INTERMEDIATE
Algebra

Third Edition

James Streeter
Late Professor of Mathematics
Clackamas Community College

Donald Hutchison
Clackamas Community College

Louis Hoelzle
Bucks County Community College

WCB
McGraw-Hill

Boston   Burr Ridge, IL   Dubuque, IA   Madison, WI   New York   San Francisco   St. Louis
Bangkok   Bogotá   Caracas   Lisbon   London   Madrid
Mexico City   Milan   New Delhi   Seoul   Singapore   Sydney   Taipei   Toronto

WCB/McGraw-Hill

A Division of The **McGraw·Hill** Companies

Intermediate Algebra

Instructor's Edition for Intermediate Algebra

This book is printed on acid-free paper.

1 2 3 4 5 6 7 8 9 0 VNH VNH 9 0 0 9 8

ISBN 0-07-063277-4 (Student Edition)

ISBN 0-07-063282-0 (Instructor's Edition)

Publisher: Tom Casson
Sponsoring editors: Jack Shira and Maggie Rogers
Marketing manager: Michelle Sala
Project manager: Eva Marie Strock
Production supervisor: Richard DeVitto
Designer: Wanda Kofax
Cover designer: Linear Design Group
Compositor: York Graphic Services, Inc.
Typeface: Times New Roman
Printer: Von Hoffmann Press, Inc.

Library of Congress Cataloging-in-Publication Data

Streeter, James (James A.)
   Intermediate algebra/James Streeter, Donald Hutchison, Louis
Hoelzle.—3rd ed.
    p. cm.
   Includes index.
   ISBN 0-07-063277-4 (student ed.).—ISBN 0-07-063282-0
(instructor's ed.)
   1. Algebra. I. Hutchison, Donald, 1948– . II. Hoelzle, Louis F.
III. Title.
QA154.2.S78   1998
512.9—dc21   97–44114
                                          CIP

http://www.mhhe.com

About the Authors

While a graduate student at the University of Washington, **James Streeter** paid for his education as a math tutor. It was here that he began to formulate the ideas that would eventually become this package. Upon graduation, he taught for 2 years at Centralia Community College. In 1968 he moved on to Clackamas Community College to become the school's first mathematics chair.

At the college, Jim recognized that he faced a very different population than the one he had tutored at the University of Washington. Jim was convinced that to reach the maximum number of these students, he would have to utilize every medium available to him. Jim opened a math lab that included CAI, original slides and tapes (which were eventually published by Harper & Row), and original worksheets and text materials. With the assistance of the people at McGraw-Hill, that package has been refined to include media and supplements that did not even exist when this project began.

Donald Hutchison spent his first 10 years of teaching working with disadvantaged students. He taught in an intercity elementary school and an intercity high school. He also worked for 2 years at Wassaic State School in New York and 2 years at the Portland Habilitation Center. He worked with both physically and mentally disadvantaged students in these two settings.

In 1982, Don was hired by Jim Streeter to teach at Clackamas Community College. In 1989, Don became Chair of the Mathematics Department at the college. It was here at Clackamas that Don discovered two things that, along with his family, form the focus for his life. Jim introduced Don to the joy of writing (with the first edition of *Beginning Algebra*), and Jack Scrivener converted him to a born-again environmentalist.

Don is also active in several professional organizations. He was a member of the ACM committee that undertook the writing of computer curriculum for the 2-year college. From 1989 to 1994 he was the Chair of the Technology in Mathematics Education Committee for AMATYC. He was President of ORMATYC from 1996 to 1998.

Louis Hoelzle has been teaching at Bucks County Community College for 27 years. In 1989, Lou became Chair of the Mathematics Department at Bucks County Community College. He has taught the entire range of courses from Arithmetic to Calculus, giving him an excellent view of the current and future needs of developmental students.

Over the past 34 years, Lou has also taught Physics courses at 4-year colleges, which has enabled him to have the perspective of the practical applications of mathematics. In addition, Lou has extensively reviewed manuscripts and written several solutions manuals for major textbooks. In these materials he has focused on writing for the student.

Lou is also active in professional organizations. He has served on the Placement and Assessment Committee for AMATYC since 1989.

This book is dedicated to the family I work with, which includes Susan, Alice, Kurt, Barry, Jack, Jacque, Janet, Kathy, Brenda, James, Medy, Larry, Dave, John, Lyle, Chuck, Liz, Joe, and Stephanie. Every member of this group has contributed to the way I teach and the way I learn. Without their advice, criticism, and friendship, this book would have no energy and no focus. Thank you all.

Don Hutchison

This book is dedicated to my grandchildren, Matthew, Moira, and Nicholas. May the brightness and curiosity in their eyes and the happiness of their lives stay with them forever. They are our future.

Louis Hoelzle

THIS SERIES IS DEDICATED TO THE MEMORY OF JAMES ARTHUR STREETER, AN ARTISAN WITH WORDS, A GENIUS WITH NUMBERS, AND A VIRTUOSO WITH PICTURES FROM 1940 UNTIL 1989.

CONTENTS

PREFACE

Statement of Philosophy

We believe that the key to learning mathematics, at any level, is active participation. When students are active participants in the learning process, they are directly involved in the development of mathematical ideas and can make connections to previously studied material. Such participation leads to understanding, success, and confidence. We developed this text with this philosophy in mind and integrated many features throughout the book to reflect this philosophy. The *Check Yourself* exercises are designed to keep the student involved and active with every page of exposition. The graphing calculator references involve the student actively in the development of mathematical ideas. We have placed great emphasis on a visual approach since it helps students see a problem and its solution in a less abstract setting than symbolic manipulation alone permits. Furthermore, the combination of graphing calculators and the function approach allows us to put a fresh face on some topics that might otherwise be all too familiar to students who are taking their third or fourth algebra class.

Early Introduction of Functions

In the study of college algebra or precalculus, the concept of function plays a central role. Since functions of various types are frequently applied in many fields of study, we believe it essential that our students understand what a function is and be able to distinguish several types of functions before they enter that next level of mathematics. These concepts, coupled with the appropriate use of graphing technology, empower students to solve many new types of equations, inequalities, and application problems. The function theme should therefore permeate an intermediate algebra course. It is a thread that can successfully tie together the traditional topics taught here.

For these reasons, we introduce the concept of function very early in the course. In Chapter 1, relations, linear functions, and their graphs are developed. In Chapter 2, students are introduced to graphical solutions for equations and inequalities. This is a new approach to old problems for almost every student. We find that this new approach has two benefits. For many students, looking at a graphical solution provides a key for understanding the concept of "solution." For other students it provides a fresh look at material that they have seen many times before. Chapters 1 and 2 are the foundation for the rest of the course.

Why We Changed Our Approach

We were certainly impressed by the AMATYC Standards. Much of what we have included—group exercises, writing exercises, data analysis, functions, elementary statistics concepts, and use of technology—comes directly from the Standards. But seeing these items in the Standards was more confirmation than inspiration. Most of these changes in our text have evolved from the way we teach the course at Clackamas Community College. We find students to be more receptive to new ideas than ever in our Intermediate Algebra courses primarily because of the degree to which this approach involves the students in their education. They are forced to become active learners.

Graphing Calculators

In this text we assume that each student holds, or has access to, a graphing calculator. We consider such machines invaluable aids in the study (and teaching!) of college mathematics. Since visualization is such an important theme in this course, the instructor will find frequent opportunities to illustrate, or encourage discovery of, important concepts via the graphing calculator. Margin notes accompanying the text also urge the student to view a particular graph on a calculator. Furthermore, with these powerful tools students are able to solve problems that involve messy, real-world data.

We do, however, view the graphing calculator as a tool (albeit a wonderful one), one of many tools to be used in the study of algebra. Our goal is the understanding of concepts, aided by a visual approach. Whether such visualization is provided by a calculator, or by a sketch based on graphing principles, is not critical. Instructors and students must continually decide whether to turn on the calculator or to make a sketch of the problem by hand. This in itself is a valuable process.

Important Features

Functions-Based Approach

Students are introduced to functions in Chapter 1 and graphical solutions in Chapter 2. This emphasis on functions and graphing is carried throughout the rest of the text in every subsequent chapter. The exposition and the check yourself exercises have been extensively revised to incorporate this approach.

Vignettes, Applications, Check Yourself Exercises

Those familiar with the Streeter series will recognize most of the pedagogical elements of this text. The chapter-opening **vignettes** provide a sketch of a real event or scenario that requires the use of the concepts of the current chapter as part of the solution. They also provide fodder for interesting **applications** in the exercise sections. The **check yourself exercises** in each chapter engage the students in practice after every example. Answers are provided at the end of the section for immediate feedback.

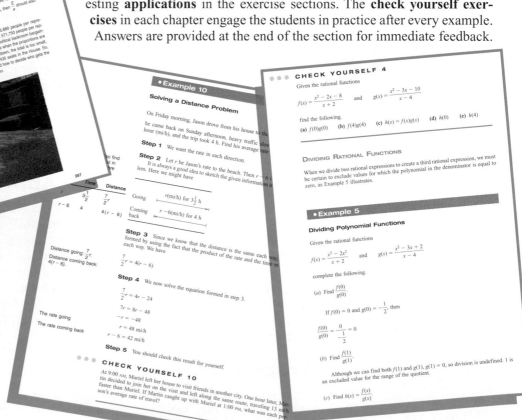

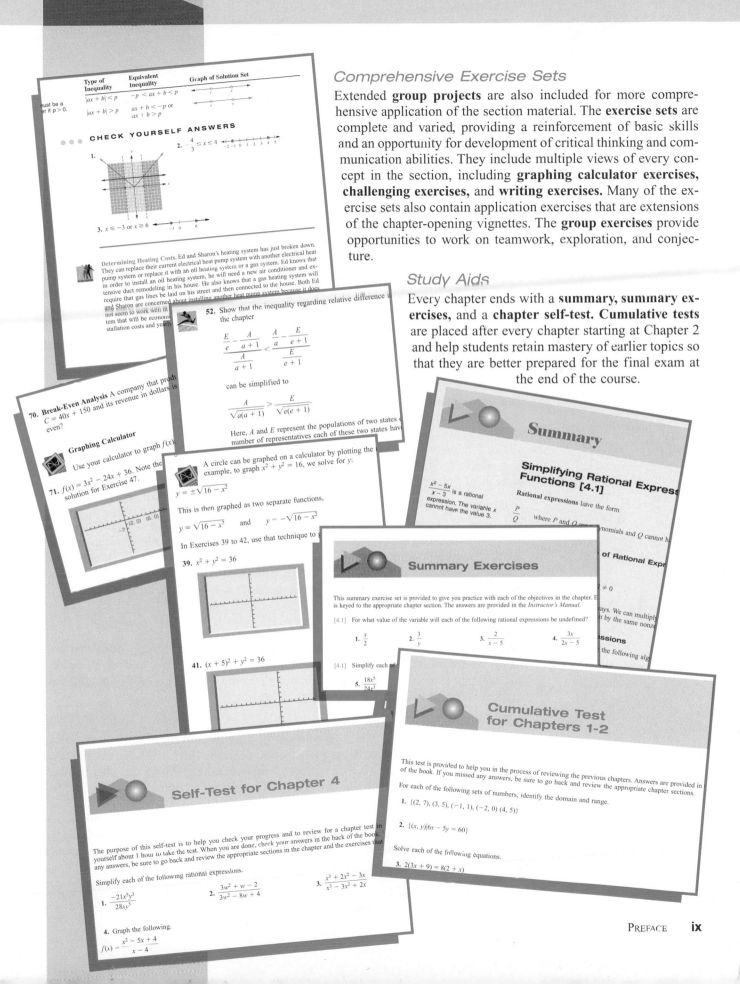

Comprehensive Exercise Sets

Extended **group projects** are also included for more comprehensive application of the section material. The **exercise sets** are complete and varied, providing a reinforcement of basic skills and an opportunity for development of critical thinking and communication abilities. They include multiple views of every concept in the section, including **graphing calculator exercises, challenging exercises,** and **writing exercises.** Many of the exercise sets also contain application exercises that are extensions of the chapter-opening vignettes. The **group exercises** provide opportunities to work on teamwork, exploration, and conjecture.

Study Aids

Every chapter ends with a **summary, summary exercises,** and a **chapter self-test. Cumulative tests** are placed after every chapter starting at Chapter 2 and help students retain mastery of earlier topics so that they are better prepared for the final exam at the end of the course.

SUPPLEMENTS

Supplements

A comprehensive set of ancillary materials for both the student and the instructor is available with this text.

Instructor's Edition

This ancillary includes answers to all exercises and tests. These answers are printed in a second color for ease of use by the instructor and are located on the appropriate pages throughout the text.

Instructor's Solutions Manual

The manual provides worked-out solutions to the even-numbered exercises in the text.

Instructor's Resource Manual

The resource manual contains multiple-choice placement tests for three levels of testing: (1) a diagnostic pretest for each chapter and three forms of multiple-choice and open-ended chapter tests; (2) two forms of multiple-choice and open-ended cumulative tests; and (3) two forms of multiple-choice and open-ended final tests. Also included is an answer section and appendixes that cover collaborative learning and the implementation of the new standards.

Print and Computerized Testing

The testing materials provide an array of formats that allow the instructor to create tests using both algorithmically generated test questions and those from a standard testbank. This testing system enables the instructor to choose questions either manually or randomly by section, question type, difficulty level, and other criteria. Testing is available for IBM, IBM-compatible, and Macintosh computers. A softcover print version of the testbank provides most questions found in the computerized version.

Streeter Video Series

The video series is completely new to this edition. It gives students additional reinforcement of the topics presented in the book. The videos were developed especially for the Streeter pedagogy, and features are tied directly to the main text's individual chapters and section objectives. The videos feature an effective combination of learning techniques, including personal instruction, state-of-the-art graphics, and real-world applications.

Multimedia Tutorial

This interactive CD-ROM is a self-paced tutorial specifically linked to the text and reinforces topics through unlimited opportunities to review concepts and practice problem solving. It requires virtually no computer training on the part of the students and supports IBM and Macintosh computers.

MathWorks

This DOS-based interactive tutorial software is available and specifically designed to accompany the Streeter pedagogy. The program supports IBM, IBM-compatible, and Macintosh computers as well as a variety of networks. MathWorks can also be used with its companion program, the Instructor's Management System, to track and record the progress of students in the class.

In addition, a number of other technology and Web-based ancillaries are under development; they will support the ever-changing technology needs in developmental mathematics. For further information about these or any supplements, please contact your local WCB/McGraw-Hill sales representative.

ACKNOWLEDGMENTS

Acknowledgments

Through three editions and a revision, the list of contributors to this text has grown steadily. Foremost are the users and the reviewers. We do not have space to individually thank each user, but suffice it to say that every comment that has reached us has helped in the evolution of this text. We do have space for thanks to each reviewer. Every person on the list has helped by questioning, correcting, or complimenting. Our thanks to each of them.

Jackie Cohen, Augusta College (GA)
Maura Corley, Henderson Community College (KY)
Jon Davidson, Southern State Community College (OH)
Kenneth Hufham, Cape Fear Community College (NC)
Robert Farinelli, Community College of Allegheny County (PA)
James Warren Fightmaster, Virginia Western Community College
Roseann Fuglio, Gloucester Community College (NJ)
Brenda Foster, Central Virginia Community College
David Freeman, Marietta College (OH)
Dorothy Gotway, University of Missouri at St. Louis
Pauline Gravelinc, State University of New York–College of Technology at Canton
John Jacobs, Massachusetts Bay Community College
Billie L. James, University of South Dakota
Norma James, New Mexico State University
Conrad Johnson, Central Florida Community College
Diane L. Johnson, Humboldt State University (CA)
Robert Malena, Community College of Allegheny County (PA)
Laurie McManus, St. Louis University
Valerie Melvin, Cape Fear Community College (NC)
Fred Monaco, State University of New York–College of Technology at Canton
Cheryl Raboin, Eastern Wyoming College
Janice Rech, University of Nebraska at Omaha
Thomas Roe, South Dakota State University
Rakesh Rustagi, Northeastern Illinois University
Debra Sample, Arkansas State University
Hugh Sanders, Georgia College
Richard Semmler, Northern Virginia Community College
Randy Taylor, Las Positas College (CA)
Carol Walker, Hinds Community College (MS)
Raymond Whaley, Waubonsee College (IL)

The staff at WCB/McGraw-Hill has had an incredible knack for providing what was needed. Sometimes it was a new direction, sometimes it was a friend, always it was time. Thanks to Tom, Maggie, Eva, Jack, and Paul. As always, we encourage prospective authors to talk with the people at WCB/McGraw-Hill. It will be a valuable use of your time.

In this age of the high-tech mathematics lab, the supplements and the supplement authors have become as important as the text itself. We are proud that our names appear on the supplements together with the following authors: John Garlow, Tarrant County Junior College; Gloria Langer; John Robert Martin, Tarrant County Junior College; and Mark Stevenson, Oakland Community College.

We must also thank all the students who have been taught, talked to, questioned, and tested. This text was created for them, this text was created by them. We also want

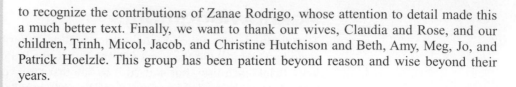

to recognize the contributions of Zanae Rodrigo, whose attention to detail made this a much better text. Finally, we want to thank our wives, Claudia and Rose, and our children, Trinh, Micol, Jacob, and Christine Hutchison and Beth, Amy, Meg, Jo, and Patrick Hoelzle. This group has been patient beyond reason and wise beyond their years.

Donald Hutchison
Louis Hoelzle

TO THE STUDENT

To The Student

Intermediate Algebra is the final step in your preparation for college transfer mathematics. In this course you will cover many new concepts and see the expansion of several familiar ones. All the topics in this text are designed to prepare you for subsequent courses in Pre-Calculus, College Algebra, or Statistics. We have provided situations and applications that will help you understand the relevance of what you are learning. As your mathematical expertise expands in succeeding courses, the applications of your skills will become even more diverse and interesting. Your ability to succeed in those later courses will very much depend on how well you understand the material covered in this text. The following suggestions are designed to help you succeed in this, and future, mathematics classes.

1. If you are in a lecture class, take the time to read the appropriate section before the lecture. Take careful notes of every example your instructor presents in class.

2. When you sit down to study, have your calculator, pencil, and paper ready. Work through the examples in the text. Do every *Check Yourself* exercise, checking your answer against the one at the end of the section. If you have difficulty, go back and reread the previous example. Make certain you understand what you are doing and why. The best test of whether you do understand a concept lies in your ability to explain that concept to a classmate. Try working together.

3. At the end of each section is a set of exercises. Work these carefully in order to check your progress on the section just completed. The answers to odd-numbered exercises are at the end of the section. If you have difficulty with any of the exercises, review the appropriate parts of the section. If your confusion is not completely cleared up, by all means do not become discouraged. Ask your instructor or an available tutor for assistance. A word of caution: Work the exercises on a regular (preferably daily) basis. Learning algebra requires active participation on your part. As is the case with the learning of any skill, the main ingredient is practice.

4. When you have completed a chapter, review by using the *Summary*. Following the summary are *Summary Exercises* for further practice. The exercises are keyed to chapter sections so you will know where to turn if you are still struggling.

5. When finished with the Summary Exercises, try the *Self-Test* that appears at the end of each chapter. Answers, with section references, are in the back of the book.

6. Finally, an important element of success in studying algebra is the process of regular review. We provide a series of *Cumulative Reviews* throughout this book. Use these tests to prepare for any midterms or finals. If it appears that you have forgotten some concepts from earlier chapters, do not worry. Go back and review the section where the idea was first explained, or the appropriate chapter summary; that is the purpose of the *Cumulative Tests*.

We hope you will find these suggestions helpful as you work through the material in this text. Best of luck in this course!

Donald Hutchison
Louis Hoelzle

CHAPTER 1

ORDERED PAIRS AND LINEAR FUNCTIONS

INTRODUCTION

Data are usually presented in one of three forms: a table, a formula, or a graph. Graphs often show trends that may not be easy to see when data are read from a table, or when a rule is given in a formula.

Economists are among the many professionals who use graphs to show connections between two sets of data. For example, an economist may use a graph to look for a connection between two different measures for the standard of living in various countries.

One way of measuring the standard of living in a country is the per capita gross domestic product (GDP). The GDP is the total value of all goods and services produced by all businesses and individuals over the course of 1 year. To find the per capita GDP, we divide that total value by the population of the country.

Other economists, including some who wrote an article in *Scientific American* in May 1993, question this method. Rather than comparing GDP among countries, they use survival rate (life expectancy) to measure the quality of life. The following table compares life expectency and GDP.

Country	Life Expectancy at Birth (in years)	Per Capita GPD (in dollars)
Botswana	59	$ 2,496
Libya	62	7,250
Sri Lanka	71	2,053
Costa Rica	75	3,760
Thailand	66	2,576
Israel	76	9,182
United States	76	17,615
Japan	78	13,135

Adapted from *The Human Development Report,* United Nations, Oxford Press: Cambridge, England, 1990.

© Wojnarowicz/The Image Works

Not many people could quickly read and interpret the information in the table. Although all the information is available, it is not as easy to see as in the following graph (called a scatter plot). This graph contains the same information, but in a form that is easier to interpret. Note that, if life expectancy is a good measure of the standard of living, productivity may not be a good measure. It does not appear to be the case that higher productivity leads to a longer life.

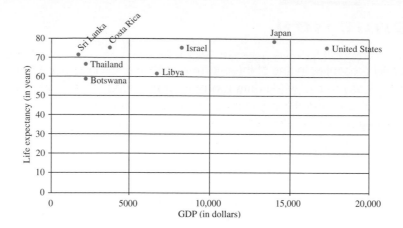

A graph such as this one helps us to visualize complex information. This kind of graph is created using ordered pairs. We will introduce ordered pairs in this first chapter and then use them throughout the book. ■

 **1.1**

Ordered Pairs and Relations

1.1 OBJECTIVES

1. Identify ordered pairs
2. Identify relations
3. Identify the domain and range of a relation

The number line extends infinitely in both directions.

Note that sets always use **braces** to enclose their contents.

ALGEBRA

Algebra is a system consisting of a set, together with operations, that follow certain properties. The word **algebra** comes from the title of a book written in the ninth century by Al-Khowarazmi. It described the techniques used in solving equations. Even after more than a 1000 years, solving equations and simplifying expressions continue to be the primary topics of an algebra course taught at this level.

In intermediate algebra, most of our work will be with the set of real numbers. These are all the numbers that can be found on a number line.

In set notation, we refer to the set of all real numbers as $\{x \mid x \in \mathbb{R}\}$, which is read, "The set of all numbers x such that x is a real number."

ORDERED PAIRS

We will usually be interested in the relationship between two numbers. Whenever this is the case, the order in which we present the two numbers will be important, which is why we will call them **ordered pairs.**

Given two related numbers, x and y, we will write the pair of numbers as (x, y). The set of all possible ordered pairs of real numbers will then be written as follows: $\{(x, y) \mid x \in \mathbb{R}, y \in \mathbb{R}\}$.

The ordered pair $(2, -3)$ is different from the ordered pair $(-3, 2)$. Whereas, the set of ordered pairs $\{(1, 3), (4, -1)\}$ is the same as the set of ordered pairs $\{(4, -1), (1, 3)\}$. Order is important in an ordered pair. Order is not important in a set.

● Example 1

Identifying Ordered Pairs

Which of the following are ordered pairs?

(a) $(2, -\pi)$ (b) $\{2, -4\}$ (c) $(1, 3, -1)$

(d) $\{(1, -5), (9, 0)\}$ (e) $2, 5$

Only (a) is an ordered pair. (b) is a set (it uses braces instead of parentheses), (c) has three numbers instead of two, (d) is a set of ordered pairs, and (e) is simply a list of two numbers.

● ● ●   CHECK YOURSELF 1

Which of the following are ordered pairs?

(a) $\left\{\dfrac{1}{2}, -3\right\}$ (b) $\left(-3, \dfrac{1}{3}\right)$ (c) $\{(5, 0)\}$ (d) $(1, -5)$ (e) $-3, 6$

RELATIONS

> A set of ordered pairs is called a **relation.**

The ordered pairs can either be listed explicitly, or implied. Examples of relations include

> $A = \{(-2, 8), (1, 2), (-3, -5)\}$
>
> $R = \{(1, 3), (1, 4), (1, 5)\}$
>
> $B = \{(x, y) | y = x\}$

Relation B is comprised of every ordered pair in which x and y are equal. This relation includes $(0, 0)$, $\left(\dfrac{1}{2}, \dfrac{1}{2}\right)$, $(-3, -3)$, and (π, π).

DOMAIN AND RANGE

In most cases, the domain is the set of all x values, and the range is the set of all y values.

> The **domain** of a relation is the set of all the first elements of the ordered pairs. The **range** of a relation is the set of all the second elements of the ordered pairs.

• Example 2

Finding the Domain and Range of a Relation

Find the domain and range for each relation.

(*a*) $A = \{(-2, 8), (1, 2), (-3, -5)\}$

The domain is the set $\{-2, 1, -3\}$. The range is the set $\{8, 2, -5\}$.

(*b*) $R = \{(1, 3), (1, 4), (1, 5)\}$

The domain is the set $\{1\}$. The range is the set $\{3, 4, 5\}$.

● ● ● **CHECK YOURSELF 2**

Find the domain and range for each relation.

(a) $A = \{(-5, 4), (-4, 7), (-4, 9)\}$ (b) $R = \{(1, 0), (3, 0), (5, 0)\}$

It is also possible to find the domain when the relation is described by rule rather than by list. When looking for the domain from a rule, the question to ask is, "What values can I substitute for x?" When looking for the range, we ask, "What values can be substituted for y?"

• Example 3

Identifying the Domain and Range

Find the domain and range for each relation.

(a) $A = \{(x, y)|x = 3\}$

The symbol $\mathbb{R}$ represents the real numbers, $\mathbb{N}$ represents the natural numbers, and $\mathbb{Z}$ represents the integers.

The domain is $\{3\}$. The description of the relation defines x to always be 3. The range is all real numbers, $\mathbb{R}$.

(b) $B = \{(x, y)|y = x\}$

The domain is the set of all real numbers, $\mathbb{R}$. The range is also the set of all real numbers, $\mathbb{R}$.

● ● ● CHECK YOURSELF 3

Find the domain and range for each relation.

(a) $A = \{(x, y)|y = -1\}$ **(b)** $B = \{(x, y)|y = x - 1\}$

● ● ● CHECK YOURSELF ANSWERS

1. (b) and **(d)** are ordered pairs.   **2. (a)** Domain $= \{-5, -4\}$, range $= \{4, 7, 9\}$; **(b)** domain $= \{1, 3, 5\}$, range $= \{0\}$.   **3. (a)** The domain is all real numbers, $\mathbb{R}$. The range is $\{-1\}$. **(b)** The domain is the set of all real numbers, $\mathbb{R}$. The range is also the set of all real numbers, $\mathbb{R}$.

1.1 Exercises

In Exercises 1 to 4, identify the ordered pairs.

1. **(a)** $(3, -5)$ **(b)** $\{7, 9\}$ **(c)** $(2, 5)$ **(d)** 5, 2 **(e)** $((3, 1), 4)$

2. **(a)** $\{7, 23\}$ **(b)** $(1, 0, (5, 6))$ **(c)** $\left(\frac{1}{2}, -1\right)$ **(d)** $[5, 6]$ **(e)** $(23, 7)$

3. **(a)** 18, 67 **(b)** $(-3, -9)$ **(c)** $\{3, 9\}$ **(d)** $(3, 7, -3)$ **(e)** $[12, 56]$

4. **(a)** $\{45, 67]$ **(b)** $(9, 3)$ **(c)** 5, 8 **(d)** $(11, -3, 9)$ **(e)** $[5, 2]$

In Exercises 5 to 20, identify the domain and range in the sets of ordered pairs.

5. {(1, 2), (3, 4), (5, 6), (7, 8), (9, 10)}

6. {(2, 3), (3, 5), (4, 7), (5, 9), (6, 11)}

7. {(1, 2), (4, 6), (3, 3), (5, 4), (6, 1)}

8. {(3, 4), (5, 7), (6, 1), (2, 2), (4, 3)}

9. {(1, 2), (1, 3), (1, 4), (1, 5), (1, 6)}

10. {(3, 4), (3, 6), (3, 8), (3, 9), (3, 10)}

11. {(1, 5), (2, 5), (3, 6), (2, 4), (4, 5)}

12. {(2, 8), (3, 9), (2, 9), (3, 8), (4, 7)}

13. {(−1, 3), (−2, 4), (−3, 5), (4, 4), (5, 6)}

14. {(−2, 4), (1, 4), (−3, 4), (5, 4), (7, 4)}

15. {(x, y)|x + 2y = 3}

16. {(x, y)|3x + 4y = 12}

17. {(x, y)|y = 5}

18. {(x, y)|y = −4}

19. {(x, y)|x = 23}

20. {(x, y)|x = −9}

21. The stock prices for a given stock over a week's time are displayed in a table. List this information as a set of ordered pairs using the day of the week as the domain.

Day	1	2	3	4	5
Price	$9\frac{1}{8}$	8	$8\frac{7}{8}$	$9\frac{1}{4}$	9

22. Food Purchases In the snack department of the local supermarket, candy costs $1.58 per pound. For 1 to 5 pounds, write the cost of candy as ordered pairs.

 In Exercises 23 to 26, write a set of ordered pairs that describes each situation. Give the domain and range of each relation.

23. The first coordinate is an integer between −3 and 3. The second coordinate is the cube of the first coordinate.

24. The first coordinate is a positive integer less than 6. The second coordinate is the sum of the first coordinate and −2.

25. The first coordinate is the number of hours worked (10, 20, 30, 40); the second coordinate is the salary at $6 per hour.

26. The first coordinate is the number of toppings on a pizza (up to 4); the second coordinate is the price of the pizza, which is $9 plus $1 per topping.

27. Explain why the ordered pair (2, 3) is different from the ordered pair (3, 2) while the set {2, 3} is the same as {3, 2}.

28. Explain the difference between the relation $\{(x, y)|x + y = 5\}$ and the relation $\{(x, y)|x + y = 5$ and x is an integer between 1 and 7}.

29. Think of four pairs of quantities that are related as are the pizza toppings and pizza prices in Exercise 26, and write a sentence explaining the relationship. Then, give four sets of ordered pairs that might represent these values. For example, "The amount of money for a tip in a restaurant depends on the amount of the bill."
Ordered pairs: (21.00, 3.15), (38.00, 5.70), (95.00, 14.25), and (15.00, 10.00).

Answers

1. (a) and **(c)**   **3. (b)**   **5.** D: {1, 3, 5, 7, 9}; R: {2, 4, 6, 8, 10}   **7.** D: {1, 3, 4, 5, 6}; R: {1, 2, 3, 4, 6}

9. D: {1}; R: {2, 3, 4, 5, 6}   **11.** D: {1, 2, 3, 4}; R: {4, 5, 6}   **13.** D: {−1, −2, −3, 4, 5}; R: {3, 4, 5, 6}

15. D: reals; R: reals   **17.** D: reals; R: {5}   **19.** D: {23}; R: reals   **21.** $\left\{\left(1, 9\frac{1}{8}\right), (2, 8), \left(3, 8\frac{7}{8}\right),\right.$

$\left.\left(4, 9\frac{1}{4}\right), (5, 9)\right\}$ **23.** {(−2, −8), (−1, −1), (0, 0), (1, 1), (2, 8)}; D: {−2, −1, 0, 1, 2}; R: {−8, −1, 0, 1, 8}

25. {(10, 60), (20, 120), (30, 180), (40, 240)}; D: {10, 20, 30, 40}; R: {60, 120, 180, 240}   **27.** In ordered pairs, the order of elements is important.   **29.**

1.2 The Cartesian Coordinate System

1.2 OBJECTIVES

1. Plot ordered pairs
2. Identify plotted points
3. Scale the axes

The development of the **coordinate system** was part of an effort to combine the knowledge of geometry with that of algebra.

In the eighteenth century, René Descartes, a French philosopher and mathematician, created a way of graphing ordered pairs. He developed a system that consists of two perpendicular number lines. The point at which the lines meet (their **intersection**) is called the **origin.** The coordinates of the origin are (0, 0).

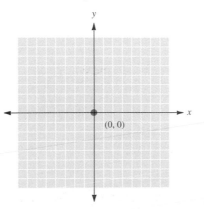

PLOTTING ORDERED PAIRS

To **plot an ordered pair,** move along the horizontal line to the point corresponding to the first number of the ordered pair. Imagine a vertical line is drawn there. Now, move along the vertical axis to the point corresponding to the second number of the ordered pair. Imagine a horizontal line is drawn there. The intersection of those two lines is the point that corresponds to the ordered pair. A dot is generally used to designate that point.

● Example 1

Plotting Ordered Pairs

Plot the point associated with each ordered pair.

(*a*) $A(2, 4)$

Move to the right on the horizontal axis 2 units. Picture a vertical line through that point. Move up the vertical axis 4 units. Picture a horizontal line through that point. Plot the point at the intersection of those two lines.

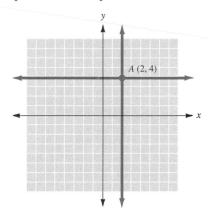

(*b*) $B(-2, 5\frac{3}{4})$

Moving two units to the left of the origin, and $5\frac{3}{4}$ units above it, we plot the point associated with $\left(-2, 5\frac{3}{4}\right)$.

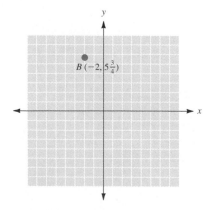

● ● ● **CHECK YOURSELF 1**

Plot the points associated with the following ordered pairs.

$A(0, 3)$ $B(-2, -4)$ $C(3, -1)$

To find the ordered pair associated with a plotted point, we reverse the process. We move vertically from the point to find the first coordinate of the ordered pair, and horizontally from the point to find the second coordinate.

●Example 2

Identifying Plotted Points

Find the ordered pair associated with each point.

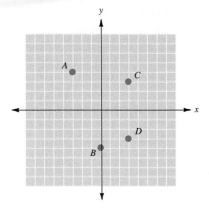

When no scale appears on the grid, we assume that each division on the axis is one unit.

From point A, a vertical line meets the axis at -3. A horizontal line meets the axis at 4. The ordered pair is $(-3, 4)$. B is associated with $(0, -4)$, C with $(3, 3)$, and D with $(3, -3)$.

● ● ● **CHECK YOURSELF 2**

Find the ordered pair associated with each point.

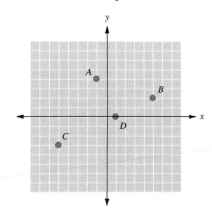

SCALING THE AXES

The number lines used in the cartesian plane are called axes (pronounced "axees"). The horizontal line is usually called the *x* axis. Note that, in all our examples, the *x* value was associated with the horizontal axis. The vertical number line is usually referred to as the *y* axis.

It is not necessary, or even desirable, to always use the same scale on both the *x* and *y* axes. For example, if we were plotting ordered pairs in which the first value represented the age of a used car and the second value represented the number of miles driven, it would be necessary to have a different scale on the two axes. If not, the following extreme case could happen.

Assume that the cars range in age from 1 to 15 years. The cars have mileage from 2000 to 150,000 miles. If we use the same scale on both axes, 0.5 in. between each two counting numbers, how large would the paper have to be on which the points were plotted? The horizontal axes would have to be 15(0.5) = 7.5 in. The vertical axis would have to be 150,000(0.5) = 75,000 in. = 6250 feet = almost 1.2 miles long!

So what do we do? We simply use a different, but clearly marked, scale on the axes. In this case, the horizontal axis could be marked in 1s, but the vertical axis would be marked in 10,000s. Additionally, all the numbers would be positive, so we really need only the upper right portion of the graph (this portion is called the **first quadrant**), in which *x* and *y* are both always positive. We could draw the graph like this:

The same decisions must be made when you are using a graphing calculator. When graphing this kind of relation on a calculator, you must decide what the appropriate **viewing window** should be.

The other three quadrants are numbered in a counterclockwise direction.

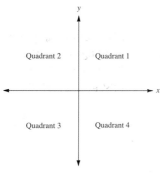

Every ordered pair is either in one of the quadrants or on one of the axes.

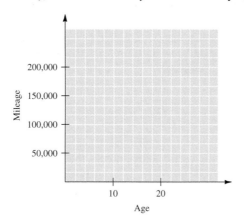

• Example 3

Scaling the Axes

A survey of residents in a large apartment building was recently taken. The following points represent ordered pairs in which the first number is the number of years of education a person has had, and the second number is their 1996 income (in thousands of dollars). Estimate, and interpret, each ordered pair represented.

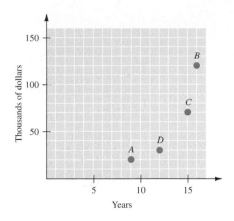

A is (9, 20), *B* is (16, 120), *C* is (15, 70), and *D* is (12, 30). Person *A* completed 9 years of education and made $20,000 in 1996. Person *B* completed 16 years of education and made $120,000 in 1996. Person *C* had 15 years education and made $70,000. Person *D* had 12 years and made $30,000.

Note that there is no obvious "relation" that would allow one to predict income from years of education, but you might suspect that in most cases, more education results in more income.

● ● ● ● **CHECK YOURSELF 3**

Each year on his son's birthday, Armand records his son's weight. The following points represent ordered pairs in which the first number represents his son's age and the second number represents his weight. For example, point *A* indicates that when his son was 1 year old, the boy weighed 14 pounds. Estimate each ordered pair represented.

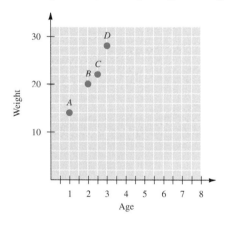

● ● ● **CHECK YOURSELF ANSWERS**

1.

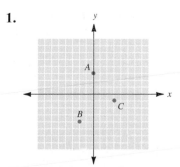

2. *A*(−1, 4), *B*(5, 2), *C*(−5, −3), and *D*(1, 0).

3. *A*(1, 14), *B*(2, 20), *C*($3\frac{1}{2}$, 22), and *D*(3, 28).

1.2 Exercises

In Exercises 1 to 12, give the quadrant in which each of the following points is located or the axis on which the point lies.

1. $(4, 5)$

2. $(-3, 2)$

3. $(-4, -3)$

4. $(2, -4)$

5. $(5, 0)$

6. $(-5, 7)$

7. $(-4, 7)$

8. $(-3, -7)$

9. $(0, -7)$

10. $(-3, 0)$

11. $\left(5\frac{3}{4}, -3\right)$

12. $\left(-2, 4\frac{5}{6}\right)$

In Exercises 13 to 26, graph the points on the accompanying graph.

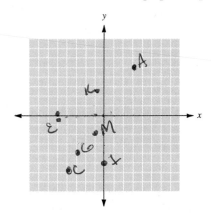

13. $A(3, 5)$

14. $B(-4, 6)$

15. $C(-4, -6)$

16. $D(5, -6)$

17. $E\left(-5, -\frac{1}{2}\right)$

18. $F(6, 0)$

19. $G(-3, -4)$

20. $H(-1, 4)$

21. $I(0, -5)$

22. $J(2, -3)$

23. $K\left(-1, \frac{5}{2}\right)$

24. $L\left(-5, \frac{3}{4}\right)$

25. $M(-1, -2)$

26. $N\left(5, -\frac{6}{5}\right)$

In Exercises 27 to 36, give the coordinates (ordered pairs) associated with the points indicated in the figure.

27. *P*   **28.** *Q*   **29.** *R*   **30.** *S*   **31.** *T*   **32.** *U*   **33.** *V*   **34.** *W*   **35.** *X*   **36.** *Y*

37. On the accompanying graph, draw the graph of the set $\{(x, y)|x$ is an integer between -3 and 3 and $y = 2\}$.

[handwritten: X = -2, -1, 0, 1, 2   Y=2]

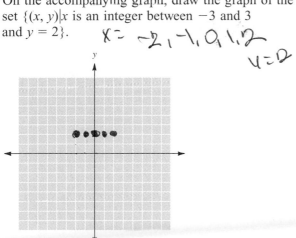

38. On the accompanying graph, draw the graph of the set $\{(x, y)|x$ is an integer between -2 and 2 and $y = x + 2\}$.

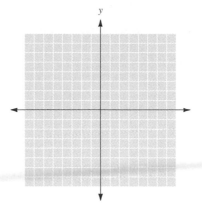

39. We mentioned that the cartesian coordinate system was named for the French philosopher and mathematician René Descartes. What philosophy book is Descartes most famous for? Use an encyclopedia as a reference.

40. What characteristic is common to all points on the *x* axis? On the *y* axis?

41. A company has kept a record of the number of items produced by an employee as the number of days on the job increases. In the following figure, points correspond to an ordered-pair relationship in which the first number represents days on the job and the second number represents the number of items produced. Estimate each ordered pair produced.

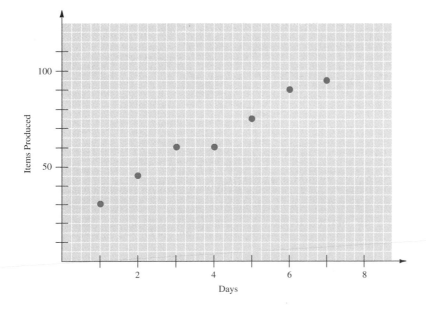

42. In the following figure, points correspond to an ordered-pair relationship between height and age in which the first number represents age and the second number represents height. Estimate each ordered pair represented.

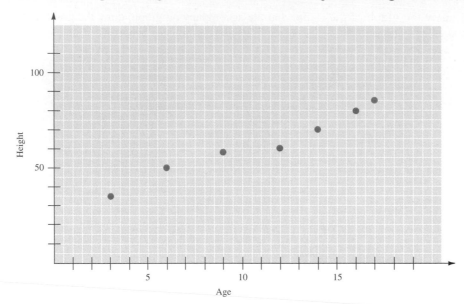

 43. Graph the points with coordinates (1, 2), (2, 3), and (3, 4). What do you observe? Give the coordinates of another point with the same property.

 44. Graph points with coordinates (−1, 3), (0, 0), and (1, −3). What do you observe? Give the coordinates of another point with the same property.

 45. Although high employment is a measure of a country's economic vitality, economists worry that periods of low unemployment will lead to inflation. Look at the following table.

Year	Unemployment Rate (%)	Inflation Rate (%)
1955	4.4	−0.4
1960	5.5	1.7
1965	4.5	1.6
1970	4.9	5.7
1975	8.5	9.1
1980	7.1	13.5
1985	7.2	3.6
1990	5.5	5.4

Plot the figures in the table with unemployment rates as the domain and inflation rates as the range. What do these plots tell you? Do higher inflation rates seem to be associated with lower unemployment rates? Explain.

Answers

1. Quadrant I   **3.** Quadrant III   **5.** x axis   **7.** Quadrant II   **9.** y axis   **11.** Quadrant IV

13 to **25.** See figure: **27.** P **29.** R **31.** T **33.** V **35.** X

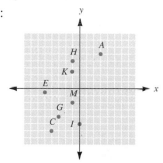

37.

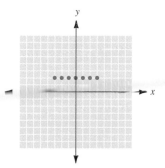

39.

41. Points are $(1, 30)$, $(2, 45)$, $(3, 60)$, $(4, 60)$, $(5, 75)$, $(6, 90)$, $(7, 95)$ **43.** The y coordinate is 1 more than the x coordinate; $(4, 5)$ **45.**

1.3 An Introduction to Functions

1.3 OBJECTIVES

1. Evaluate expressions
2. Evaluate functions
3. Identify the domain and range of a function
4. Use the vertical line test to identify a function

Variables can be used to represent unknown real numbers. Together with the operations of addition, subtraction, multiplication, division, and exponentiation, these numbers and variables form expressions such as:

$$3 + 5 \qquad 7x - 4 \qquad x^2 - 3x - 10 \qquad x^4 - 2x^2 + 3x + 4$$

Four different actions can be taken with expressions. We can:

1. Substitute values for the variable(s) and **evaluate the expression.**
2. Rewrite an expression as some simpler equivalent expression. This rewriting is called **simplifying the expression.**
3. Set two expressions equal to each other and **solve for the stated variable.**
4. Set two expressions equal to each other and **graph the equation.**

Throughout this book, everything we do will involve one of these four actions. We now focus on the first item, evaluating expressions.

EVALUATING EXPRESSIONS

Expressions can be evaluated for an indicated value of the variable(s), as Example 1 illustrates.

• Example 1

Evaluating Expressions

Evaluate the expression $x^4 - 2x^2 + 3x + 4$ for the indicated value of x.

(*a*) $x = 0$

Substituting 0 for x in the expression yields:

$$(0)^4 - 2(0)^2 + 3(0) + 4 = 0 - 0 + 0 + 4$$
$$= 4$$

(*b*) $x = 2$

Substituting 2 for x in the expression yields:

$$(2)^4 - 2(2)^2 + 3(2) + 4 = 16 - 8 + 6 + 4$$
$$= 18$$

(*c*) $x = -1$

Substituting -1 for x in the expression yields:

$$(-1)^4 - 2(-1)^2 + 3(-1) + 4 = 1 - 2 - 3 + 4$$
$$= 0$$

● ● ● CHECK YOURSELF 1

Evaluate the expression $2x^3 - 3x^2 + 3x + 1$ for the indicated value of x.

(a) $x = 0$ **(b)** $x = 1$ **(c)** $x = -2$

Function Notation

We could design a machine whose function would be to crank out the value of an expression for each given value of x. We could call this machine something simple such as f, our **function notation.** Our machine might look like this.

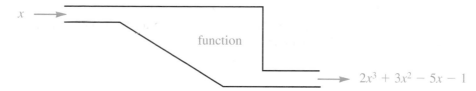

For example, when we put -1 into the machine, the machine would substitute -1 for x in the expression, and 5 would come out the other end because

$$2(-1)^3 + 3(-1)^2 - 5(-1) - 1 = -2 + 3 + 5 - 1 = 5$$

In fact, the idea of the function machine is very useful in mathematics. Your graphing calculator can be used as a function machine. You can enter the expression into the calculator as Y_1 and then evaluate Y_1 for different values of x.

Generally, in mathematics, we do not write $Y_1 = 2x^3 + 3x^2 - 5x - 1$. Instead, we write $f(x) = 2x^3 + 3x^2 - 5x - 1$, which is read as "$f$ of x is equal to . . . ." Instead of calling f a function machine, we say that f is a function of x. The greatest benefit of this notation is that it lets us easily note the input value of x along with the output of the function. Instead of "Evaluate Y_1 for $x = 4$" we say "Find $f(4)$."

• Example 2

Evaluating Expressions with Function Notation

Given $f(x) = x^3 - 3x^2 + x + 5$, find the following.

(a) $f(0)$

Substituting 0 for x in the expression on the right, we get

$$(0)^3 - 3(0)^2 + (0) + 5 = 5$$

(b) $f(-3)$

Substituting -3 for x in the expression on the right, we get

$$(-3)^3 - 3(-3)^2 + (-3) + 5 = -27 - 27 - 3 + 5$$
$$= -52$$

(c) $f\left(\dfrac{1}{2}\right)$

Substituting $\dfrac{1}{2}$ for x in the expression on the right, we get

$$\left(\frac{1}{2}\right)^3 - 3\left(\frac{1}{2}\right)^2 + \left(\frac{1}{2}\right) + 5 = \frac{1}{8} - 3\left(\frac{1}{4}\right) + \frac{1}{2} + 5$$

$$= \frac{1}{8} - \frac{3}{4} + \frac{1}{2} + 5$$

$$= \frac{1}{8} - \frac{6}{8} + \frac{4}{8} + 5$$

$$= -\frac{1}{8} + 5$$

$$= 4\frac{7}{8} \text{ or } \frac{39}{8}$$

● ● ● ●   **CHECK YOURSELF 2**

Given $f(x) = 2x^3 - x^2 + 3x - 2$, find the following.

(a) $f(0)$ **(b)** $f(3)$ **(c)** $f\left(-\dfrac{1}{2}\right)$

FUNCTIONS AND ORDERED PAIRS

In this book, we are interested in a special kind of relation called a **function.** We talked about a function machine, and we used function notation. As expressed already, the idea is that each x we put in (our input) gives us a unique output. No matter how many times we put that value for x into the machine, the output is always the same. This concept is the central idea of a function.

> A *function* is a set of ordered pairs (a relation) in which no two first coordinates are equal.

Given a function f, the pair of numbers $(x, f(x))$ is very significant. We always write them in that order, hence the name *ordered pairs*. In Example 2(a), we saw that, given $f(x) = x^3 - 3x^2 + x + 5$, $f(0) = 5$, which meant that the ordered pair $(0, 5)$ was associated with the function. The ordered pair consists of the x value first and the function value at that x (the $f(x)$) second.

● Example 3

Finding Ordered Pairs

Given the function $f(x) = 2x^2 - 3x + 5$, find the ordered pair $(x, f(x))$ associated with each given value for x.

(a) $x = 0$

$f(0) = 5$, so the ordered pair is $(0, 5)$.

(b) $x = -1$

$f(-1) = 2(-1)^2 - 3(-1) + 5 = 10$. The ordered pair is $(-1, 10)$.

(c) $x = \dfrac{1}{4}$

$f\left(\dfrac{1}{4}\right) = 2\left(\dfrac{1}{16}\right) - 3\left(\dfrac{1}{4}\right) + 5 = \dfrac{35}{8}$. The ordered pair is $\left(\dfrac{1}{4}, \dfrac{35}{8}\right)$.

● ● ● ● **CHECK YOURSELF 3**

Given $f(x) = 2x^3 - x^2 + 3x - 2$, find the ordered pair associated with each given value of x.

(a) $x = 0$ **(b)** $x = 3$ **(c)** $x = -\dfrac{1}{2}$

FUNCTIONS AND DOMAIN AND RANGE

We saw the idea of a function machine as a model that allows us to put in a value for x and get out a value that is a function of x. These two values, x and $f(x)$, have a relationship that can be represented as an ordered pair.

A similar type of relationship is used in every field in which mathematics is applied.

- The physicist looks for the relationship that uses a planet's mass to predict its gravitational pull.
- The economist looks for the relationship that uses the tax rate to predict the employment rate.
- The business marketer looks for the relationship that uses an item's price to predict the number that will be sold.
- The college board looks for the relationship between tuition costs and the number of students enrolled at the college.
- The biologist looks for the relationship that uses temperature to predict a body of water's nutrient level.

In each of these examples, a researcher matches an item from the given set (the *domain*) with an item from the related set (the *range*.) Each pairing becomes an ordered pair.

FUNCTIONS AND RELATIONS

In Section 1.1, we looked at examples of a *relation,* which is a set of ordered pairs. Above, we mentioned the relationship between a planet's mass and its gravitational pull. This relationship is an example of a function. There cannot be two different gravitational pulls associated with the same planet. If you know a planet's mass, you can find its gravitational pull.

Every set of ordered pairs defines a relation, but not every set of ordered pairs defines a function, as illustrated in Example 4.

• Example 4

Identifying a Function

For each table of values below, decide whether the relation is a function.

(a)

x	y
−2	1
−1	1
1	3
2	3

(b)

x	y
−5	−2
−1	3
−1	6
2	9

(c)

x	y
−3	1
−1	0
0	2
2	4

Part (*a*) represents a function. No element of the domain (*x*) is matched with two different elements of the range (*y*). Part (*b*) is not a function because −1 is matched with two different range elements, 3 and 6. Part (*c*) is a function.

● ● ●   **CHECK YOURSELF 4**

For each table of values below, decide whether the relation is a function.

(a)

x	y
−3	0
−1	1
1	2
3	3

(b)

x	y
−2	−2
−1	−2
1	3
2	3

(c)

x	y
−2	0
−1	1
0	2
0	3

We defined a function in terms of ordered pairs. A set of ordered pairs can be specified in several ways; here are the most common.

> **1.** We can present the ordered pairs in a list or table, as in Example 4.
>
> **2.** We can give a rule or equation that will generate the ordered pairs.
>
> **3.** We can use a graph to indicate the ordered pairs. The graph can show distinct ordered pairs, or it can show all the ordered pairs on a line or curve.

VERTICAL LINE TEST

Let's look at a graph of the ordered pairs from Example 4 to introduce the **vertical line test,** which is a graphic test for identifying a function.

(a)

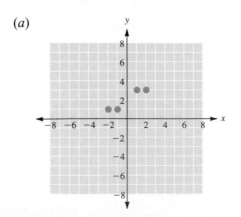

(b)

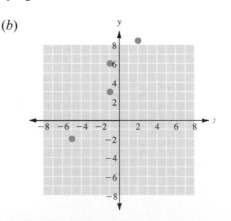

(c)

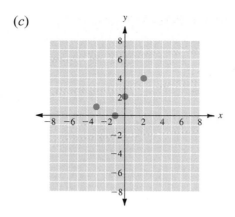

Notice that in the graphs of relations (*a*) and (*c*), there is no vertical line that can pass through two different points of the graph. In relation (*b*), a vertical line can pass through the two points that represent the ordered pairs $(-1, 3)$ and $(-1, 6)$. This leads to the following definition.

Vertical Line Test

If no vertical line can pass through two or more points in the graph of a relation, then the relation is a function.

● Example 5

Identifying a Function

For each set of ordered pairs, plot the related points on the provided axes. Then use the vertical line test to determine which of the sets is a function.

(*a*) $\{(0, -1), (2, 3), (2, 6), (4, 2), (6, 3)\}$

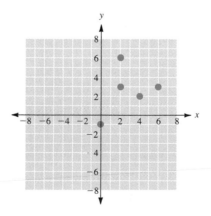

Since a vertical line can be drawn through the points $(2, 3)$ and $(2, 6)$, the relation does not pass the vertical line test. This is not a function.

(*b*) {(1, 1), (2, 0), (3, 3), (4, 3), (5, 3)}

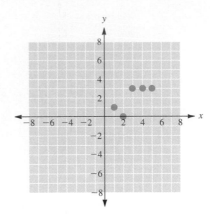

This is a function. Although a horizontal line can be drawn through several points, no vertical line passes through more than one point.

● ● ●   **CHECK YOURSELF 5**

For each set of ordered pairs, plot the related points. Then use the vertical line test to determine which of the sets is a function.

(a) {(−2, 4), (−1, 4), (0, 4), (1, 3), (5, 5)}
(b) {(−3, −1), (−1, −3), (1, −3), (1, 3)}

The vertical line test can be used to determine whether a graph is the graph of a function.

●Example 6

Identifying a Function

Which of the following graphs represents the graph of a function?

(*a*)

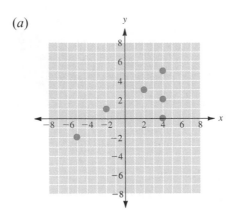

(*b*)

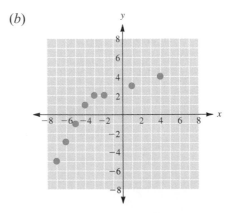

(*c*)

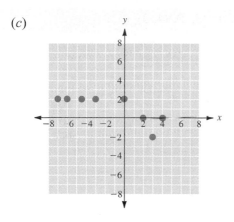

Part (*a*) is not a function, part (*b*) is a function, and part (*c*) is a function.

● ● ● **CHECK YOURSELF 8**

Which of the following graphs represents the graph of a function?

(a)

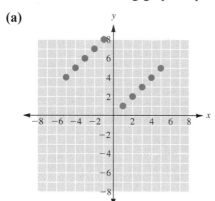

(b)

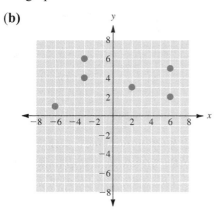

(c)

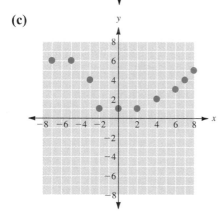

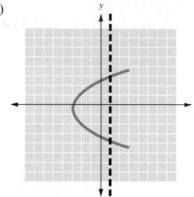

•Example 7

Identifying a Function

Which of the following graphs represents the graph of a function?

(a)

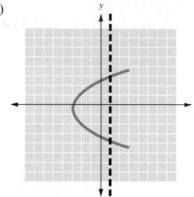

(b)

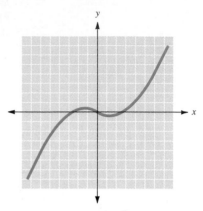

Curves, like the number line, are made up of a continuous set of points.

(c)

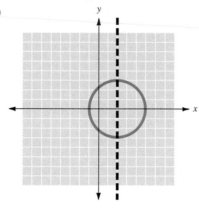

Part (*a*) is not a function; it does not pass the vertical line test. Part (*b*) is a function because it passes the vertical line test. Part (*c*) is not a function.

● ● ● **CHECK YOURSELF 7**

Which of the following graphs represents the graph of a function?

(a)

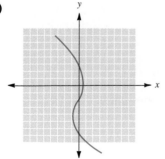

(b)

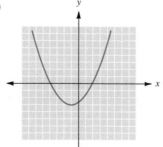

(c)

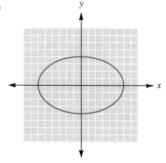

● ● ● **CHECK YOURSELF ANSWERS**

1. (a) 1, **(b)** 3, **(c)** −33.　　**2. (a)** −2, **(b)** 52, **(c)** −4.　　**3. (a)** (0, −2),

(b) (3, 52), **(c)** $(-\frac{1}{2}, -4)$.　**4. (a)** Is a function, **(b)** is a function, **(c)** is not a function.

5. (a) Is a function, **(b)** is not a function.

6. (a) Is a function, **(b)** is not a function, **(c)** is a function.

7. (a) Is not a function, **(b)** is a function, **(c)** is not a function.

1.3 Exercises

In Exercises 1 to 10, evaluate each function for the value specified.

1. $f(x) = x^2 - x - 2$; find **(a)** $f(0)$, **(b)** $f(-2)$, and **(c)** $f(1)$.

2. $f(x) = x^2 - 7x + 10$; find **(a)** $f(0)$, **(b)** $f(5)$, and **(c)** $f(-2)$.

3. $f(x) = 3x^2 + x - 1$; find **(a)** $f(-2)$, **(b)** $f(0)$, and **(c)** $f(1)$.

4. $f(x) = -x^2 - x - 2$; find **(a)** $f(-1)$, **(b)** $f(0)$, and **(c)** $f(2)$.

5. $f(x) = x^3 - 2x^2 + 5x - 2$; find **(a)** $f(-3)$, **(b)** $f(0)$, and **(c)** $f(1)$.

6. $f(x) = -2x^3 + 5x^2 - x - 1$; find **(a)** $f(-1)$, **(b)** $f(0)$, and **(c)** $f(2)$.

7. $f(x) = -3x^3 + 2x^2 - 5x + 3$; find **(a)** $f(-2)$, **(b)** $f(0)$, and **(c)** $f(3)$.

8. $f(x) = -x^3 + 5x^2 - 7x - 8$; find **(a)** $f(-3)$, **(b)** $f(0)$, and **(c)** $f(2)$.

9. $f(x) = 2x^3 + 4x^2 + 5x + 2$; find **(a)** $f(-1)$, **(b)** $f(0)$, and **(c)** $f(1)$.

10. $f(x) = -x^3 + 2x^2 - 7x + 9$; find **(a)** $f(-2)$, **(b)** $f(0)$, and **(c)** $f(2)$.

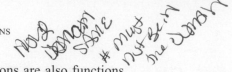

In Exercises 11 to 18, determine which of the relations are also functions.

11. {(1, 6), (2, 8), (3, 9)} *function*

12. {(2, 3), (3, 4), (5, 9)}

13. {(−1, 4), (−2, 5), (−3, 7)} *function*

14. {(−2, 1), (−3, 4), (−4, 6)}

15. {(1, 3), (1, 2), (1, 1)} *not function*

16. {(2, 4), (2, 5), (3, 6)}

17. {(−1, 1), (2, 1), (2, 3)} *not function*

18. {(2, −1), (3, 4), (3, −1)}

In Exercises 19 to 24, decide whether the relation is a function in each table of values.

19.

x	y
3	1
−2	4
5	3
−7	4

yes

20.

x	y
−2	3
1	4
5	6
2	−1

21.

x	y
2	3
4	2
2	−5
−6	−3

no

22.

x	y
1	5
3	−6
1	−5
−2	−9

23.

x	y
−1	2
3	6
6	2
−9	4

yes

24.

x	y
4	−6
2	3
−7	1
−3	−6

In Exercises 25 to 30, for each set of ordered pairs, plot the related points on the graph. Then use the vertical line test to determine which sets are functions.

25. {(−3, 1), (−1, 2), (−2, 3), (1, 4)}

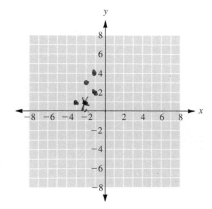

26. {(2, 2), (1, 1), (3, 3), (4, 5)}

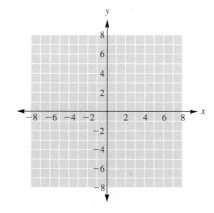

27. $\{(-1, 1), (2, 2), (3, 4), (5, 6)\}$

28. $\{(1, 4), (-1, 5), (0, 2), (2, 3)\}$

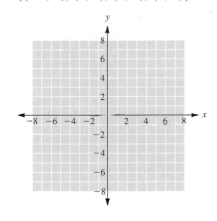

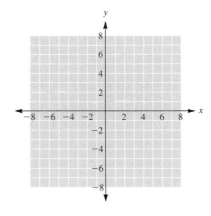

29. $\{(1, 2), (1, 3), (2, 1), (3, 1)\}$

30. $\{(-1, 1), (3, 4), (-1, 2), (5, 3)\}$

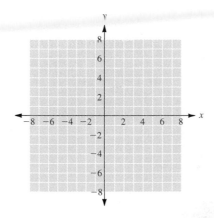

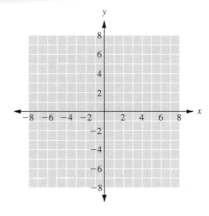

 Your graphing calculator can be used to evaluate a function for a specific value of x. If $f(x) = 3x^2 - 7$, and you wish to find $f(-3)$,

    1. Use the $\boxed{Y =}$ key to enter $Y_1 = 3x^2 - 7$.

    2. Select $\boxed{\text{TABLE}}$ ($\boxed{2^{\text{nd}}}$ $\boxed{\text{GRAPH}}$), and choose -3 for x.

    3. The table will give you a value of 11 for Y_1.

Use that technique to evaluate the functions in Exercises 31 to 34.

31. $f(x) = 3x^2 - 5x + 7$; find **(a)** $f(-5)$, **(b)** $f(5)$, and **(c)** $f(12)$.

32. $f(x) = 4x^3 - 7x^2 + 9$; find **(a)** $f(-6)$, **(b)** $f(6)$, and **(c)** $f(10)$.

33. $f(x) = 3x^4 - 6x^3 + 2x^2 - 17$; find **(a)** $f(-3)$, **(b)** $f(4)$, and **(c)** $f(7)$.

34. $f(x) = 5x^7 + 8x^4 - 9x^2 - 13$; find **(a)** $f(-4)$, **(b)** $f(-3)$, and **(c)** $f(2)$.

For Exercises 35 to 42, use the vertical line test to determine whether the graphs represent a function.

35.

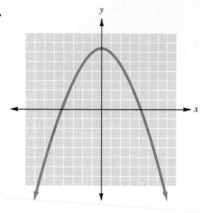

36.

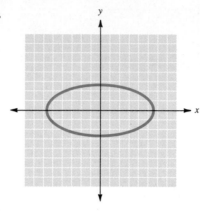

37.

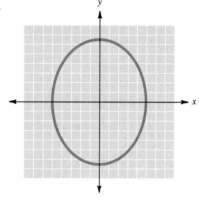

38.

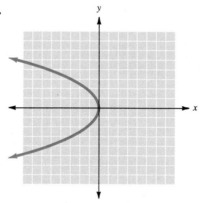

39.

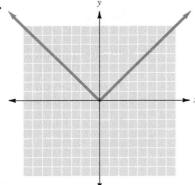

40.

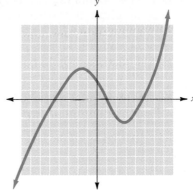

41.

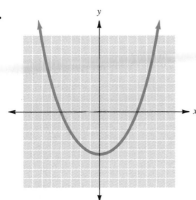

42.

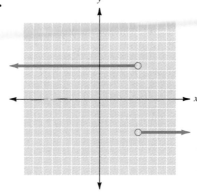

43. Consider the following graph.

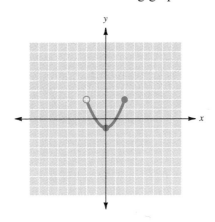

(a) Identify the domain and range of the relation whose graph is given.

(b) Does this graph represent a function? Explain your answer.

(c) How do you use the graph to determine the domain and range of the relation it represents?

 44. Are all relations functions? Are all functions relations? Explain your answer.

45. The following table shows the average hourly earnings for blue collar workers from 1947 to 1993. These figures are given in "real" wages, which means that the *purchasing* power of the money is given rather than the actual dollar amount. In other words, the amount earned for 1947 is not the actual amount listed here; in fact, it was much lower. The amount you see here is the amount in dollars that 1947 earnings could buy in 1947 compared to what 1993 wages could buy in 1993.

Year	Average Hourly Earnings (in 1993 dollars)
1947	$ 6.75
1967	10.67
1973	12.06
1979	12.03
1982	11.61
1989	11.26
1991	10.95
1993	10.83

Make a cartesian coordinate graph of this data, using the year as the domain and the hourly earnings as the range. You will have to decide how to set up the axes so that the data all fit on the graph nicely. (*Hint*: Do not start the year at 0!) In complete sentences, answer the following questions: What are the trends that you notice from reading the table? What additional information does the graph show? Is this relation a function? Why or why not?

Answers

1. (a) −2, (b) 4, (c) −2   **3.** (a) 9, (b) −1, (c) 3   **5.** (a) −62, (b) −2, (c) 2   **7.** (a) 45, (b) 3, (c) −75

9. (a) −1, (b) 2, (c) 13   **11.** Function   **13.** Function   **15.** Not a function   **17.** Not a function

19. Function   **21.** Not a function   **23.** Function   **25.**

27.

Function

Function

29.

Not a Function

31. 107, 57, 379   **33.** 406, 399, 5226   **35.** Function

37. Not a function   **39.** Function   **41.** Function   **43.** (a) D: −2 < x ≤ 2; R: −1 ≤ y ≤ 2;
(b) Yes; (c) Answers will vary   **45.**

Properties of Functions

1.4 OBJECTIVES

1. Find the sum or difference of two functions
2. Find the product of two functions
3. Find the quotient of two functions
4. Find the domain of sum or difference of functions

The profit that a company makes on an item is determined by subtracting the cost of making the item from the total revenue the company receives from selling the item. This is an example of **combining functions.** It can be written as

$$P(x) = R(x) - C(x)$$

Many applications of functions involve the combining of two or more component functions. In this section, we will look at several properties that allow for the addition, subtraction, multiplication, and division of functions.

> The **sum of two functions** f and g is written as $f + g$ and can be defined as
>
> $$f + g: (f + g)(x) = f(x) + g(x)$$
>
> for every value of x that is in the domain of both functions f and g.

> The **difference of two functions** f and g is written as $f - g$ and can be defined as
>
> $$f - g: (f - g)(x) = f(x) - g(x)$$
>
> for every value of x that is in the domain of both functions f and g.

● Example 1

Finding the Sum or Difference of Two Functions

Given the functions $f(x) = 2x - 1$ and $g(x) = -3x + 4$

(a) Find $f + g$.

(b) Find $f - g$.

(c) Find $(f + g)(2)$.

Solution

(a) $(f + g)(x) = f(x) + g(x)$

$$= (2x - 1) + (-3x + 4) = -x + 3$$

(b) $(f - g)(x) = f(x) - g(x)$

$$= (2x - 1) - (-3x + 4) = 5x - 5$$

(*c*) If we use the definition of the sum of two functions, we find that

$$(f + g)(2) = f(2) + g(2)$$
$$= 3 + (-2) = 1$$

As an alternative, we could use part (*a*) and say

$$(f + g)(x) = -x + 3$$

Therefore,

$$(f + g)(2) = -2 + 3$$
$$= 1$$

● ● ● **CHECK YOURSELF 1**

Given the functions $f(x) = -2x - 3$ and $g(x) = 5x - 1$,

(a) Find $f + g$.　　　　**(b)** Find $f - g$.　　　　**(c)** Find $(f + g)(2)$.

In defining the sum of two functions, we indicated that the domain was determined by the domain of both functions. We will find the domain in Example 2.

● Example 2

Finding the Domain of the Sum or Difference of Functions

Given $f(x) = 2x - 4$ and $g(x) = \dfrac{1}{x}$,

(*a*) Find $f + g$.

(*b*) Find the domain of $f + g$.

Solution

(*a*) $f + g = (2x - 4) + \dfrac{1}{x} = 2x - 4 + \dfrac{1}{x}$

(*b*) The domain of $f + g$ is the set of all numbers in the domain of f and also in the domain of g. The domain of f consists of all real numbers. The domain of g consists of all real numbers except 0 because we cannot divide by 0. The domain of $f + g$ is the set of all real numbers except 0.

● ● ● **CHECK YOURSELF 2**

Given $f(x) = -3x + 1$ and $g(x) = \dfrac{1}{(x - 2)}$,

(a) Find $f + g$.　　　　　　　　**(b)** Find the domain of $f + g$.

From earlier work in algebra, you should recall that the product of two binomials,

$(x + a)(x + b) =$

$x^2 + bx + ax + ab$

> The **product of two functions** f and g is written as $f \cdot g$ and can be defined as
>
> $f \cdot g: (f \cdot g)(x) = f(x) \cdot g(x)$
>
> for every value of x that is in the domain of both functions f and g.

• Example 3

Finding the Product of Two Functions

Given $f(x) = x - 1$ and $g(x) = x + 5$, find $f \cdot g$.

$f \cdot g = f(x) \cdot g(x) = (x - 1)(x + 5) = x^2 + 5x \ \ x - 5 = x^2 + 4x - 5$

●●● **CHECK YOURSELF 3**

Given $f(x) = x - 3$ and $g(x) = x + 2$, find $f \cdot g$.

The final operation on functions that we will look at involves the division of two functions.

> The **quotient of two functions** f and g is written as $f \div g$ and can be defined as
>
> $f \div g: (f \div g)(x) = f(x) \div g(x)$
>
> for every value of x that is in the domain of both functions f and g, where $g \neq 0$.

• Example 4

Finding the Quotient of Two Functions

Given $f(x) = x - 1$ and $g(x) = x + 5$,

(a) Find $f \div g$.

(b) Find the domain of $f \div g$.

Solution

(a) $f \div g = f(x) \div g(x) = (x - 1) \div (x + 5) = \dfrac{x - 1}{x + 5}$

(b) The domain is the set of all real numbers except -5 because $g(-5) = 0$, and division by 0 is undefined.

● ● ● **CHECK YOURSELF 4**

Given $f(x) = x - 3$ and $g(x) = x + 2$,

(a) Find $f \div g$. **(b)** Find the domain for $f \div g$.

● ● ● **CHECK YOURSELF ANSWERS**

1. (a) $3x - 4$, **(b)** $-7x - 2$, **(c)** 2. **2. (a)** $-3x + 1 + \dfrac{1}{(x-2)}$,

(b) $D = \{x | x \neq 2\}$. **3.** $(x - 3)(x + 2) = x^2 - x - 6$. **4. (a)** $\dfrac{(x-3)}{(x+2)}$,
(b) $D = \{x | x \neq -2\}$.

1.4 Exercises

In Exercises 1 to 8, find **(a)** $f + g$, **(b)** $f - g$, **(c)** $(f + g)(3)$, and **(d)** $(f - g)(2)$.

1. $f(x) = -4x + 5 \quad g(x) = 7x - 4$ **2.** $f(x) = 9x - 3 \quad g(x) = -3x + 5$

3. $f(x) = 8x - 2 \quad g(x) = -5x + 6$ **4.** $f(x) = -7x + 9 \quad g(x) = 2x - 1$

5. $f(x) = x^2 + x - 1 \quad g(x) = -3x^2 - 2x + 5$ **6.** $f(x) = -3x^2 - 2x + 5 \quad g(x) = 5x^2 + 3x - 6$

7. $f(x) = -x^3 - 5x + 8 \quad g(x) = 2x^2 + 3x - 4$ **8.** $f(x) = 2x^3 + 3x^2 - 5 \quad g(x) = -4x^2 + 5x - 7$

In Exercises 9 to 14, find **(a)** $f + g$ and **(b)** the domain of $f + g$.

9. $f(x) = -9x + 11 \quad g(x) = 15x - 7$ **10.** $f(x) = -11x + 3 \quad g(x) = 8x - 5$

11. $f(x) = 3x + 2 \quad g(x) = \dfrac{1}{x - 2}$ **12.** $f(x) = -2x + 5 \quad g(x) = \dfrac{3}{x + 1}$

13. $f(x) = x^2 + x - 5$ $g(x) = \dfrac{2}{3x + 1}$

14. $f(x) = 3x^2 - 5x + 1$ $g(x) = \dfrac{-2}{2x - 3}$

In Exercises 15 to 20, find **(a)** $f \cdot g$, **(b)** $\dfrac{f}{g}$, and **(c)** the domain of $\dfrac{f}{g}$.

15. $f(x) = 2x - 1$ $g(x) = x - 3$

16. $f(x) = x + 3$ $g(x) = x + 4$

17. $f(x) = 3x + 2$ $g(x) = 2x - 1$

18. $f(x) = -3x + 5$ $g(x) = -x + 2$

19. $f(x) = x - 1$ $g(x) = x^2 + x + 1$

20. $f(x) = x + 2$ $g(x) = x^2 - 2x + 4$

In business, the profit, $P(x)$, obtained from selling x units of a product is equal to the revenue, $R(x)$, minus the cost, $C(x)$. In Exercises 21 and 22, find the profit, $P(x)$, for selling x units.

21. $R(x) = 25x$ $C(x) = x^2 + 4x + 50$

22. $R(x) = 20x$ $C(x) = x^2 + 2x + 30$

The velocity, $V(t)$, of a freely falling object is the sum of two functions: the initial velocity, V_0, with which it is thrown, and the acceleration, $a(t)$, which is the change in velocity due to gravity, such that

$$V(t) = V_0 + a(t)$$

In Exercises 23 and 24, find the velocity at any time.

23. $V_0 = 10$m/s $a(t) = -4.9t^2$

24. $V_0 = 64$ft/s^2 $a(t) = -16t^2$

Answers

1. (a) $3x + 1$, **(b)** $-11x + 9$, **(c)** 10, **(d)** -13 **3. (a)** $3x + 4$, **(b)** $13x - 8$, **(c)** 13, **(d)** 18

5. (a) $-2x^2 - x + 4$, **(b)** $4x^2 + 3x - 6$, **(c)** -17, **(d)** 16 '   **7. (a)** $-x^3 + 2x^2 - 2x + 4$,

(b) $-x^3 - 2x^2 - 8x + 12$; **(c)** -11, **(d)** -20 **9. (a)** $6x + 4$, **(b)** all reals   **11. (a)** $3x + 2 + \dfrac{1}{x - 2}$,

(b) $\{x | x \neq 2\}$ **13. (a)** $x^2 + x - 5 + \dfrac{2}{3x + 1}$, **(b)** all reals $\neq -\dfrac{1}{3}$ **15. (a)** $(2x - 1)(x - 3) = 2x^2 - 7x + 3$,

(b) $\dfrac{2x - 1}{x - 3}$, **(c)** $\{x | x \neq 3\}$ **17. (a)** $6x^2 + x - 2$, **(b)** $\dfrac{3x + 2}{2x - 1}$, **(c)** $\left\{ x | x \neq \dfrac{1}{2} \right\}$

19. (a) $(x - 1)(x^2 + x + 1) = x^3 - 1$, **(b)** $\dfrac{x - 1}{x^2 + x + 1}$, **(c)** all reals   **21.** $P(x) = -x^2 + 21x - 50$

23. $V = 10 - 4.9t^2$

1.5 The Graph of a Linear Equation

In previous algebra classes you solved equations in one variable such as

$$2x - 1 = x + 2$$

Solving such equations required finding the value for x that made the equation a true statement. In this case, the value was $x = 3$ because

$$2(3) - 1 = (3) + 2$$

This is a true statement because each side of the equation is equal to 5; no other value for x makes the statement true. Thus, the solution can be written three ways: as $x = 3$; $\{x | x = 3\}$, which is read "x, such that x is equal to 3"; or simply $\{3\}$, which is the set that contains only the number 3.

What if we have an equation in two variables, such as $x + 2y = 6$? The solution set is defined in a similar fashion.

Solution Set for an Equation in Two Variables

The **solution set** for an equation in two variables is the set containing all ordered pairs of real numbers (x, y) that will make the equation a true statement.

The solution set for an equation in two variables is a set of ordered pairs. Typically, there will be an infinite number of ordered pairs that make an equation a true statement. We can find some of these ordered pairs by substituting a value for x, then solving the remaining equation for y. We will use that technique in Example 1.

● Example 1

Finding Ordered Pair Solutions

Find three ordered pairs that are solutions for each equation.

(a) $3x + y = 6$

We will pick three values for x, set up a table for ordered pairs, and then determine the related value for y.

x	y
-1	
0	
1	

Substituting -1 for x, we get

$$3(-1) + y = 6$$
$$-3 + y = 6$$
$$y = 9$$

The ordered pair $(-1, 9)$ is a solution to the equation $3x + y = 6$.
Substituting 0 for x, we get

$$3(0) + y = 6$$
$$0 + y = 6$$
$$y = 6$$

The ordered pair $(0, 6)$ is a solution to the equation $3x + y = 6$.
Substituting 1 for x, we get

$$3(1) + y = 6$$
$$3 + y = 6$$
$$y = 3$$

To indicate the set of all solutions to the equation, we write

$\{(x, y) \mid 3x + y = 6\}$

The ordered pair $(1, 3)$ is a solution to the equation $3x + y = 6$.
Completing the table gives us the following:

x	y
-1	9
0	6
1	3

(b) $2x - y = 1$

Let's try a different set of values for x. We will use the following table.

x	y
-5	
0	
5	

Substituting -5 for x, we get

$$2(-5) - y = 1$$
$$-10 - y = 1$$
$$-y = 11$$
$$y = -11$$

The ordered pair $(-5, -11)$ is a solution to the equation $2x - y = 1$.
Substituting 0 for x, we get

$$2(0) - y = 1$$
$$0 - y = 1$$
$$-y = 1$$
$$y = -1$$

Again, the set of all solutions is

$\{(x, y) \mid 2x - y = 1\}$

The ordered pair $(0, -1)$ is a solution to the equation $2x - y = 1$.

Substituting 5 for x, we get

$2(5) - y = 1$

$10 - y = 1$

$-y = -9$

$y = 9$

The ordered pair $(5, 9)$ is a solution to the equation $2x - y = 1$.

Completing the table gives us the following:

x	y
-5	-11
0	-1
5	9

● ● ● CHECK YOURSELF 1

Find three ordered pairs that are solutions for each equation.

(a) $2x - y = 6$ **(b)** $3x + y = 2$

The graph of the solution set of an equation in two variables, usually called the **graph of the equation,** is the set of all points with coordinates (x, y) that satisfy the equation.

In this section, we are primarily interested in a particular kind of equation in x and y and the graph of that equation. The equations we refer to involve x and y to the first power, and they are called **linear equations.**

Linear Equations

An equation of the form

$ax + by = c$

where a and b cannot both be zero, is called the **standard form for a line.** Its graph is always a straight line.

● Example 2

Graphing by Plotting Points

Graph the equation

$x + y = 5$

Since two points determine a straight line, technically two points are all that are needed to graph the equation. You may want to locate at least one other point as a check of your work.

If you first rewrite an equation so that y is isolated on the left side, it can be easily entered and graphed with a graphing calculator. In this case, graph the equation

$$y = -x + 5$$

This is a linear equation in two variables. To draw its graph, we can begin by assigning values to x and finding the corresponding values for y. For instance, if $x = 1$, we have

$$1 + y = 5$$
$$y = 4$$

Therefore, $(1, 4)$ satisfies the equation and is in the graph of $x + y = 5$.

Similarly, $(2, 3)$, $(3, 2)$, and $(4, 1)$ are in the graph. Often these results are recorded in a table of values, as shown below. We then plot the points determined and draw a straight line through those points.

$x + y = 5$

x	y
1	4
2	3
3	2
4	1

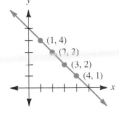

Every point on the graph of the equation $x + y = 5$ has coordinates that satisfy the equation, and every point with coordinates that satisfy the equation lies on the line.

● ● ● CHECK YOURSELF 2

Graph the equation $2x - y = 6$.

An algorithm is a sequence of steps that solve a problem.

The following algorithm summarizes our first approach to **graphing a linear equation in two variables.**

> **To Graph a Linear Equation**
>
> **STEP 1** Find at least three solutions for the equation, and write your results in a table of values.
> **STEP 2** Graph the points associated with the ordered pairs found in step 1.
> **STEP 3** Draw a straight line through the points plotted above to form the graph of the equation.

Two particular points are often used in graphing an equation because they are very easy to find. The *x intercept* of a line is the x coordinate of the point where the line crosses the x axis. If the x intercept exists, it can be found by setting $y = 0$ in the equation and solving for x. The *y intercept* is the y coordinate of the point where the line crosses the y axis. If the y intercept exists, it is found by letting $x = 0$ and solving for y.

Solving for y, we get

$$y = \frac{1}{2}x - 3$$

To graph this result on your calculator, you can enter

$$Y_1 = (1 \div 2)\, x - 3$$

using the $\boxed{X, T, \theta, n}$ key for x.

Graphing by the Intercept Method

Use the intercepts to graph the equation

$$x - 2y = 6$$

To find the x intercept, let $y = 0$.

$$x - 2 \cdot 0 = 6$$
$$x = 6$$

To find the y intercept, let $x = 0$.

$$0 - 2y = 6$$
$$-2y = 6$$
$$y = -3$$

Graphing the intercepts and drawing the line through those intercepts, we have the desired graph.

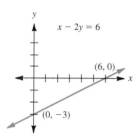

● ● ● ● **CHECK YOURSELF 3**

Graph, using the intercept method.

$$4x + 3y = 12$$

The following algorithm summarizes the steps of graphing a straight line by the **intercept method.**

Graphing by the Intercept Method

STEP 1 Find the x intercept. Let $y = 0$, and solve for x.
STEP 2 Find the y intercept. Let $x = 0$, and solve for y.
STEP 3 Plot the two intercepts determined in steps 1 and 2.

STEP 4   Draw a straight line through the intercepts.

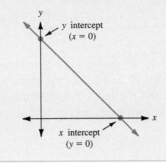

When can the intercept method not be used? Some straight lines have only one intercept. For instance, the graph of $x + 2y = 0$ passes through the origin. In this case, other points must be used to graph the equation.

● Example 4

Graph the equation

$$y = -\frac{1}{2}x$$

Note that the line passes through the origin.

Graphing a Line That Passes Through the Origin

Graph $x + 2y = 0$.

Letting $y = 0$ gives

$$x + 2 \cdot 0 = 0$$
$$x = 0$$

Thus $(0, 0)$ is a solution, and the line has only one intercept.
We continue by choosing any other convenient values for x. If $x = 2$:

$$2 + 2y = 0$$
$$2y = -2$$
$$y = -1$$

So $(2, -1)$ is a solution. You can easily verify that $(4, -2)$ is also a solution. Again, plotting the points and drawing the line through those points, we have the desired graph.

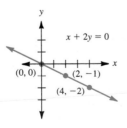

● ● ●   **CHECK YOURSELF 4**

Graph the equation $x - 3y = 0$.

In Section 1.1, we defined the terms *domain* and *range*. Recall that the domain of a relation is the set of all the first elements in the ordered pairs. The range is the set of all the second elements. Recall that a line is the graph of a set of ordered pairs. In Example 5, we will examine the domain and range for a line.

● Example 5

Finding the Domain and Range for a Line

Find the domain and range for the relation described by the equation

$x + y = 5$

We can analyze the domain and range either graphically or algebraically. First, we will look at a graphical analysis. From Example 2, let's look at the graph of the equation.

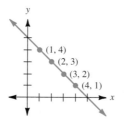

The graph continues forever at both ends. For every value of x, there is an associated point on the line. Therefore, the domain (D) is the set of all real numbers. In set builder notation, we write

$D = \{x | x \in \mathbb{R}\}$

This is read, "The domain is the set of every x that is a real number."

To find the range (R), we look at the graph to see what values are associated with y. Note that every y is associated with some point. The range is written as

$R = \{y | y \in \mathbb{R}\}$

This is read, "The range is the set of every y that is a real number."

Let's find the domain and range for the same relation by using an algebraic analysis. Look at the following equation.

$x + y = 5$

To determine the domain, we need to find every value of x that allows us to solve for y. That combination will result in an ordered pair (x, y). The set of all those x values is the domain of the relation.

We can find a value for y for *any* real value of x. For example, if $x = -5$,

$$-5 + y = 5$$
$$y = 10$$

The ordered pair $(-5, 10)$ is part of the relation. As in our graphical analysis, the domain is

$$D = \{x | x \in \mathbb{R}\}$$

and the range is

$$R = \{y | y \in \mathbb{R}\}$$

● ● ● **CHECK YOURSELF 5**

Find the domain and range for the relation described by the following equation.

$$x - y = 4$$

Two types of equations are worthy of special attention. Their graphs are lines that are parallel to the x or y axis, and the equations are special cases of the general form

$$ax + by = c$$

in which either $a = 0$ or $b = 0$.

Vertical or Horizontal Lines

1. A line with an equation of the form

$$y = k$$

is horizontal (parallel to the x axis).
2. A line with an equation of the form

$$x = h$$

is vertical (parallel to the y axis).

Example 6 illustrates both cases.

Since part (*a*) is a function, it can be graphed on your calculator. Part (*b*) is not a function and cannot be graphed on your calculator.

● **Example 6**

Graphing Horizontal and Vertical Lines

(*a*) Graph the line with equation

$$y = 3$$

You can think of the equation in the equivalent form

$$0 \cdot x + y = 3$$

Note that any ordered pair of the form (__, 3) will satisfy the equation. Since x is multiplied by 0, y will always be equal to 3.

For instance, $(-2, 3)$ and $(5, 3)$ are on the graph. The graph, a horizontal line, is shown below.

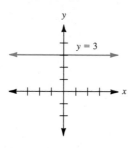

The domain for a horizontal line is every real number. The range is a single y value. We write

$$D = \{x | x \in \mathbb{R}\} \quad \text{and} \quad R = \{3\}$$

(*b*) Graph the line with equation

$$x = -2$$

In this case, you can think of the equation in the equivalent form

$$x + 0 \cdot y = -2$$

Now any ordered pair of the form $(-2, \_\_)$ will satisfy the equation. Examples are $(-2, -1)$ and $(-2, 5)$. The graph, a vertical line, is shown below.

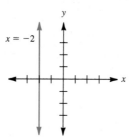

● ● ● **CHECK YOURSELF 6**

Graph each equation.

(a) $y = -3$ **(b)** $x = 5$

Many applications involve linear relationships between variables, and the methods of this section can be used to picture or graph those relationships.

In Section 1.2, we mentioned that the axes may have different scales. Example 7 is such an application involving a linear equation.

● Example 7

Adjusting the Scale for the *x* and *y* Axes

A car rental agency advertises daily rates for a midsized automobile at \$20 per day plus 10¢ per mile. The cost per day C and the distance driven in miles s are then related by the following linear equation:

$$C = 0.10s + 20 \qquad\qquad (1)$$

Graph the relationship between C and s.

First, we proceed by finding three points on the graph.

s	C
0	20
100	30
200	40

So as the distance s varies from 0 to 200 mi, the cost C changes from \$20 to \$40. To draw a "reasonable" graph, it makes sense to choose a different scale for the horizontal (or s) axis than for the vertical (or C) axis.

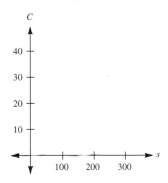

Before you graph this function on your calculator, adjust the scales on both axes. This is done from the WINDOW menu.

We have chosen units of 100 for the s axis and units of 10 for the C axis. The graph can then be completed, as shown below.

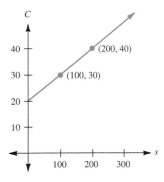

Note that the graph of equation (1) does not extend beyond the first quadrant because of the nature of our problem, in which solutions are only realistic when $s \geq 0$.

● ● ● **CHECK YOURSELF 7**

A salesperson's monthly salary S is based on a fixed salary of $1200 plus 8% of all monthly sales x. The linear equation relating S and x is

$$S = 0.08x + 1200$$

Graph the relationship between S and x. *Hint:* Find the monthly salary for sales of $0, $10,000, and $20,000.

● ● ● **CHECK YOURSELF ANSWERS**

1. (a) Answers will vary, but could include $(0, -6)$.   **(b)** Answers will vary, but could include $(0, 2)$.   **2.**

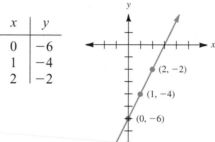

x	y
0	−6
1	−4
2	−2

3.

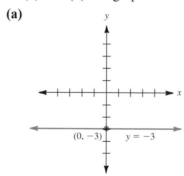

4.

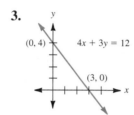

5. $D = \{x | x \in \mathbb{R}\}$ and $R = \{y | y \in \mathbb{R}\}$

6. (a) and **(b)** see graphs.

(a)

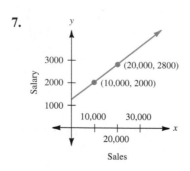

(b)

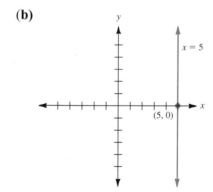

7.

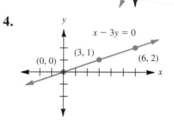

Exercise and Age Aerobic exercise requires that your heartbeat must be at a certain rate for 12 minutes or more for full physical benefit. To determine the proper heart rate for healthy persons, we start with the age of the person. To compute the proper heart rate, start with the number 220 and subtract the person's age. Then multiply by 0.70. The result is the target aerobic heart rate, the rate to maintain during exercise.

1. Write a formula for the relation between a person's age (A) and the person's target aerobic heart rate (R).
2. Using at least 10 different ages, construct a table of target heart rates by age.
3. Draw a graph of this table of values.
4. What are reasonable limits for the person's age that you would use with your formula? Would it make sense to use $A = 2$? Or $A = 150$? In other words, what is a reasonable domain for A?
5. What are the benefits of aerobic exercise over other types of exercise?
6. List some different types of exercise that are nonaerobic. Describe the differences between the two different types of exercise.

1.5 Exercises

In Exercises 1 to 8, find three ordered pairs that are solutions to the given equations.

1. $2x + y = 5$ **2.** $3x + y = 7$ **3.** $7x - y = 8$ **4.** $5x - y = 3$

5. $4x + 5y = 20$ **6.** $2x + 3y = 6$ **7.** $3x + y = 0$ **8.** $2x - y = 0$

In Exercises 9 to 26, graph each of the equations.

9. $x + y = 6$

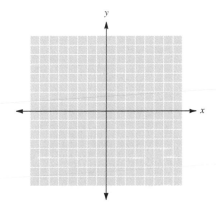

10. $x - y = 6$

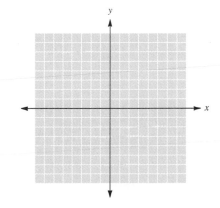

11. $y = x - 2$

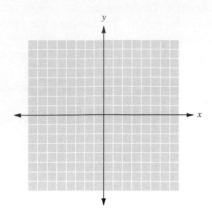

12. $y = x + 5$

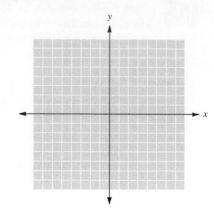

13. $y = x + 1$

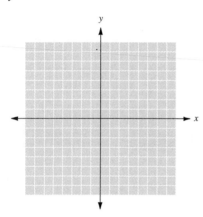

14. $y = 2x + 2$

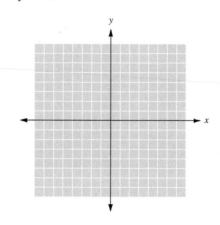

15. $y = -2x + 1$

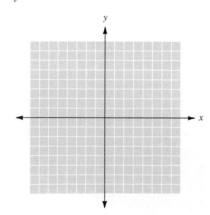

16. $y = -3x + 1$

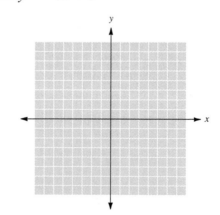

17. $y = \dfrac{1}{2}x - 3$

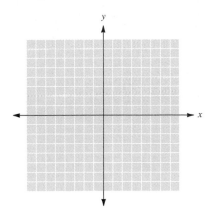

18. $y = 2x - 4$

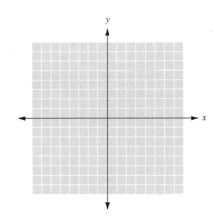

19. $y = -x - 3$

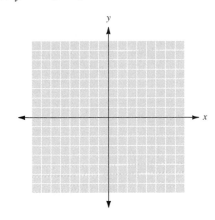

20. $y = -2x - 4$

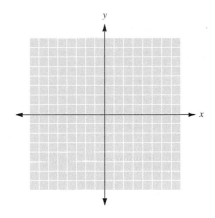

21. $x + 2y = 0$

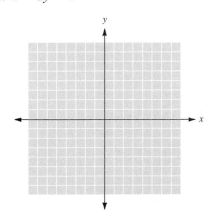

22. $x - 2y = 0$

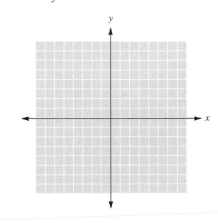

23. $x = 4$

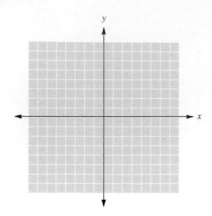

24. $x = -4$

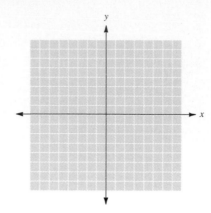

25. $y = 4$

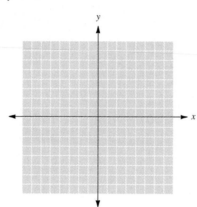

26. $y = -6$

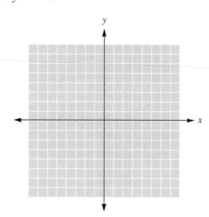

In Exercises 27 to 38, find the x and y intercepts and then graph each equation.

27. $x - 2y = 4$

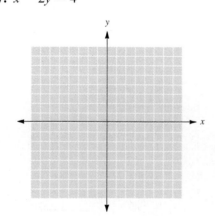

28. $x + 3y = 6$

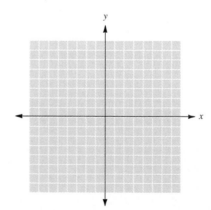

29. $2x - y = 6$

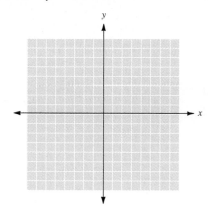

30. $3x + 2y = 12$

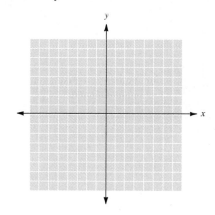

31. $2x + 5y = 10$

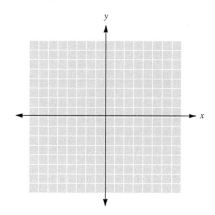

32. $2x - 3y = 6$

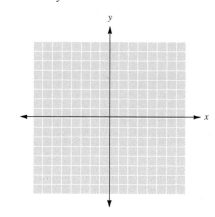

33. $5x - 6y = 0$

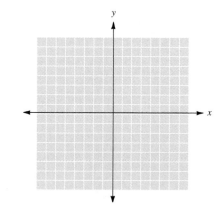

34. $2x + 7y = 0$

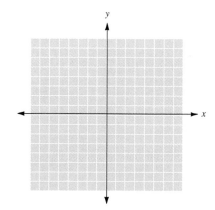

35. $x + 4y + 8 = 0$

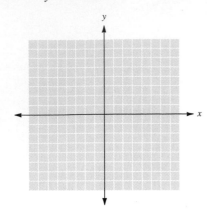

36. $2x - y + 6 = 0$

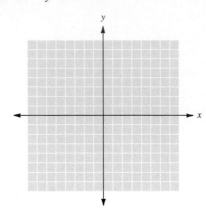

37. $8x = 4y$

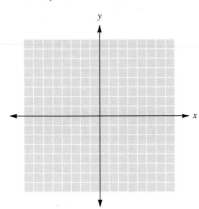

38. $6x = -7y$

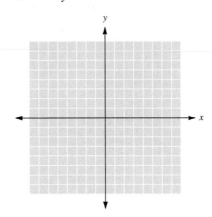

In Exercises 39 to 46, find the domain and range of each of the relations.

39. $3x + 2y = 4$

40. $5x - 4y = 20$

41. $6x + 2y = 18$

42. $-x + 5y = 8$

43. $x = 4$

44. $2x - 10 = 0$

45. $y = 3$

46. $3y + 12 = 0$

47. Consumer Affairs A car rental agency charges \$12 per day and 8¢ per mile for the use of a compact automobile. The cost of the rental C and the number of miles driven per day s are related by the equation

$C = 0.08 \, s + 12$

Graph the relationship between C and s. Be sure to select appropriate scaling for the C and s axes.

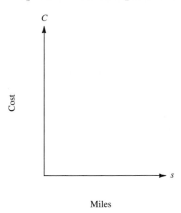

48. **Checking Account Charges** A bank has the following structure for charges on checking accounts. The monthly charges consists of a fixed amount of $8 and an additional charge of 5¢ per check. The monthly cost of an account C and the number of checks written per month n are related by the equation

$$C = 0.05n + 8$$

Graph the relationship between C and n.

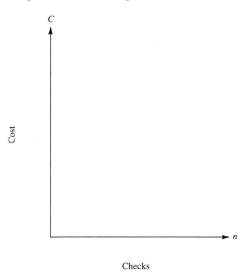

49. **Tuition Charges** A college has tuition charges based on the following pattern. Tuition is $35 per credit-hour plus a fixed student fee of $75.

(a) Write a linear equation that shows the relationship between the total tuition charge T and the number of credit-hours taken h.
(b) Graph the relationship between T and h.

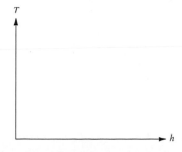

50. Weekly Salary A salesperson's weekly salary is based on a fixed amount of $200 plus 10% of the total amount of weekly sales.

   (a) Write an equation that shows the relationship between the weekly salary S and the amount of weekly sales x (in dollars).
   (b) Graph the relationship between S and x.

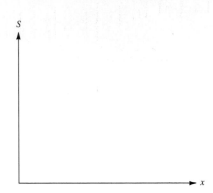

 For Exercises 51 to 58, select a window that allows you to see both the x and y intercepts on your calculator. If that is not possible, explain why not.

51. $x + y = 40$ **52.** $x - y = 80$

53. $2x + 3y = 900$ **54.** $5x - 8y = 800$

55. $y = 5x + 90$ **56.** $y = 3x - 450$

57. $y = 30x$ **58.** $y = 200$

Two distinct lines in the plane either are parallel or they intersect. In Exercises 59 to 62, graph each pair of equations on the same set of axes, and find the point of intersection, where possible.

59. $x + y = 6$ **60.** $y = x + 3$
     $x - y = 4$ $y = -x + 1$

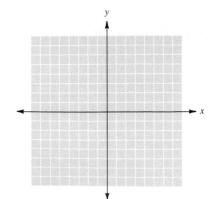 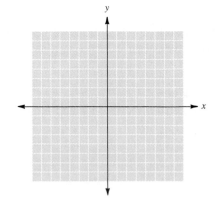

61. $y = 2x$

$y = x + 1$

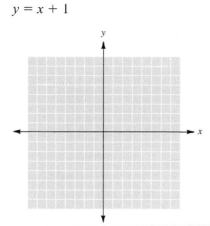

62. $2x + y = 3$

$2x + y = 5$

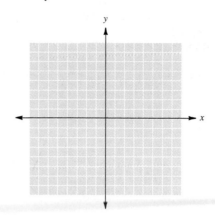

63. Graph $y = x$ and $y = 2x$ on the same set of axes. What do you observe?

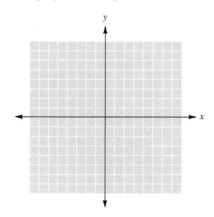

64. Graph $y = 2x + 1$ and $y = -2x + 1$ on the same set of axes. What do you observe?

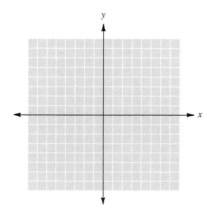

65. Graph $y = 2x$ and $y = 2x + 1$ on the same set of axes. What do you observe?

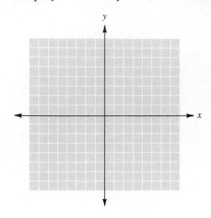

66. Graph $y = 3x + 1$ and $y = 3x - 1$ on the same set of axes. What do you observe?

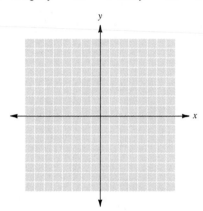

67. Graph $y = 2x$ and $y = -\dfrac{1}{2}x$ on the same set of axes. What do you observe?

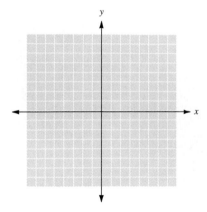

68. Graph $y = \dfrac{1}{3}x + \dfrac{7}{3}$ and $y = -3x + 2$ on the same set of axes. What do you observe?

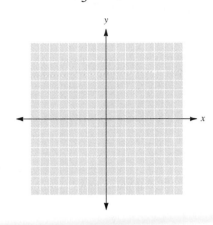

Use your graphing utility to graph each of the following equations.

69. $y = -3$

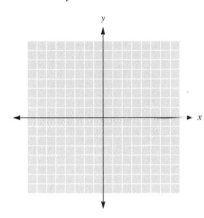

70. $y = 2$

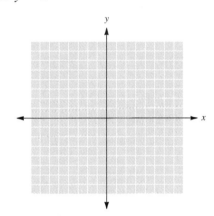

71. $y = 3x - 1$

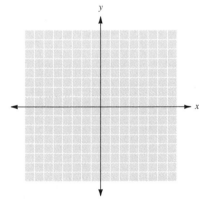

72. $y = -2x + 2$

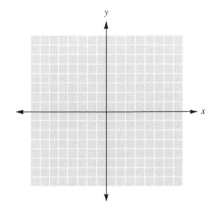

Answers

Answers

1. $\{(0, 5), (1, 3), (-1, 7)\}$ **3.** $\{(0, -8), (1, -1), (-1, -15)\}$ **5.** $\left\{(0, 4), (5, 0), \left(-1, \dfrac{24}{5}\right)\right\}$

7. $\{(0, 0), (1, -3), (-1, 3)\}$

9. $x + y = 6$

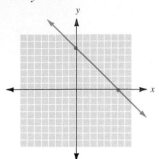

11. $y = x - 2$

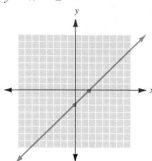

13. $y = x + 1$

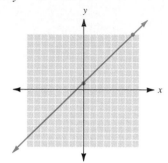

15. $y = -2x + 1$

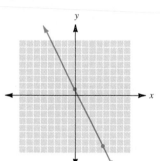

17. $y = \dfrac{1}{2}x - 3$

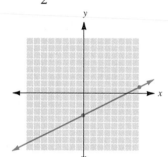

19. $y = -x - 3$

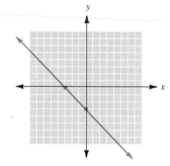

21. $x + 2y = 0$

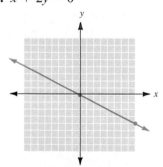

23. $x = 4$

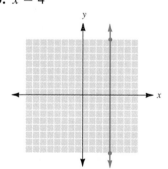

25. $y = 4$

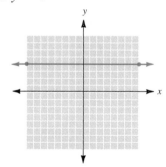

27. $x - 2y = 4$; y intercept $(0, -2)$; x intercept $(4, 0)$

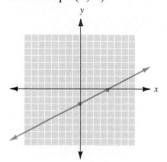

29. $2x - y = 6$; y intercept $(0, -6)$; x intercept $(3, 0)$

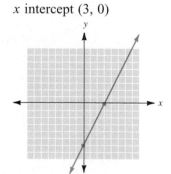

31. $2x + 5y = 10$; y intercept $(0, 2)$; x intercept $(5, 0)$

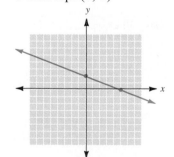

33. $5x - 6y = 0$;
intercepts: $(0, 0)$

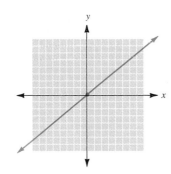

35. $x + 4y + 8 = 0$;
y intercept $(0, -2)$;
x intercept $(-8, 0)$

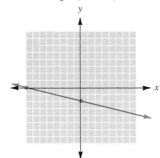

37. $8x = 4y$;
intercept: $(0, 0)$

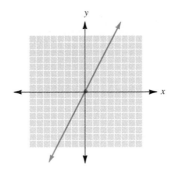

39. D: $\{x|x = \mathbb{R}\}$; R: $\{y|y = \mathbb{R}\}$

41. D: $\{x|x = \mathbb{R}\}$; R = $\{y|y = \mathbb{R}\}$

43. D: $\{4\}$; R: $\{y|y = \mathbb{R}\}$

45. D: $\{x|x = \mathbb{R}\}$; R: $\{3\}$

47. $C = 0.08s + 12$

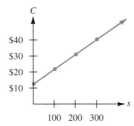

49. **(a)** $T = 35h + 75$ and **(b)** see graph

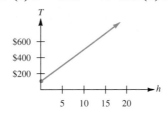

51. X max = 40, Y max = 40

53. X max = 450, Y max = 300

55. X min = -18, Y max = 90

57. The only intercept is $(0, 0)$

59. Intersection: $(5, 1)$

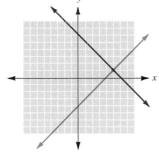

61. Intersection: $(1, 2)$

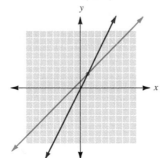

63. The line corresponding to $y = 2x$ is steeper than that corresponding to $y = x$.

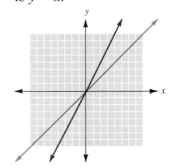

65. The two lines appear to be parallel.

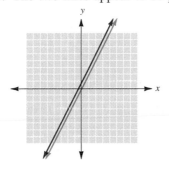

67. The lines appear to be perpendicular.

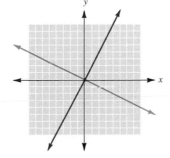

69.

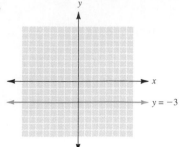

71.

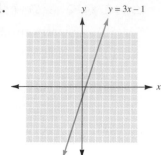

1.6 Reading Values from Graphs

1.6 OBJECTIVES

1. Given *x*, find the function value on the graph
2. Given a function value, find the related *x* value
3. Find the *x* and *y* intercepts from a graph

READING VALUES FROM GRAPHS

In Section 1.2, we learned to read the coordinates of a point by drawing a vertical line from the point to the *x* axis to find the *x* coordinate and then drawing a horizontal line from the point to the *y* axis to find the *y* coordinate. A graph of a curve (including a graph of a straight line) is actually the graph of an infinite number of connected points. Finding the coordinates of any point on a curve is exactly the same as finding the coordinates of a point.

Keep in mind that although we usually say something like, "Find the coordinates of the point . . . ," every time we read a graph we are able to only *estimate* the coordinates.

•Example 1

Reading Values from a Graph

Find the coordinates of the labeled points.

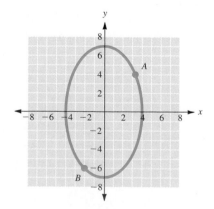

Point *A* has an *x* coordinate of 3 and a *y* coordinate of 4. Point *A* represents the ordered pair (3, 4). Point *B* represents the ordered pair $(-2, -6)$.

⬤ ⬤ ⬤ **CHECK YOURSELF 1**

Find the coordinates of the labeled points.

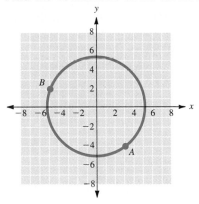

READING FUNCTION VALUES FROM GRAPHS

If a graph is the graph of a function, then every ordered pair (x, y) can be thought of as $(x, f(x))$.

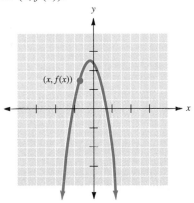

For a specific value of x, let's call it a, we can find $f(a)$ with the following algorithm.

STEP 1	Draw a vertical line through a on the x axis.
STEP 2	Find the point of intersection of that line with the graph.
STEP 3	Draw a horizontal line through the graph at that point.
STEP 4	Find the intersection of the horizontal line with the y axis.
STEP 5	$f(a)$ is that y value.

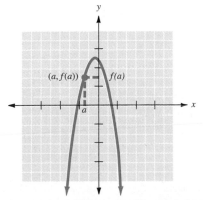

Example 2 illustrates this algorithm.

• Example 2

Finding the Function Value on a Graph Given *x*

Consider the following graph of the function *f*. Use the graph to estimate *f*(2).

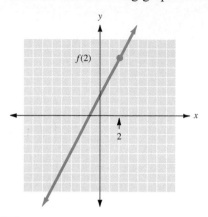

f(2) is a *y* value. It is the *y* value that is paired with an *x* value of 2. Locate the number 2 on the *x* axis, draw a vertical line to the graph of the function, and then draw a horizontal line to the *y* axis, as shown below.

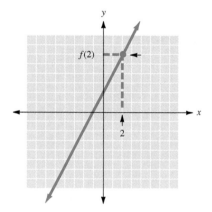

The coordinates of the point are (2, *f*(2)). The *y* value of the point is *f*(2). Read the *y* value of this point on the *y* axis. It appears that *f*(2), the *y* value of the point, is approximately 6. Therefore, *f*(2) = 6.

● ● ● **CHECK YOURSELF 2**

Using the graph of the function *f* in Example 2, estimate each of the following.

(a) *f*(1) **(b)** *f*(−1) **(c)** *f*(−3)

In the preceding problem, you were given the *x* value and asked to find the corresponding function value or *y* value. Now you will do the opposite operation. You will be given the function value and then asked to find the corresponding *x* value(s). Consider Example 3.

● Example 3

Finding the *x* Value from a Graph Given the Function Value

Use the following graph of the function f to find all values of x such that $f(x) = -5$.

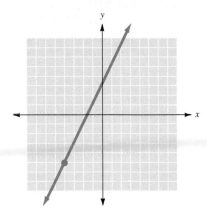

This time -5 is a function value, or y value. Locate -5 on the y axis, and draw a horizontal line to the graph of the function, followed by a vertical line to the x axis, as shown below.

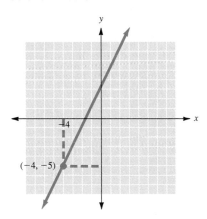

The solution of $f(x) = -5$ is $x = -4$. In particular, $f(-4) = -5$.

● ● ●   CHECK YOURSELF 3

Use the following graph to find all values of x such that

(a) $f(x) = 1,$ **(b)** $f(x) = 7,$ **(c)** $f(x) = -1.$

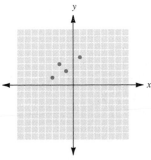

If given the function value, we can find the associated x value by using the following algorithm.

> **Finding x Values from Function Values**
>
> **STEP 1**   Find the given function value on the y axis.
> **STEP 2**   Draw a horizontal line through that point.
> **STEP 3**   Find every point on the graph that intersects the horizontal line.
> **STEP 4**   Draw a vertical line through each of those points of intersection.
> **STEP 5**   The x value(s) are each point of intersection of the vertical lines and the x axis.

READING X AND Y INTERCEPTS FROM GRAPHS

Among the most important values that can be read from graphs are the values of the x and y intercepts.

•Example 4

Finding x and y Intercepts

Find the x and y intercepts from the graph.

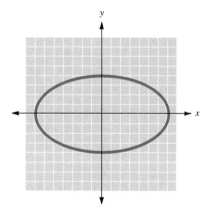

The x intercept is the value of x for which $y = 0$. It is the x value of any point on the graph that touches the x axis. This graph touches the x axis at $(7, 0)$ and also at $(-7, 0)$. The x intercepts are $(7, 0)$ and $(-7, 0)$.

The y intercept is the value of y when $x = 0$. It is the y value of any point that touches the y axis. This graph touches the y axis at $(0, 4)$ and $(0, -4)$. The y intercepts are $(0, 4)$ and $(0, -4)$.

CHECK YOURSELF 4

Find the x and y intercepts from the graph.

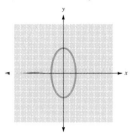

CHECK YOURSELF ANSWERS

1. (a) $(3, -4)$, **(b)** $(-5, 2)$.   **2. (a)** $f(1) = 3$, **(b)** $f(-1) = 0$, **(c)** $f(-3) = -4$.
3. (a) $x = 1, 5$; **(b)** $x = 0, 6$; **(c)** $x = 3$.   **4.** x int: $(-2, 0)$, $(2, 0)$; y int: $(0, -4)$, $(0, 4)$.

1.6 Exercises

In Exercises 1 to 12, find the coordinates of the labeled points. Assume that each small square is a 1-unit square.

1.

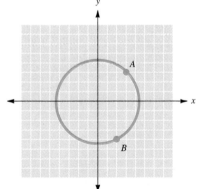

2.

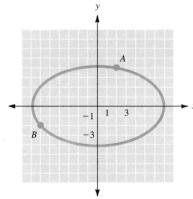

3.

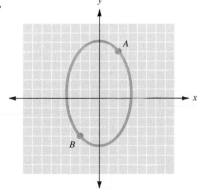

4.

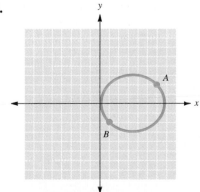

5.

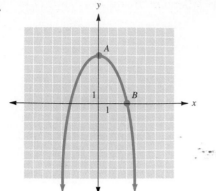

6.

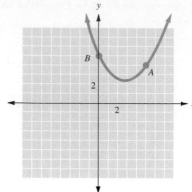

7.

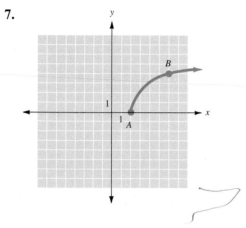

8.

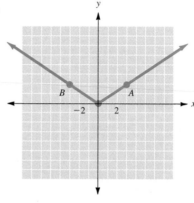

9.

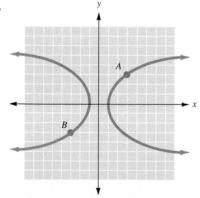

10.

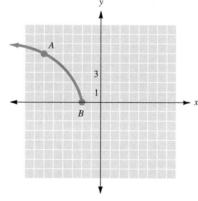

11.

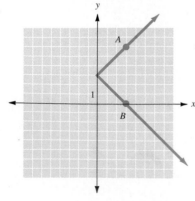

12.

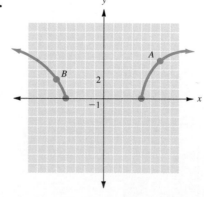

In Exercises 13 to 20, use the graph of the function to estimate each of the following values: **(a)** $f(1)$, **(b)** $f(-1)$, **(c)** $f(0)$, **(d)** $f(3)$, and **(e)** $f(-2)$.

13.

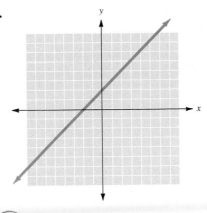

14.

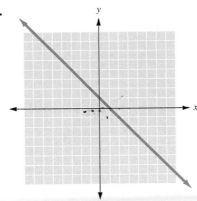

15.

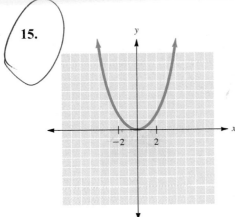

16.

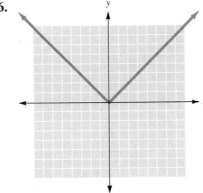

17.

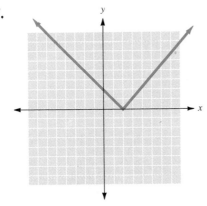

18.

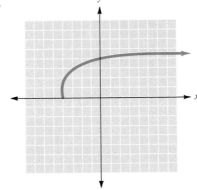

19.

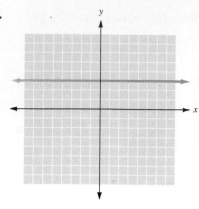

20.

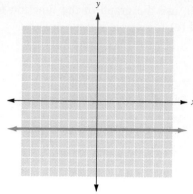

In Exercises 21 to 28, use the graph of $f(x)$ to find all values of x such that **(a)** $f(x) = -1$, **(b)** $f(x) = 0$, and **(c)** $f(x) = 2$.

21.

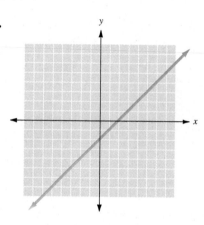

22.

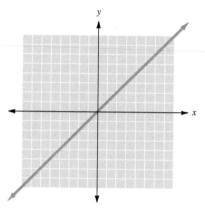

23.

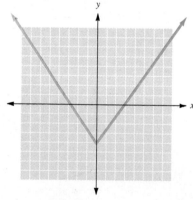

24.

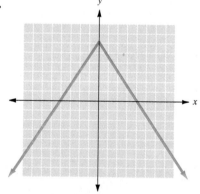

25.

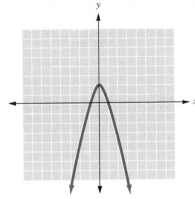

26.

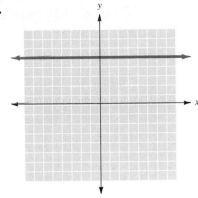

27.

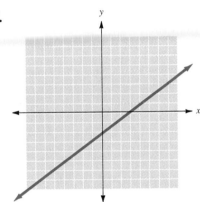

28.

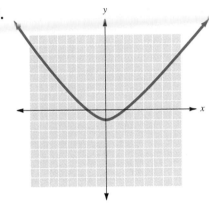

29. Your friend Sam Weatherby is a salesperson and has just been offered two new jobs, one for $600 a month plus 9% of the amount of all his sales over $20,000. The second job offer is $1000 a month plus 3% of the amount of his sales.

Sam and his spouse are about to have a child, and he feels that he has to make $2500 a month just to make ends meet. He has called you to ask for your help in deciding which job to take. To help him picture his options, graph both offers on the same graph, and add a graph of the income of $2500.

Next, write an explanation that answers these questions: How much does he have to sell in each position to earn $2500? When is the first offer better? When is the second better? What sales would he have made to make less than $2000 in each position? Which job should he take?

Answers

1. A (3, 3), B (2, -4)   **3.** A (2, 5), B (-2, -4)   **5.** A (0, 5), B (3, 0)   **7.** A (2, 0), B (6, 4)

9. A (3, 3), B (-3, -3)   **11.** A (3, 6), B (3, 0)   **13.** (a) 3, (b) 1, (c) 2, (d) 5, (e) 0   **15.** (a) 1, (b) 1,

(c) 0, (d) 6, (e) 4   **17.** (a) 1, (b) 3, (c) 2, (d) 1, (e) 4   **19.** (a) 3, (b) 3, (c) 3, (d) 3, (e) 3   **21.** (a) 1,

(b) 2, (c) 4   **23.** (a) 2, -2, (b) 3, -3, (c) 4.5, -4.5 **25.** (a) -1.5, 1.5, (b) 1, -1, (c) 0   **27.** (a) 2,

(b) 3, (c) 6   **29.**

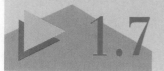

1.7 The Slope of a Line

1.7 OBJECTIVES

1. Find the slope of a line
2. Find the slopes of parallel and perpendicular lines
3. Find the slope of a line given an equation
4. Find the slope given a graph
5. Graph linear equations using the slope of a line

FINDING THE SLOPE

On the coordinate system below, plot a point, any point.

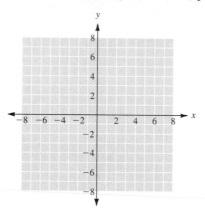

How many different lines can you draw through that point? Hundreds? Thousands? Millions? Actually, there is no limit to the number of different lines that pass through that point.

On the coordinate system below, plot two distinct points.

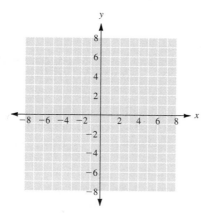

Now, how many different (straight) lines can you draw through those points? Only one! The two points were enough to define the line.

In Section 1.8, we will see how we can find the equation of a line if we are given two of its points. The first part of finding that equation is finding the **slope** of the line, which is a way of describing the steepness of a line.

To define a formula for slope, choose any two distinct points on the line, say, P with coordinates (x_1, y_1) and Q with coordinates (x_2, y_2). As we move along the line from P to Q, the x value, or coordinate, changes from x_1 to x_2. That change in x, also called the **horizontal change,** is $x_2 - x_1$. Similarly, as we move from P to Q, the corresponding change in y, called the **vertical change,** is $y_2 - y_1$. The *slope* is then defined as the ratio of the vertical change to the horizontal change. The letter m is used to represent the slope, which we now define on page 71.

The difference, $x_2 - x_1$, is often called the **run.** The difference, $y_2 - y_1$, is the **rise.** So the slope can be thought of as "rise over run."

Note that $x_1 \neq x_2$ or $x_2 - x_1 \neq 0$ ensures that the denominator is nonzero, so that the slope is defined. It also means the line cannot be vertical.

Slope of a Line

The *slope* of a line through two distinct points $P(x_1, y_1)$ and $Q(x_2, y_2)$ is given by

$$m = \frac{\text{change in } y}{\text{change in } x} = \frac{y_2 - y_1}{x_2 - x_1} \tag{1}$$

where $x_1 \neq x_2$.

Let's look at some examples using the definition.

● Example 1

Finding the Slope Through Two Points

Find the slope of the line through the points $(-3, 2)$ and $(3, 5)$.

Let $(x_1, y_1) = (-3, 2)$ and $(x_2, y_2) = (3, 5)$. From the definition we have

$$m = \frac{5 - 2}{3 - (-3)} = \frac{3}{6} = \frac{1}{2}$$

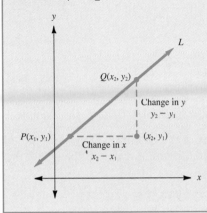

Note that if the pairs are reversed, so that

$$(x_1, y_1) = (3, 5) \qquad \text{and} \qquad (x_2, y_2) = (-3, 2)$$

then we have

$$m = \frac{2 - 5}{-3 - 3} = \frac{-3}{-6} = \frac{1}{2}$$

The slope in either case is the same.

The work here suggests that no matter which point is chosen as (x_1, y_1) or (x_2, y_2), the slope formula will give the same result. Simply stay with your choice once it is made, and use the same order of subtraction in the numerator and the denominator.

● ● ● **CHECK YOURSELF 1**

Find the slope of the line through the points $(-2, -1)$ and $(1, 1)$.

The slope indicates both the direction of a line and its steepness. First, we will compare the steepness of two examples.

•Example 2

Finding the Slope

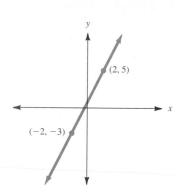

Find the slope of the line through $(-2, -3)$ and $(2, 5)$.
   Again, by equation (1),

$$m = \frac{5 - (-3)}{2 - (-2)} = \frac{8}{4} = 2$$

Compare the lines in Examples 1 and 2. In Example 1 the line has slope $\frac{1}{2}$. The slope here is 2. Now look at the two lines. Do you see the idea of slope as measuring steepness? The greater the absolute value of the slope, the steeper the line.

● ● ● ●   **CHECK YOURSELF 2**

Find the slope of the line through the points $(-1, 2)$ and $(2, 7)$. Draw a sketch of this line and the line in the Check Yourself 1 exercise on the same coordinate axes. Compare the lines and the two slopes.

The sign of the slope indicates in which direction the line tilts, as Example 3 illustrates.

•Example 3

Finding the Slope

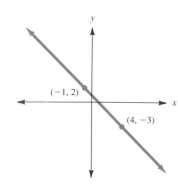

Find the slope of the line through the points $(-1, 2)$ and $(4, -3)$.
   We see that

$$m = \frac{-3 - 2}{4 - (-1)} = \frac{-5}{5} = -1$$

Now the slope is negative.
   Comparing this with our previous examples, we see that

1. In Examples 1 and 2, the lines were rising from left to right, and the slope was **positive.**
2. In this example, the line is falling from left to right, and the slope is **negative.**

● ● ●   **CHECK YOURSELF 3**

Find the slope of the line through the points $(-2, 5)$ and $(4, -1)$.

Let's continue by looking at the slopes of lines in two particular cases.

● Example 4

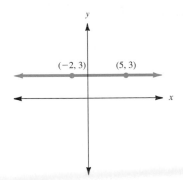

Finding the Slope of a Horizontal Line

Find the slope of the line through $(-2, 3)$ and $(5, 3)$.

$$m = \frac{3 - 3}{5 - (-2)} = \frac{0}{7} = 0$$

The slope of the line is 0. Note that the line is parallel to the x axis and $y_2 - y_1 = 0$. *The slope of any horizontal line will be 0.*

● ● ● ● CHECK YOURSELF 4

Find the slope of the line through the points $(-2, -4)$ and $(3, -4)$.

● Example 5

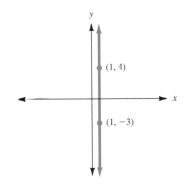

Finding the Slope of a Vertical Line

Find the slope of the line through the points $(1, -3)$ and $(1, 4)$.

$$m = \frac{4 - (-3)}{1 - 1} = \frac{7}{0}$$

Here the line is parallel to the y axis, and $x_2 - x_1$ (the denominator of the slope formula) is 0. Since division by 0 is undefined, we say that the slope is **undefined,** as will be the case for *any vertical line.*

Be very careful not to confuse a slope of 0 (in the case of a horizontal line) with an undefined slope (in the case of a vertical line).

● ● ● ● CHECK YOURSELF 5

Find the slope of the line through the points $(2, -3)$ and $(2, 7)$.

Here is a summary of our work in the previous examples.

> **1.** If the slope of a line is *positive,* the line is rising from left to right.
> **2.** If the slope of a line is *negative,* the line is falling from left to right.
> **3.** If the slope of a line is 0, the line is *horizontal.*
> **4.** If the slope of a line is undefined, the line is *vertical.*

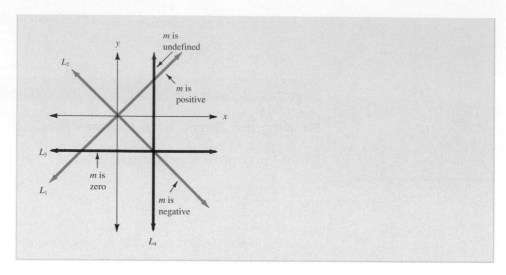

There are two more important results regarding the slope. Recall from geometry that two distinct lines in the plane either intersect at a point or never intersect. Two lines in the plane that do not intersect are called **parallel lines.** It can be shown that two distinct parallel lines will always have the same slope, and we can state the following.

This means that if the lines are parallel, then their slopes are equal. Conversely, if the slopes are equal, then the lines are parallel.

Mathematicians use the symbol ⇔ to represent "if and only if."

Slopes of Parallel Lines

For nonvertical lines L_1 and L_2, if line L_1 has slope m_1 and line L_2 has slope m_2, then

L_1 is parallel to L_2 if and only if $m_1 = m_2$

Note: All vertical lines are parallel to each other.

• Example 6

Parallel Lines

Are lines L_1 through $(2, 3)$ and $(4, 6)$ and L_2 through $(-4, 2)$ and $(0, 8)$ parallel, or do they intersect?

$$m_1 = \frac{6 - 3}{4 - 2} = \frac{3}{2}$$

Unless, of course, L_1 and L_2 are actually the *same line.* In this case, a quick sketch will show that the lines are distinct.

$$m_2 = \frac{8 - 2}{0 - (-4)} = \frac{6}{4} = \frac{3}{2}$$

Since the slopes of the lines are equal, the lines are parallel. They do *not* intersect.

● ● ● **CHECK YOURSELF 6**

Are lines L_1 through $(-2, -1)$ and $(1, 4)$ and L_2 through $(-3, 4)$ and $(0, 8)$ parallel, or do they intersect?

Two lines are perpendicular if they intersect at right angles. Also, if two lines (which are not vertical or horizontal) are perpendicular, their slopes are the negative reciprocals of each other. We can then state the following result for perpendicular lines.

SLOPES OF PERPENDICULAR LINES

For nonvertical lines L_1 and L_2, if line L_1 has slope m_1 and line L_2 has slope m_2, then

L_1 is perpendicular to L_2 if and only if $m_1 = -\dfrac{1}{m_2}$

or, equivalently,

$m_1 \cdot m_2 = -1$

Note: Horizontal lines are perpendicular to vertical lines.

• Example 7

Perpendicular Lines

Are lines L_1 through points $(-2, 3)$ and $(1, 7)$ and L_2 through points $(2, 4)$ and $(6, 1)$ perpendicular?

$$m_1 = \frac{7 - 3}{1 - (-2)} = \frac{4}{3}$$

Note:

$$\left(\frac{4}{3}\right)\left(-\frac{3}{4}\right) = -1$$

$$m_2 = \frac{1 - 4}{6 - 2} = -\frac{3}{4}$$

Since the slopes are negative reciprocals, the lines are perpendicular.

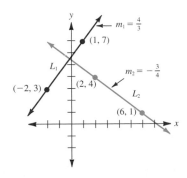

• • • CHECK YOURSELF 7

Are lines L_1 through points $(1, 3)$ and $(4, 1)$ and L_2 through points $(-2, 4)$ and $(2, 10)$ perpendicular?

FINDING THE SLOPE FROM AN EQUATION OR GRAPH

Given the equation of a line, we can also find its slope, as Example 8 illustrates.

• Example 8

Finding the Slope from an Equation

Note: Let's try solving the original equation for y:

$3x + 2y = 6$

$2y = 3x + 6$

$y = -\dfrac{3}{2}x + 3$

Consider the coefficient of x. What do you observe?

Find the slope of the line with equation $3x + 2y = 6$.

First, find any two points on the line. In this case, $(2, 0)$ and $(0, 3)$, the x and y intercepts, will work and are easy to find. From the slope formula,

$$m = \frac{0 - 3}{2 - 0} = \frac{-3}{2} = -\frac{3}{2}$$

The slope of the line with equation $3x + 2y = 6$ is $-\dfrac{3}{2}$.

• • • CHECK YOURSELF 8

Find the slope of the line with equation $3x - 4y = 12$.

We can find the slope of a graphed line by identifying two points on the line. We will use that technique in Example 9.

• Example 9

Finding the Slope from a Graph

Determine the slope of the line from its graph.

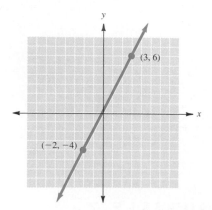

We can choose any two points to determine the slope of the line. Here we pick $(3, 6)$ and $(-2, -4)$. We find the slope by the usual method.

$$m = \frac{6 - (-4)}{3 - (-2)} = \frac{10}{5} = 2$$

The slope of the line is 2.

● ● ● **CHECK YOURSELF 9**

Determine the slope of the line from its graph.

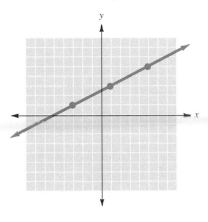

     The slope of a line can also be useful in graphing a line. In Example 10, the slope of a line is used in sketching its graph.

● Example 10

Graphing a Line with a Given Slope

Suppose a line has slope $\frac{3}{2}$ and passes through the point (5, 2). Graph the line.

     First, locate the point (5, 2) in the coordinate system. Now, since the slope, $\frac{3}{2}$, is the ratio of the change in y to the change in x, move 2 units to the right in the x direction and then 3 units up in the y direction. This determines a second point, here (7, 5), and we can draw our graph.

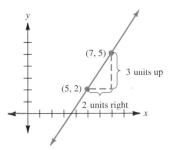

● ● ● **CHECK YOURSELF 10**

Graph the line with slope $-\frac{3}{4}$ that passes through the point (2, 3). *Hint:* Consider the x change as 4 units and the y change as -3 units (down).

Since, given a point on a line and its slope, we can graph the line, we also should be able to write its equation. That is, in fact, the case, as we will see in Section 1.8.

● ● ● CHECK YOURSELF ANSWERS

1. $m = \dfrac{2}{3}$.　　**2.** $m = \dfrac{5}{3}$.　　**3.** $m = -1$.　　**4.** 0.　　**5.** Undefined.

6. The lines intersect.　　**7.** The lines are perpendicular.　　**8.** $m = \dfrac{3}{4}$.　　**9.** $\dfrac{1}{2}$.

10.

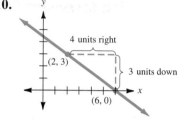

1.7 Exercises

In Exercises 1 to 12, find the slope (if it exists) of the line determined by the following pairs of points. Sketch each line so that you can compare the slopes.

1. $(2, 3)$ and $(4, 7)$

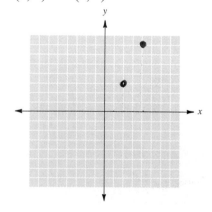

2. $(-1, 2)$ and $(5, 3)$

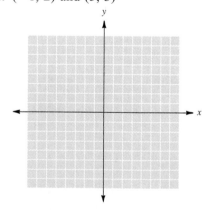

3. $(2, -3)$ and $(-2, -5)$

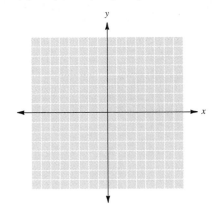

4. $(0, 0)$ and $(5, 7)$

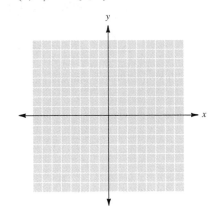

5. $(2, 5)$ and $(-3, 5)$

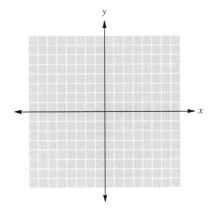

6. $(-2, -4)$ and $(5, 3)$

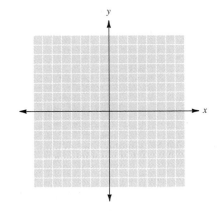

7. $(-1, 4)$ and $(-1, 7)$

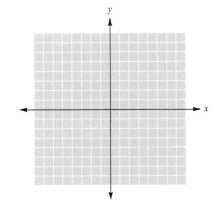

8. $(4, 2)$ and $(-2, 5)$

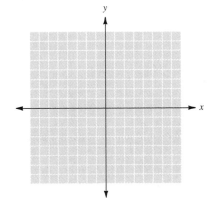

9. $(8, -3)$ and $(-2, -5)$

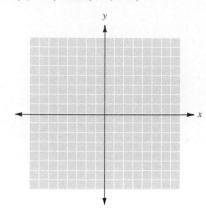

10. $(4, -3)$ and $(-2, 7)$

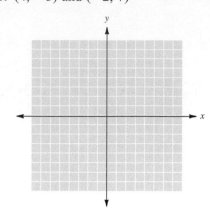

11. $(-4, -3)$ and $(2, -7)$

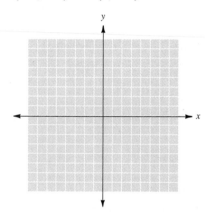

12. $(3, 6)$ and $(3, -4)$

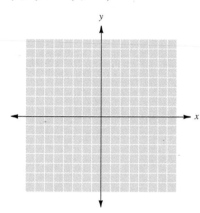

In Exercises 13 to 20, find the slope of the line determined by each equation.

13. $y = -3x - \dfrac{1}{2}$ **14.** $y = \dfrac{1}{4}x + 3$ **15.** $y + \dfrac{1}{2}x = 2$ **16.** $2y - 3x + 5 = 0$

17. $2x - 3y = 6$ **18.** $x + 4y = 4$ **19.** $3x + 4y = 12$ **20.** $x - 3y = 9$

In Exercises 21 to 26, are the pairs of lines parallel, perpendicular, or neither?

21. L_1 through $(-2, -3)$ and $(4, 3)$; L_2 through $(3, 5)$ and $(5, 7)$

22. L_1 through $(-2, 4)$ and $(1, 8)$; L_2 through $(-1, -1)$ and $(-5, 2)$

23. L_1 through $(8, 5)$ and $(3, -2)$; L_2 through $(-2, 4)$ and $(4, -1)$

24. L_1 through $(-2, -3)$ and $(3, -1)$; L_2 through $(-3, 1)$ and $(7, 5)$

25. L_1 with equation $x - 3y = 6$; L_2 with equation $3x + y = 3$

26. L_1 with equation $x + 2y = 4$; L_2 with equation $2x + 4y = 5$

27. Find the slope of any line parallel to the line through points $(-2, 3)$ and $(4, 5)$.

28. Find the slope of any line perpendicular to the line through points $(0, 5)$ and $(-3, -4)$.

29. A line passing through $(-1, 2)$ and $(4, y)$ is parallel to a line with slope 2. What is the value of y?

30. A line passing through $(2, 3)$ and $(5, y)$ is perpendicular to a line with slope $\dfrac{3}{4}$. What is the value of y?

If points P, Q, and R are collinear (lie on the same line), the slope of the line through P and Q must equal the slope of the line through Q and R. In Exercises 31 to 36, use the slope concept to determine whether the sets of points are collinear.

31. $P(-2, -3)$, $Q(3, 2)$, and $R(4, 3)$

32. $P(-5, 1)$, $Q(-2, 4)$, and $R(4, 9)$

33. $P(0, 0)$, $Q(2, 4)$, and $R(-3, 6)$

34. $P(-2, 5)$, $Q(-5, 2)$, and $R(1, 12)$

35. $P(2, 4)$, $Q(-3, -6)$, and $R(-4, 8)$

36. $P(-1, 5)$, $Q(2, -4)$, and $R(-2, 8)$

In Exercises 37 to 44, graph the lines through each of the specified points having the given slope.

37. $(0, 1)$, $m = 3$

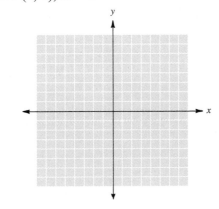

38. $(0, -2)$, $m = -2$

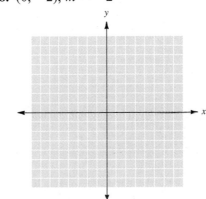

39. $(3, -1)$, $m = 2$

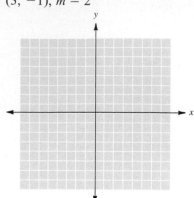

40. $(2, -3)$, $m = -3$

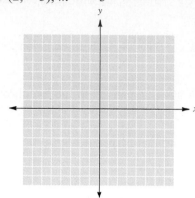

41. $(2, 3)$, $m = \dfrac{2}{3}$

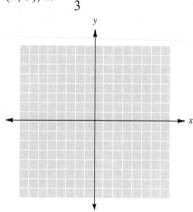

42. $(-2, 1)$, $m = -\dfrac{3}{4}$

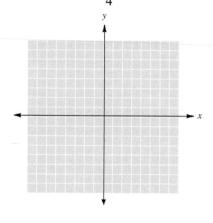

43. $(4, 2)$, $m = 0$

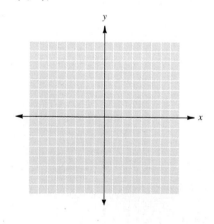

44. $(3, 0)$, m is undefined

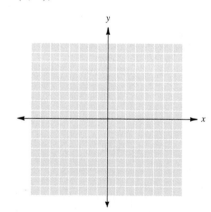

45. On the same graph, sketch lines with slope 2 through each of the following points: $(-1, 0)$, $(2, 0)$, and $(5, 0)$.

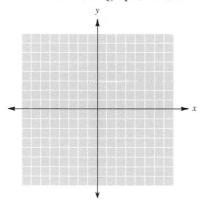

46. On the same graph, sketch one line with slope $\dfrac{1}{3}$ and one line with slope -3, having both pass through point $(2, 3)$.

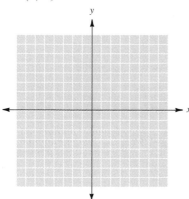

A four-sided figure (quadrilateral) is a parallelogram if the opposite sides have the same slope. If the adjacent sides are perpendicular, the figure is a rectangle. In Exercises 47 to 50, for each quadrilateral *ABCD*, determine whether it is a parallelogram; then determine whether it is a rectangle.

47. $A(0, 0)$, $B(2, 0)$, $C(2, 3)$, $D(0, 3)$

48. $A(-3, 2)$, $B(1, -7)$, $C(3, -4)$, $D(-1, 5)$

49. $A(0, 0)$, $B(4, 0)$, $C(5, 2)$, $D(1, 2)$

50. $A(-3, -5)$, $B(2, 1)$, $C(-4, 6)$, $D(-9, 0)$

ises 51 to 54, solve each equation for y, then use your graphing utility to graph each equation.

51. $2x + 5y = 10$

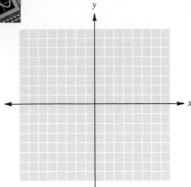

52. $5x - 3y = 12$

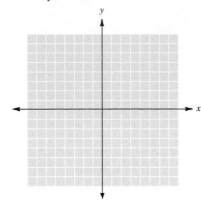

53. $x + 7y = 14$

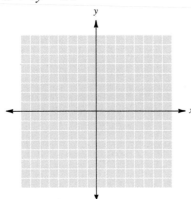

54. $-2x - 3y = 9$

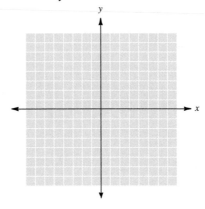

In Exercises 55 to 62, use the graph to determine the slope of the line.

55.

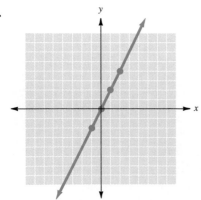

56.

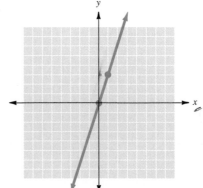

57.

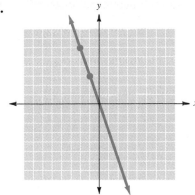

58.

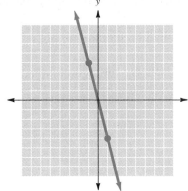

59.

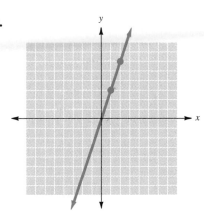

60.

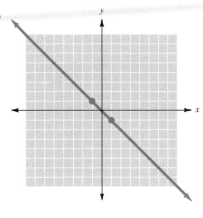

61.

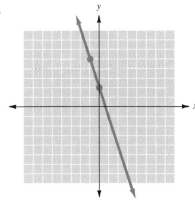

62.

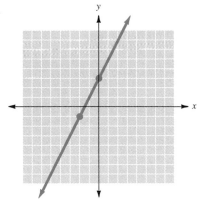

63. Consumer Affairs In 1960, the cost of a soft drink was 20¢. By 1995, the cost of the same soft drink had risen to 95¢. During this time period, what was the annual rate of change of the cost of the soft drink?

64. Science On a certain February day in Philadelphia, the temperature at 6:00 AM was 10°F. By 2:00 PM the temperature was up to 26°F. What was the hourly rate of temperature change?

65. Business In a certain business, the fixed costs are $750, and each item costs $15 to produce. Find the total cost required to produce 75 items.

66. Construction The rise-to-run ratio used to determine the steepness of the roof on a house is 4:5. Determine the maximum height of the attic if the house is 25 feet long.

Answers

1. 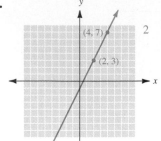 2

3. 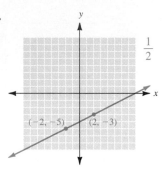 $\frac{1}{2}$

5. 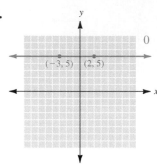 0

7. 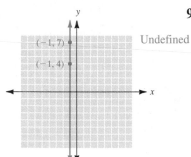 Undefined

9. 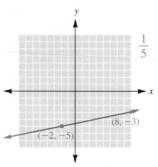 $\frac{1}{5}$

11. 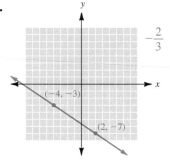 $-\frac{2}{3}$

13. -3 **15.** $-\frac{1}{2}$ **17.** $\frac{2}{3}$ **19.** $-\frac{3}{4}$ **21.** Parallel **23.** Neither **25.** Perpendicular

27. $\frac{1}{3}$ **29.** 12 **31.** Collinear **33.** Not collinear **35.** Not collinear

37.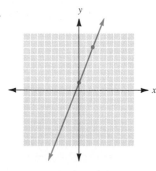

39.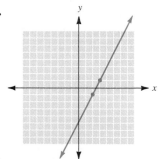

41.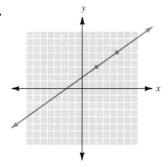

43.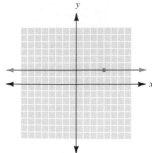

45.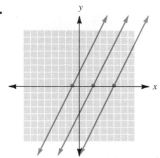

47. Parallelogram, rectangle

49. Parallelogram, not a rectangle

51.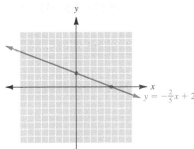

53.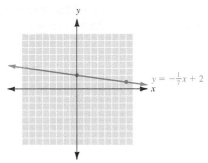

55. 2   **57.** -3 **59.** 3

61. -3 **63.** 2.14¢ per year   **65.** $1875

1.8 Forms of Linear Equations

1.8 OBJECTIVES

1. Write the equation of a line given its slope and y intercept
2. Write the equation of a line given a slope and any point on the line
3. Graph the equation of a line
4. Write the equation of a line given a point and a slope
5. Write the equation of a line given two points
6. Express the equation of a line as a linear function
7. Write an equation as a function
8. Graph a linear function

Just as m is used for the slope, b is used for the y intercept.

The special form

$$ax + by = c$$

where a and b cannot both be zero, is called the **standard form for a linear equation.** In Section 1.7, we determined the slope of a line from two ordered pairs. In this section, we will look at several forms for a linear equation. For the first of these forms, we use the concept of slope to write the equation of a line.

First, suppose we know the y intercept of a line L and its slope m. Since b is the y intercept, we know that the point with coordinates $(0, b)$ is on the line. Let $P(x, y)$ be any other point on that line. Using $(0, b)$ as (x_1, y_1) and (x, y) as (x_2, y_2) in the slope formula, we have

$$m = \frac{y - b}{x - 0} \tag{1}$$

or

$$m = \frac{y - b}{x} \tag{2}$$

Multiplying both sides of equation (2) by x gives

$$mx = y - b$$

or

$$y = mx + b \tag{3}$$

Equation (3) will be satisfied by any point on line L, including $(0, b)$. It is called the **slope-intercept form** for a line, and we can state the following general result.

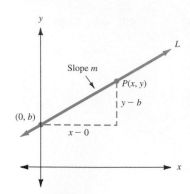

Slope-Intercept Form for the Equation of a Line

The equation of a line with y intercept b and with slope m can be written as
$$y = mx + b$$

The slope-intercept form for the equation of a line is the most convenient form for entering an equation into the calculator.

The x coefficient is 2; the y intercept is 3.

• Example 1

Finding the Equation of a Line

Write the equation of the line with slope 2 and y intercept 3.

Here $m = 2$ and b (the y intercept) $= 3$. Applying the slope-intercept form, we have

$$y = 2x + 3$$

as the equation of the specified line.

It is easy to see that whenever a linear equation is written in slope-intercept form (that is, solved for y), then the slope of the line is simply the x coefficient and the y intercept is given by the constant.

● ● ● **CHECK YOURSELF 1**

Write the equation of the line with slope $-\dfrac{2}{3}$ and y intercept -3.

Note that the slope-intercept form now gives us a second (and generally more efficient) means of finding the slope of a line whose equation is written in standard form. Recall that we determined two specific points on the line and then applied the slope formula. Now, rather than using specific points, we can simply solve the given equation for y to rewrite the equation in the slope-intercept form and identify the slope of the line as the x coefficient.

• Example 2

Finding the Slope and y Intercept of a Line

Find the slope and y intercept of the line with equation

$$2x + 3y = 3$$

To write the equation in slope-intercept form, we solve for y.

$$2x + 3y = 3$$

$$3y = -2x + 3 \qquad \text{Subtract } 2x \text{ from both sides.}$$

$$y = -\frac{2}{3}x + 1 \qquad \text{Divide by 3.}$$

We now see that the slope of the line is $-\dfrac{2}{3}$ and the y intercept is 1.

● ● ● ● **CHECK YOURSELF 2**

Find the slope and y intercept of the line with equation

$3x - 4y = 8$

We can also use the slope-intercept form to determine whether the graphs of given equations will be parallel, intersecting, or perpendicular lines.

● **Example 3**

Verifying That Two Lines Are Perpendicular

Show that the graphs of $3x + 4y = 4$ and $-4x + 3y = 12$ are perpendicular lines. First, we solve each equation for y.

Two lines are perpendicular if their slopes are negative reciprocals, so

$m_1 = -\dfrac{1}{m_2}$

$3x + 4y = 4$

$\qquad 4y = -3x + 4$

$\qquad\quad y = -\dfrac{3}{4}x + 1$ \hfill (4)

$-4x + 3y = 12$

$\qquad\quad 3y = 4x + 12$

$\qquad\quad\; y = \dfrac{4}{3}x + 4$ \hfill (5)

We now look at the product of the two slopes: $-\dfrac{3}{4} \cdot \dfrac{4}{3} = -1$. Any two lines whose slopes have a product of -1 are perpendicular lines. These two lines are perpendicular.

● ● ● ● **CHECK YOURSELF 3**

Show that the graphs of the equations

$-3x + 2y = 4 \qquad$ and $\qquad 2x + 3y = 9$

are perpendicular lines.

The slope-intercept form can also be used in graphing a line, as Example 4 illustrates.

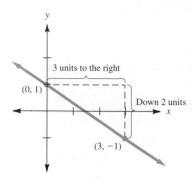

We treat $-\dfrac{2}{3}$ as $\dfrac{-2}{+3}$ to move over 3 units and down 2 units.

• Example 4

Graphing the Equation of a Line

Graph the line $2x + 3y = 3$.

In Example 2, we found that the slope-intercept form for this equation was

$$y = -\frac{2}{3}x + 1$$

To graph the line, plot the y intercept at $(0, 1)$. Now, since the slope m is equal to $-\dfrac{2}{3}$, from $(0, 1)$ we move to the right 3 units and then *down* 2 units, to locate a second point on the graph of the line, here $(3, -1)$. We can now draw a line through the two points to complete the graph.

CHECK YOURSELF 4

Graph the line with equation

$$3x - 4y = 8$$

Hint: You worked with this equation in Check Yourself 2.

The following algorithm summarizes the use of graphing with the slope-intercept form.

The desired form for the equation is

$y = mx + b$

Graphing by Using the Slope-Intercept Form

STEP 1 Write the original equation of the line in slope-intercept form.

STEP 2 Determine the slope m and the y intercept b.

STEP 3 Plot the y intercept at $(0, b)$.

STEP 4 Use m (the change in y over the change in x) to determine a second point on the desired line.

STEP 5 Draw a line through the two points determined above to complete the graph.

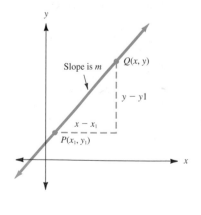

Often in mathematics it is useful to be able to write the equation of a line, given its slope and *any* point on the line. We will now derive a third special form for a line for this purpose.

Suppose a line has slope m and passes through the known point $P(x_1, y_1)$. Let $Q(x, y)$ be any other point on the line. Once again we can use the definition of slope and write

$$m = \frac{y - y_1}{x - x_1} \tag{6}$$

Multiplying both sides of equation (6) by $x - x_1$, we have

$$m(x - x_1) = y - y_1$$

or

$$y - y_1 = m(x - x_1) \qquad (7)$$

Equation (7) is called the **point-slope form** for the equation of a line, and all points lying on the line [including (x_1, y_1)] will satisfy this equation. We can state the following general result.

The equation of a line with undefined slope passing through the point (x_1, y_1) is given by $x = x_1$.

> **Point-Slope Form for the Equation of a Line**
>
> The equation of a line with slope m that passes through point (x_1, y_1) is given by
>
> $$y - y_1 = m(x - x_1)$$

● Example 5

Finding the Equation of a Line

Write the equation for the line that passes through point $(3, -1)$ with a slope of 3.
    Letting $(x_1, y_1) = (3, -1)$ and $m = 3$ in point-slope form, we have

$$y - (-1) = 3(x - 3)$$

or

$$y + 1 = 3x - 9$$

We can write the final result in slope-intercept form as

$$y = 3x - 10$$

● ● ● **CHECK YOURSELF 5**

Write the equation of the line that passes through point $(-2, 4)$ with a slope of $\dfrac{3}{2}$. Write your result in slope-intercept form.

___

Since we know that two points determine a line, it is natural that we should be able to write the equation of a line passing through two given points. Using the point-slope form together with the slope formula will allow us to write such an equation.

• Example 6

Finding the Equation of a Line

Write the equation of the line passing through (2, 4) and (4, 7).
First, we find m, the slope of the line. Here

$$m = \frac{7 - 4}{4 - 2} = \frac{3}{2}$$

Note: We could just as well have chosen to let

$(x_1, y_1) = (4, 7)$

The resulting equation will be the same in either case. Take time to verify this for yourself.

Now we apply the point-slope form with $m = \frac{3}{2}$ and $(x_1, y_1) = (2, 4)$:

$$y - 4 = \frac{3}{2}(x - 2)$$

$$y - 4 = \frac{3}{2}x - 3 \qquad \text{Write the result in slope-intercept form.}$$

$$y = \frac{3}{2}x + 1$$

• • •   CHECK YOURSELF 6

Write the equation of the line passing through $(-2, 5)$ and $(1, 3)$. Write your result in slope-intercept form.

A line with slope zero is a horizontal line. A line with an undefined slope is vertical. Example 7 illustrates the equations of such lines.

• Example 7

Finding the Equation of a Line

(*a*) Find the equation of a line passing through $(7, -2)$ with a slope of zero.
We could find the equation by letting $m = 0$. Substituting into the slope-intercept form, we can solve for the y intercept b.

$$y = mx + b$$
$$-2 = 0(7) + b$$
$$-2 = b$$

So,

$$y = 0x - 2 \qquad y = -2$$

It is far easier to remember that any line with a zero slope is a horizontal line and has the form

$$y = b$$

The value for b will always be the y coordinate for the given point.

(b) Find the equation of a line with undefined slope passing through $(4, -5)$.

A line with undefined slope is vertical. It will always be of the form $x = a$, where a is the x coordinate for the given point. The equation is

$x = 4$

● ● ● ● **CHECK YOURSELF 7**

(a) Find the equation of a line with zero slope that passes through point $(-3, 5)$.
(b) Find the equation of a line passing through $(-3, -6)$ with undefined slope.

Alternate methods for finding the equation of a line through two points do exist and have particular significance in other fields of mathematics, such as statistics. Example 8 shows such an alternate approach.

● Example 8

Finding the Equation of a Line

Write the equation of the line through points $(-2, 3)$ and $(4, 5)$.
First, we find m, as before.

$$m = \frac{5 - 3}{4 - (-2)} = \frac{2}{6} = \frac{1}{3}$$

We now make use of the slope-intercept equation, but in a slightly different form.
Since $y = mx + b$, we can write

$b = y - mx$

We substitute these values because the line must pass through $(-2, 3)$

Now, letting $x = -2$, $y = 3$, and $m = \frac{1}{3}$, we can calculate b.

$$b = 3 - \left(\frac{1}{3}\right)(-2)$$

$$= 3 + \frac{2}{3} = \frac{11}{3}$$

With $m = \frac{1}{3}$ and $b = \frac{11}{3}$, we can apply the slope-intercept form, to write the equation of the desired line. We have

$$y = \frac{1}{3}x + \frac{11}{3}$$

● ● ● ● **CHECK YOURSELF 8**

Repeat the Check Yourself 6 exercise, using the technique illustrated in Example 8.

We now know that we can write the equation of a line once we have been given appropriate geometric conditions, such as a point on the line and the slope of that line. In some applications, the slope may be given not directly but through specified parallel or perpendicular lines.

• Example 9

Finding the Equation of a Line

Find the equation of the line passing through $(-4, -3)$ and parallel to the line determined by $3x + 4y = 12$.

First, we find the slope of the given parallel line, as before.

$$3x + 4y = 12$$

$$4y = -3x + 12$$

The slope of the given line is $-\dfrac{3}{4}$

$$y = -\frac{3}{4}x + 3$$

Now, since the slope of the desired line must also be $-\dfrac{3}{4}$, we can use the point-slope form to write the required equation.

The line must pass through $(-4, -3)$, so let $(x_1, y_1) = (-4, -3)$

$$y - (-3) = -\frac{3}{4}[x - (-4)]$$

This simplifies to

$$y = -\frac{3}{4}x - 6$$

and we have our equation in slope-intercept form.

● ● ●   **CHECK YOURSELF 9**

Find the equation of the line passing through $(5, 4)$ and perpendicular to the line with equation $2x - 5y = 10$. *Hint:* Recall that the slopes of perpendicular lines are negative reciprocals of each other.

There are many applications of our work with linear equations in various fields. The following is just one of many typical examples.

• Example 10

An Application of a Linear Function

In producing a new product, a manufacturer predicts that the number of items produced x and the cost in dollars C of producing those items will be related by a linear equation.

Suppose that the cost of producing 100 items will be $5000 and the cost of producing 500 items will be $15,000. Find the linear equation relating x and C.

Solution

To solve this problem, we must find the equation of the line passing through points (100, 5000) and (500, 15,000).

Even though the numbers are considerably larger than we have encountered thus far in this section, the process is exactly the same.

First, we find the slope:

$$m = \frac{15,000 - 5000}{500 - 100} = \frac{10,000}{400} = 25$$

We can now use the point-slope form as before to find the desired equation.

$$C - 5000 = 25(x - 100)$$

$$C - 5000 = 25x - 2500$$

$$C = 25x + 2500$$

To graph the equation we have just derived, we must choose the scaling on the x and C axes carefully to get a "reasonable" picture. Here we choose increments of 100 on the x axis and 2500 on the C axis since those seem appropriate for the given information.

Note how the change in scaling "distorts" the slope of the line.

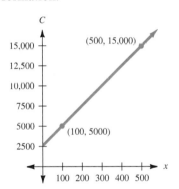

CHECK YOURSELF 10

A company predicts that the value in dollars V and the time that a piece of equipment has been in use t are related by a linear equation. If the equipment is valued at $1500 after 2 years and at $300 after 10 years, find the linear equation relating t and V.

We used the term *function* several times in this chapter. We identified functions, looked at a function machine, used function notation, and found the domain and range for a function. But how does all this relate to the equations in two variables that we studied in this chapter? When is y the same as $f(x)$?

Anytime we solved a linear equation for y, such as

$$y = 3x - 2$$

then y was a function of x. The x is considered the **independent variable,** and the y is considered the **dependent variable.** This means that y changes because x has changed. Let's look at some examples of variables that are related and determine which is the dependent variable.

• Example 11

Identifying the Dependent Variable

From each pair, identify which variable is dependent on the other.

(*a*) The age of a car and its resale value.
   The value depends on the age, so we would assign the age of the car the independent variable (*x*) and the value the dependent variable (*y*).

(*b*) The amount of interest earned in a bank account and the amount of time the money has been in the bank.
   The interest depends on the time, so interest is the dependent variable (*y*) and time is the independent variable (*x*).

If you think about it, you will see that time will be the independent variable in most ordered pairs. Most everything depends on time rather than the reverse.

(*c*) The number of cigarettes one has smoked and the chances of dying from a smoking-related disease.
   The number of cigarettes is the independent variable (*x*), and dying from a smoking-related disease is the dependent variable (*y*).

● ● ●   **CHECK YOURSELF 11**

From each pair, identify which variable is dependent on the other.

(a) The number of credits taken and the amount of tuition paid.
(b) The temperature of a cup of coffee and the length of time since it was poured.

We indicated earlier in this chapter that any nonvertical line could be represented as a function. In cases in which *y* is a function of *x*, we can rewrite the equation in function form.

• Example 12

Writing Equations as Functions

Rewrite each linear equation as a function of *x*.

(*a*) $y = 3x - 4$

   This can be rewritten as

$$f(x) = 3x - 4$$

(*b*) $2x - 3y = 6$

   We must first solve the equation for *y* (recall that this will give us the slope-intercept form).

$$-3y = -2x + 6$$

$$y = \frac{2}{3}x - 2$$

This can be rewritten as

$$f(x) = \frac{2}{3}x - 2$$

CHECK YOURSELF 12

Rewrite each equation as a function of x.

(a) $y = -2x + 5$ **(b)** $3x + 5y = 15$

The process of finding the graph of a linear function is identical to the process of finding the graph of a linear equation.

● Example 13

Graphing a Linear Function

Graph the function

$$f(x) = 3x - 5$$

We could use the slope and y intercept to graph the line, or we can find three points (the third is just a check point) and draw the line through them. We will do the latter.

$$f(0) = -5 \qquad f(1) = -2 \qquad f(2) = 1$$

We will use the three points $(0, -5)$, $(1, -2)$, and $(2, 1)$ to graph the line.

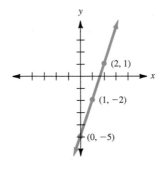

CHECK YOURSELF 13

Graph the function

$$f(x) = 5x - 3$$

One benefit of having a function written in $f(x)$ form is that it makes it fairly easy to substitute values for x. In Example 13, we substituted the values 0, 1, and 2. Sometimes it is useful to substitute nonnumeric values for x.

• Example 14

Substituting Nonnumeric Values for x

Let $f(x) = 2x + 3$. Evaluate f as indicated.

(a) $f(a)$

Substituting a for x in our equation, we see that

$f(a) = 2a + 3$

(b) $f(2 + h)$

Substituting $2 + h$ for x in our equation, we get

$f(2 + h) = 2(2 + h) + 3$

Distributing the 2, then simplifying, we have

$f(2 + h) = 4 + 2h + 3$

$\qquad = 2h + 7$

● ● ● **CHECK YOURSELF 14**

Let $f(x) = 4x - 2$. Evaluate f as indicated.

(a) $f(b)$ **(b)** $f(4 + h)$

● ● ● **CHECK YOURSELF ANSWERS**

1. $y = -\dfrac{2}{3}x - 3$.   **2.** $m = \dfrac{3}{4}$ and $b = -2$.   **3.** $m_1 = \dfrac{3}{2}$ and $m_2 = -\dfrac{2}{3}$;

$(m_1)(m_2) = -1$.   **4.** 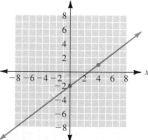   **5.** $y = \dfrac{3}{2}x + 7$.

6. $y = -\dfrac{2}{3}x + \dfrac{11}{3}$.   **7. (a)** $y = 5$, **(b)** $x = -3$.   **8.** $y = -\dfrac{2}{3}x + \dfrac{11}{3}$.

9. $y = -\dfrac{5}{2}x + \dfrac{33}{2}$.   **10.** $V = -150t + 1800$.   **11. (a)** Tuition is dependent

on credits taken. **(b)** The temperature is dependent on the time since the coffee was

poured.   **12. (a)** $f(x) = -2x + 5$, **(b)** $f(x) = -\dfrac{3}{5}x + 3$.

13.

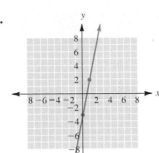

14. **(a)** $4b - 2$, **(b)** $4h + 14$.

1.8 Exercises

In Exercises 1 to 8, match the graph with one of these equations: **(a)** $y = 2x$; **(b)** $y = x + 1$; **(c)** $y = -x + 3$; **(d)** $y = 2x + 1$; **(e)** $y = -3x - 2$; **(f)** $y = \dfrac{2}{3}x + 1$; **(g)** $y = -\dfrac{4}{3}x + 1$; and **(h)** $y = -4x$.

1.

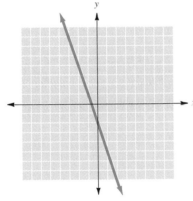

2.

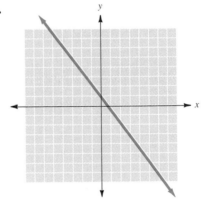

3.

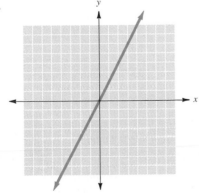

4.

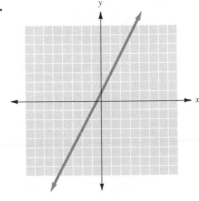

5.

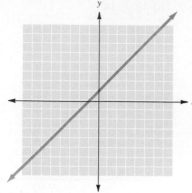

6.

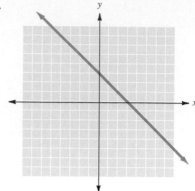

7.

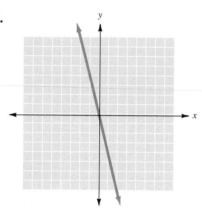

8.

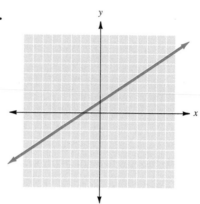

In Exercises 9 to 20, write each equation in slope-intercept form. Give its slope and y intercept.

9. $x + y = 5$

10. $2x + y = 3$

11. $2x - y = -2$

12. $x + 3y = 6$

13. $x + 3y = 9$

14. $4x - y = 8$

15. $2x - 3y = 6$

16. $3x - 4y = 12$

17. $2x - y = 0$

18. $3x + y = 0$

19. $y + 3 = 0$

20. $y - 2 = 0$

In Exercises 21 to 38, write the equation of the line passing through each of the given points with the indicated slope. Give your results in slope-intercept form, where possible.

21. $(0, 2)$, $m = 3$

22. $(0, -4)$, $m = -2$

23. $(0, 2)$, $m = \dfrac{3}{2}$

24. $(0, -3)$, $m = -2$

25. $(0, 4)$, $m = 0$

26. $(0, 5)$, $m = -\dfrac{3}{5}$

27. $(0, -5)$, $m = \dfrac{5}{4}$

28. $(0, -4)$, $m = -\dfrac{3}{4}$

29. $(1, 2)$, $m = 3$

30. $(-1, 2)$, $m = 3$

31. $(-2, -3)$, $m = -3$

32. $(1, -4)$, $m = -4$

33. $(5, -3)$, $m = \dfrac{2}{5}$

34. $(4, 3)$, $m = 0$

35. $(2, -3)$, m is undefined

36. $(2, -5)$, $m = \dfrac{1}{4}$

37. $(5, 0)$, $m = -\dfrac{4}{5}$

38. $(-3, 0)$, m is undefined

In Exercises 39 to 48, write the equation of the line passing through each of the given pairs of points. Write your result in slope-intercept form, where possible.

39. $(2, 3)$ and $(5, 6)$

40. $(3, -2)$ and $(6, 4)$

41. $(-2, -3)$ and $(2, 0)$

42. $(-1, 3)$ and $(4, -2)$

43. $(-3, 2)$ and $(4, 2)$

44. $(-5, 3)$ and $(4, 1)$

45. $(2, 0)$ and $(0, -3)$

46. $(2, -3)$ and $(2, 4)$

47. $(0, 4)$ and $(-2, -1)$

48. $(-4, 1)$ and $(3, 1)$

In Exercises 49 to 58, write the equation of the line L satisfying the given geometric conditions.

49. L has slope 4 and y intercept -2.

50. L has slope $-\dfrac{2}{3}$ and y intercept 4.

51. L has x intercept 4 and y intercept 2.

52. L has x intercept -2 and slope $\dfrac{3}{4}$.

53. L has y intercept 4 and a 0 slope.

54. L has x intercept -2 and an undefined slope.

55. L passes through point $(3, 2)$ with a slope of 5.

56. L passes through point $(-2, -4)$ with a slope of $-\dfrac{3}{2}$.

57. L has y intercept 3 and is parallel to the line with equation $y = 3x - 5$.

58. L has y intercept -3 and is parallel to the line with equation $y = \frac{2}{3}x + 1$.

In Exercises 59 to 70, write the equation of each line in function form.

59. L has y intercept 4 and is perpendicular to the line with equation $y = -2x + 1$.

60. L has y intercept 2 and is parallel to the line with equation $y = -1$.

61. L has y intercept 3 and is parallel to the line with equation $y = 2$.

62. L has y intercept 2 and is perpendicular to the line with equation $2x - 3y = 6$.

63. L passes through point $(-3, 2)$ and is parallel to the line with equation $y = 2x - 3$.

64. L passes through point $(-4, 3)$ and is parallel to the line with equation $y = -2x + 1$.

65. L passes through point $(3, 2)$ and is parallel to the line with equation $y = \frac{4}{3}x + 4$.

66. L passes through point $(-2, -1)$ and is perpendicular to the line with equation $y = 3x + 1$.

67. L passes through point $(5, -2)$ and is perpendicular to the line with equation $y = -3x - 2$.

68. L passes through point $(3, 4)$ and is perpendicular to the line with equation $y = -\frac{3}{5}x + 2$.

69. L passes through $(-2, 1)$ and is parallel to the line with equation $x + 2y = 4$.

70. L passes through $(-3, 5)$ and is parallel to the x axis.

71. Geometry Find the equation of the perpendicular bisector of the segment joining $(-3, -5)$ and $(5, 9)$. *Hint:* First determine the midpoint of the segment. The midpoint would be $\left(\frac{-3 + 5}{2}, \frac{-5 + 9}{2} \right) = (1, 2)$. The perpendicular bisector passes through that point and is perpendicular to the line segment connecting the points.

72. Geometry Find the equation of the perpendicular bisector of the segment joining $(-2, 3)$ and $(8, 5)$.

73. Science A temperature of 10°C corresponds to a temperature of 50°F. Also 40°C corresponds to 104°F. Find the linear equation relating F and C.

74. Business In planning for a new item, a manufacturer assumes that the number of items produced, x, and the cost in dollars, C, of producing these items are related by a linear equation. Projections are that 100 items will cost $10,000 to produce and that 300 items will cost $22,000 to produce. Find the equation that relates C and x.

75. Business Mike bills a customer at the rate of $35 per hour plus a fixed service call charge of $50.

(a) Write an equation that will allow you to compute the total bill for any number of hours, x, that it takes to complete a job.

(b) What will the total cost of a job be if it takes 3.5 hours to complete?

(c) How many hours would a job have to take if the total bill were $160.25?

76. Business Two years after an expansion, a company had sales of $42,000. Four years later the sales were $102,000. Assuming that the sales in dollars, S, and the time, t, in years are related by a linear equation, find the equation relating S and t.

In Exercises 77 to 80, without graphing, compare and contrast the slopes and intercepts of the equations.

77. (a) $y = 2x + 1$, **(b)** $y = 2x - 5$, **(c)** $y = 2x$

78. (a) $y = 3x + 2$, **(b)** $y = \dfrac{3}{4}x + 2$, **(c)** $y = -2x + 2$

79. (a) $y = 4x + 5$, **(b)** $y - -4x + 5$, **(c)** $y = 4x - 5$

80. (a) $y = 3x$, **(b)** $y = \dfrac{1}{3}x$, **(c)** $y = -3x$

In Exercises 81 to 86, use your graphing utility to graph the following.

81. $3x - 5y = 30$

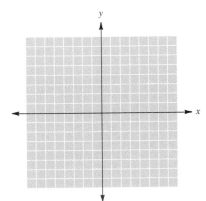

82. $2x + 7y = 14$

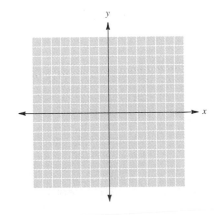

83. The line with slope $\dfrac{2}{3}$ and y intercept at 7.

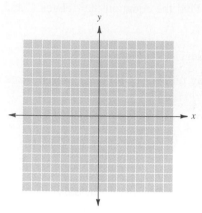

84. The line with slope $-\dfrac{1}{5}$ and y intercept at 3.

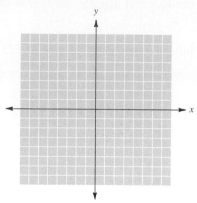

85. The line with slope n passing through the point $(1, 5)$.

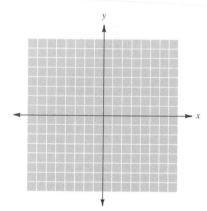

86. The line with slope $\sqrt{2}$ passing through the point $(2, -2)$.

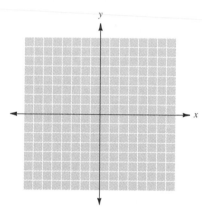

In Exercises 87 to 94, use the graph to determine the slope and y intercept of the line.

87.

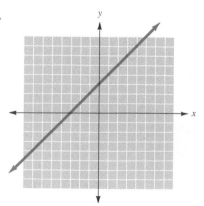

88.

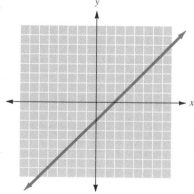

89.

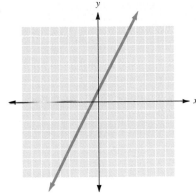

90.

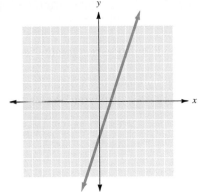

91.

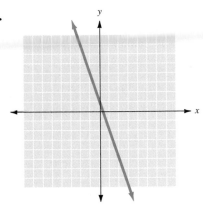

92.

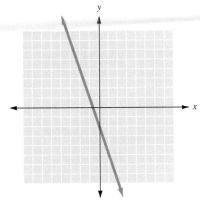

93.

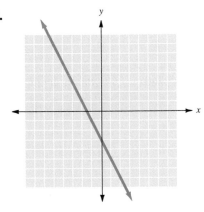

94.

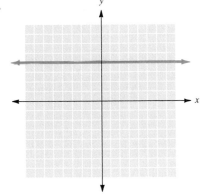

In Exercises 95 to 100, from each pair, identify which variable is dependent and which is independent.

95. The amount of a phone bill and the length of the call.

96. The cost of filling a car's gas tank and the size of the tank.

97. The height of a ball thrown in the air and the time in the air.

98. The amount of penalty on an unpaid tax bill and the length of the time unpaid.

99. The length of time needed to graduate from college and the number of credits taken per semester.

100. The amount of snowfall in Boston and the length of the winter.

In Exercises 101 to 110, rewrite each equation as a function of x.

101. $y = -3x + 2$

102. $y = 5x + 7$

103. $y = 4x - 8$

104. $y = -7x - 9$

105. $3x + 2y = 6$

106. $4x + 3y = 12$

107. $-2x + 6y = 9$

108. $-3x + 4y = 11$

109. $-5x - 8y = -9$

110. $4x - 7y = -10$

In Exercises 111 to 116, graph the functions.

111. $f(x) = 3x + 7$

112. $f(x) = -2x - 5$

113. $f(x) = -2x + 7$

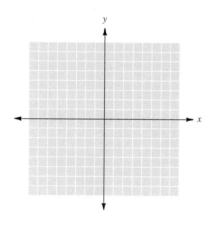

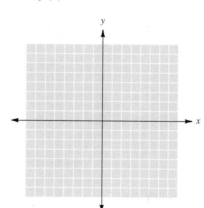

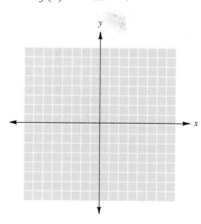

114. $f(x) = -3x + 8$

115. $f(x) = -x - 1$

116. $f(x) = -2x - 5$

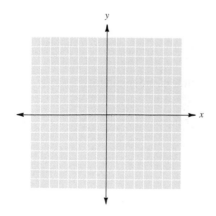

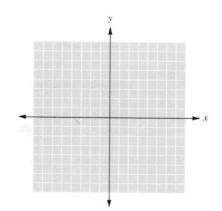

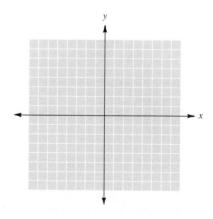

In Exercises 117 to 122, if $f(x) = 4x - 3$, find the following.

117. $f(5)$ **118.** $f(0)$ **119.** $f(4)$

120. $f(-1)$ **121.** $f(-4)$ **122.** $f(\frac{1}{2})$

In Exercises 123 to 128, if $f(x) = 5x - 1$, find the following.

123. $f(a)$ **124.** $f(2r)$ **125.** $f(x + 1)$

126. $f(a - 2)$ **127.** $f(x + h)$ **128.** $\dfrac{f(x + h) - f(x)}{h}$

In Exercises 129 to 132, if $g(x) = -3x + 2$, find the following.

129. $g(m)$ **130.** $g(5n)$ **131.** $g(x + 2)$ **132.** $g(s - 1)$

In Exercises 133 to 136, let $f(x) = 2x + 3$.

133. Find $f(1)$ **134.** Find $f(3)$

135. Form the ordered pairs $(1, f(1))$ and $(3, f(3))$.

136. Write the equation of the line passing through the points determined by the ordered pairs in Exercise 135.

Answers

1. (e)   **3. (a)**   **5. (b)**   **7. (h)**   **9.** $y = -x + 5$, $m = -1$, y int $= 5$ **11.** $y = 2x + 2$, $m = 2$,

y int $= 2$ **13.** $y = -\dfrac{1}{3}x + 3$, $m = -\dfrac{1}{3}$, y int $= 3$ **15.** $y = \dfrac{2}{3}x - 2$, $m = \dfrac{2}{3}$, y int $= -2$

17. $y = 2x$, $m = 2$, y int $= 0$ **19.** $y = -3$, $m = 0$, y int $= -3$ **21.** $y = 3x + 2$

23. $y = \dfrac{3}{2}x + 2$ **25.** $y = 4$ **27.** $y = \dfrac{5}{4}x - 5$ **29.** $y = 3x - 1$ **31.** $y = -3x - 9$ **33.** $y = \dfrac{2}{5}x - 5$

35. $x = 2$ **37.** $y = -\dfrac{4}{5}x + 4$ **39.** $y = x + 1$ **41.** $y = \dfrac{3}{4}x - \dfrac{3}{2}$ **43.** $y = 2$ **45.** $y = \dfrac{3}{2}x - 3$

47. $y = \dfrac{5}{2}x + 4$ **49.** $y = 4x - 2$ **51.** $y = -\dfrac{1}{2}x + 2$ **53.** $y = 4$ **55.** $y = 5x - 13$ **57.** $y = 3x + 3$

59. $f(x) = \dfrac{1}{2}x + 4$ **61.** $f(x) = 3$ **63.** $f(x) = 2x + 8$ **65.** $f(x) = \dfrac{4}{3}x - 2$ **67.** $f(x) = \dfrac{1}{3}x - \dfrac{11}{3}$

69. $f(x) = \dfrac{1}{2}x$ **71.** $f(x) = -\dfrac{4}{7}x + \dfrac{18}{7}$ **73.** F $-\dfrac{9}{5}$C I 32   **75. (a)** $C = 35x + 50$, **(b)** \$172.50,

(c) 3.14 h   **77.** Same slope but different y int

79. (a) and **(b)** have the same y intercept but **(a)** rises while **(b)** falls both at the same rate. **(a)** and **(c)** have the same slope but different y int.

81.

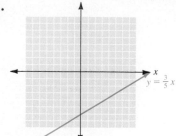

$y = \frac{3}{5}x$

83.

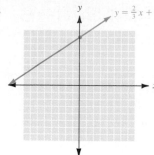

$y = \frac{2}{3}x + 7$

85.

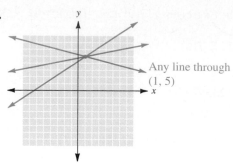

Any line through (1, 5)

87. Slope = 1; y int = 3   **89.** Slope = 2; y int = 1   **91.** Slope = −3; y int = 1

93. Slope = −2; y int = −3   **95.** Independent: length of call; dependent: amount of bill

97. Independent: time in air; dependent: height of ball   **99.** Independent: number of credits; dependent: time

to graduate   **101.** $f(x) = -3x + 2$ **103.** $f(x) = 4x - 8$ **105.** $f(x) = -\frac{3}{2}x + 3$ **107.** $f(x) = \frac{1}{3}x + \frac{3}{2}$

109. $f(x) = \frac{-5}{8}x + \frac{9}{8}$ **111.**

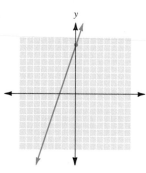

113.

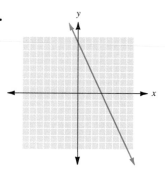

115.

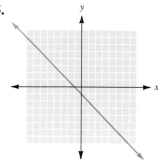

117. 17   **119.** 13   **121.** −19   **123.** $5a - 1$ **125.** $5x + 4$

127. $5x + 5h - 1$ **129.** $-3m + 2$ **131.** $-3x - 4$ **133.** 5

135. (1, 5), (3, 9)

1.9 Scatter Plots and Prediction Lines

1.9 OBJECTIVES

1. Create a scatter plot
2. Use a TI-82 or TI-83 to graph a scatter plot
3. Use a TI-82 or TI-83 to find the equation of a prediction line

A **scatter plot** is a graph of a set of ordered pairs. Scatter plots help us see the relationship between two sets of data. For example, the following graph represents the relationship between the number of wins and the number of losses that a professional football team might have in a full season. We can use this graph to determine the number of losses that a team with 10 wins would have.

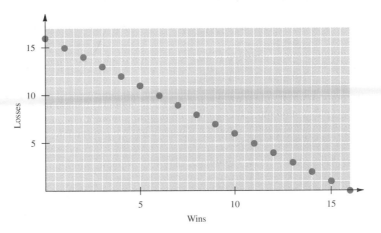

The ordered pair (10, 6) indicates that a team with 10 wins would have 6 losses. Notice that the ordered pairs form a perfect line with slope -1.

The set of ordered pairs graphed below show the relationship between the number of miles driven and the amount of gas purchased the last 10 times that Allie filled her gas tank. Notice that the points almost form a straight line.

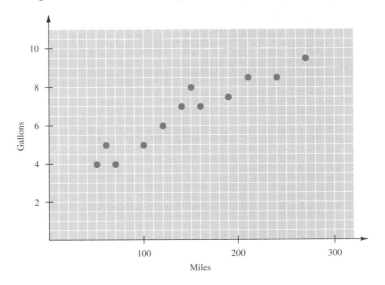

A *prediction line* is a line that gives us a "reasonable" estimation for *y* when we have a given *x*. We will reserve definition of the word reasonable for future mathematics classes.

Suppose that you were asked to estimate the amount of gas Allie will need to buy if she drives 250 miles. Even though there is no ordered pair associated with 250 miles, you can comfortably guess that Allie will need about 9 gallons of gas. You arrived at this answer by noting that the points fell in a fairly straight line, and you estimated where that line would be when the *x* was 250. Essentially, you created a **prediction line,** which is a line that is used to estimate the *y* value when you are given a value for *x*.

In this section, we will develop a technique (using a graphing calculator) that will allow us to find a prediction line. The first step in finding a prediction line is to create and sketch a scatter plot, as Example 1 illustrates.

• Example 1

Creating a Scatter Plot

Carlos kept the following chart next to his treadmill. Create a scatter plot for the ordered pairs.

Minutes	Miles
53	6.4
48	5.7
55	6.8
30	4.5
40	5.2
62	7.0
35	4.9
50	6.0
65	7.2

Each combination of minutes and miles makes an ordered pair. The first ordered pair is (53, 6.4). The scatter plot is the graph of all 10 ordered pairs.

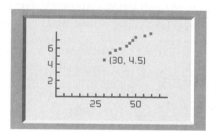

• • • CHECK YOURSELF 1

Whitney keeps track of use of the Xerox machine in the library. She created the following chart:

Month	School Days	Duplication Count
September	9	1230
October	21	3268
December	8	1124
January	15	2253
February	19	2872
March	17	2597
April	21	3410
May	22	3502
June	10	1470

Create a scatter plot for the ordered pairs.

USING A GRAPHING CALCULATOR

For the remainder of this section, we will assume that you have a TI-82 (or TI-83). We will use those calculators to graph and to find the equation of the prediction line.

• Example 2

Using a TI to Graph the Scatter Plot

The challenge of this example is to have you graph the scatter plot for the data in Example 1 on the display screen of your TI-82 (or TI-83). You must first enter the data into the calculator. The following algorithm will help you accomplish that.

> 1. Press STAT.
> 2. Choose 4 (ClrLst).
> 3. Press L₁ (2nd "1"), comma (,), and L₂ (2nd "2").
> 4. Press ENTER.
> 5. Press STAT.
> 6. Choose 1 (Edit).
> 7. Enter the numbers in the minutes column. Press ENTER after each number.
> 8. When all 10 numbers have been entered, use the right arrow to go to List 2.
> 9. Enter the 10 numbers in the miles column in list.
> 10. When all numbers have been entered, press 2ND QUIT.

Now we are ready to create a scatter plot with the data. Use the following algorithm:

> 1. Press 2ND STAT PLOT (the Y = key).
> 2. Choose 1.
> 3. Use the left arrow and ENTER to select ON.
> 4. Choose the first graph type by highlighting it and pushing ENTER.
> 5. Push ZOOM and select 9 (ZoomStat).

You should now have the scatter plot displayed on your calculator screen. It will look very much like the scatter plot you saw in Example 1. If you wish to adjust the windows so that the x and y axes are included, push WINDOW and set X min and Y min to 0. You can then press the GRAPH key.

● ● ● CHECK YOURSELF 2

Use your graphing calculator to create a scatter plot for the data in Check Yourself 1.

One reason we create a scatter plot is so that we can see whether a pattern exists. This pattern can then be used to make predictions about other possible ordered pairs. In Example 1, you may have wondered how many miles Carlos could jog in 60 minutes. Even though there was no ordered pair that began with 60, you could use the pattern of the graph to reasonably predict the number of miles he might jog. It is important that you recognize that if a prediction is any good at all, it is good only within the domain of the data points. The **domain of the data points** is the set of values between the minimum and maximum values for the independent variable. For example, we used the scatter plot to predict Carlos' miles for a time (58 minutes) that fell between the minimum (25 minutes) and the maximum (65 minutes.) What if we had predicted his mileage for a workout time of 584 minutes? He certainly could not have kept running for that long. His mileage would have probably been the mileage to the local hospital, but he would not have been running there!

To ensure that our predictions are based on reason and not guesswork, we find an equation for our prediction line. Although this can be done by hand, it is a long (and somewhat complicated) process. Instead, we will use the calculator to find the equation for the prediction line.

• Example 3

Finding the Equation of a Prediction Line

We now want to use the calculator to find the equation of the prediction line. As before, the data must be in List 1 and List 2. Since the data for Carlos' workouts are there already, we will find the equation for that prediction line.

> **1.** Press $\boxed{\text{STAT}}$.
> **2.** Use the right arrow to choose $\boxed{\text{CALC}}$.
> **3.** Select LinReg.
> **4.** Press $\boxed{\text{ENTER}}$.

The calculator has given you the equation of the prediction line in slope-intercept form. Instead of writing $y = mx + b$, it has written $y = ax + b$. The values for a and b are below the equation. Rounded to the nearest hundredth, the values are $a = 0.09$ and $b = 1.44$. The equation of the prediction line is $y = 0.09x + 1.44$.

● ● ● **CHECK YOURSELF 3**

Use your graphing calculator to find the prediction equation from Check Yourself 1.

Example 4 ties together all that we learned in this section.

• Example 4

Finding and Using the Prediction Equation

Kareem kept excellent financial records in 1997. The chart on page 113 represents 9 different weeks of net pay and the amount he spent on food that week.

Net	Food Expenses
175	61
280	115
225	96
237	101
193	88
145	47
210	90
276	121
123	58

Use the data to complete the following:

(*a*) Sketch a scatter plot.
(*b*) Find the prediction equation.
(*c*) Draw the prediction line on the scatter plot.
(*d*) Predict how much he would have spent on food in a week in which he netted $250.
(*e*) Explain why you would not predict how much he would have spent on food in a week in which he netted $2000.

Solution

(*a*) Following the algorithm from Example 3, we enter the data into List 1 (net earnings) and List 2 (food expenses). Then we create the scatter plot, which should look something like this.

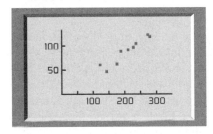

(*b*) From the algorithm in Example 4, we find the equation of the prediction line to be $y = 0.46x - 8.34$.
(*c*) Adding the prediction line to the scatter plot, we get the following graph.

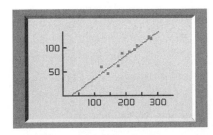

(*d*) If $x = \$250$, we replace x in the equation for the prediction line with 250, which gives us $y = 0.46(250) - 8.34 = 106.66$. Rounded to the nearest dollar, we get $107.
(*e*) The domain does not include $2000, so our prediction equation would not be valid for that amount.

● ● ●   **CHECK YOURSELF 4**

Emanuel has been curious about the relationship between the amount of time people spend studying and the grade they get on a test. He got the entire Intermediate Algebra class to keep track of the number of hours they studied for the midterm. After the test they gave him a sheet of paper that showed the ordered pair (hours, score). The following chart represents 10 of those ordered pairs.

Hours Studied	Test Score
10	73
19	82
28	85
12	88
5	71
3	48
21	92
34	98
9	65
11	75

Use the data to complete the following:

(a) Sketch a scatter plot.
(b) Find the prediction equation.
(c) Draw the prediction line on the scatter plot.
(d) Predict the test score for someone who studied 25 hours.
(e) Explain why you would not predict a test score for someone who studied 400 hours.

The clue for how good a predictor the line will be is given by the *"r"* value that appears each time you look for the equation. The closer *r* is to 1 (or negative 1) the better the prediction equation is.

In closing this section, we must note that not all ordered pairs can be made to find useful prediction lines. Every case that we looked at in this section started with a scatter plot that lent itself nicely to drawing a line through (or at least close to) the points. What if the points are scattered in no particular pattern? For example, what if we collected ordered pairs that had students' scores on their last math test paired with their scores on their last English test? In that case, we probably would not bother even looking for the prediction equation.

● ● ●   **CHECK YOURSELF ANSWERS**

1. and 2.

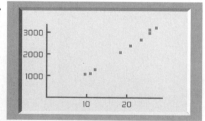

3. $y = 170x - 270$.

4. (a)

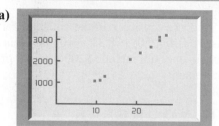

(b) $y = 1.2x + 60$,

(c)

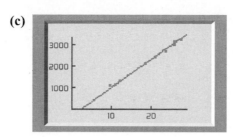

(d) 89, and **(e)** Not in the domain.

1.9 Exercises

1. In a local industrial plant, the number of work-hours in safety training and the number of work-hours lost due to accidents have been recorded for 10 divisions.

Division	No. of Work-Hours in Safety Training	No. of Work-Hours Lost from Accidents
1	10	80
2	15	75
3	20	72
4	25	70
5	30	60
6	40	53
7	45	50
8	50	48
9	60	42
10	65	35

(a) Create a scatter plot.
(b) Find the prediction equation.
(c) Draw the prediction line on the scatter plot.
(d) Predict how many work-hours would be lost due to accidents if 35 hours of safety training were given.

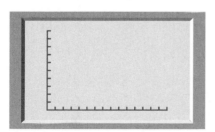

2. In a statistics class, the mid-term and final exam scores were collected for 10 students. Each exam was worth a total of 100 points.

Mid-Term Exam Scores	Final Exam Scores
71	80
79	85
84	88
76	81
62	75
93	90
88	87
91	96
68	82
77	83

(a) Create a scatter plot.
(b) Find the prediction equation.
(c) Draw the prediction line on the scatter plot.
(d) Predict the final exam grade for Derek if he scored an 87 on the mid-term.

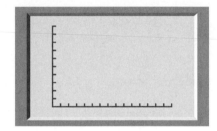

3. A rental car agency has collected data relating the number of miles traveled and the total cost in dollars.

Miles Traveled (in thousands)	Cost (in $)
2	60
6	100
10	200
14	275
8	175
5	90
12	290
16	400
3	75
18	450
21	475

(a) Create a scatter plot.
(b) Find the prediction equation.
(c) Draw the prediction line on the scatter plot.
(d) How much would repairs cost after **(i)** 9000 miles and **(ii)** 20,000 miles?
(e) Why would you not predict maintenance costs after traveling 35,000 miles?

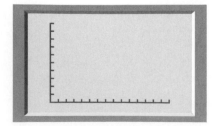

4. A math placement test was given to all entering freshmen at Bucks County Community College. The placement test scores and the score on the first test were recorded for students in a college algebra class.

Placement Test Scores (max. of 40)	First Test Score
25	78
18	75
30	88
14	65
10	62
32	85
12	68
16	73
22	78
27	82
38	93

(a) Create a scatter plot.
(b) Find the prediction equation.
(c) Draw the prediction line on the scatter plot.
(d) Predict the score on the first college algebra test if the placement test score is **(i)** 20 and **(ii)** 35.

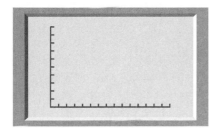

5. Students claim they can tell the cost of a textbook by the thickness of the book. They picked 10 books of roughly the same height and weight. The following data were collected.

Thickness (in cm)	Cost (in $)
1.0	44
0.8	43
3.0	53
2.4	50
1.6	46
1.9	48
0.5	42
1.2	45
3.2	54

(a) Create a scatter plot.
(b) Find the prediction equation.
(c) Draw the prediction line on the scatter plot.
(d) How much would a book that is 2.8 cm thick cost?
(e) Why would you not want to predict the cost of a book 5 cm thick?

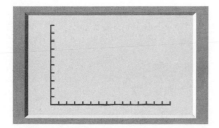

6. The following table shows the IQ of 12 students along with their cumulative Grade Point Average (GPA) after 4 years of college.

IQ	GPA
117	3.2
93	2.6
102	2.9
110	3.1
88	2.4
75	1.9
107	3.1
111	3.2
120	3.5
95	2.7
115	3.4
99	2.9

(a) Create a scatter plot.
(b) Find the prediction equation.
(c) Draw the prediction line on the scatter plot.
(d) Predict the GPA of a student with an IQ of 115.
(e) Why would you not want to predict the GPA of a student with an IQ of 140?

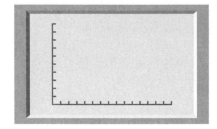

7. The following table gives the median income for households in the United States headed by persons under 25 years of age. Median income is the income that is right in the middle: 50% of the families made this income or more, and 50% made this income or less.

When income is measured over a period of years, the dollar amounts must be related to the value in a particular year in order to account for inflation. In this chart, the dollar amounts are in equivalent 1993 dollars. In other words, the income shown for 1967 is not the amount listed here but an equivalent amount that it would buy in 1993 dollars.

Year	Median Family Income for Families with Householders Under Age 25 (in 1993 $)
1967	$23,263
1973	24,534
1979	25,416
1989	19,885
1992	16,014
1993	17,440

Adapted from Lawrence Mishel and Jared Bernstein, *The State of Working America 1994–95.*

(a) What do you notice about these figures by looking at the table?
(b) To get a better picture of what has happened to income for this age group, draw a graph of income versus time in years, plotting a point for each pair of figures given here. Scale the x axis in years since 1965 and the y axis in thousands of dollars.
(c) What does this scatter plot tell you that you did not notice when you looked at the table?
(d) Now, using a ruler, draw a straight line that comes as close to as many points as possible to get a "line of best fit." When you have drawn the line, read a couple of points on the line and calculate the slope of your line of best fit. What does the slope tell you about the trend in incomes for this age group?
(e) Now, read the y intercept from your graph and write a linear equation for your line of best fit. This line, although it may not actually touch any of the points, is a description of the general trend in income for households headed by people under 25. Use your equation to project what the median income will be in 2000 if this trend continued.

Answers

1. (a) and (c) see graph, (b) $y = -0.8x + 87$, (d) 59

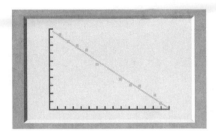

3. (a) and (c) see graph, (b) $y = 24x - 15$, (d) $201, $465, (e) 35,000 is outside the domain

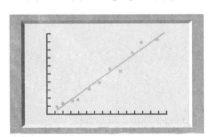

5. (a) and (c) see graph, (b) $y = 4.5x + 39$, (d) $51.60, (e) the domain does not include 5 cm, so our prediction equation is not valid for that size.

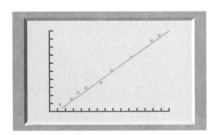

7.

Summary

Ordered Pairs and Relations [1.1]

The set {(1, 4), (2, 5), (1, 6)} is a relation.

The domain is {1, 2}.

The range is {4, 5, 6}.

Ordered Pair
Given two related values, x and y, we write the pair of values as (x, y).
Relation
A set of ordered pairs.
Domain
The set of all first elements of a relation.
Range
The set of all second elements of a relation.

The Cartesian Coordinate System [1.2]

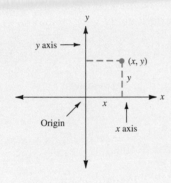

The **cartesian coordinate system** allows us to establish a one-to-one correspondence between points in the plane and ordered pairs of real numbers.

$$(x, y)$$
x coordinate y coordinate

To **graph** (or **plot**) a point (x, y) in the plane:

1. Start at the origin.
2. Move to the right or left along the x axis according to the value of the x coordinate.
3. Move up or down and parallel to the y axis according to the value of the y coordinate.

It is not always desirable to use the same scale on both the x and y axes. In these situations, we use a different marked scale. This is called scaling the axes.

An Introduction to Functions [1.3]

{(1, 2), (2, 3), (3, 4)} is a function.
{(1, 2), (2, 3), (2, 4)} is *not* a function.

A **function** is set of ordered pairs (a relation) in which no two first coordinates are equal.

The set of points in the plane that correspond to ordered pairs in a relation or function is called the *graph* of that relation or function.

A useful means of determining whether a graph represents a relation that is also a function is called the **vertical-line test**.

If a vertical line meets the graph of a relation in two or more points, the relation is *not* a function.

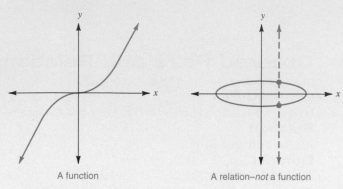

A function   A relation–*not* a function

Properties of Functions [1.4]

The **sum of two functions** f and g is written $f + g$. It is defined as

$$f + g: (f + g)(x) = f(x) + g(x)$$

The **difference of two functions** f and g is written $f - g$. It is defined as

$$f - g: (f - g)(x) = f(x) - g(x)$$

The **product of two functions** f and g is written $f \cdot g$. It is defined as

$$f \cdot g: (f \cdot g)(x) = f(x) \cdot g(x)$$

The **quotient of two functions** f and g is written $f \div g$. It is defined as

$$f \div g: (f \div g)(x) = f(x) \div g(x)$$

(1, 3) is a solution for

$3x + 2y = 9$

because

$3 \cdot 1 + 2 \cdot 3 = 9$

is a true statement.

The Graph of a Linear Equation [1.5]

The **solution set** for an equation in two variables is the set containing all ordered pairs of real numbers (x, y) that will make the equation a true statement.

An equation of the form

$$ax + by = c \tag{1}$$

where a and b cannot both be zero, is a linear equation in two variables. The graph of such an equation is always a *straight line*. An equation in form (1) is called the **standard form of the equation of a line.**

To graph a linear equation:

1. Find at least three solutions for the equation, and write your results in a table of values.
2. Graph the points associated with the ordered pairs found in step 1.
3. Draw a straight line through the points plotted above to form the graph of the equation.

A second approach to graphing linear equations uses the **x and y intercepts** of the line. The x intercept is the x coordinate of the point where the line intersects the x axis. The y intercept is the y coordinate of the point where the line intersects the y axis.

To graph

$y = 2x - 3$

$(0, -3)$, $(1, -1)$, and $(2, 1)$ are solutions.

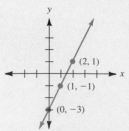

To graph by the intercept method:

1. Find the x intercept. Let $y = 0$ and solve for x.
2. Find the y intercept. Let $x = 0$ and solve for y.
3. Plot the two intercepts determined in steps 1 and 2.
4. Draw a straight line through the intercepts.

Vertical or Horizontal Lines

1. A line with an equation of the form

$$y = k$$

   is horizontal (parallel to the x axis).
2. A line with an equation of the form

$$x = h$$

   is vertical (parallel to the y axis).

Reading Values from Graphs [1.6]

For a specific value of x, let's call it a, we can find $f(a)$ with the following algorithm.

1. Draw a vertical line through a on the x axis.
2. Find the point of intersection of that line with the graph.
3. Draw a horizontal line through the graph at that point.
4. Find the intersection of the horizontal line with the y axis.
5. $f(a)$ is that y value.

If given the function value, one finds the x value associated with it as follows.

1. Find the given function value on the y axis.
2. Draw a horizontal line through that point.
3. Find every point on the graph that intersects the horizontal line.
4. Draw a vertical line through each of those points of intersection.
5. The x value(s) are each point of intersection of the vertical lines and the x axis.

The Slope of a Line [1.7]

The **slope** of a line gives a numerical measure of the direction and steepness, or inclination, of the line. The slope m of a line containing the distinct points in the plane (x_1, y_1) and (x_2, y_2) is given by

$$m = \frac{y_2 - y_1}{x_2 - x_1} \qquad x_2 \neq x_1$$

$y = \frac{2}{3}x + 4$ is in slope-intercept form. The slope m is $\frac{2}{3}$, and the y intercept is 4.

The slopes of two nonvertical parallel lines are equal. The slopes of two nonvertical perpendicular lines are the negative reciprocals of each other.

If line l has slope $m = -2$ and passes through $(-2, 3)$, its equation is

$$y - 3 = -2[x - (-2)]$$
$$y - 3 = -2(x + 2)$$
$$y - 3 = -2x - 4$$
$$y = -2x - 1$$

Forms of Linear Equations [1.8]

There are two useful special forms for the equation of a line. The **slope-intercept form** of the equation of a line is $y = mx + b$, where the line has slope m and intercept b. The **point-slope form** of the equation of a line is $y - y_1 = m(x - x_1)$, where the line has slope m and passes through the point (x_1, y_1). And $x = x_1$ is the equation of a line through (x, y) with undefined slope.

Scatter Plots and Prediction Lines [1.9]

The following algorithm allows you to use your TI-82 or TI-83 to plot a scatter plot.

1. Press STAT .
2. Choose 4 (ClrLst).
3. Press L₁ (2nd "1"), comma (,) , and L₂ (2nd "2").
4. Press ENTER .
5. Press STAT .
6. Choose 1 (Edit).
7. Enter numbers in the minutes column. Press ENTER after each number.
8. When all 10 numbers have been entered, use the right arrow to go to List 2.
9. Enter the 10 numbers in the miles column in list.
10. When all numbers have been entered, press 2nd QUIT .

To create a scatter plot with the data, use the following algorithm:

1. Press 2nd STAT PLOT (the Y=key).
2. Choose 1 .
3. Use the left arrow and ENTER to select ON .
4. Choose the first graph type by highlighting it and pushing ENTER .
5. Push ZOOM and select 9 (ZoomStat).

If you wish to adjust the windows so that the x and y axes are included, push WINDOW and set X min and Y min to 0. You can then press the GRAPH key.

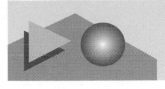

Summary Exercises

This summary exercise set is provided to give you practice with each of the objectives in the chapter. Each exercise is keyed to the appropriate chapter section. The answers are provided in the *Instructor's Manual*.

[1.1]   In Exercises 1 and 2, identify which are ordered pairs.

1. (a) $(2, 1)$ **(b)** $\{3, 4\}$ **(c)** $1, 4$ **(d)** $(-4, -3)$ **(e)** $((3, 2), 5)$

2. (a) $\{-1, 4\}$ **(b)** $6, 8$ **(c)** $(3, 4)$ **(d)** $\{(3, -1), 4\}$ **(e)** $(-2, 5)$

In Exercises 3 to 10, for each set of ordered pairs, identify the domain and range.

3. $\{(3, 5), (4, 6), (1, 2), (8, 1), (7, 3)\}$

4. $\{(-1, 3), (-2, 5), (3, 7), (1, 4), (2, -2)\}$

5. $\{(1, 3), (1, 5), (1, 7), (1, 9), (1, 10)\}$

6. $\{(2, 4), (-1, 4), (-3, 4), (1, 4), (6, 4)\}$

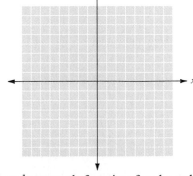

7. $\{(x, y) \mid x + y - 4\}$

8. $\{(x, y) \mid 3x + 2y = 6\}$

9. $\{(x, y) \mid y = 5\}$

10. $\{(x, y) \mid x = 8\}$

[1.2]   In Exercises 11 to 18, graph the following points in the cartesian coordinate system.

11. $A(2, -3)$
12. $B(3, 5)$
13. $C(0, 5)$
14. $D(-2, -6)$
15. $E(-4, 1)$
16. $F(-6, 0)$
17. $G(4, -5)$
18. $H(0, -2)$

[1.3]   In Exercises 19 to 24, evaluate each function for the value specified.

19. $f(x) = x^2 - 3x + 5$; find **(a)** $f(0)$, **(b)** $f(-1)$, and **(c)** $f(1)$.

20. $f(x) = -2x^2 + x - 7$; find **(a)** $f(0)$, **(b)** $f(2)$, and **(c)** $f(-2)$.

21. $f(x) = x^3 - x^2 - 2x + 5$; find **(a)** $f(-1)$, **(b)** $f(0)$, and **(c)** $f(2)$.

22. $f(x) = -x^2 + 7x - 9$; find **(a)** $f(-3)$, **(b)** $f(0)$, and **(c)** $f(1)$.

23. $f(x) = 3x^2 - 5x + 1$; find **(a)** $f(-1)$, **(b)** $f(0)$, and **(c)** $f(2)$.

24. $f(x) = -x^3 + 3x - 5$; find **(a)** $f(2)$, **(b)** $f(0)$, and **(c)** $f(1)$.

In Exercises 25–30, determine which relations are also functions.

25. $\{(1, 3), (2, 4), (5, -1), (-1, 3)\}$ *function*   **26.** $\{(-2, 4), (3, 6), (1, 5), (0, 1)\}$

27. $\{(1, 2), (0, 4), (1, 3), (2, 5)\}$ *not a function*   **28.** $\{(1, 3), (2, 3), (3, 3), (4, 3)\}$

29.

x	y
-3	2
-1	1
0	3
1	4
3	5

30.

x	y
-1	3
0	2
1	3
2	4
3	5

In Exercises 31 to 34, use the vertical line test to determine whether the graphs represent a function.

31.

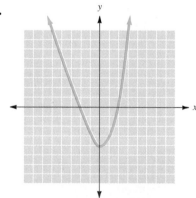

32.

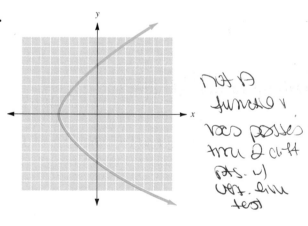

not a function, does passes thru 2 diff pts. w/ vert. line test

33.

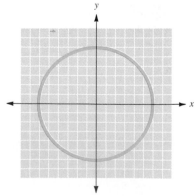

34.

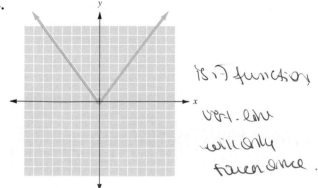

is a function vert. line will only touch once.

[1.4] In Exercises 35 to 38, find **(a)** $f + g$, **(b)** $f - g$, **(c)** $(f + g)(2)$, **(d)** $(f - g)(1)$, and **(e)** the domain of $f + g$.

35. $f(x) = 3x - 8$ $g(x) = -9x - 15$

36. $f(x) = -8x + 14$ $g(x) = 2x + 3$

37. $f(x) = -9x + 5$ $g(x) = \dfrac{3}{x - 8}$

38. $f(x) = 2x + 7$ $g(x) = \dfrac{-4x}{x - 9}$

In Exercises 39 to 42, find **(a)** $f \cdot g$, **(b)** $\dfrac{f}{g}$, and **(c)** the domain of $\dfrac{f}{g}$.

39. $f(x) = 7x + 3$ $g(x) = x - 2$

40. $f(x) = -8x - 3$ $g(x) = x + 11$

41. $f(x) = -9x + 12$ $g(x) = 3x - 5$

42. $f(x) = 5x + 3$ $g(x) = 4x + 12$

[1.5] In Exercises 43 to 52, find the x and y intercepts and graph the equations.

43. $x + y = 7$

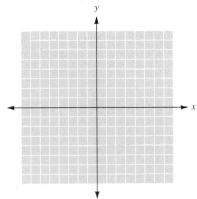

44. $x - y = 3$

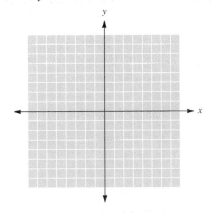

45. $3x + 2y = 10$

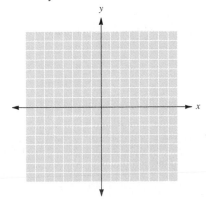

46. $5x - 2y = 10$

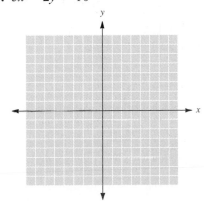

47. $4x - 3y = 12$

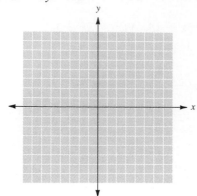

48. $5x + 4y = 40$

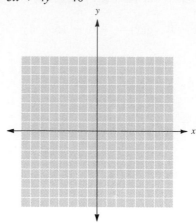

49. $2x + 7y = 14$

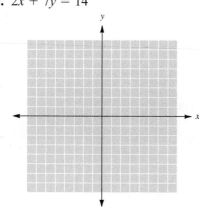

50. $3x - 9 = 0$

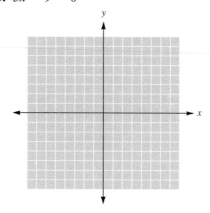

51. $16 - 2x = 0$

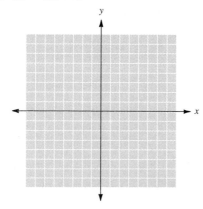

52. $2x = 18y$

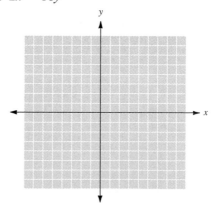

In Exercises 53 to 56, find the domain and range of each relation.

53. $4y - 5y = 40$

54. $-3x + 15y = 45$

55. $2x + 18 = 0$

56. $7y - 28 = 0$

[1.6]   In Exercises 57 to 60, find the coordinates of the labeled points.

57.

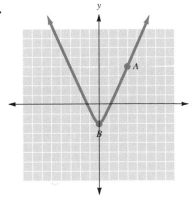

58.

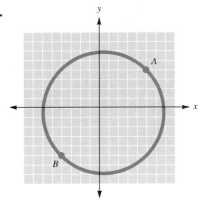

59.

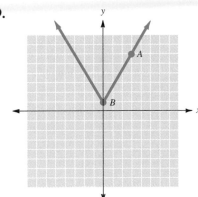

60.

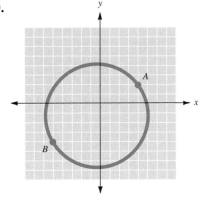

In Exercises 61 to 64, use the graph to estimate **(a)** $f(-2)$, **(b)** $f(0)$, and **(c)** $f(2)$.

61.

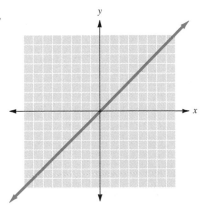

62.

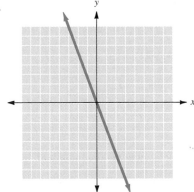

63.

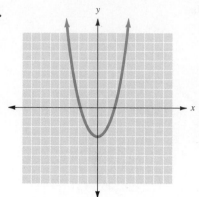

64.

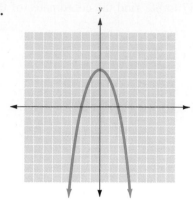

In Exercises 65 to 68, use the graph of $f(x)$ to find all values of x such that **(a)** $f(x) = -1$, **(b)** $f(x) = 0$, and **(c)** $f(x) = 1$.

65.

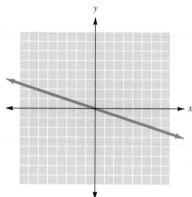

66.

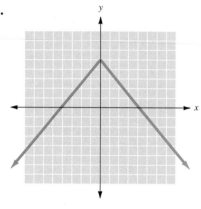

67.

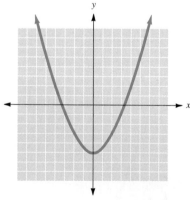

68.

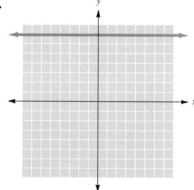

[1.7]   In Exercises 69 to 76, find the slope (if it exists) of the line determined by the following pairs of points.

69. (3, 2) and (5, 8)   **70.** (−2, 5) and (1, −1)   **71.** (2, −4) and (3, 5)

72. (−3, −4) and (3, 0)   **73.** (4, −3) and (4, 4)   **74.** (4, −2) and (−2, −2)

75. $(2, -4)$ and $(-1, -3)$

76. $(5, -2)$ and $(5, 3)$

In Exercises 77 and 78, find the slope of the line determined by the equations.

77. $3x + 2y = 6$

78. $x - 4y = 8$

In Exercises 79 to 82, are the pairs of lines parallel, perpendicular, or neither?

79. L_1 through $(-3, -2)$ and $(1, 3)$
L_2 through $(0, 3)$ and $(4, 8)$

80. L_1 through $(-4, 1)$ and $(2, -3)$
L_2 through $(0, -3)$ and $(2, 0)$

81. L_1 with equation $x + 2y = 6$
L_2 with equation $x + 3y = 9$

82. L_1 with equation $4x - 6y = 18$
L_2 with equation $2x - 3y = 6$

[1.8] In Exercises 83 to 92, write the equation of the line passing through the following points with the indicated slope. Give your results in slope-intercept form, where possible.

83. $(0, -5)$, $m = \dfrac{2}{3}$

84. $(0, -3)$, $m = 0$

85. $(2, 3)$, $m = 3$

86. $(4, 3)$, m is undefined

87. $(3, -2)$, $m = \dfrac{5}{3}$

88. $(-2, -3)$, $m = 0$

89. $(-2, -4)$, $m = -\dfrac{5}{2}$

90. $(-3, 2)$, $m = -\dfrac{4}{3}$

91. $\left(\dfrac{2}{3}, -5\right)$, $m = 0$

92. $\left(-\dfrac{5}{2}, -1\right)$, m is undefined

In Exercises 93 to 100, write the equation of the line L satisfying the following sets of geometric conditions.

93. L passes through $(-3, -1)$ and $(3, 3)$.

94. L passes through $(0, 4)$ and $(5, 3)$.

95. L has slope $\dfrac{3}{4}$ and y intercept 3.

96. L passes through $(4, -3)$ with a slope of $-\dfrac{5}{4}$.

97. L has y intercept -4 and is parallel to the line with equation $3x - y = 6$.

98. L passes through $(3, -2)$ and is perpendicular to the line with equation $3x - 5y = 15$.

99. L passes through $(2, -1)$ and is perpendicular to the line with equation $3x - 2y = 5$.

100. L passes through the point $(-5, -2)$ and is parallel to the line with equation $4x - 3y = 9$.

In Exercises 101 to 104, use the graph to determine the slope and y intercept of the line.

101.

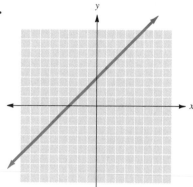

102.

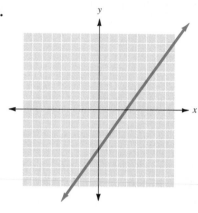

103.

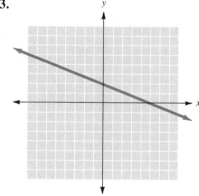

104.

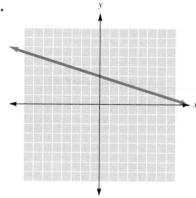

[1.9]   A public relations firm has developed a table that relates the number of years of experience and annual salary.

No. of Years	Salary (in thousands)
1	20
2	24
15	48
11	37
9	32
6	29

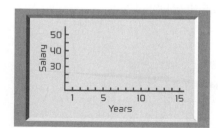

105. Create a scatter plot.

106. Find the prediction equation.

107. Draw the prediction line on the scatter plot.

108. What should the salary be for an individual with 13 years of experience?

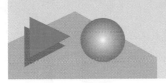
The purpose of this self-test is to help you check your progress and to review for a chapter test in class. Allow yourself about 1 hour to take the test. When you are done, check your answers in the back of the book. If you missed any answers, be sure to go back and review the appropriate sections in the chapter and the exercises that are provided.

1. Identify which of the following are ordered pairs.

    (a) $\{-1, 3\}$ **(b)** $(1, 4)$ **(c)** $[2, 4]$ **(d)** $5, 6$ **(e)** $\{(2, 1), (4, 5)\}$

2. For each of the following sets of ordered pairs, identify the domain and range.

    (a) $\{(1, 6), (-3, 5), (2, 1), (4, -2), (3, 0)\}$ **(b)** $\{(x, y) \mid 4x + 5y = 20\}$

3. Let $f(x) = -11x + 7$ and $g(x) = 4x - 9$. Find **(a)** $f + g$, **(b)** $f - g$, and **(c)** the domain of $f + g$.

4. Let $f(x) = 4x - 7$ and $g(x) = x + 9$. Find **(a)** $f \cdot g$, **(b)** $\dfrac{f}{g}$, and **(c)** the domain of $\dfrac{f}{g}$.

Write the equation of the line L that satisfies the given set of geometric conditions.

5. L passes through $(2, 3)$ and $(4, 7)$.

6. L has slope $\dfrac{2}{3}$ and y intercept of -4.

7. L passes through $(-1, 2)$ and has slope of -3.

8. L has y intercept -2 and is parallel to $4x - y = 8$.

9. If $f(x) = 3x^3 + 2x^2 - 5x + 7$, find **(a)** $f(-2)$, **(b)** $f(0)$, and **(c)** $f(1)$

In each of the following, determine which relations are functions.

10. $\{(1, 2), (3, 2), (-1, 4), (2, 4)\}$ **11.** $\{3, 1), (2, 0), (-1, 4), (2, 4)\}$

12.

x	y
1	3
4	2
-1	-3
5	1

Use the vertical line test to determine whether the following graphs represent functions.

13.

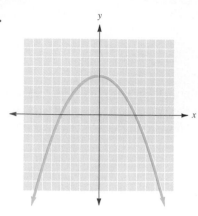

14.

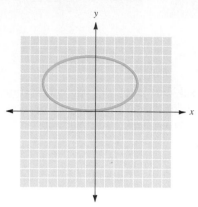

Find the x and y intercepts and graph each of the following.

15. $7x + 5y = 35$

16. $x - 3y = 12$

17. Find the domain and range of the relation $-8x + 16y = 48$.

Find the coordinates of the labeled points.

18.

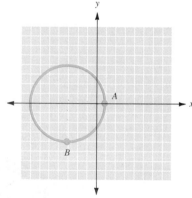

19.

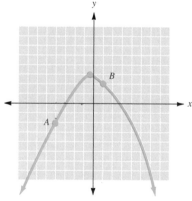

In the following, use the graph to estimate **(a)** $f(-1)$, **(b)** $f(0)$, and **(c)** $f(1)$.

20.

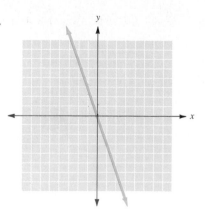

21.

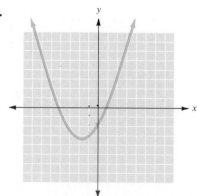

In the following, use the graph of $f(x)$ to find all values of x such that **(a)** $f(x) = -2$, **(b)** $f(x) = 0$, and **(c)** $f(x) = 2$.

22.

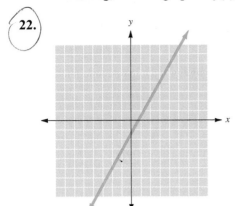

23.

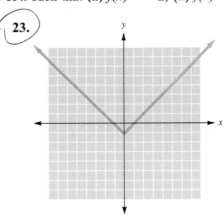

A study was made to determine the relation between weekly advertising expenditures and sales. The following data were recorded.

Advertising Costs	Sales
$40	$385
$20	$400
$25	$395
$20	$365
$30	$475
$50	$440
$40	$490
$20	$420
$50	$560
$40	$525
$25	$480
$50	$510

24. Create a scatter plot.

25. Find the prediction equation and draw the prediction line on the scatter plot.

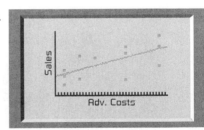

EQUATIONS AND INEQUALITIES

INTRODUCTION

Quality control is exercised in nearly all manufacturing processes. Samples of the product are taken at each stage, and various measurements are made to see if the product fits within given specifications. How often this procedure is done is determined by what is being made, the accuracy needed to produce the product, and the standards set by a specific industry or government.

In the pharmaceutical-making process, great caution must be exercised to ensure that the medicines and drugs are pure and contain precisely what is indicated on the label. Guaranteeing such purity is a task the quality control division of the pharmaceutical company assumes.

A lab technician working in quality control must run a series of tests on samples of every ingredient, even simple ingredients such as salt (NaCl). One such test is a measure of how much weight is lost as a sample is dried. The technician must set up a 3-hour procedure that involves cleaning and drying bottles and stoppers and then weighing them while they are empty and again when they contain samples of the substance to be heated and dried. At the end of the procedure, to compute the percentage of weight loss from drying, the technician uses the formula

$$L = \frac{W_g - W_f}{W_g - T} \cdot 100$$

where L = percentage loss in drying
W_g = weight of container and sample
W_f = weight of container and sample after drying process completed
T = weight of empty container

The pharmaceutical company may have a standard of acceptability for this substance. For instance, the substance may not be acceptable if the loss of weight from drying is greater than 10%. The technician would then use the following inequality to calculate acceptable weight loss:

$$10 \geq \frac{W_g - W_f}{W_g - T} \cdot 100$$

Such inequalities are more useful when solved for one of the variables, here W_f or T. In this chapter, you will learn how to solve such an inequality. ■

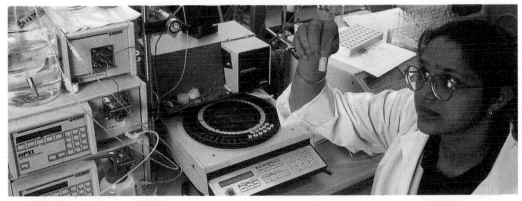

© Hank Morgan/Science Source

2.1 Solving Linear Equations in One Variable

2.1 OBJECTIVES

1. Identify linear equations
2. Combine like terms to solve an equation
3. Solve equations that involve fractions
4. Solve distance applications
5. Find a break-even point

IDENTIFYING LINEAR EQUATIONS

We begin this chapter by considering one of the most important tools of mathematics—the equation. The ability to recognize and solve various types of equations and inequalities is probably the most useful algebraic skill you will learn, and we will continue to build on the methods developed here throughout the remainder of the text. To start, let's describe what we mean by an equation.

An **equation** is a mathematical statement in which two expressions represent the same quantity. An equation has three parts:

$$\underbrace{5x + 6}_{\text{Left side}} \underset{\substack{\uparrow \\ \text{Equals} \\ \text{sign}}}{=} \underbrace{2x - 3}_{\text{Right side}}$$

The equation simply says that the expression on the left and the expression on the right represent the same quantity.

In this chapter, we will work with a particular kind of equation.

> Linear equations are also called **first-degree equations** because the highest power of the variable is the first power, or first degree.

A **linear equation in one variable** is any equation that can be written in the form

$$ax + b = 0$$

where a and b are any real numbers

$$a \neq 0$$

> We also say the solution *satisfies* the equation.

The **solution** of an equation in one variable is any number that will make the equation a true statement. The **solution set** for such an equation is simply the set consisting of all solutions.

• Example 1

Checking Solutions

Solve for x.

$$5x + 6 = 2x - 3$$

The solution is -3 because replacing x with -3 gives

> We use the question mark over the equals sign when we are checking to see if the statement is true.

$$5(-3) + 6 \overset{?}{=} 2(-3) - 3$$
$$-15 + 6 \overset{?}{=} -6 - 3$$
$$-9 = -9 \qquad \text{A true statement}$$

138

● ● ●   **CHECK YOURSELF 1**

Verify that 7 is a solution for this equation.

$5x - 15 = 2x + 6$

Solving linear equations in one variable will require using **equivalent equations.**

> Two equations are *equivalent* if they have the same solution set.

For example, the three equations

$$5x + 5 = 2x - 4 \qquad 3x = -9 \qquad x = -3$$

You can easily verify this by replacing x with −3 in each equation.

are all equivalent because they all have the same solution set, $\{-3\}$. Note that replacing x with -3 will give a true statement in the third equation, but it is not as clear that -3 is a solution for the other two equations. This leads us to an equation-solving strategy of *isolating* the variable, as is the case in the equation $x = -3$.

To form equivalent equations that will lead to the solution of a linear equation, we need two properties of equality: addition and multiplication. The addition property is defined here.

> **Addition Property of Equality**
>
> If $a = b$
>
> then $a + c = b + c$

Adding the same quantity to both sides of an equation gives an equivalent equation, which holds true whether c is positive or negative.

Since subtraction can always be defined in terms of addition,

$$a - c = a + (-c)$$

The addition property also allows us to *subtract* the same quantity from both sides of an equation.

The multiplication property is defined here.

> **Multiplication Property of Equality**
>
> If $a = b$
>
> then $ac = bc$ where $c \neq 0$

Multiplying both sides of an equation by the same nonzero quantity gives an equivalent equation.

Since division can be defined in terms of multiplication,

$$\frac{a}{c} = a \cdot \frac{1}{c} \qquad c \neq 0$$

The multiplication property allows us to *divide* both sides of an equation by the same nonzero quantity.

Let's see how these properties are applied to find the solution of a linear equation.

• Example 2

Applying the Properties of Equality

Solve for x.

$$3x - 5 = 4 \tag{1}$$

We start by using the addition property to add 5 to both sides of the equation.

$$3x - 5 + 5 = 4 + 5$$
$$3x = 9 \tag{2}$$

Why did we add 5? We added 5 because it is the *opposite* of −5, and the resulting equation will have the variable term on the left and the constant term on the right.

Now we want to get the x term alone on the left with a coefficient of 1 (we call this *isolating* the x). To do this, we use the multiplication property and multiply both sides by $\dfrac{1}{3}$.

We choose $\dfrac{1}{3}$ because $\dfrac{1}{3}$ is the *reciprocal* of 3 and

$$\frac{1}{3} \cdot 3 = 1$$

$$\frac{1}{3}(3x) = \frac{1}{3}(9)$$

$$\left(\frac{1}{3} \cdot 3\right)(x) = 3$$

$$\text{So, } x = 3 \tag{3}$$

We could also use set builder notation. We write $\{x \mid x = 3\}$, which is read, "Every x such that x equals three." We will use both notations throughout the text.

In set notation, we write $\{3\}$, which represents the set of all solutions. No other value of x makes the original equation true.

Since any application of the addition or multiplication properties leads to an equivalent equation, equations (1), (2), and (3) in Example 2 all have the same solution, 3.

To check this result, we can replace x with 3 in the original equation:

$$3(3) - 5 \stackrel{?}{=} 4$$
$$9 - 5 \stackrel{?}{=} 4$$
$$4 = 4 \qquad \text{A true statement.}$$

You may prefer a slightly different approach in the last step of the solution above. From equation (2),

$$3x = 9$$

The multiplication property can be used to *divide* both sides of the equation by 3. Then,

$$\frac{3x}{3} = \frac{9}{3}$$
$$x = 3$$

Of course, the result is the same.

● ● ● ● **CHECK YOURSELF 2**

Solve for x.

$4x - 7 = 17$

The steps involved in using the addition and multiplication properties to solve an equation are the same if more terms are involved in an equation.

• Example 3

Applying the Properties of Equality

Solve for x.

$5x - 11 = 2x - 7$

Our objective is to use the properties of equality to isolate x on one side of an equivalent equation. We begin by adding 11 to both sides.

Again, adding 11 leaves us with the constant term on the right.

$5x - 11 + 11 = 2x - 7 + 11$

$$5x = 2x + 4$$

We continue by adding $-2x$ to (or subtracting $2x$ from) both sides.

If you prefer, write

$5x - 2x = 2x - 2x + 4$

Again:

$3x = 4$

$5x + (-2x) = 2x + (-2x) + 4$

$$3x = 4$$

To isolate x, we now multiply both sides by $\dfrac{1}{3}$.

This is the same as dividing both sides by 3. So

$\dfrac{3x}{3} = \dfrac{4}{3}$

$x - \dfrac{4}{3}$

$\dfrac{1}{3}(3x) = \dfrac{1}{3}(4)$

$$x = \frac{4}{3}$$

In set notation, we write $\left\{\dfrac{4}{3}\right\}$. We leave it to you to check this result by substitution.

● ● ● ● **CHECK YOURSELF 3**

Solve for x.

$7x - 12 = 2x - 9$

Both sides of an equation should be simplified as much as possible *before* the addition and multiplication properties are applied. If like terms are involved on one side (or on both sides) of an equation, they should be combined before an attempt is made to isolate the variable. Example 4 illustrates this approach.

• Example 4

Applying the Properties of Equality with Like Terms

Solve for x.

Note the like terms on the left and right sides of the equation.

$8x + 2 - 3x = 8 + 3x + 2$

Here we combine the like terms $8x$ and $-3x$ on the left and the like terms 8 and 2 on the right as our first step. We then have

$5x + 2 = 3x + 10$

We can now solve as before.

$5x + 2 - 2 = 3x + 10 - 2$ Subtract 2 from both sides.

$\qquad 5x = 3x + 8$

Then,

$5x - 3x = 3x - 3x + 8$ Subtract $3x$ from both sides.

$\qquad 2x = 8$

$\qquad \dfrac{2x}{2} = \dfrac{8}{2}$ Divide both sides by 2.

$\qquad x = 4 \qquad \text{or} \qquad \{4\}$

The solution is 4, which can be checked by returning to the *original equation*.

● ● ●   **CHECK YOURSELF 4**

Solve for x.

$7x - 3 - 5x = 10 + 4x + 3$

If parentheses are involved on one or both sides of an equation, the parentheses should be removed by applying the distributive property as the first step. Like terms should then be combined before an attempt is made to isolate the variable. Consider Example 5.

• Example 5

Applying the Properties of Equality with Parentheses

Solve for x.

$x + 3(3x - 1) = 4(x + 2) + 4$

First, apply the distributive property to remove the parentheses on the left and right sides.

$$x + 9x - 3 = 4x + 8 + 4$$

Combine like terms on each side of the equation.

$$10x - 3 = 4x + 12$$

Recall that to isolate the x, we must get x alone on the left side with a coefficient of 1.

Now, isolate variable x on the left side.

$$10x - 3 + 3 = 4x + 12 + 3 \qquad \text{Add 3 to both sides.}$$

$$10x = 4x + 15$$

$$10x - 4x = 4x - 4x + 15 \qquad \text{Subtract } 4x \text{ from both sides.}$$

$$6x = 15$$

$$\frac{6x}{6} = \frac{15}{6} \qquad \text{Divide both sides by 6.}$$

$$x = \frac{5}{2} \qquad \text{or} \qquad \left\{\frac{5}{2}\right\}$$

The solution is $\dfrac{5}{2}$. Again, this can be checked by returning to the original equation.

CHECK YOURSELF 5

Solve for x.

$$x + 5(x + 2) = 3(3x - 2) + 18$$

The LCM of a set of denominators is also called the **lowest common denominator (LCD)**.

To solve an equation involving fractions, the first step is to multiply both sides of the equation by the **least common multiple (LCM)** of all denominators in the equation. This will clear the equation of fractions, and we can proceed as before.

● Example 6

Applying the Properties of Equality with Fractions

Solve for x.

$$\frac{x}{2} - \frac{2}{3} = \frac{5}{6}$$

First, multiply each side by 6, the least common multiple of 2, 3, and 6.

$$6\left(\frac{x}{2} - \frac{2}{3}\right) = 6\left(\frac{5}{6}\right)$$

$$6\left(\frac{x}{2}\right) - 6\left(\frac{2}{3}\right) = 6\left(\frac{5}{6}\right)$$ Apply the distributive law.

$$\overset{3}{\cancel{6}}\left(\frac{x}{\underset{1}{\cancel{2}}}\right) - \overset{2}{\cancel{6}}\left(\frac{2}{\underset{1}{\cancel{3}}}\right) = \overset{1}{\cancel{6}}\left(\frac{5}{\underset{1}{\cancel{6}}}\right)$$ Simplify.

Next, isolate the variable x on the left side.

The equation is now cleared of fractions.

$$3x - 4 = 5$$
$$3x = 9$$
$$x = 3 \quad \text{or} \quad \{3\}$$

The solution, 3, can be checked as before by returning to the original equation.

● ● ●   **CHECK YOURSELF 6**

Solve for x.

$$\frac{x}{4} - \frac{4}{5} = \frac{19}{20}$$

Be sure that the distributive property is applied properly so that *every term* of the equation is multiplied by the LCM.

● Example 7

Applying the Properties of Equality with Fractions

Solve for x.

$$\frac{2x - 1}{5} + 1 = \frac{x}{2}$$

First, multiply each side by 10, the LCM of 5 and 2.

$$10\left(\frac{2x - 1}{5} + 1\right) = 10\left(\frac{x}{2}\right)$$

$$\overset{2}{\cancel{10}}\left(\frac{2x - 1}{\underset{1}{\cancel{5}}}\right) + 10(1) = \overset{5}{\cancel{10}}\left(\frac{x}{\underset{1}{\cancel{2}}}\right)$$ Apply the distributive property on the left. Reduce.

$$2(2x - 1) + 10 = 5x$$
$$4x - 2 + 10 = 5x$$
$$4x + 8 = 5x$$ Next, isolate x. Here we isolate x on the right side.
$$8 = x \quad \text{or} \quad \{8\}$$

The solution for the original equation is 8.

● ● ● ● **CHECK YOURSELF 7**

Solve for x.

$$\frac{3x + 1}{4} - 2 = \frac{x + 1}{3}$$

Thus far, we have considered only equations of the form $ax + b = 0$, where $a \neq 0$. If we allow the possibility that $a = 0$, two additional equation forms arise. The resulting equations can be classified into three types depending on the nature of their solutions.

CONDITIONAL EQUATIONS, IDENTITIES, AND CONTRADICTIONS

1. An equation that is true for only particular values of the variable is called a **conditional equation.** Here the equation can be written in the form

   $$ax + b = 0$$

   where $a \neq 0$. This case was illustrated in all our previous examples and exercises.
2. An equation that is true for all possible values of the variable is called an **identity.** In this case, *both a and b are 0,* so we get the equation $0 = 0$. This will be the case if both sides of the equation reduce to the same expression (a true statement).
3. An equation that is never true, no matter what the value of the variable, is called a **contradiction.** For example, if a is 0 but b is nonzero, we end up with something like $4 = 0$. This will be the case if both sides of the equation reduce to a false statement.

Example 8 illustrates the second and third cases.

● Example 8

Identities and Contradictions

(a) Solve for x.

$$2(x - 3) - 2x = -6$$

Apply the distributive property to remove the parentheses.

$$2x - 6 - 2x = -6$$

$$-6 = -6 \qquad \text{A } \textit{true} \text{ statement.}$$

See the definition of an identity, above. By adding 6 to both sides of this equation, we have $0 = 0$.

Since the two sides reduce to the true statement $-6 = -6$, the original equation is an *identity,* and the solution set is the set of all real numbers.

(b) Solve for x.

$$3(x + 1) - 2x = x + 4$$

Again, apply the distributive property.

$$3x + 3 - 2x = x + 4$$

$$x + 3 = x + 4$$

$$3 = 4 \qquad \text{A } \textit{false} \text{ statement.}$$

See the definition of a contradiction, above. Subtracting 3 from both sides, we have 0 = 1.

Since the two sides reduce to the false statement $3 = 4$, the original equation is a contradiction. There are no values of the variable that can satisfy the equation. The solution set has nothing in it. We call this the **empty set** and write $\{\}$ or $\varnothing$.

● ● ● **CHECK YOURSELF 8**

Determine whether each of the following equations is a conditional equation, an identity, or a contradiction.

(a) $2(x + 1) - 3 = x$ **(b)** $2(x + 1) - 3 = 2x + 1$ **(c)** $2(x + 1) - 3 = 2x - 1$

An **algorithm** is a step-by-step process for problem solving.

An organized step-by-step procedure is the key to an effective equation-solving strategy. The following algorithm summarizes our work in this section and gives you guidance in approaching the problems that follow.

> **Solving Linear Equations in One Variable**
>
> **STEP 1** Multiply both sides of the equation by the LCM of any denominators, to clear the equation of fractions.
> **STEP 2** Remove any grouping symbols by applying the distributive property.
> **STEP 3** Combine any like terms that appear on either side of the equation.
> **STEP 4** Apply the addition property of equality to write an equivalent equation with the variable term on *one side* of the equation and the constant term on the *other side*.
> **STEP 5** Apply the multiplication property of equality to write an equivalent equation with the variable isolated on one side of the equation.
> **STEP 6** Check the solution in the *original* equation.
>
> *Note:* If the equation derived in step 5 is always true, the original equation was an *identity*. If the equation is always false, the original equation was a *contradiction*.

When you are solving an equation for which a calculator is recommended, it is often easiest to do all calculations as the last step.

● Example 9

Evaluating Expressions Using a Calculator

Solve the following equation for x.

$$\frac{185(x - 3.25) + 1650}{500} = 159.44$$

Following the steps of the algorithm, we get

$$185(x - 3.25) + 1650 = 159.44 \cdot 500 \qquad \text{Multiply by the LCM.}$$

$$185x - 185 \cdot 3.25 + 1650 = 159.44 \cdot 500 \qquad \text{Remove parentheses.}$$

$$185x = 159.44 \cdot 500 + 185 \cdot 3.25 - 1650 \qquad \text{Apply the addition property.}$$

$$x = \frac{159.44 \cdot 500 + 185 \cdot 3.25 - 1650}{185} \qquad \text{Isolate the variable.}$$

Now, remembering to insert parentheses around the numerator, we use a calculator to simplify the expression on the right.

$$x = 425.25 \qquad \text{or} \qquad \{425.25\}$$

CHECK YOURSELF 9

Solve the following equation for x.

$$\frac{2200(x + 17.5) - 1550}{75} = 2326$$

SOLVING APPLICATION PROBLEMS

We are now ready to use our equation-solving skills to reach an important goal—that of solving application problems. The process is easier if we regularly follow these five basic steps, which you may have encountered in previous coursework.

> **Solving Applications**
>
> **STEP 1** Read the problem carefully to determine the unknown quantities.
> **STEP 2** Choose a variable to represent the unknown. Express all other unknowns in terms of this variable.
> **STEP 3** Translate the problem to the language of algebra to form an equation.
> **STEP 4** Solve the equation, and answer the question of the original problem.
> **STEP 5** Verify your solution by returning to the original problem.

Our first two applications fall into the category of **uniform-motion problems.** Uniform motion means that the speed of an object does not change over a certain distance of time. To solve these problems, we will need a relationship between the distance traveled, represented by d, the rate (or speed) of travel, r, and the time of that travel, t. In general, the relationship for the distance traveled d, rate r, and time t, is expressed as

$$d \quad = \quad r \quad \cdot \quad t$$
$$\uparrow \qquad \uparrow \qquad \uparrow$$
$$\text{Distance} \quad \text{Rate} \quad \text{Time}$$

The solution for uniform-motion problems always involves a relationship between the distance, rate, or time. That relationship is then used to form the necessary equation in step 3.

Consider Example 10.

• Example 10

Solving a Distance Problem

On Friday morning, Jason drove from his house to the beach in $3\frac{1}{2}$ hours (h). When he came back on Sunday afternoon, heavy traffic slowed his speed by 6 miles per hour (mi/h), and the trip took 4 h. Find his average rate in each direction.

Step 1 We want the rate in each direction.

Step 2 Let r be Jason's rate to the beach. Then $r - 6$ is his return rate.

It is always a good idea to sketch the given information in a uniform-motion problem. Here we might have

Some students also find using a table helpful in motion problems. Here we have

Rate	Time	Distance
r	$3\frac{1}{2}$	$\frac{7}{2}r$
$r - 6$	4	$4(r - 6)$

Going $\qquad \xrightarrow{\quad r\text{(mi/h) for } 3\frac{1}{2}\text{ h}\quad}$

Coming back $\xleftarrow{\quad r - 6\text{(mi/h) for 4 h}\quad}$

Step 3 Since we know that the distance is the same each way, our equation is formed by using the fact that the product of the rate and the time must be the same each way. We have

Distance going: $\frac{7}{2}r$.

Distance coming back: $4(r - 6)$.

$$\frac{7}{2}r = 4(r - 6)$$

Step 4 We now solve the equation formed in step 3.

$$\frac{7}{2}r = 4r - 24$$

$$7r = 8r - 48$$

$$-r = -48$$

The rate going

$$r = 48 \text{ mi/h}$$

The rate coming back

$$r - 6 = 42 \text{ mi/h}$$

Step 5 You should check this result for yourself.

● ● ● **CHECK YOURSELF 10**

At 9:00 AM, Muriel left her house to visit friends in another city. One hour later, Martin decided to join her on the visit and left along the same route, traveling 13 mi/h faster than Muriel. If Martin caught up with Muriel at 1:00 PM, what was each person's average rate of travel?

In the previous example and exercise, the unknown quantity was the rate of travel. Let's consider another variation of the motion problem in which the time of travel is the unknown.

• Example 11

Solving a Distance Problem

At noon, Linda decides to leave Las Vegas for Los Angeles, driving at 50 mi/h. At 1 PM, Lou leaves Los Angeles for Las Vegas, driving at 55 mi/h along the same route. If the two cities are 260 mi apart, at what time will Linda and Lou meet?

Step 1 Here the time until they meet is the unknown.

Lou left 1 h later!

Step 2 Let t be the time of Linda's travel until they meet. Then $t - 1$ is the time of Lou's travel.

Again, a sketch of the given information is a good idea.

You might want to use a table such as that in Example 10.

```
Los                55 mi/h for (t − 1) h   50 mi/h for t h   Las
Angeles         ─────────────────►│◄───────────────    Vegas
                         Lou                    Linda
                └──────────  260 mi  ──────────────┘
```

Step 3 Again, since distance is the product of rate and time, from the sketch of step 2, we have

Linda's distance: $50t$

Lou's distance: $55(t - 1)$

Since the sum of those distances must be 260 mi, we can write

$$50t + 55(t - 1) = 260$$

Step 4 Solving the equation as before yields

$$t = 3 \text{ h}$$

Finally, since Linda left Las Vegas at noon, the two will meet at 3 PM.

Step 5 Again, we leave the checking of this result to you. Verify that the sum of the distances that the two travel is 260 mi.

● ● ● ● **CHECK YOURSELF 11**

At 7 AM, a freight train leaves a city, traveling west, averaging 42 mi/h. Two hours later, an express train leaves the same station, traveling east at 68 mi/h. At what time will the two trains be exactly 359 mi apart?

To complete this section, we consider an application from business. But we will need some new terminology. The total cost of manufacturing an item consists of two types of costs. The **fixed cost,** sometimes called the **overhead,** includes costs such as product design, rent, and utilities. In general, this cost is constant and does not change with the number of items produced. The **variable cost,** which is a cost per item, includes costs such as material, labor, and shipping. The variable cost depends on the number of items being produced.

A typical cost equation might be

$$C = 3.50x + 5000$$

Variable cost   Fixed cost

where total cost, C, equals variable cost times the number of items produced, x, plus the fixed cost.

The total **revenue** is the income the company makes. It is calculated as the product of the selling price of the item and the number of items sold. A typical revenue equation might be

$$R = 7.50x$$

Selling price   Number of
per item   items sold

where total revenue equals an item's selling price times the number sold.

The **break-even point** is that point at which the revenue equals the cost (the company would exactly break even without a profit or a loss).

Let's apply these concepts in Example 12.

• Example 12

Finding the Break-Even Point

A firm producing videocassette tapes finds that its fixed cost is $5000 per month and that its variable cost is $3.50 per tape. The cost of producing x tapes is then given by

$$C = 3.50x + 5000$$

The firm can sell the tapes at $7.50 each, so the revenue from selling x tapes is

$$R = 7.50x$$

Find the break-even point.

Since the break-even point is that point where the revenue equals the cost, or $R = C$, from our given equations we have

$$7.50x = \underline{3.50x + 5000}$$

Revenue   Cost

Solving as before gives

$4x = 5000$

$x = 1250$

The firm will break even (no profit or loss) by producing and selling exactly 1250 tapes each month.

CHECK YOURSELF 12

A firm producing lawn chairs has fixed costs of $525 per week. The variable cost is $8.50 per chair, and the revenue per chair is $15.50. This means that the cost equation is

$C = 8.50x + 525$

and the revenue equation is

$R = 15.50x$

Find the break-even point.

CHECK YOURSELF ANSWERS

1. $5(7) - 15 \stackrel{?}{=} 2(7) + 6$ **2.** 6.   **3.** $\dfrac{3}{5}$.   **4.** -8.   **5.** $-\dfrac{2}{3}$.   **6.** 7.
   $35 - 15 \stackrel{?}{=} 14 + 6$
   $20 = 20$
   A true statement.
7. 5.   **8. (a)** Conditional; **(b)** contradiction; **(c)** identity.   **9.** 62.5.
10. Muriel: 39 mi/h; Martin: 52 mi/h.   **11.** 11:30 AM.   **12.** 75 chairs.

Density and Body Fat. You have probably heard the question, "Which weighs more, a ton of feathers or a ton of bricks?" The usual response of "bricks" is, of course, incorrect. A ton of anything weighs 2000 pounds. Why do people miss this question so often? We know from experience that bricks are heavier than feathers. However, what we mean is that bricks have a much higher *density* than feathers (they weigh more per unit volume). A cubic meter of bricks weighs far more than a cubic meter of feathers.

We compare the density of a substance to the density of water. Water has a density of 1 g/cm^3. Things denser than water (such as body lean) sink, whereas things less dense (such as body fat) float.

The concept of density is important in analyzing the significance of the percentage of body fat one has. There are many ways to measure, or estimate, one's percentage of body fat. Most health clubs and many college health classes offer the opportunity to find this measure. In his book *Fit or Fat*, Covert Bailey suggests that people concerned with their fitness should take and record their percentage of body fat every 6 months. By multiplying this percentage times your weight, you can get the actual pounds of fat in your body.

However, we are looking for a more useful measure. Body density, D, can be calculated using the following formula.

$$D = \frac{1}{\dfrac{A}{a} + \dfrac{B}{b}}$$

where A = proportion of body fat
B = proportion of lean body tissue $(1 - A)$
a = density of fat body tissue in grams/cubic cm (approximately 0.9 g/cm³)
b = density of lean body tissue in grams/cubic cm (approximately 1.1 g/cm³)

Use this formula to compute the following.

1. Substituting $1 - A$ for B, 0.9 for a, and 1.1 for b, solve the formula so that D is a function of A.
2. Find the ordered pairs associated with body fat proportions of 0.1, 0.15, 0.2, 0.25, and 0.3. Use these points to graph the function.
3. Use the graph and a little research to determine a reasonable domain and range for the graph. For example, is it reasonable that someone would have a body fat proportion of 0.9?
4. Use other resources (health professionals, health clubs, the Internet, and the library) to find another measure of body density. How do the two methods compare?

2.1 Exercises

In Exercises 1 to 14, solve each equation, and check your results. Express each answer in set notation.

1. $5x - 8 = 17$

2. $4x + 9 = -11$

3. $8 - 7x = -41$

4. $-7 - 4x = 21$

5. $7x - 5 = 6x + 6$

6. $9x + 4 = 8x - 3$

7. $8x - 4 = 3x - 24$

8. $5x + 2 = 2x - 5$

9. $7x - 4 = 2x + 26$

10. $11x - 3 = 4x - 31$

11. $4x - 3 = 1 - 2x$

12. $8x + 5 = -19 - 4x$

13. $2x + 8 = 7x - 37$

14. $3x - 5 = 9x + 22$

In Exercises 15 to 32, simplify and then solve each equation. Express your answer in set notation.

15. $5x - 2 + x = 9 + 3x + 10$

16. $5x + 5 - x = -7 + x - 2$

17. $7x - 3 - 4x = 5 + 5x - 13$

18. $8x - 3 - 6x = 7 + 5x + 17$

19. $5x = 3(x - 6)$

20. $2(x - 15) = 7x$

21. $5(8 - x) = 3x$

22. $7x = 7(6 - x)$

23. $2(2x - 1) = 3(x + 1)$

24. $3(3x - 1) = 4(3x + 1)$

25. $8x - 3(2x - 4) = 17$

26. $7x - 4(3x + 4) = 9$

27. $7(3x + 4) = 8(2x + 5) + 13$

28. $-4(2x - 1) + 3(3x + 1) = 9$

29. $9 - 4(3x + 1) = 3(6 - 3x) - 9$

30. $13 - 4(5x + 1) = 3(7 - 5x) - 15$

31. $5 - 2[x - 2(x - 1)] = 55 - 4[x - 3(x + 2)]$

32. $7 - 5[x - 3(x + 2)] = 25 - 2[x - 2(x - 3)]$

In Exercises 33 to 46, clear fractions and then solve each equation. Express your answer in set notation.

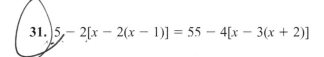

33. $\dfrac{2x}{3} - \dfrac{5}{3} = 3$

34. $\dfrac{3x}{4} + \dfrac{1}{4} = 4$

35. $\dfrac{x}{6} + \dfrac{x}{5} = 11$

36. $\dfrac{x}{6} - \dfrac{x}{8} = 1$

37. $\dfrac{2x}{3} - \dfrac{x}{4} = \dfrac{5}{2}$

38. $\dfrac{5x}{6} + \dfrac{2x}{3} = \dfrac{5}{6}$

39. $\dfrac{x}{5} - \dfrac{x - 7}{3} = \dfrac{1}{3}$

40. $\dfrac{x}{6} + \dfrac{3}{4} = \dfrac{x - 1}{4}$

41. $\dfrac{5x - 3}{4} - 2 = \dfrac{x}{3}$

42. $\dfrac{6x - 1}{5} - \dfrac{2x}{3} = 3$

43. $\dfrac{2x + 3}{5} - \dfrac{2x - 1}{3} = \dfrac{8}{15}$

44. $\dfrac{3x}{5} - \dfrac{3x - 1}{2} = \dfrac{11}{10}$

45. $0.5x - 6 = 0.2x$

46. $0.7x - 7 = 0.3x - 5$

In Exercises 47 to 56, classify each equation as a conditional equation, an identity, or a contradiction.

47. $3(x - 1) = 2x + 3$

48. $2(x + 3) = 2x + 6$

49. $3(x - 1) = 3x + 3$

50. $2(x + 3) = x + 5$

51. $3(x - 1) = 3x - 3$

52. $2(x + 3) = 3x + 5$

53. $3x - (x - 3) = 2(x + 1) + 2$

54. $5x - (x + 4) = 4(x - 2) + 4$

55. $\dfrac{x}{2} - \dfrac{x}{3} = \dfrac{x}{6}$

56. $\dfrac{3x}{4} - \dfrac{2x}{3} = \dfrac{x}{6}$

   57. What is the common characteristic of equivalent equations?

58. What is meant by a *solution* to a linear equation?

59. Define **(a)** identity and **(b)** contradiction.

60. Why does the multiplication property of equality not include multiplying both sides of the equation by 0?

Label Exercises 61 to 66 true or false.

61. Adding the same value to both sides of an equation creates an equivalent equation.

62. Multiplying both sides of an equation by 0 creates an equivalent equation.

63. To clear an equation of fractions, we multiply both sides by the GCF of the denominator.

64. The multiplication property of equations allows us to divide both sides by the same nonzero quantity.

65. Some equations have more than one solution.

66. No matter what value is substituted for *x*, the expressions on either side of the equals sign have the same value.

Solve the following applications.

67. Speed. On her way to a business meeting, Kim took the freeway, and the trip took 3 h. Returning, she decided to take a side road, and her speed along that route averaged 9 mi/h slower than on the freeway. If Kim's return trip took $3\frac{1}{2}$ h and the distance driven was the same each way, find her average speed in each direction.

68. Speed. Beth was required to make a cross-country flight in training for her pilot's license. When she flew from her home airport, a steady 30-mi/h wind was behind her, and the first leg of the trip took 5 h. When she returned against the same wind, the flight took 7 h. Find the plane's speed in still air and the distance traveled on each leg of the flight.

69. Speed. Craig was driving on a 220-mi trip. For the first 3 h he traveled at a steady speed. At that point, realizing that he would be late to his destination, he increased his speed by 10 mi/h for the remaining 2 h of the trip. What was his driving speed for each portion of the trip?

70. Distance. Robert can drive to work in 45 min, whereas if he decides to take the bus, the same trip takes 1 h 15 min. If the average rate of the bus is 16 mi/h slower than his driving rate, how far does he travel to work?

71. **Time.** At 9 AM, Tom left Boston for Baltimore, traveling at 45 mi/h. One hour later, Andrea left Baltimore for Boston, traveling at 50 mi/h along the same route. If the cities are 425 mi apart, at what time did Tom and Andrea meet?

72. **Time.** A passenger bus left a station at 1 PM, traveling north at an average rate of 50 mi/h. One hour later, a second bus left the same station, traveling south at a rate of 55 mi/h. At what time will the two buses be 260 mi apart?

73. **Distance.** On Tuesday, Malia drove to a conference and averaged 54 mi/h for the trip. When she returned on Thursday, road construction slowed her average speed by 9 mi/h. If her total driving time was 11 h, what was her driving time each way, and how far away from her home was the conference?

74. **Time.** At 8:00 AM, Robert left on a trip, traveling at 45 mi/h. One-half hour later, Laura discovered that Robert forgot his luggage and left along the same route, traveling at 54 mi/h, to catch up with him. When did Laura catch up with Robert?

75. **Business.** A firm producing gloves finds that its fixed cost is $4000 per week and its variable cost is $2.50 per pair. The revenue is $6.50 per pair of gloves, so that cost and revenue equations are, respectively,

    $$C = 2.50x + 4000 \quad \text{and} \quad R = 6.50x$$

    Find the break-even point for the firm.

76. **Business.** A company that produces calculators determines that its fixed cost is $8820 per month. The variable cost is $70 per calculator: the revenue is $105 per calculator. The cost and revenue equations, respectively, are given by

    $$C = 70x + 8820 \quad \text{and} \quad R = 105x$$

    Find the number of calculators the company must produce and sell in order to break even.

77. **Business.** A firm that produces scientific calculators has fixed costs of $1260 per week and variable costs of $6.50 per calculator. If the company can sell the calculators for $13.50, find the break-even point.

78. **Business.** A publisher finds that the fixed costs associated with a new paperback are $18,000. Each book costs $2 to produce and will sell for $6.50. Find the publisher's break-even point.

79. **Business.** An important economic application involves supply and demand. The number of units of a commodity that manufacturers are willing to **supply,** S, is related to the market price, p. A typical supply equation is

    $$S = 40p - 285 \tag{1}$$

    (Generally the supply increases as the price increases.)

The number of units that consumers are willing to buy, D, is called the **demand,** and it is also related to the market price. A typical demand equation is

$$D = -45p + 1500 \qquad (2)$$

(Generally the demand decreases as the price increases.)

The price where the supply and demand are equal (or $S = D$) is called the **equilibrium price** for the commodity. The supply and demand equations for a certain model portable radio are given in equations (1) and (2). Find the equilibrium price for the radio.

80. Business. The supply and demand equations for a certain type of computer modem are

$$S = 25p - 2500 \quad \text{and} \quad D = -40p + 5300$$

Find the equilibrium price for the modem.

81. Business. You find a new bicycle that you like, and you plan to ride it for exercise several days a week. You are also happy to find that this very model is on sale for 22% off. You speak to the salesclerk, who begins writing up the sale by first adding on your state's 7.8% sales tax. He then takes off the 22%. "No!", you object, "You should take the 22% off first and then add the sales tax." The salesclerk says he is sorry, but he has been instructed to first calculate the amount of tax. Who is correct? Defend your position using algebra.

Answers

1. $\{5\}$ **3.** $\{7\}$ **5.** $\{11\}$ **7.** $\{-4\}$ **9.** $\{6\}$ **11.** $\left\{\dfrac{2}{3}\right\}$ **13.** $\{9\}$ **15.** $\{7\}$ **17.** $\left\{\dfrac{5}{2}\right\}$

19. $\{-9\}$ **21.** $\{5\}$ **23.** $\{5\}$ **25.** $\left\{\dfrac{5}{2}\right\}$ **27.** $\{5\}$ **29.** $\left\{-\dfrac{4}{3}\right\}$ **31.** $\{-13\}$ **33.** $\{7\}$

35. $\{30\}$ **37.** $\{6\}$ **39.** $\{15\}$ **41.** $\{3\}$ **43.** $\left\{\dfrac{3}{2}\right\}$ **45.** $\{20\}$ **47.** Conditional

49. Contradiction   **51.** Identity   **53.** Contradiction   **55.** Identity   **57.**   **59.**

61. True   **63.** False   **65.** True   **67.** 63 mi/h going; 54 mi/h returning   **69.** 40 mi/h for 3 h; 50 mi/h for 2 h   **71.** 2:00 PM   **73.** 5 h going; 6h returning, 270 mi   **75.** 1000 pairs of gloves   **77.** 180 calculators   **79.** $21   **81.**

 **2.2**

Graphical Solutions to Equations in One Variable

2.2 OBJECTIVES

1. Rewrite a linear equation in one variable as $f(x) = g(x)$
2. Find the point of intersection of $f(x)$ and $g(x)$
3. Interpret the point of intersection of $f(x)$ and $g(x)$
4. Solve a linear equation in one variable by writing it as the functional equality $f(x) = g(x)$

In Section 2.1, you learned to solve linear equations in one variable algebraically. In this section, we will look at graphical solutions for the same type of equations.

Recall that a solution to a linear equation is a value for the variable that makes the equation a true statement. In this section, we will look at a graphical method for finding the solution to a linear equation. This method will be particularly useful in the next section, when we discuss linear inequalities.

In our first example, we will look at the solution to a simple linear equation. The method may seem unnecessarily complicated, but remember that we are learning the method so we can solve more complex equations and inequalities later.

•Example 1

A Graphical Solution to a Linear Equation

Graphically solve the following equation.

$2x - 6 = 0$

Step 1 Let each side of the equation represent a function of x.

$f(x) = 2x - 6$

$g(x) = 0$

Step 2 Graph the two functions on the same set of axes.

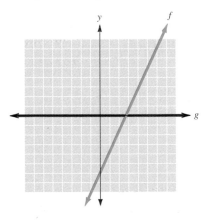

Step 3 Find the intersection of the two graphs. This intersection represents the solution to the original equation.

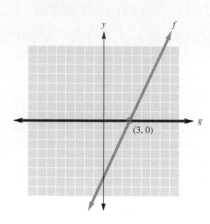

The two lines intersect on the x axis at the point $(3, 0)$. We are looking for the x value at the point of intersection, which is 3.

● ● ●   **CHECK YOURSELF 1**

Graphically solve the following equation.

$$-3x + 6 = 0$$

The same three-step process is used for solving any equation. In Example 2, we look for an intersection that is *not* on the x axis.

●**Example 2**

A Graphical Solution to a Linear Equation

Graphically solve the following equation.

$$2x - 6 = -3x + 4$$

Step 1   Let each side of the equation represent a function of x.

$$f(x) = 2x - 6$$
$$g(x) = -3x + 4$$

Step 2   Graph the two functions on the same set of axes.

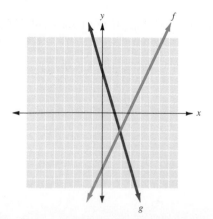

Step 3 Find the intersection of the two graphs. This intersection represents the solution to the original equation.

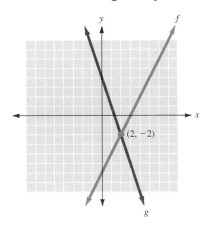

If the solution was an ordered pair, the solution set would be {(2, −2)}, but we are looking for only the *x* value.

The two lines intersect at the point $(2, -2)$. Again, the solution is the *x* value of the ordered pair, so the solution is $x = 2$ and the solution set is $\{2\}$. But what is the significance of the *y* value of the ordered pair? Note that, when you substitute 2 for *x* in the two functions, both yield -2.

● ● ● **CHECK YOURSELF 2**

Graphically display the solution to the following equation.

$$-3x - 4 = 2x - 1$$

The following algorithm summarizes our work in finding a graphical solution for an equation.

Finding a Graphical Solution for an Equation

STEP 1 Let each side of the equation represent a function of *x*.
STEP 2 Graph the two functions on the same set of axes.
STEP 3 Find the intersection of the two graphs. The *x* value at this intersection represents the solution to the original equation.

Linear equations are often first solved by algebraic means. The graph of the equation can then be used to check the solution. This concept is illustrated in Example 3.

● **Example 3**

Solving Linear Equations Algebraically and Graphically

Solve the linear equation algebraically, then graphically display the solution.

$$2(x + 3) = -3x - 4$$

Begin by using the distributive property to rid the left side of parentheses.

$$2x + 6 = -3x - 4$$
$$5x + 6 = -4$$
$$5x = -10$$
$$x = -2$$

The solution set is $\{-2\}$.

To graphically display the solution, let

$$f(x) = 2(x + 3)$$
$$g(x) = -3x - 4$$

Graphing both lines, we get

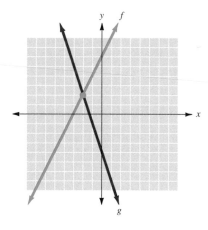

The point of intersection appears to be $(-2, 2)$, which confirms that -2 is a reasonable solution to the equation

$$2(x + 3) = -3x - 4$$

● ● ●   **CHECK YOURSELF 3**

First solve the linear equation algebraically, then graphically display the solution.

$$3(x - 2) = -4x + 1$$

In Example 4, we turn to a business application.

● Example 4

Solving a Business Application

A manufacturer can produce and sell x items per week at the following cost.

$$C(x) = 30x + 800$$

The revenue from selling those items is given by

$$R(x) = 110x$$

Find the break-even point, which is the number of units at which the revenue equals the cost.

We form a linear system from the given equations.

$$C(x) = 30x + 800 \qquad\qquad\qquad (1)$$

$$R(x) = 110x \qquad\qquad\qquad\qquad (2)$$

Finding the break-even point requires that revenue equal cost, $R = C$. From equation (2), we can substitute $110x$ for C in equation (1). We then have

$$110x = 30x + 800$$

or

$$110x - 30x = 800$$
$$80x = 800$$
$$x = 10$$

The solution set is $\{10\}$.

For $x = 10$ units, the cost (and the revenue) is $1100. This system is illustrated below.

Note that to the right of the break-even point, the revenue line is above the cost line.

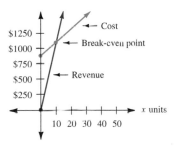

Note that if the company sells more than 10 units, it makes a profit since the revenue exceeds the cost.

● ● ● CHECK YOURSELF 4

A manufacturer can produce and sell x items per week at a cost

$$C(x) = 30x + 1800$$

The revenue from selling those items is given by

$$R(x) = 120x$$

Find the break-even point.

● ● ● **CHECK YOURSELF ANSWERS**

1. $f(x) = -3x + 6$
$g(x) = 0$

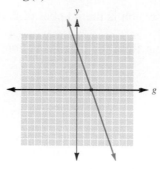

Solution $\{2\}$

2. $f(x) = -3x - 4$
$g(x) = 2x - 1$

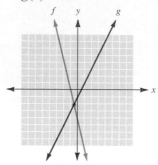

Solution $\left\{-\dfrac{3}{5}\right\}$

3. $3(x - 2) = -4x + 1$
$3x - 6 = -4x + 1$
$7x = 7$
$x = 1$

Graphically,
$f(x) = 3(x - 2)$
$g(x) = -4x + 1$

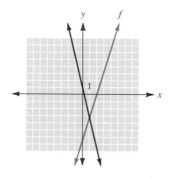

4. $\{20\}$

2.2 Exercises

Solve Exercises 1 to 8, and check your answer graphically.

1. $2x - 8 = 0$

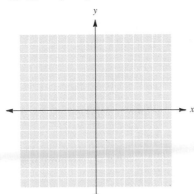

2. $4x + 12 = 0$

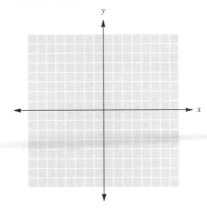

3. $7x - 7 = 0$

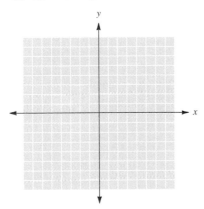

4. $2x - 6 = 0$

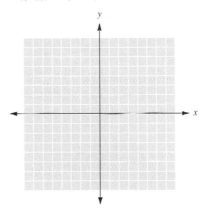

5. $5x - 8 = 2$

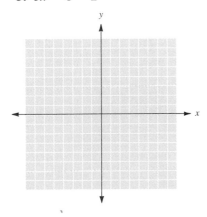

6. $4x + 5 = -3$

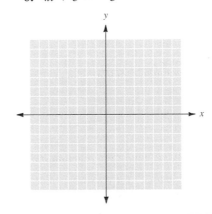

7. $2x - 3 = 7$

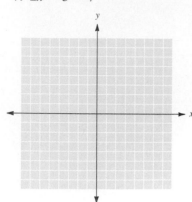

8. $5x + 9 = 4$

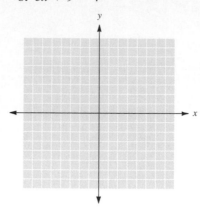

In Exercises 9 to 16, solve the linear equations and graphically display the solution.

9. $4x - 2 = 3x + 1$

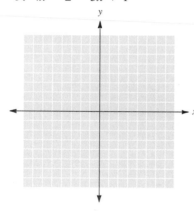

10. $6x + 1 = x + 6$

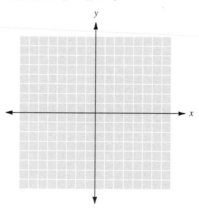

11. $\dfrac{7}{5}x - 3 = \dfrac{3}{10}x + \dfrac{5}{2}$

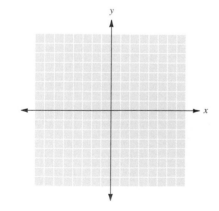

12. $2x - 3 = 3x - 2$

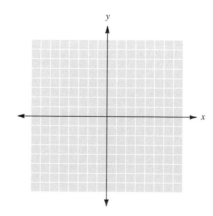

13. $3(x - 1) = 4x - 5$

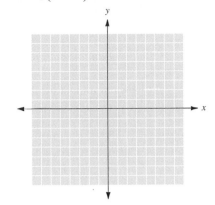

14. $2(x + 1) = 5x - 7$

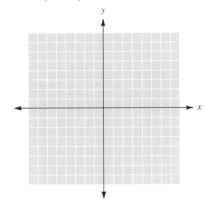

15. $7\left(\dfrac{1}{5}x - \dfrac{1}{7}\right) = x + 1$

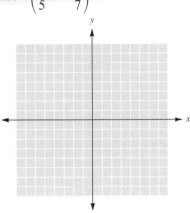

16. $2(3x - 1) = 12x + 4$

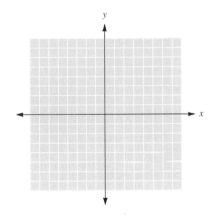

17. The following graph represents the rates that two different car rental agencies charge. The x axis represents the number of miles driven (in hundreds of miles), and the y axis represents the total charge. How would you use this graph to decide which agency to use?

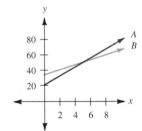

18. **Business.** A firm producing flashlights finds that its fixed cost is $2400 per week, and its variable cost is $4.50 per flashlight. The revenue is $7.50 per flashlight, so the cost and revenue equations are, respectively,

$$C(x) = 4.50x + 2400 \qquad \text{and} \qquad R(x) = 7.50x$$

Find the break-even point for the firm (the point at which the revenue equals the cost).

19. **Business.** A company that produces portable television sets determines that its fixed cost is $8750 per month. The variable cost is $70 per set, and the revenue is $105 per set. The cost and revenue equations, respectively, are given by

$$C(x) = 70x + 8750 \qquad \text{and} \qquad R(x) = 105x$$

Find the number of sets the company must produce and sell in order to break even.

20. Graphs can be used to solve distance, time, and rate problems because graphs make pictures of the action.

   (a) Consider Exercise 74 in Section 2.1: "Robert left on a trip, traveling at 45 mi/h. One-half hour later, Laura discovered that Robert forgot his luggage and so she left along the same route, traveling at 54 mi/h, to catch up with him. When did Laura catch up with Robert?" How could drawing a graph help solve this problem? If you graph Robert's distance as a function of time and Laura's distance as a function of time, what does the slope of each line correspond to in the problem?

   (b) Use a graph to solve this problem: Marybeth and Sam left her mother's house to drive home to Minneapolis along the interstate. They drove an average of 60 mi/h. After they had been gone for $\frac{1}{2}$ h, Marybeth's mother realized they had left their laptop computer. She grabbed it, jumped into her car, and pursued the two at 70 mi/h. Marybeth and Sam also noticed the missing computer, but not until 1 h after they had left. When they noticed that it was missing, they slowed to 45 mi/h while they considered what to do. After driving for another $\frac{1}{2}$ h, they turned around and drove back toward the home of Marybeth's mother at 65 mi/h. Where did they meet? How long had Marybeth's mother been driving when they met?

   (c) Now that you have become experts at this, try solving this problem by drawing a graph. It will require that you think about the slope and perhaps make several guesses when drawing the graphs. If you ride your new bicycle to class, it takes you 1.2 h. If you drive, it takes you 40 min. If you drive in traffic an average of 15 mi/h faster than you can bike, how far away from school do you live? Write an explanation of how you solved this problem by using a graph.

21. Graphing. The family next door to you is trying to decide which health maintenance organization (HMO) to join. One parent has a job with health benefits for the employee only, but the rest of the family can be covered if the employee agrees to a payroll deduction. The choice is between The Empire Group, which would cost the family $185 per month for coverage and $25.50 for each office visit, and Group Vitality, which costs $235 per month and $4.00 for each office visit.

(a) Write an equation showing total yearly costs for each HMO. Graph the cost per year as a function of the number of visits, and put both graphs on the same axes.

(b) Write a note to the family explaining when The Empire Group would be better and when Group Vitality would be better. Explain how they can use your data and graph to help make a good decision. What other issues might be of concern to them?

Answers

1.

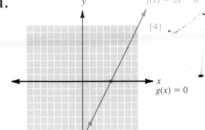

3.

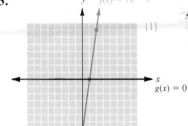

5.

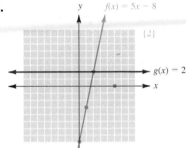

7.

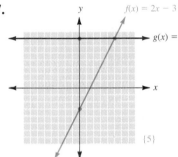

9.

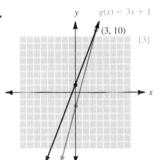

11.

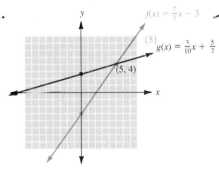

13.

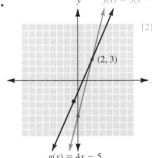

15.

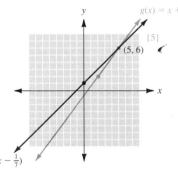

17. Always use the lower graph to determine the cheaper cost.   **19.** 250 sets   **21.**

Graphing Linear Inequalities in One Variable

2.3

2.3 OBJECTIVE

Solve linear inequalities in one variable graphically

In Section 2.2, we looked at the graphical solution to a linear equation. In this section, we will use the graphs of linear functions to determine the solution to a linear inequality.

Linear inequalities in one variable, x, are obtained from linear equations by replacing the symbol for equality ($=$) with one of the inequality symbols ($<$, $>$, $\leq$, $\geq$).

The general form for a linear inequality in one variable is

$$x < a$$

where the symbol $<$ can be replaced with $>$, $\leq$, or $\geq$. Examples of linear inequalities in one variable include

$$x \geq -3 \qquad 2x + 5 > 7 \qquad 2x - 3 \leq 5x + 6$$

Recall that the solution set for an equation is the set of all values for the variable (or ordered pair) that make the equation a true statement. Similarly, the solution set for an inequality is the set of all values that make the inequality a true statement. Example 1 looks at the two-dimensional graph of an equation in one variable.

• Example 1

Graphing the Solution Set to an Inequality

Graph the solution set to the inequality

$$2x + 5 > 7$$

First, rewrite the inequality as a comparison of two functions. Here, $f(x) > g(x)$, where $f(x) = 2x + 5$ and $g(x) = 7$.

Now graph the two functions on a single set of axes.

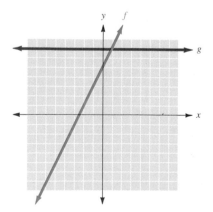

Next, draw a vertical dotted line through each point of intersection of the two functions. In this case, there will be a vertical line through the point (1, 7).

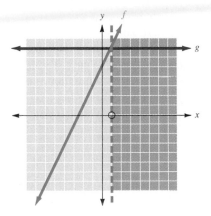

The solution set is every x value that results in $f(x)$ being greater than $g(x)$, which is every x value to the right of the dotted line.

The solution set will be all the x values that make the original statement, $2x + 5 > 7$, true.

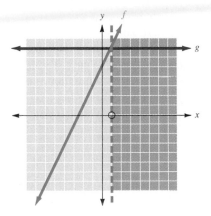

Finally, we express the solution set in set builder notation

$$\{x \mid x > 1\}$$

which is read as, "The set of all x such that x is greater than 1."

● ● ● CHECK YOURSELF 1

Graph the solution set to the inequality $3x - 2 < 4$.

In Example 1, the function $g(x) = 4$ resulted in a horizontal line. In Example 2, we see that the same method works when comparing any two functions.

● Example 2

Graphing the Solution Set to an Inequality

Graph the solution set to the inequality

$$2x - 3 \geq 5x$$

First, rewrite the inequality as a comparison of two functions. Here, $f(x) \geq g(x)$, where $f(x) = 2x - 3$ and $g(x) = 5x$.

Now graph the two functions on a single set of axes.

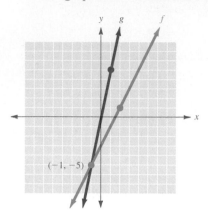

As in Example 1, draw a vertical line through each point of intersection of the two functions. In this case, there will be a vertical line through the point $(-1, -5)$. In this case, the line is included (greater than or *equal to*), so the line is solid, not dotted.

Again, we need to fill in every x value that makes the statement true. In this case, that is every x for which the line representing $f(x)$ is above or intersects the line representing $g(x)$. That is the region in which $f(x)$ is greater than or equal to $g(x)$. We shade the region to the left of the line, but we also want to include the x value on the line, so we make it a solid circle.

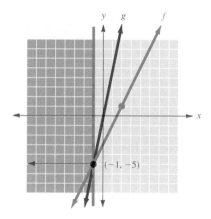

Finally, we express the solution in set builder notation. We see that the solution is every ordered pair in which x is -1, or less, so we write

$$\{x \mid x \leq -1\}$$

● ● ● **CHECK YOURSELF 2**

Graph the solution set to the inequality

$$3x + 2 \geq -2x - 8$$

The following algorithm summarizes our work in this section.

Finding the Solution for an Inequality in One Variable

STEP 1   Rewrite the inequality as a comparison of two functions.

$$f(x) < g(x) \qquad f(x) > g(x) \qquad f(x) \leq g(x) \qquad f(x) \geq g(x)$$

STEP 2   Graph the two functions on a single set of axes.

STEP 3   Draw a vertical line through each point of intersection of the two functions. Use a dotted line if equality is not included ($<$ or $>$). Use a solid line if equality is included ($\leq$ or $\geq$).

STEP 4   Shade the x values that make the inequality a true statement.

STEP 5   Write the solution in set builder notation.

● ● ●   CHECK YOURSELF ANSWERS

1.

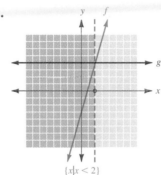

$\{x | x < 2\}$

2.

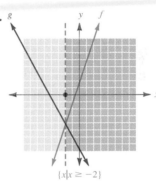

$\{x | x \geq -2\}$

2.3 Exercises

In Exercises 1 to 16, graph the solution set to each inequality.

1. $2x < 8$

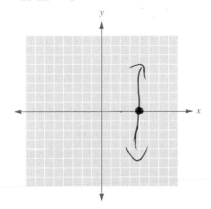

2. $-x < 4$

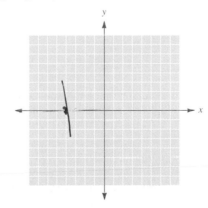

3. $\dfrac{x+3}{2} < -1$

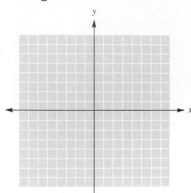

4. $\dfrac{-3x+3}{4} > -3$

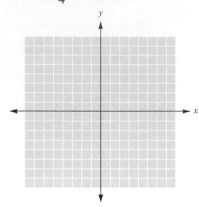

5. $6x \geq 6$

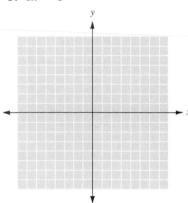

6. $-3x \leq 6$

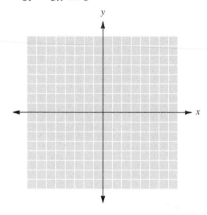

7. $7x - 7 < -2x + 2$

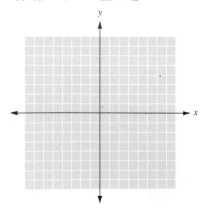

8. $7x + 2 > x - 4$

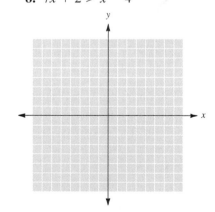

9. $\dfrac{14x + 4}{3} > 2(4x - 1)$

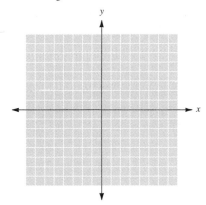

10. $2(3x + 1) < 4(x + 1)$

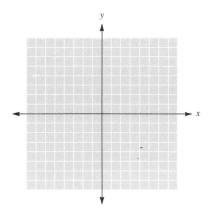

11. $6(1 + x) \geq 2(3x - 5)$

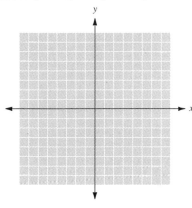

12. $2(x - 5) \geq 2x - 1$

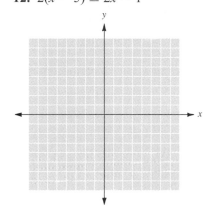

13. $7x > \dfrac{9x - 5}{2}$

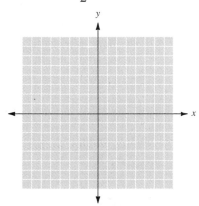

14. $-4x - 12 < x + 8$

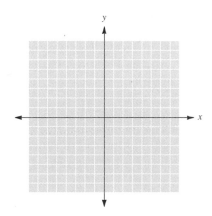

15. $4x - 6 \leq 2x - 2(5x - 12)$

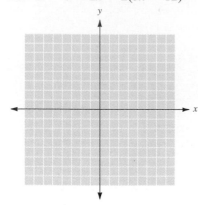

16. $5x + 3 > 2(4 - x) + 7x$

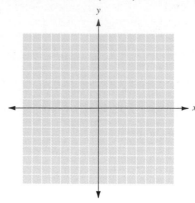

In Exercises 17 to 22, solve the following applications.

17. Business. The cost to produce x units of wire is $C(x) = 50x + 5000$, and the revenue generated is $R(x) = 60x$. Find all values of x for which the product will at least break even.

18. Business. Find the values of x for which a product will at least break even if the cost is $C(x) = 85x + 900$ and the revenue is given by $R(x) = 105x$.

19. Car Rental. Tom and Jean went to Salem, Massachusetts, for 1 week. They needed to rent a car, so they checked out two rental firms. Wheels, Inc. wanted $28 per day with no mileage fee. Downtown Edsel wanted $98 per week and 14¢ per mile. Set up inequalities to express the rates of the two firms, and then decide when each deal should be taken.

20. Mileage. A fuel company has a fleet of trucks. The annual operating cost per truck is $C(x) = 0.58x + 7800$, where x is the number of miles traveled by a truck per year. What number of miles will yield an operating cost that is less than $25,000?

21. Wedding. Eileen and Tom are having their wedding reception at the Warrington Fire Hall. They can spend at the most $3000 for the reception. If the hall charges a $250 cleanup fee plus $25 per person, find the largest number of people they can invite.

22. Tuition. A nearby college charges annual tuition of $6440. Meg makes no more than $1610 per year in her summer job. What is the smallest number of summers that she must work in order to make enough for 1 year's tuition?

23. Graphing. Explain to a relative how a graph is helpful in solving each inequality below. Be sure to include the significance of the point at which the lines meet (or what happens if the lines do not meet).

(a) $3x - 2 < 5$ (b) $3x - 2 \leq 4 - x$ (c) $4(x - 1) \geq 2 + 4x$

24. College. Look at the data here about enrollment in college. Assume that the changes occurred at a constant rate over the years. Make one linear graph for men and one for women, but on the same set of axes. What conclusions could you draw from reading the graph?

Year	No., in millions, of men in the U.S. enrolled in college	No., in millions, of women in the U.S. enrolled in college
1960	2.3	1.2
1991	6.4	7.8

Answers

1.

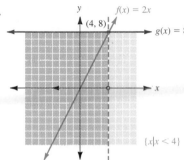

3.

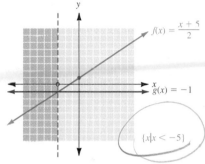

5.

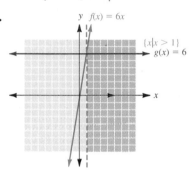

7.

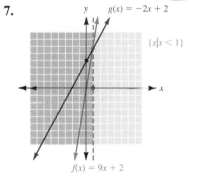

9.

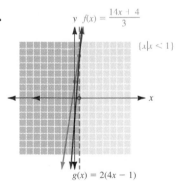

11.

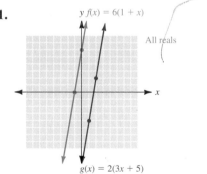

13.

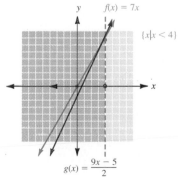

15.

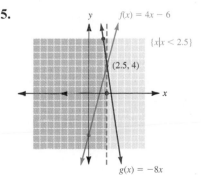

17. $x \geq 500$

1

19. If miles are under 700, Downtown Edsel; if over 700, Wheels, Inc.; $W = \$28 \times 7 = \$196$; $DE = 98 + 0.14x$ (x is number of miles)   **21.** 110 people   **23.**

Absolute Value Equations

Equations may contain absolute value notation in their statements. In this section, we will look at both graphical and algebraic solutions to statements that include absolute values. First, we will review the concept of absolute value.

The **absolute value** of a signed number is the distance from that signed number to 0. Because absolute value is a distance, it is always positive. Formally, we say

> The absolute value of a number x is given by
>
> $$|x| = \begin{cases} -x \text{ if } x < 0 \\ x \text{ if } x \geq 0 \end{cases}$$

● Example 1

Finding the Absolute Value of a Number

Find the absolute value for each expression.

(a) $|-3|$ (b) $|7 - 2|$ (c) $|-7 - 2|$

Solution

(a) Because $-3 < 0$, $|-3| = -(-3) = 3$.

(b) $|7 - 2| = |5|$ because $5 \geq 0$, $|5| = 5$.

(c) $|-7 - 2| = |-9|$ because $-9 < 0$, $|-9| = -(-9) = 9$.

● ● ●   CHECK YOURSELF 1

Find the absolute value for each expression.

(a) $|12|$ **(b)** $|-9 + 5|$ **(c)** $|-3 - 4|$

Graph the function

$y = |x|$ as

$Y_1 = \text{abs}(x)$

To look at graphical solutions, we must first look at the graph of an absolute value function. We will start by looking at the graph of the function $f(x) = |x|$. All other graphs of absolute value functions are variations of this graph.

The graph can be found using a graphing calculator (most graphing calculators use $\boxed{\text{abs}}$ to represent the absolute value). We will develop the graph from a table of values.

x	$f(x) = |x|$
-3	3
-2	2
-1	1
0	0
1	1
2	2

Plotting these ordered pairs, we see a pattern emerge. The graph is like a large V that has its vertex at the origin. The slope of the line to the right of 0 is 1, and the slope of the line to the left of 0 is -1.

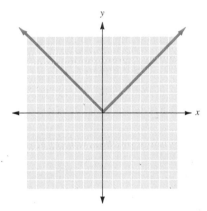

Let us now see what happens to the graph when we add or subtract some constant inside the parentheses.

● Example 2

Graphing an Absolute Value Function

Graph each function.

$f(x) = |x - 3|$

Would be entered as

$Y_1 = \text{abs}(x - 3)$

(a) $f(x) = |x - 3|$

Again, we start with a table of values.

x	$f(x)$
-2	5
-1	4
0	3
1	2
2	1
3	0
4	1
5	2

Then, we plot the points associated with the set of ordered pairs.

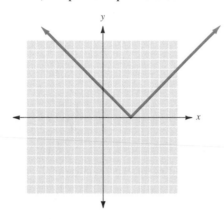

The graph of the function $f(x) = |x - 3|$ is the same shape as the graph of the function $f(x) = |x|$; it has just shifted to the right 3 units.

(b) $f(x) = |x + 1|$

We begin with a table of values.

x	$f(x)$
-2	1
-1	0
0	1
1	2
2	3
3	4

Then we graph.

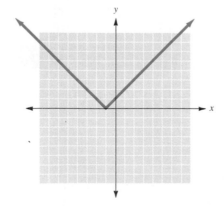

Note that the graph of $f(x) = |x + 1|$ is the same shape as the graph of the function $f(x) = |x|$, except that it is shifted 1 unit to the left.

● ● ● **CHECK YOURSELF 2**

Graph each function.

(a) $f(x) = |x - 2|$ **(b)** $f(x) = |x + 3|$

We can summarize what we have discovered about the horizontal shift of the graph of an absolute value function.

> The graph of the function $f(x) = |x - a|$ will be the same shape as the graph of $f(x) = |x|$ except that the graph will be shifted a units
>
> to the right if a is positive
> to the left if a is negative

If a is negative, $x - a$ will be x plus some positive number.

We will now use these methods to solve equations that contain an absolute value expression.

● **Example 3**

Solving an Absolute Value Equation Graphically

Graphically, find the solution set for the equation.

$$|x - 3| = 4$$

We graph the function associated with each side of the equation.

$$f(x) = |x - 3| \quad \text{and} \quad g(x) = 4$$

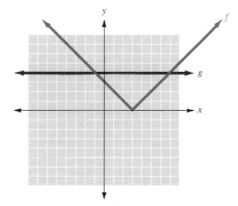

Then, we draw a vertical line through each of the intersection points.

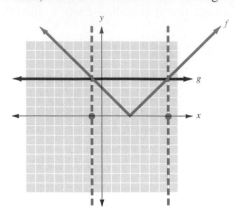

Looking at the x values of the two vertical lines, we find the solutions to the original equation. There are two x values that make the statement true: -1 and 7. The solution set is $\{-1, 7\}$.

● ● ● **CHECK YOURSELF 3**

Graphically find the solution set for the equation.

$$|x - 2| = 3$$

We can also solve absolute value equations algebraically. To do so, we use the following rule.

Absolute Value Equations, Property 1

If $\quad |x| = p$

then $\quad x = p \quad$ or $\quad x = -p$

Let's use this definition to find an algebraic solution to the same equation we solved in Example 3.

● Example 4

Solving an Absolute Value Equation Algebraically

Algebraically solve the equation

$$|x - 3| = 4$$

From Property 1 above, we know that the expression inside the absolute value signs, $(x - 3)$, must either equal 4 or -4. We set up two equations and solve them both.

CAUTION

Be Careful! A common mistake is to solve *only* the equation $(x - 3) = 4$. You must solve *both* equivalent equations to find the **two** required solutions.

$$(x - 3) = 4 \qquad (x - 3) = -4$$
$$x - 3 = 4 \qquad x - 3 = -4 \qquad \text{Add 3 to both sides of the equation.}$$
$$x = 7 \qquad\qquad x = -1$$

We arrive at the same solution set, $\{-1, 7\}$.

● ● ● CHECK YOURSELF 4

Algebraically find the solution set for the equation.

$$|x - 2| = 3$$

We will use the algebraic method to solve subsequent examples in this section. You can use your graphing calculator to check these solutions.

● Example 5

Solving an Absolute Value Equation

Solve for x.

$$|3x - 2| = 4$$

From Property 1, we know that $|3x - 2| = 4$ is equivalent to the equations

$$3x - 2 = 4 \qquad \text{or} \qquad 3x - 2 = -4 \qquad \text{Divide by 2.}$$
$$3x = 6 \qquad\qquad\qquad 3x = -2 \qquad \text{Divide by 3.}$$
$$x = 2 \qquad\qquad\qquad x = -\frac{2}{3}$$

Remember to check your solutions with your calculator.

The solution set is $\left\{-\dfrac{2}{3}, 2\right\}$. These solutions are easily checked by replacing x with $-\dfrac{2}{3}$ and 2 in the original absolute value equation.

● ● ● CHECK YOURSELF 5

Solve for x.

$$|4x + 1| = 9$$

An equation involving absolute value may have to be rewritten before you can apply Property 1. Consider Example 6.

• Example 6

Solving an Absolute Value Equation

Solve for x.

$$|2 - 3x| + 5 = 10$$

To use Property 1, we must first isolate the absolute value on the left side of the equation. This is easily done by subtracting 5 from both sides for the result

$$|2 - 3x| = 5$$

We can now proceed as before by using Property 1.

$$2 - 3x = 5 \qquad \text{or} \qquad 2 - 3x = -5 \qquad \text{Subtract 2.}$$
$$-3x = 3 \qquad\qquad\qquad -3x = -7 \qquad \text{Divide by } -3.$$
$$x = -1 \qquad\qquad\qquad x = \frac{7}{3}$$

Check.

The solution set is $\left\{-1, \dfrac{7}{3}\right\}$.

● ● ● **CHECK YOURSELF 6**

Solve for x.

$$|5 - 2x| - 4 = 7$$

In some applications, there is more than one absolute value in an equation. Consider an equation of the form

$$|x| = |y|$$

Since the absolute values of x and y are equal, x and y are the same distance from 0, which means they are either *equal* or *opposite in sign*. This leads to a second general property of absolute value equations.

> **Absolute Value Equations, Property 2**
>
> If $\quad |x| = |y|$
> then $\quad x = y$ or $x = -y$

Let's look at an application of this second property in Example 7.

•Example 7

Solving Equations with Two Absolute Value Expressions

Solve for x.

$$|3x - 4| = |x + 2|$$

By Property 2, we can write

$3x - 4 = x + 2$	or	$3x - 4 = -(x + 2)$
		$3x - 4 = -x - 2$ Add 4 to both sides.
$3x = x + 6$		$3x = -x + 2$ Isolate x.
$2x = 6$		$4x = 2$ Divide by 2.
$x = 3$		$x = \dfrac{1}{2}$

Check.

The solution set is $\left\{\dfrac{1}{2}, 3\right\}$.

● ● ● CHECK YOURSELF 7

Solve for x.

$$|4x - 1| = |x + 5|$$

● ● ● CHECK YOURSELF ANSWERS

1. (a) 12, **(b)** 4, **(c)** 7.

2. (a)

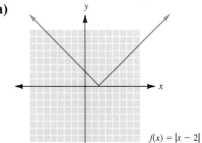

$f(x) = |x - 2|$

(b)

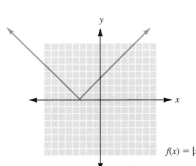

$f(x) = |x + 3|$

3.

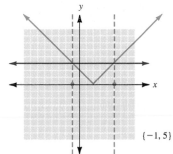

$\{-1, 5\}$

4. $|x - 2| = 3$

$x - 2 = 3$	or	$x - 2 = -3$
$x = 5$		$x = -1$

$\{-1, 5\}$

5. $\left\{-\dfrac{5}{2}, 2\right\}$ **6.** $\{-3, 8\}$ **7.** $\left\{2, -\dfrac{4}{5}\right\}$

Assessing Piston Design. Combustion engines get their power from the force exerted by burning fuel on a piston inside a chamber. The piston is forced down out of the cylinder by the force of a small explosion caused by burning fuel mixed with air. The piston in turn moves a piston rod, which transfers the motion to the work of the engine. The rod is attached to a flywheel, which pushes the piston back into the cylinder to begin the process all over. Cars usually have four to eight of these cylinders and pistons. It is crucial that the piston and the cylinder fit well together, with just a thin film of oil separating the sides of the piston and the sides of the cylinder. When these are manufactured, the measurements for each part must be accurate. But, there is always some error. How much error is a matter for the engineers to set and for the quality control department to check.

Suppose the diameter of the cylinder is meant to be 7.6 cm, and the engineer specifies that this part must be manufactured to within 0.1 mm of that measurement. This figure is called the **tolerance.** As parts come off the assembly line, someone in quality control takes samples and measures the cylinders and the pistons. Given this information, complete the following.

1. Write an absolute value statement about the diameter, d_c, of the cylinder.
2. If the diameter of the piston is to be 7.59 cm with a tolerance of 0.1 mm, write an absolute value statement about the diameter, d_p, of the piston.
3. Investigate all the possible ways these two parts will fit together. If the two parts have to be within 0.1 mm of each other for the engine to run well, is there a problem with the way the parts may be paired together? Write your answer and use a graph to explain.
4. Accuracy in machining the parts is expensive, so the tolerance should be close enough to make sure the engine runs correctly, but not so close that the cost is prohibitive. If you think a tolerance of 0.1 is too large, find another that you think would work better. If it is too small, how much can it be enlarged and still have the engine run according to design? (That is, so $|d_c - d_p| \leq 0.1$ mm.) Write the tolerance using absolute value signs. Explain your reasoning if you think a tolerance of 0.1 mm is not workable.
5. After you have decided on the appropriate tolerance for these parts, think about the quality control engineer's job. Hazard a few educated opinions to answer these questions: How many parts should be pulled off the line and measured? How often? How many parts can reasonably be expected to be outside the expected tolerance before the whole line is shut down and the tools corrected?

2.4 Exercises

In Exercises 1 to 14, find the absolute value for each expression.

1. $|15|$

2. $|21|$

3. $|-15|$

4. $|-18|$

5. $|8 - 3|$

6. $|35 - 23|$

7. $|-12 + 8|$

8. $|-23 - 11|$

9. $|-12 - 19|$

10. $|-19 - 27|$

11. $-|-19| + |-27|$

12. $|-13| - |-12|$

13. $|-13| + |12|$

14 $|-13 - 12|$

In Exercises 15 to 20, graph each function.

15. $f(x) = |x - 3|$

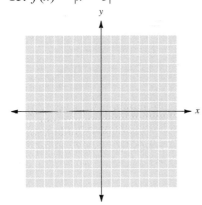

16. $f(x) = |x + 2|$

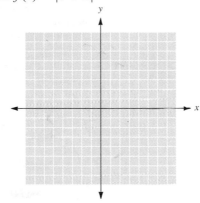

17. $f(x) = |x + 3|$

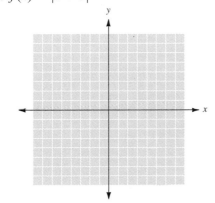

18. $f(x) = |x - 4|$

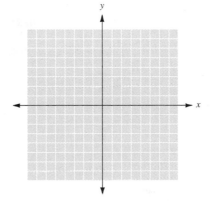

19. $f(x) = |x - (-3)|$

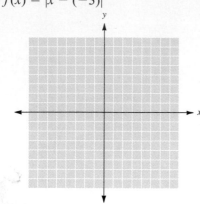

20. $f(x) = |x - (-5)|$

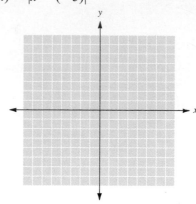

In Exercises 21 to 26, solve the equations graphically.

21. $|x| = 3$

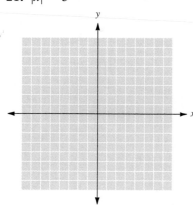

22. $|x| = 5$

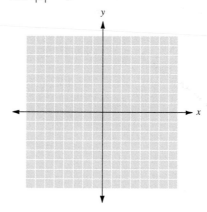

23. $|x - 2| = \dfrac{7}{2}$

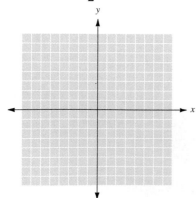

24. $|x - 5| = 3$

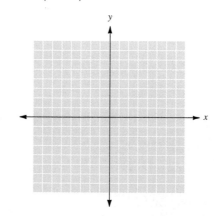

25. $|x + 2| = 4$

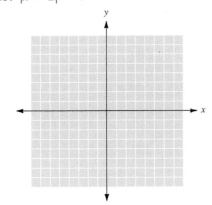

26. $|x + 4| = 2$

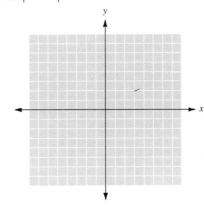

In Exercises 27 to 46, solve the equations algebraically, if possible.

27. $|x| = 5$

28. $|x| = 6$

29. $|x - 1| = 4$

30. $|x - 2| = 6$

31. $|2x - 1| = 6$

32. $|3x - 2| = 8$

33. $3|2x - 6| = 9$

34. $2|3x - 5| = 12$

35. $|5x - 2| = -3$

36. $|4x - 9| = -8$

37. $|x - 2| + 3 = 5$

38. $|x + 5| - 2 = 5$

39. $8 - |x - 4| = 5$

40. $10 - |2x + 1| = 3$

41. $|2x - 1| = |x + 3|$

42. $|3x + 1| = |2x - 3|$

43. $|5x - 2| = |2x - 4|$

44. $|5x + 2| = |x - 3|$

45. $|7x - 3| = |2x + 7|$

46. $|x - 2| = |2 - x|$

In Exercises 47 to 50, determine the function represented by each graph.

47.

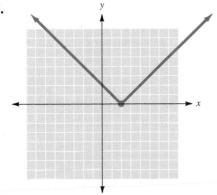

48.

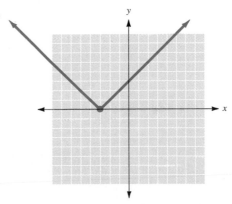

49.

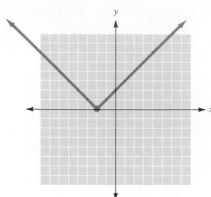

50.

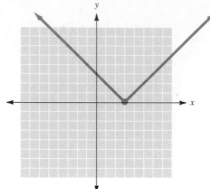

51. Distance. Two towns, Hope and Suntown, are 21 miles apart on a straight stretch of East-West highway. If Hope is at milepost A and Suntown is at milepost B, we have the following absolute value statement

$$|A - B| = 21$$

Solve the following.

(a) Another town, Newtown, is to be situated so that the distance from this town to Suntown is 12 miles. If Newtown is at milepost C, write a statement using absolute value for the distance from B to C and another statement for the distance from A to C. Be sure to consider all possibilities. Compare your statements with other students' statements.

(b) If the third town does not have to be on the highway but still has to be 12 miles from Suntown, draw a sketch of where it might possibly be. In this case, use inequalities to write a statement about how far Newtown is from Hope. *Note:* Since the town may not be on the highway, C cannot be used to express distance, and you will need another variable.

Answers

1. 15   **3.** 15   **5.** 5   **7.** 4   **9.** 31   **11.** 8   **13.** 25

15.

17.

19.

21.

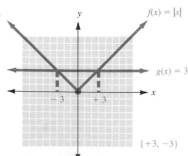

$f(x) = |x|$
$g(x) = 3$
-3 $+3$
$\{+3, -3\}$

23. $f(x) = |x - 2|$

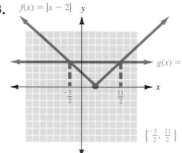

$g(x) = \frac{7}{2}$
$-\frac{3}{2}$ $\frac{11}{2}$
$\left\{ \frac{-3}{2}, \frac{11}{2} \right\}$

25.

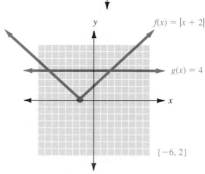

$f(x) = |x + 2|$
$g(x) = 4$
$\{-6, 2\}$

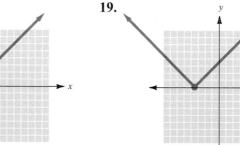

27. $\{5, -5\}$　　**29.** $\{5, -3\}$　　**31.** $\{3.5, -2.5\}$　　**33.** $\{1.5, 4.5\}$　　**35.** No solution　　**37.** $\{0, 4\}$

39. $\{7, 1\}$　　**41.** $\left\{-\dfrac{2}{3}, 4\right\}$　　**43.** $\left\{-\dfrac{2}{3}, \dfrac{6}{7}\right\}$　　**45.** $\left\{2, -\dfrac{4}{9}\right\}$　　**47.** $f(x) = |x - 2|$　　**49.** $f(x) = |x + 2|$

51.

2.5　**Absolute Value Inequalities**

2.5 OBJECTIVES

1. Solve absolute value inequalities in one variable graphically
2. Solve absolute value inequalities in one variable algebraically

In Section 2.4, we looked at a graphical solution to an absolute value equation, and then we learned to solve the same equations algebraically. In this section, we will follow the same process for absolute value inequalities.

Absolute value inequalities in one variable, x, are obtained from absolute value equations by replacing the symbol for equality ($=$) with one of the inequality symbols ($<, >, \le, \ge$).

The general form for an absolute value inequality in one variable is

$$|x - a| < b$$

where the symbol $<$ can be replaced with $>$, $\le$, or $\ge$. Examples of absolute value inequalities in one variable include

$$|x| < 6 \qquad |x - 4| \ge 2 \qquad |3x - 5| \le 8$$

Example 1 will demonstrate a graphical solution to an absolute value inequality.

● Example 1

Solving an Absolute Value Inequality Graphically

Graphically solve

$$|x| < 6$$

As we did in previous sections, we begin by letting each side of the inequality represent a function. Here

$$f(x) = |x| \qquad \text{and} \qquad g(x) = 6$$

Now we graph both functions on the same set of axes.

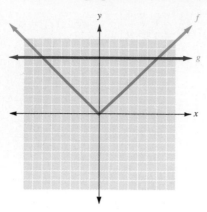

We next draw a dotted line (equality is not included) through the points of inter-section of the two graphs.

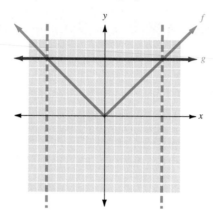

The solution set is any value of x for which the graph of $f(x)$ is below the graph of $g(x)$.

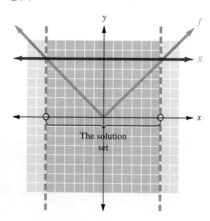

In set notation, we write $\{x \mid -6 < x < 6\}$.

● ● ●   CHECK YOURSELF 1

Graphically solve the inequality

$$|x| > 3$$

The graphical solution to Example 1 suggests the following general statement.

Absolute Value Inequalities, Property 1

For any positive number p, if

$$|x| < p$$

then

$$-p < x < p$$

Let's look at an application of Property 1 in solving an absolute value inequality.

• Example 2

Solving Absolute Value Inequalities

Solve and graph the solution set of

$$|2x - 3| < 5$$

With Property 1 we can *translate* an absolute value inequality to an inequality *not* containing an absolute value, which can be solved by our earlier methods.

From Property 1, we know that the given absolute value inequality is equivalent to the double inequality

$$-5 < 2x - 3 < 5$$

Solve as before.

$$-5 < 2x - 3 < 5 \qquad \text{Add 3 to all three parts.}$$
$$-2 < 2x < 8 \qquad \text{Divide by 2.}$$
$$-1 < x < 4$$

The solution set is

$$\{x \mid -1 < x < 4\}$$

Note that the solution is an open interval on the number line.

The graph is shown below.

$$
\begin{array}{ccccccccc}
-2 & -1 & 0 & 1 & 2 & 3 & 4 & 5
\end{array}
$$

● ● ● CHECK YOURSELF 2

Solve and graph the solution set.

$$|3x - 4| \leq 8$$

We know that the solution set for the absolute value inequality

$$|x| < 4$$

consists of those numbers whose distance from the origin is *less than* 4. Now what about the solution set for

$$|x| > 4$$

It must consist of those numbers whose distance from the origin is *greater than* 4. The solution set is pictured below.

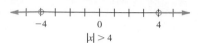

$$|x| > 4$$

The solution set can be described by the compound inequality

$$x < -4 \qquad \text{or} \qquad x > 4$$

and this suggests the following general statement.

Absolute Value Inequalities, Property 2

For any positive number p, if

$$|x| > p$$

then

$$x < -p \qquad \text{or} \qquad x > p$$

Let's apply Property 2 to the solution of an absolute value inequality.

• Example 3

Solving Absolute Value Inequalities

Solve and graph the solution set of

$$|5x - 2| > 8$$

From Property 2, we know that the given absolute inequality is equivalent to the compound inequality

Again we *translate* the absolute value inequality to the compound inequality *not* containing an absolute value.

$$5x - 2 < -8 \qquad \text{or} \qquad 5x - 2 > 8$$

Solving as before, we have

$$
\begin{array}{llll}
5x - 2 < -8 & \text{or} & 5x - 2 > 8 & \text{Add 2.} \\
5x < -6 & & 5x > 10 & \text{Divide by 5.} \\
x < -\dfrac{6}{5} & & x > 2 &
\end{array}
$$

You could describe the solution set as

$$\left\{x \,\middle|\, x < -\frac{6}{5}\right\} \text{ or } \{x \mid x > 2\}$$

The solution set is $\left\{x \,\middle|\, x < -\dfrac{6}{5} \text{ or } x > 2\right\}$, and the graph is shown below.

```
 ←—+——+—⊙—+——+—+——⊙—+——+——→
  -3 -2 -1  0  1  2  3  4
```

CHECK YOURSELF 3

Solve and graph the solution set.

$$|3 - 2x| \geq 9$$

The following chart summarizes our discussion of absolute value inequalities.

As before, p must be a positive number if $p > 0$.

Type of Inequality	Equivalent Inequality	Graph of Solution Set
$\lvert ax + b \rvert < p$	$-p < ax + b < p$	 ←———⊙————⊙———→ r s
$\lvert ax + b \rvert > p$	$ax + b < -p$ or $ax + b > p$	 ←———⊙ ⊙———→ r s

CHECK YOURSELF ANSWERS

1.

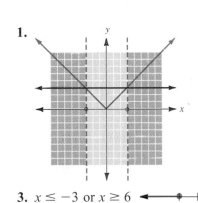

2. $-\dfrac{4}{3} \leq x \leq 4$

```
 ←+—●—+——+——+——+——+——●—+——→
  -2 -1  0  1  2  3  4  5
```

3. $x \leq -3$ or $x \geq 6$

```
 ←———————●———+———————●———→
        -3   0       6
```

Determining Heating Costs. Ed and Sharon's heating system has just broken down. They can replace their current electrical heat pump system with another electrical heat pump system or replace it with an oil heating system or a gas system. Ed knows that in order to install an oil heating system, he will need a new air conditioner and extensive duct remodeling in his house. He also knows that a gas heating system will require that gas lines be laid on his street and then connected to the house. Both Ed and Sharon are concerned about installing another heat pump system because it does not seem to work well in their cold Northeast climate. They both want to install a system that will be economical in the long run. Ed did some research and collected installation costs and yearly operational costs These data appear in the following chart.

Type of Heating	Installation Costs	Operational Costs (per year)
Reinstall an electric system	$4,200	$900
Convert to oil system	$8,500	$550
Convert to gas system	$13,000	$425

1. Use 5-year intervals to construct a table that shows the total cost of heating over a 35-year period for the three heating systems.
2. Determine the average yearly heating costs for each of the three systems after **(a)** 5 years, **(b)** 10 years, **(c)** 15 years, **(d)** 20 years, **(e)** 25 years, **(f)** 30 years, and **(g)** 35 years.
3. The total cost of each system is a function composed of the operational costs multiplied by the number of years of operation added to the installation costs. For each of the three systems, develop a function that represents the total cost.
4. Graph the three straight lines developed in Exercise 3 on the same coordinate axis.
5. Using the graphs in Exercise 4, estimate the following:
   (a) When does the oil system become cheaper than the electric system?
   (b) When does the gas system become cheaper than the electric system?
   (c) When does the gas system become cheaper than the oil system?
6. Using the functions developed in Exercise 3, determine the following:
   (a) When does the oil system become cheaper than the electric system?
   (b) When does the gas system become cheaper than the electric system?
   (c) When does the gas system become cheaper than the oil system?
7. Do the following:
   (a) Contact your local electric and gas utility and obtain the current rate schedule as well as the cost of installing a new heating system that would be adequate for a 2000-square ft home with 8-ft ceilings.
   (b) Contact a local oil company and obtain the current prices as well as installation costs of a new heating system that would be adequate for a 2000-square ft home with 8-ft ceilings.
   (c) Use the information from **(a)** and **(b)** and construct a table similar to that developed in Exercise 1.
   (d) Develop a function that represents the total heating costs for each of the three systems based on the information you obtained in **(a)** and **(b)**.
   (e) Based solely on the costs associated with the heating systems, which one would you choose to install in your house?
8. In addition to cost, what other factors would you consider in choosing a new heating system for your home?
9. Considering all the factors involved, decide what type of heating system you would pick for a new home you are building, and write a paragraph explaining your decision.

2.5 Exercises

In Exercises 1 to 12, solve each inequality graphically.

1. $|x| < 4$

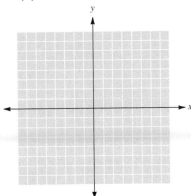

2. $|x| < 6$

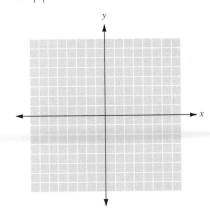

3. $|x| \geq 5$

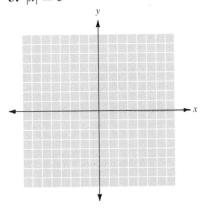

4. $|x| \geq 2$

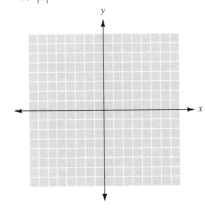

5. $|x - 3| < 4$

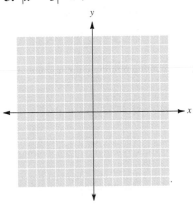

6. $|x - 1| < 5$

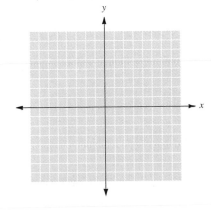

7. $|x - 2| \geq 5$

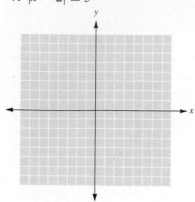

8. $|x + 2| > 4$

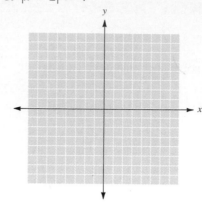

9. $|x + 1| \leq 5$

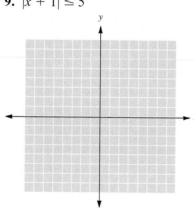

10. $|x + 4| > 1$

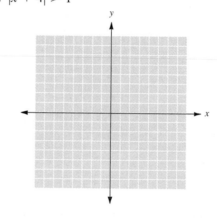

11. $|x + 2| \geq -2$

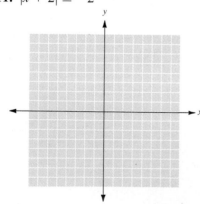

12. $|x - 4| > -1$

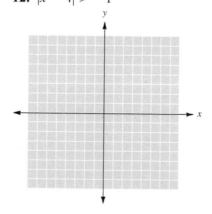

In Exercises 13 to 32, solve each inequality algebraically. Graph the solution set.

13. $|x| < 5$

14. $|x| > 3$

15. $|x| \geq 7$

16. $|x| \leq 4$

17. $|x - 4| > 2$

18. $|x + 5| < 3$

19. $|x + 6| \leq 4$

20. $|x - 7| > 5$

21. $|3 - x| > 5$

22. $|5 - x| < 3$

23. $|x - 7| < 0$

24. $|x + 5| \geq 0$

25. $|2x - 5| < 3$

26. $|3x - 1| > 8$

27. $|3x + 4| \geq 5$

28. $|2x + 3| \leq 9$

29. $|5x - 3| > 7$

30. $|6x - 5| < 13$

31. $|2 - 3x| \leq 11$

32. $|3 - 2x| > 11$

In Exercises 33 to 40, use absolute value notation to write an inequality that represents each sentence.

33. x is within 3 units of 0 on the number line.

34. x is within 4 units of 0 on the number line.

35. x is at least 5 units from 0 on the number line.

36. x is at least 2 units from 0 on the number line.

37. x is less than 7 units from -2 on the number line.

38. x is more than 6 units from 4 on the number line.

39. x is at least 3 units from -4 on the number line.

40. x is at most 3 units from 4 on the number line.

Answers

1.

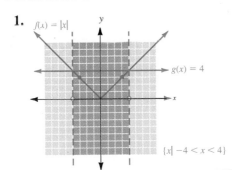

$\{x| -4 < x < 4\}$

3.

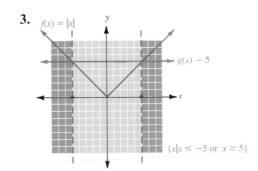

$\{x|x \leq -5 \text{ or } x \geq 5\}$

5.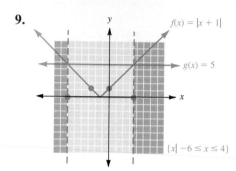

$\{x| -1 < x < 7\}$

7.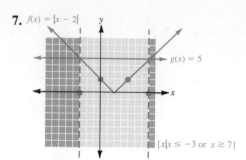

$\{x|x \le -3 \text{ or } x \ge 7\}$

9.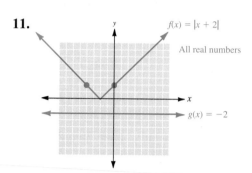

$f(x) = |x + 1|$

$g(x) = 5$

$\{x| -6 \le x \le 4\}$

11.

$f(x) = |x + 2|$

All real numbers

$g(x) = -2$

13. $-5 < x < 5$

−5 0 5

15. $x \le -7 \text{ or } x \ge 7$

−7 0 7

17. $x < 2 \text{ or } x > 6$

0 2 6

19. $-10 \le x \le -2$

−10 −2 0

21. $x < -2 \text{ or } x > 8$

−2 0 8

23. No solution

25. $|2x - 5| < 3$

0 1 4

27. $|3x + 4| \ge 5$

−3 0 $\frac{1}{3}$

29. $|5x - 3| > 7$

$-\frac{4}{5}$ 0 2

31. $|2 - 3x| \le 11$

−3 0 $\frac{13}{3}$

33. $|x| < 3$ **35.** $|x| \ge 5$ **37.** $|x - (-2)| < 7$ **39.** $|x - (-4)| \ge 3$

Summary

Solving Linear Equations in One Variable [2.1]

$5x - 6 = 3x + 2$ is a linear equation. The variable appears only to the first power.

A **linear equation in one variable** is any equation that can be written in the form

$$ax + b = 0$$

where a and b are any real numbers and $a \neq 0$.

To solve a linear equation means to find its solution. A **solution** of an equation in one variable is any number that will make the equation a true statement. The **solution set** consists of all solutions.

Two equations are **equivalent** if they have the same solution set.

Forming a sequence of equivalent equations that will lead to the solution of a linear equation involves two properties of equality.

The solution for the equation above is 4 since

$$5 \cdot 4 - 6 \overset{?}{=} 3 \cdot 4 + 2$$
$$14 = 14$$

is a true statement.
$5x - 6 = 3x + 2$ and $2x = 8$ are equivalent equations. Both have 4 as the solution.

If $x - 3 = 7$,
then $x - 3 + 3 = 7 + 3$.

Addition Property of Equality If $a = b$, then $a + c = b + c$. In words, adding the same quantity to both sides of an equation gives an equivalent equation.

If $2x = 8$, then $\frac{1}{2}(2x) = \frac{1}{2}(8)$.

Multiplication Property of Equality If $a = b$, then $ac = bc$, $c \neq 0$. In words, multiplying both sides of an equation by the same nonzero quantity gives an equivalent equation.

The above properties are applied in the following algorithm.

Solving Linear Equations in One Variable

Solve

$$\frac{x + 1}{5} - \frac{x}{4} = \frac{1}{20}$$

$$20\left(\frac{x + 1}{5}\right) - 20\left(\frac{x}{4}\right)$$
$$= 20\left(\frac{1}{20}\right)$$

$4(x + 1) - 5x = 1$

STEP 1 Multiply both sides of the equation by the LCM of all denominators, to clear the equation of fractions.

Remove grouping symbols.

$4x + 4 - 5x = 1$

STEP 2 Remove any grouping symbols by applying the distributive property.

Combine like terms.

$-x + 4 = 1$

STEP 3 Combine like terms that appear on either side of the equation.

Subtract 4.

$-x = -3$

STEP 4 Apply the addition property of equality to write an equivalent equation with the variable term on *one side* of the equation and the constant term on the *other side*.

Divide by -1.

$x = 3$

STEP 5 Apply the multiplication property of equality to write an equivalent equation with the variable isolated on one side of the equation.

To check:

STEP 6 Check the solution in the *original* equation.

$$\frac{3 + 1}{5} - \frac{3}{4} \overset{?}{=} \frac{1}{20}$$

199

To solve $2x - 6 = 8x$,
let $f(x) = 2x - 6$
$g(x) = 8x$
then, graph both lines.

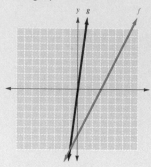

The intersection occurs when $x = -1$.

To solve $2x - 6 > 8x$,
let $f(x) = 2x - 6$
$g(x) = 8x$
then graph both lines.

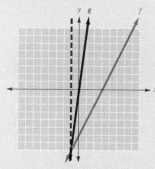

Draw a dotted line where they intersect.

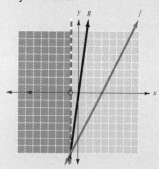

The solution is $x < -1$.

$|2x - 5| = 7$ is equivalent to

$2x - 5 = -7$ or

$2x - 5 = 7$

so

$x = -1$ or $x = 6$

Graphical Solutions to Equations in One Variable [2.2]

Finding a Graphical Solution for an Equation

STEP 1 Let each side of the equation represent a function of x.
STEP 2 Graph the two functions on the same set of axes.
STEP 3 Find the intersection of the two graphs. The x value at this intersection represents the solution to the original equation.

Graphing Linear Inequalities in One Variable [2.3]

Finding the Solution for an Inequality in One Variable

STEP 1   Rewrite the inequality as a comparison of two functions.

$$f(x) < g(x) \qquad f(x) > g(x) \qquad f(x) \leq g(x) \qquad f(x) \geq g(x)$$

STEP 2   Graph the two functions on a single set of axes.
STEP 3   Draw a vertical line through each point of intersection of the two functions. Use a dotted line if equality is not included ($<$ or $>$). Use a solid line if equality is included ($\leq$ or $\geq$).
STEP 4   Shade the x values that make the inequality a true statement.
STEP 5   Write the solution in set builder notation.

Absolute Value Equations [2.4]

Finding a Graphical Solution for an Absolute Value Equation

STEP 1 Let each side of the equation represent a function of x.
STEP 2 Graph the two functions on the same set of axes.
STEP 3 Find the intersection of the two graphs. The x value(s) at these intersection points represent the solutions to the original equation.

To solve absolute value equations, the following property is applied.

Absolute Value Equations

Property 1 For any positive number p, if

$$|x| = p$$

then

$$x = -p \qquad \text{or} \qquad x = p$$

Property 2 If $|x| = |y|$

then

$$x = y \qquad \text{or} \qquad x = -y$$

Absolute Value Inequalities [2.5]

Absolute Value Inequalities

$|3x - 5| < 7$ is equivalent to

Property 1 For any positive number p, if

$-7 < 3x - 5 < 7$

$$|x| < p$$

This yields

$-2 < 3x < 12$

then

$-\dfrac{2}{3} < x < 4$

$$-p < x < p$$

To solve this form of inequality, translate to the equivalent double inequality and solve as before.

$|2 - 5x| \geq 12$ is equivalent to

Property 2 For any positive number p, if

$2 - 5x \leq -12 \qquad \text{or}$

$$|x| > p$$

$2 - 5x \geq 12$

then

This yields

$x \geq \dfrac{14}{5} \qquad \text{or} \qquad x \leq -2$

$$x < -p \qquad \text{or} \qquad x > p$$

To solve this form of inequality, translate to the equivalent compound inequality and solve as before.

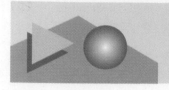

Summary Exercises

This summary exercise set is provided to give you practice with each of the objectives in the chapter. Each exercise is keyed to the appropriate chapter section. The answers are provided in the *Instructor's Manual*.

[2.1] Solve the following equations algebraically. Express your answer in set builder notation.

1. $4x - 5 = 23$

2. $7 - 3x = -8$

3. $5x + 2 = 6 - 3x$

4. $7x - 3 = 2x + 12$

5. $2x - 7 = 9x - 35$

6. $5 - 3x = 2 - 6x$

7. $7x - 3 + 2x = 5 + 6x + 4$

8. $2x + 5 - 4x = 3 - 6x + 10$

9. $3(x - 5) = x + 1$

10. $4(2x - 1) = 6x + 5$

11. $7x - 3(x - 2) = 30$

12. $8x - 5(x + 3) = -10$

13. $7(3x + 1) - 13 = 8(2x + 3)$

14. $3(2x - 5) - 2(x - 3) = 11$

15. $\dfrac{2x}{3} - \dfrac{x}{4} = 5$

16. $\dfrac{3x}{4} - \dfrac{2x}{5} = 7$

17. $\dfrac{x}{2} - \dfrac{x + 1}{3} = \dfrac{1}{6}$

18. $\dfrac{x + 1}{5} - \dfrac{x - 6}{3} = \dfrac{1}{3}$

19. Lisa left Friday morning, driving on the freeway to visit friends for the weekend. Her trip took 4 h. When she returned on Sunday, heavier traffic slowed her average speed by 6 mi/h, and the trip took $4\frac{1}{2}$ h. What was her average speed in each direction, and how far did she travel each way?

20. A bicyclist started on a 132-mi trip and rode at a steady rate for 3 h. He began to tire at that point and slowed his speed by 4 mi/h for the remaining 2 h of the trip. What was his average speed for each part of the journey?

21. At noon, Jan left her house, jogging at an average rate of 8 mi/h. Two hours later, Stanley left on his bicycle along the same route, averaging 20 mi/h. At what time will Stanley catch up with Jan?

22. At 9 AM, David left New Orleans for Tallahassee, averaging 47 mi/h. Two hours later, Gloria left Tallahassee for New Orleans along the same route, driving 5 mi/h faster than David. If the two cities are 391 mi apart, at what time will David and Gloria meet?

23. A firm producing running shoes finds that its fixed costs are $3900 per week, and its variable cost is $21 per pair of shoes. If the firm can sell the shoes for $47 per pair, how many pairs of shoes must be produced and sold each week for the company to break even?

24. For a certain type of scientific calculator, the supply available at price p (in dollars) is given by

$$S = 40p - 200$$

The demand for that same calculator at price p is

$$D = -80p + 1720$$

Find the equilibrium price (where the supply equals the demand).

[2.2]   Solve the following equations graphically.

25. $3x - 6 = 0$

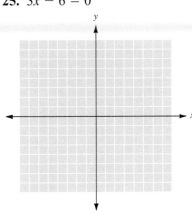

26. $4x + 3 = 7$

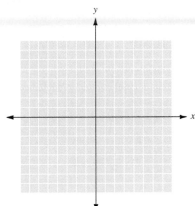

27. $3x + 5 = x + 7$

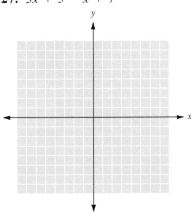

28. $4x - 3 = x - 6$

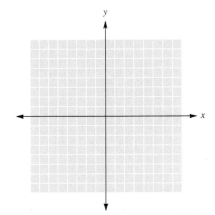

29. $\dfrac{6x - 1}{2} = 2(x - 1)$

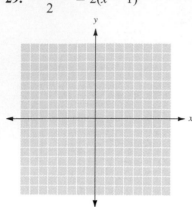

30. $3x + 2 = 2x - 1$

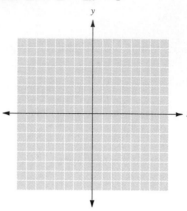

31. $3(x - 2) = 2(x - 1)$

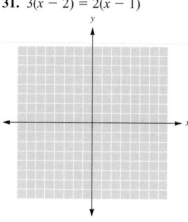

32. $3(x + 1) + 3 = -7(x + 2)$

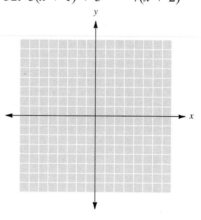

[2.3]   Graph the solution set to the following inequalities.

33. $5 < 2x + 1$

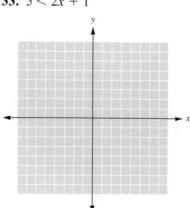

34. $-3 \geq -2x + 3$

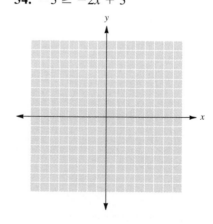

35. $3x + 2 \geq 6 - x$

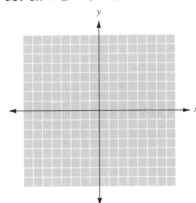

36. $3x - 5 < 15 - 2x$

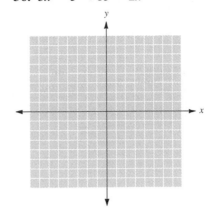

37. $x + 6 < -2x - 3$

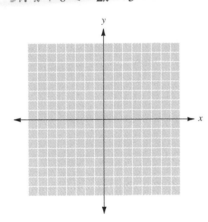

38. $4x - 3 \geq 7 - x$

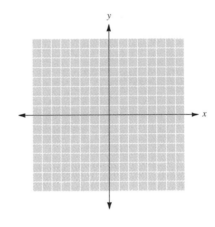

39. $x \geq -3 - 5x$

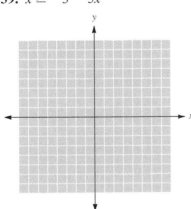

40. $x < 4 + 2x$

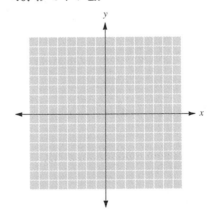

41. The cost to produce x units of a product is $C = 100x + 6000$, and the revenue is $R = 500x$. Find all values of x for which the product will at least break even.

42. For a particular line of lamps, a store has average monthly costs of $C = 55x + 180$ and corresponding revenue of $R = 100x$, where x is the number of units sold. How many units must be sold for the store to break even on the lamps for the month?

[2.4] Solve the following equations graphically.

43. $|x + 3| = 5$

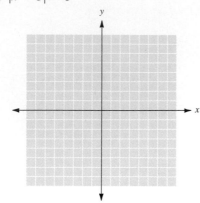

44. $|3x - 2| = 7$

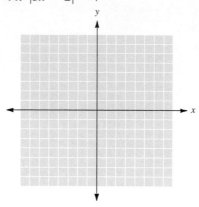

45. $|2 - x| = 3$

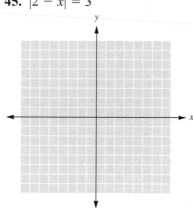

46. $|4 - x| = 2$

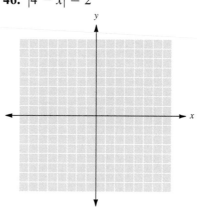

Solve the following equations algebraically, if possible.

47. $|2x + 1| = 5$

48. $|3x - 2| = 7$

49. $|2x - 5| = 3$

50. $|4x + 1| = 9$

51. $|4x - 3| = 13$

52. $|6 - 2x| = 10$

53. $|-5x + 7| = 17$

54. $|-3x + 5| = 9$

55. $|7x - 1| = -3$

56. $|2x + 5| = -1$

57. $|2x + 1| - 3 = 6$

58. $7 - |x - 3| = 5$

59. $|3x - 1| = |x + 5|$

60. $|x - 5| = |x + 3|$

[2.5]   Solve the following inequalities graphically.

61. $|x| < 7$

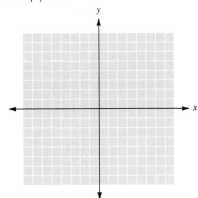

62. $|x| \leq 9$

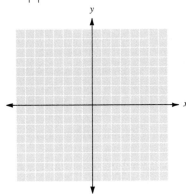

63. $|x - 1| > 6$

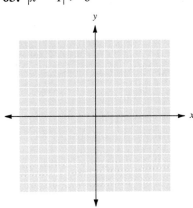

64. $|x + 3| \leq 4$

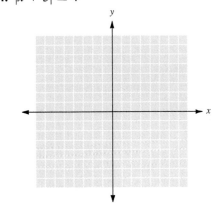

65. $|x + 5| \geq 2$

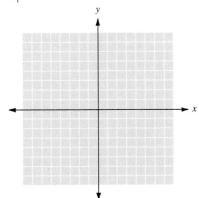

66. $|x - 1| < 8$

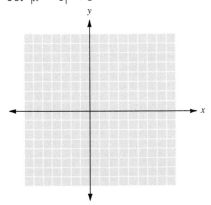

Solve the following inequalities algebraically. Graph the solution set.

67. $|x| \leq 3$

68. $|x + 3| > 5$

69. $|x - 7| > 4$

70. $|3 - x| < 6$

71. $|2x + 7| \geq 5$

72. $|3x - 1| \leq 4$

73. $|3x + 4| < 11$

74. $|5x + 2| \geq 12$

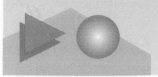

Self-Test for Chapter 2

The purpose of this self-test is to help you check your progress and to review for a chapter test in class. Allow yourself about 1 hour to take the test. When you are done, check your answers in the back of the book. If you missed any answers, be sure to go back and review the appropriate sections in the chapter and the exercises that are provided.

Solve each of the following equations algebraically. Express your answer in set notation.

1. $7 - 5x = 3$

2. $5x - 3(x - 5) = 19$

3. $6x - (-4x - 5) = -2(-4x - 5)$

4. $8x - \{5 - (4x - 9)\} = 7x$

Solve the following applications.

5. At 10 AM, Sandra left her house on a business trip and drove an average of 45 mi/h. One hour later, Adam discovered that Sandra had left her briefcase behind, and he began driving at 55 mi/h along the same route. When will Adam catch up with Sandra?

6. A company that manufactures boxed mints sells each box for $3. The company incurs a cost of $1.50 per box with a total fixed cost of $30,000. How many boxes of mints must be sold for the company to break even?

Solve the following equations graphically.

7. $4x - 7 = 5$

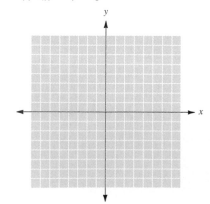

8. $6 - x = 4(x - 1)$

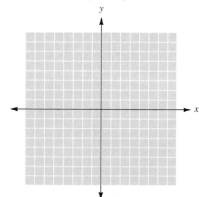

9. $-8x + 11 = 2x - 9$

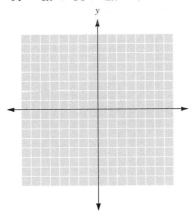

10. $6(x - 1) = -3(x - 4)$

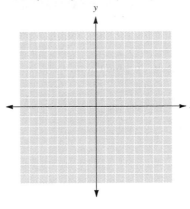

Graph the solution set to the following inequalities.

11. $5x - 3 < 7$

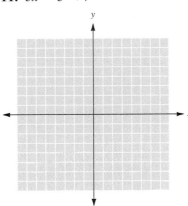

12. $2x - 1 \leq 3(x - 1)$

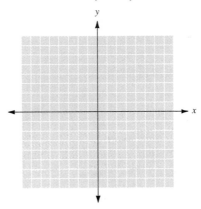

Solve the following equations graphically.

13. $|4x + 3| = 7$

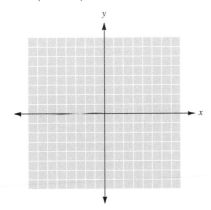

14. $|2x - 5| = 9$

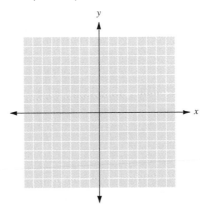

Solve the following equations algebraically.

15. $|5x - 6| - 3 = -1$

16. $|4x - 8| - 11 = 5$

Solve the following inequalities graphically.

17. $|x| \leq 3$

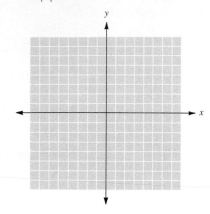

18. $|x + 1| \geq 4$

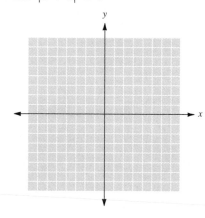

Solve the following inequalities algebraically. Graph the solution set.

19. $|x - 1| < 4$

20. $|2x + 1| \geq 7$

This test is provided to help you in the process of reviewing the previous chapters. Answers are provided in the back of the book. If you missed any answers, be sure to go back and review the appropriate chapter sections.

For each of the following sets of numbers, identify the domain and range.

1. $\{(2, 7), (3, 5), (-1, 1), (-2, 0) (4, 5)\}$

2. $\{(x, y)|6x - 5y = 60\}$

Solve each of the following equations.

3. $2(3x + 9) = 8(2 + x)$

4. $\dfrac{x - 3}{2} + \dfrac{2x + 3}{4} = \dfrac{5}{12}$

5. If $f(x) = 3x^2 + 5x - 9$, find $f(-1)$

In each of the following, determine which relations are functions.

6. $\{(1, 2), (-1, 2), (3, 4), (5, 6)\}$

7.

x	y
-3	0
-2	1
-1	5
6	3

8. Use the vertical line test to determine whether the following graph represents a function.

(a)

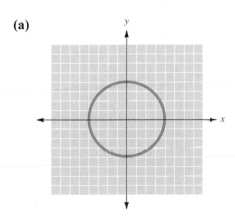

(b)

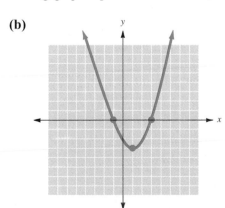

9. Find the x and y intercepts and graph the equation.

(a) $2x - 3y = 12$

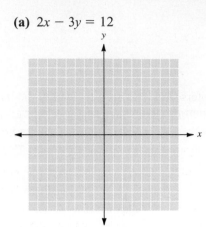

(b) $3x + 7y = 21$

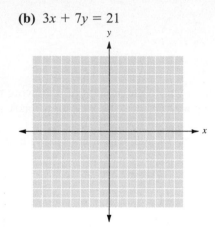

10. Find the domain and range of the relation $7x - 5y = 11$.

11. Use the given graph to estimate **(a)** $f(-3)$, **(b)** $f(0)$, and **(c)** $f(3)$.

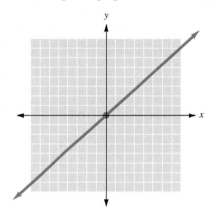

12. Find the slope of the line determined by the points $(-3, 6)$ and $(1, -2)$.

13. Write the equation of the line passing through the point $(1, 5)$ with a slope equal to -2. Write your results in slope intercept form.

14. If $f(x) = 3x + 5$ and $g(x) = -7x - 2$, find **(a)** $f + g$, **(b)** $f - g$, **(c)** $f \cdot g$, and **(d)** $\dfrac{f}{g}$.

Solve the following equations.

15. $2x - 3(x + 2) = 4(5 - x) + 7$

16. $|x - 4| = 5$

17. $4(x - 5) = -6(x - 10)$

Solve the following inequalities.

18. $-8(2 - x) \geq 16$

19. $|2x - 6| < 4$

20. $3|2x - 5| > 3$

CHAPTER 3

POLYNOMIALS

INTRODUCTION

When developing aircraft, autos, and boats, engineers use computer design programs. To ensure that the smooth, curved shapes created by the design fit together when manufactured, engineers use "polynomial splines." These splines are also useful for civil engineers when designing tunnels and highways. These splines help in the design of a roadway, ensuring that changes in direction and altitude occur smoothly and gradually. In some cases, roads make transitions from, say, a valley floor to a mountain pass while covering a distance of only a few kilometers. For example: a road passes through a valley at 75-meters altitude and then climbs through some hills, reaching an altitude of 350 meters before descending again. The following graph shows the change in altitude for 22 km of the roadway.

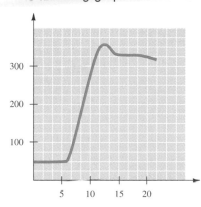

The road seems very steep in this graph, but remember that the y axis is the altitude measured in meters, and the x axis is horizontal distance measured in kilometers. To get a true feeling for the vertical change, the horizontal axis would have to be stretched by a factor of 1000.

Based on measurements of the distance, altitude, and change in the slope taken at intervals along the planned path of the road, formulas are developed that model the roadway for short sections. The formulas are then pieced together to form a model of the road over several kilometers. Such formulas are found by using algebraic methods to solve systems of linear equations. The resulting splines can be linear, quadratic, or even cubic equations. Here are some of the splines that could be fit together to create the roadway needed in our graph; y is the altitude measured in meters and x is the horizontal distance, measured in kilometers.

First 5 km: $y = 75 + 0.15x$

Next 7.5 km: $y = 21.5975x^3 + 40.726x^2 - 287.296x + 693.521$

Final 5.5 km: $y = 20.290101x^3 + 14.9427x^2 - 255.559x + 1771.94$

By entering these equations in your calculator, you can see part of the design for the road. Be certain that you adjust your graphing window appropriately.

Polynomials are useful in many areas. In this chapter, you will learn how to solve problems involving polynomials. ◼

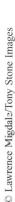

© Lawrence Migdale/Tony Stone Images

213

Positive Integer Exponents and the Scientific Calculator

3.1 OBJECTIVES

1. Use properties of exponents
2. Evaluate expressions with integer exponents
3. Use a calculator to evaluate expressions

DOUBLING

Take a sheet of paper and fold it in half. Fold the resulting paper in half; continue this process for as many folds as you can. Is it possible to get past the seventh fold? Why not? If the sheet of paper were much larger, could you then continue past the seventh fold? No matter how large the paper you start with, the answer is always no. Let's see why this is true.

The first fold doubles the thickness of the paper, so the second fold is like folding two sheets together. Each subsequent fold again doubles the thickness of the paper. The following table will help you see the result of eight folds.

Fold Number	Paper Thickness
1	2 sheets
2	4
3	8
4	16
5	32
6	64
7	128
8	256

The eighth fold is the equivalent of folding 256 sheets of paper. How large would the paper have to be for you to fold 256 sheets? Let's assume that you can fold a 1-foot-square stack of paper 256 sheets thick. How large would the original sheet have to be to equal 1 foot across after seven folds? The next table works the problem backward.

Fold Number	Paper size
8	1 foot by 1 foot
7	2 feet by 1 foot
6	2 feet by 2 feet
5	4 feet by 2 feet
4	4 feet by 4 feet
3	8 feet by 4 feet
2	8 feet by 8 feet
1	16 feet by 8 feet

The original paper would have had to have been 16 feet long and 8 feet wide to end up with a 1-foot square for the eighth fold.

EXPONENTS

Measuring the paper thickness in the paper folding exercise is an example of repeated multiplication. Exponents are a shorthand form for writing repeated multiplication. Instead of writing

$$2 \cdot 2 \cdot 2 \cdot 2 \cdot 2 \cdot 2 \cdot 2$$

214

we write

2^7

Instead of writing

$a \cdot a \cdot a \cdot a \cdot a$

we write

We call *a* the **base** of the expression and 5 the **exponent**, or **power**.

a^5

which we read as "*a* to the fifth power."

In general, for any real number *a* and any natural number *n*,

$$a^n = \underbrace{a \cdot a \cdots a}_{n \text{ factors}}$$

An expression of this type is said to be in **exponential form.**

● Example 1

Using Exponential Notation

Write each of the following, using exponential notation.

(*a*) $5y \cdot 5y \cdot 5y = (5y)^3$

(*b*) $w \cdot w \cdot w \cdot w = w^4$

● ● ● **CHECK YOURSELF 1**

Write each of the following, using exponential notation.

(a) $3z \cdot 3z \cdot 3z$ **(b)** $x \cdot x \cdot x \cdot x$

Let's consider what happens when we multiply two expressions in exponential form with the same base.

We expand the expressions and apply the associative property to regroup.

$$a^4 \cdot a^5 = \underbrace{(a \cdot a \cdot a \cdot a)}_{4 \text{ factors}}\underbrace{(a \cdot a \cdot a \cdot a \cdot a)}_{5 \text{ factors}}$$

$$= \underbrace{a \cdot a \cdot a \cdot a \cdot a \cdot a \cdot a \cdot a \cdot a}_{9 \text{ factors}}$$

$$= a^9$$

Notice that the product is simply the base taken to the power that is the sum of the two original exponents. This leads us to our first property of exponents.

> **Product Rule for Exponents**
>
> For any nonzero real number a and positive integers m and n,
>
> $$a^m \cdot a^n = \underbrace{(a \cdot a \cdots a)}_{m \text{ factors}}\underbrace{(a \cdot a \cdots a)}_{n \text{ factors}}$$
>
> $$= \underbrace{a \cdot a \cdots a}_{m + n \text{ factors}}$$
>
> $$= a^{m+n}$$

This is our *first property of exponents*

$a^m \cdot a^n = a^{m+n}$

Example 2 illustrates the product rule for exponents.

● Example 2

Using the Product Rule

Simplify each expression.

(a) $b^4 \cdot b^6 = b^{4+6} = b^{10}$

(b) $(2a)^3 \cdot (2a)^4 = (2a)^{3+4} = (2a)^7$

(c) $(-2)^5(-2)^4 = (-2)^{5+4} = (-2)^9$

(d) $(10^7)(10^{11}) = (10)^{7+11} = (10)^{18}$

● ● ● CHECK YOURSELF 2

Simplify each expression.

(a) $(5b)^6(5b)^5$ (b) $(-3)^4(-3)^3$ (c) $10^8 \cdot 10^{12}$ (d) $(xy)^2(xy)^3$

Applying the commutative and associative properties of multiplication, we know that a product such as

$2x^3 \cdot 3x^2$

can be rewritten as

$(2 \cdot 3)(x^3 \cdot x^2)$

or as

$6x^5$

We expand on these ideas in Example 3.

• Example 3

Using Properties of Exponents

Using the product rule for exponents together with the commutative and associative properties, simplify each expression.

Multiply the coefficients and *add* the exponents by the product rule. With practice you will *not need to write* the regrouping step.

(a) $(x^4)(x^2)(x^3)(x) = x^{10}$

(b) $(3x^4)(5x^2) = (3 \cdot 5)(x^4 \cdot x^2) = 15x^6$

(c) $(2x^5y)(9x^3y^4) = (2 \cdot 9)(x^5 \cdot x^3)(y \cdot y^4) = 18x^8y^5$

(d) $(-3x^2y^2)(-2x^4y^3) = (-3 \cdot -2)(x^2 \cdot x^4)(y^2 \cdot y^3) = 6x^6y^5$

● ● ● ● **CHECK YOURSELF 3**

Simplify each expression.

(a) $(x)(x^5)(x^3)$ **(b)** $(7x^5)(2x^2)$ **(c)** $(-2x^3y)(x^2y^2)$ **(d)** $(-5x^3y^2)(-x^2y^3)$

Now consider the quotient

$$\frac{a^6}{a^4}$$

If we write this in expanded form, we have

$$\frac{\overbrace{a \cdot a \cdot a \cdot a \cdot a \cdot a}^{6 \text{ factors}}}{\underbrace{a \cdot a \cdot a \cdot a}_{4 \text{ factors}}}$$

This can be reduced to

Divide the numerator and denominator by the four common factors of a.

Note that $\dfrac{a}{a} = 1$, where $a \neq 0$.

$$\frac{\not{a} \cdot \not{a} \cdot \not{a} \cdot \not{a} \cdot a \cdot a}{\not{a} \cdot \not{a} \cdot \not{a} \cdot \not{a}} \qquad \text{or} \qquad a^2$$

This means that

$$\frac{a^6}{a^4} = a^2$$

This leads us to our second property of exponents.

Quotient Rule for Exponents

In general, for any real number a ($a \neq 0$) and positive integers m and n,

This is our *second property of exponents*. We write $a \neq 0$ to avoid division by 0.

$$\frac{a^m}{a^n} = a^{m-n}$$

Example 4 illustrates this rule.

● Example 4

Using Properties of Exponents

Simplify each expression.

(a) $\dfrac{x^{10}}{x^4} = x^{10-4} = x^6$

Subtract the exponents, applying the quotient rule.

Note that $a^1 = a$; there is no need to write the exponent 1 because it is understood.

(b) $\dfrac{a^8}{a^7} = a^{8-7} = a$

(c) $\dfrac{63w^8}{7w^5} = 9w^{8-5} = 9w^3$

Divide the coefficients and subtract the exponents.

(d) $\dfrac{-32a^4b^5}{8a^2b} = -4a^{4-2}b^{5-1} = -4a^2b^4$

Divide the coefficients and subtract the exponents for *each* variable.

(e) $\dfrac{10^{16}}{10^6} = 10^{16-6} = 10^{10}$

● ● ● CHECK YOURSELF 4

Simplify each expression.

(a) $\dfrac{y^{12}}{y^5}$ (b) $\dfrac{x^9}{x^8}$ (c) $\dfrac{45r^8}{-9r^6}$ (d) $\dfrac{49a^6b^7}{7ab^3}$ (e) $\dfrac{10^{13}}{10^5}$

The calculator can be used to evaluate expressions that contain exponents. The way you tell the calculator that a number, or variable, is to be raised to a power is to follow the base with the caret key, ⌐∧⌐, then give the exponent.

To evaluate 3^4, the sequence is

⌐3⌐ ⌐∧⌐ ⌐4⌐.

Pressing ⌐enter⌐ gives the simplified result, ⌐81⌐.

● Example 5

Evaluating Expressions with Exponents

Use your calculator to evaluate each expression.

(a) 5^5

Entering

⌐5⌐⌐∧⌐⌐5⌐

we get ⌐3125⌐.

(b) $2^7 \cdot 3^4$

Entering

(2 ^ 7) (3 ^ 4)

we get 10,368 .

(c) $(7x^5)(2x^2)$, where $x = 2$.

Entering

(7 × 2 ^ 5) (2 × 2 ^ 2)

yields 1792 .

● ● ● **CHECK YOURSELF 5**

Use your calculator to evaluate each expression.

(a) 8^4 **(b)** $3^5 \cdot 2^4$ **(c)** $(7x^5)(2x^7)$, where $x = 3$

What happens when a product, such as $(2x)$, is raised to a power? We use the product-power rule.

Product-Power Rule for Exponents
In general, for any real numbers a and b ($a \neq 0$, $b \neq 0$) and positive integer n,
$(ab)^n = a^n b^n$

This is our third property of exponents.

Example 6 illustrates this rule.

● **Example 6**

Using the Product-Power Rule

Simplify each expression

(a) $(2x)^3 = 2^3 x^3 = 8x^3$

(b) $(-3x)^4 = (-3)^4 x^4 = 81x^4$

● ● ● **CHECK YOURSELF 6**

Simplify each expression.

(a) $(3x)^3$ **(b)** $(-2x)^4$

Now, consider the statement

$(3^2)^3$

This could be expanded to

$$(3^2)(3^2)(3^2)$$

and then expanded again to

$$(3)(3)(3)(3)(3)(3) = 3^6$$

This leads us to the power rule for exponents

This is our *fourth property of exponents*.

Power Rule for Exponents

In general, for any real number a ($a \neq 0$) and positive integers m and n,

$$(a^m)^n = a^{mn}$$

● Example 7

Using the Power Rule for Exponents

Simplify each expression.

(a) $(2^5)^3 = 2^{15}$

(b) $(x^2)^4 = x^8$

(c) $(2x^3)^3 = 2^3 x^9 = 8x^9$

● ● ● **CHECK YOURSELF 7**

Simplify each expression.

(a) $(3^4)^3$ **(b)** $(x^2)^6$ **(c)** $(3x^3)^4$

In Example 8, we will use the rules of exponents to evaluate some functions.

● Example 8

Evaluating Functions

Given the function $f(x) = -2x^3 + 3x^2 - 2x$, find the following.

(a) $f(2) = -2(2)^3 + 3(2)^2 - 2(2) = -2(8) + 3(4) - 2(2) = -16 + 12 - 4 = -8$

(b) $f(2a) = -2(2a)^3 + 3(2a)^2 - 2(2a)$

$$= -2(8a^3) + 3(4a^2) - 2(2a)$$

$$= -16a^3 + 12a^2 - 4a$$

● ● ● CHECK YOURSELF 8

Given the function $f(x) = 3x^3 - 3x^2 - 4x$, find

(a) $f(2)$.   **(b)** $f(2a)$.

● ● ● CHECK YOURSELF ANSWERS

1. (a) $(3z)^3$; **(b)** x^4.   **2. (a)** $(5b)^{11}$; **(b)** $(-3)^7$; **(c)** 10^{20}; **(d)** $(xy)^5$.
3. (a) x^9; **(b)** $14x^7$; **(c)** $-2x^5y^3$; **(d)** $5x^5y^5$.   **4. (a)** y^7; **(b)** x; **(c)** $-5r^2$;
(d) $7a^5b^4$; **(e)** 10^8.   **5. (a)** 4096; **(b)** 3888; **(c)** 30,618.
6. (a) $27x^3$; **(b)** $16x^4$.   **7. (a)** 3^{12}; **(b)** x^{12}; **(c)** $81x^{12}$.
8. (a) 4; **(b)** $24a^3 - 12a^2 - 8a$.

3.1 Exercises

In Exercises 1 to 16, simplify cach expression.

1. $x^4 \cdot x^5$ **2.** $x^7 \cdot x^9$ **3.** $x^5 \cdot x^3 \cdot x^2$ **4.** $x^8 \cdot x^4 \cdot x^7$

5. $3^5 \cdot 3^2$ **6.** $(-3)^4(-3)^6$ **7.** $(-2)^3(-2)^5$ **8.** $4^3 \cdot 4^4$

9. $4 \cdot x^2 \cdot x^4 \cdot x^7$ **10.** $3 \cdot x^3 \cdot x^5 \cdot x^8$ **11.** $\left(\dfrac{1}{2}\right)^2\left(\dfrac{1}{2}\right)^3\left(\dfrac{1}{2}\right)$ **12.** $\left(-\dfrac{1}{3}\right)^4\left(-\dfrac{1}{3}\right)\left(-\dfrac{1}{3}\right)^5$

13. $(-2)^2(-2)^3(x^4)(x^5)$ **14.** $(-3)^4(-3)^2(x)^2(x)^6$ **15.** $(2x)^2(2x)^3(2x)^4$ **16.** $(-3x)^3(-3x)^5(-3x)^7$

In Exercises 17 to 28, use the product rule of exponents together with the commutative and associative properties to simplify the products.

17. $(x^2y^3)(x^4y^2)$ **18.** $(x^4y)(x^2y^3)$ **19.** $(x^3y^2)(x^4y^2)(x^2y^3)$

20. $(x^2y^3)(x^3y)(x^4y^2)$ **21.** $(2x^4)(3x^3)(-4x^3)$ **22.** $(2x^3)(-3x)(-4x^4)$

23. $(5x^2)(3x^3)(x)(-2x^3)$ **24.** $(4x^2)(2x)(x^2)(2x^3)$ **25.** $(5xy^3)(2x^2y)(3xy)$

26. $(-3xy)(5x^2y)(-2x^3y^2)$ **27.** $(x^2yz)(x^3y^5z)(x^4yz)$ **28.** $(xyz)(x^8y^3z^6)(x^2yz)(xyz^4)$

In Exercises 29 to 36, use the quotient rule of exponents to simplify each expression.

29. $\dfrac{x^{10}}{x^7}$

30. $\dfrac{b^{23}}{b^{18}}$

31. $\dfrac{x^7 y^{11}}{x^4 y^3}$

32. $\dfrac{x^5 y^9}{xy^4}$

33. $\dfrac{x^5 y^4 z^2}{xy^2 z}$

34. $\dfrac{x^8 y^6 z^4}{x^3 yz^3}$

35. $\dfrac{21x^4 y^5}{7xy^2}$

36. $\dfrac{48x^6 y^6}{12x^3 y}$

 In Exercises 37 to 48, use your calculator to evaluate each expression.

37. 4^3 **38.** 5^7 **39.** $(-3)^4$ **40.** $(-4)^5$

41. $2^3 \cdot 2^5$ **42.** $3^4 \cdot 3^6$

43. $(3x^2)(2x^4)$, where $x = 2$ **44.** $(4x^3)(5x^4)$, where $x = 3$

45. $(2x^4)(4x^2)$, where $x = -2$ **46.** $(3x^5)(2x^3)$, where $x = -3$

47. $(-2x^3)(-3x^5)$, where $x = 2$ **48.** $(-3x^2)(-4x^4)$, where $x = 4$

 In Exercises 49 to 56, use a calculator to evaluate the functions.

49. Given $f(x) = -2x^3 + 7x^2 - 3x + 1$, find $f(-2)$, and $f(2)$.

50. Given $f(x) = -3x^3 - 6x^2 + 5x - 2$, find $f(-2)$, and $f(2)$.

51. Given $f(x) = 2x^4 + 3x^3 - 2x^2 + 7x + 4$, find $f(3)$, and $f(-3)$.

52. Given $f(x) = 4x^4 - 2x^3 + x^2 - 6x - 1$, find $f(2)$, and $f(-1)$.

53. Given $f(x) = -2x^4 + 3x^3 - x^2 - x + 2$, find $f(1)$, $f(0)$, and $f(-2)$.

54. Given $f(x) = 3x^3 - 2x^2 + x - 10$, find $f(0)$, $f(-2)$, and $f(2)$.

55. Given $f(x) = -2x^4 + 6x^3 - 7x^2 + 15$, find $f(2)$, $-f(-3)$, and $f(0)$.

56. Given $f(x) = -3x^4 + 7x^2 - 5x - 10$, find $f(4)$, $f(0)$, and $-f(-4)$.

In Exercises 57 to 78, simplify each expression.

57. $(-3x)(5x^5)$

58. $(-5x^2)(-2x^2)$

59. $(2x)^3$

60. $(-3x)^3$

61. $(x^3)^7$

62. $(-x^3)^5$

63. $(3x)(-2x)^3$

64. $(2x)(-3x)^3$

65. $(2x^3)^5$

66. $(-3x^2)^3$

67. $(-2x^2)^3(3x^2)^3$

68. $(-3x^2)^2(5x^2)^2$

69. $(2x^3)(5x^3)^2$

70. $(3x^3)^2(x^2)^4$

71. $(2x^3)^4(3x^4)^2$

72. $(3x^2)^3(2x^5)^3$

73. $\left(\dfrac{2x^5}{y^3}\right)^2$

74. $\left(\dfrac{2x^5}{3x^8}\right)^3$

75. $(-8x^2y)(-3x^4y^5)^4$

76. $(5x^5y)^2(-3x^3y^4)^3$

77. $\left(\dfrac{3x^4y^9}{2x^2y^7}\right)\left(\dfrac{x^6y^3}{x^3y^2}\right)^2$

78. $\left(\dfrac{6x^5y^4}{5xy}\right)\left(\dfrac{x^3y^2}{xy^3}\right)^3$

79. Business. The value, P, of a savings account that compounds interest annually is given by the formula

$$P = A(1 + r)^t$$

where A = original amount
r = interest rate in decimal form
t = time in years

Find the amount of money in the account after 8 years if $2000 was invested initially at 5% compounded annually.

80. Business. Using the formula for compound interest in Exercise 79, determine the amount of money in the account if the original investment is doubled.

81. You have learned rules for working with exponents when multiplying, dividing, and raising an expression to a power.

(a) Explain each rule in your own words. Give numerical examples.
(b) Is there a rule for raising a *sum* to a power? That is, does $(a + b)n = an + bn$? Use numerical examples to explain why this is true in general or why it is not. Is it always true or always false?

82. Work with another student to investigate the rate of inflation. The rate of inflation has been about 3% from 1990 to 1997. This means that the value of the goods that you could buy for $1 in 1990 would cost 3% more in 1991, 3% more than that in 1992, etc. If a movie ticket cost $5.50 in 1990, what would it cost today if movie tickets just kept up with inflation? Make a table and figure out when the cost of a movie ticket would be double what it was in 1990. How many years would it take for the cost to double again? Do this same exercise, but consider inflation to be 6% a year. If the inflation rate were 6%, what price would you pay for a movie ticket in the year 2020, a little more than 20 years from now?

83. Prime numbers—numbers that cannot be factored into factors other than 1 and the number—have fascinated people for centuries. (In this century, prime numbers have found practical uses in security codes and encrypting systems.) Over the centuries, several formulas have been proposed for generating prime numbers. One such formula is in the form $2n - 1$. Try some positive integer values for n and see if you get a prime number when the expression is evaluated. Do all values of n give a prime number? Do *any* values of n give a prime number? Make a table and see if you can discern a pattern of values for n that can be depended on to produce prime numbers. Write your conjecture in a complete sentence.

Answers

1. x^9 **3.** x^{10} **5.** 3^7 **7.** $(-2)^8$ **9.** $4x^{13}$ **11.** $\left(\dfrac{1}{2}\right)^6$ **13.** $(-2)^5 x^9$ **15.** $(2x)^9$ **17.** $x^6 y^5$

19. $x^9 y^7$ **21.** $-24x^{10}$ **23.** $-30x^9$ **25.** $30x^4 y^5$ **27.** $x^9 y^7 z^3$ **29.** x^3 **31.** $x^3 y^8$ **33.** $x^4 y^2 z$

35. $3x^3 y^3$ **37.** 64   **39.** 81   **41.** 256   **43.** 384   **45.** 512   **47.** 1536   **49.** 51, 7

51. 250, 46   **53.** 1, 2, −56   **55.** 3, −372, 15   **57.** $-15x^6$ **59.** $8x^3$ **61.** x^{21} **63.** $-24x^4$

65. $32x^{15}$ **67.** $-216x^{12}$ **69.** $50x^9$ **71.** $144x^{20}$ **73.** $\dfrac{4x^{10}}{y^6}$ **75.** $-648x^{18} y^{21}$ **77.** $\dfrac{3x^8 y^4}{2}$

79. $2954.91   **81.**   **83.**

Adding and Subtracting Polynomials

3.2 OBJECTIVES

1. Use the language of polynomials
2. Evaluate $f(x) + g(x)$ for a given x
3. Add and subtract polynomial functions

In Chapter 1, we looked at a class of functions called *linear functions*. In this section, we examine polynomial functions. We begin by defining some important words.

> A **term** is a number or the product of a number and one or more variables, raised to a power.

• Example 1

Identifying Terms

Which of the following are terms?

$$5x^3 \qquad \frac{7}{x} \qquad 2x^2 + 3x \qquad 4xy$$

$5x^3$, $\dfrac{7}{x}$, and $4xy$ are all terms. $2x^2 + 3x$ is not a term; it is the sum of two terms.

● ● ● CHECK YOURSELF 1

Which of the following are terms?

(a) $\dfrac{5x}{3y}$ (b) $4x^3 - 2y$ (c) $2x^3y^2$ (d) x^7

If terms contain exactly the same variables raised to the *same powers*, they are called **like terms**. Examples include $6s$ and $7s$, $4x^2$ and $9x^2$, and $7xy^2z^3$ and $10xy^2z^3$. The following are *not* like terms

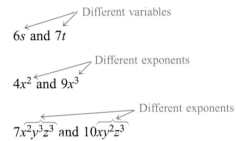

● Example 2

Identifying Like Terms

For each of the following pairs of terms, decide whether they are like terms.

(*a*) $5x^3$ and $5x^2$

(*b*) $3xy$ and $\sqrt{2}\, xy$

(*c*) $4xy^2z$ and $9xy^2z$

(*d*) $2xy^2$ and $7x^2y$

(*e*) $3x^2y$ and $2x^2z$

Solution

(*a*) Not like terms—different exponents.

(*b*) Like terms.

(*c*) Like terms.

(*d*) Not like terms—different exponents.

(*e*) Not like terms—different variables.

● ● ●   **CHECK YOURSELF 2**

For each of the following pairs of terms, decide whether they are like terms.

(a) $2ab^2c$ and $3ab^2d$

(b) $4xy^3$ and $\frac{1}{2}xy^3$

(c) $5x^2y^3z^2$ and $7x^3y^2z^2$

(d) $3x$ and $4xy$

A **polynomial** consists of one or more terms in which the only allowable exponents are the whole numbers, 0, 1, 2, 3, . . . and so on. The terms are connected by addition or subtraction signs.

Certain polynomials occur enough that they are given special names according to the number of terms that they have.

The prefix "mono" means 1.

A polynomial with one term is called a **monomial.** For example,

$7x^3 \qquad 2x^2y^3 \qquad 4xy \qquad -12$

are all monomials. But,

$\dfrac{9}{x} \qquad 5\sqrt{x}$

are not monomials because the exponent in the first term is -1 and the variable in the second term is under the radical (we will see in Chapter 5 that the exponent in this case is $\left(\dfrac{1}{2}\right)$.

The prefix "bi" means 2 and "tri" means 3.

A polynomial with two terms is a **binomial.** A polynomial with three terms is called a **trinomial.**

The numerical factor in a polynomial is called the **numerical coefficient,** or more simply the **coefficient,** of that term. For example, in the polynomials

$8x^4 \qquad 9x^3y^4 \qquad 6x^2y^7 \qquad -10xy$

the numerical coefficients are 8, 9, 6, and -10.

● **Example 3**

Classifying Polynomials

Remember that $\dfrac{1}{x} = x^{-1}$

Which of the following are polynomials? Classify the polynomials as monomial, binomial, or trinomial.

(a) $5x^2y$

(b) $3m + 5n$

(c) $4a^3 + 3a - 2$

(d) $5y^2 - \dfrac{2}{x}$

Solution

(a) Monomial.

(b) Binomial.

(c) Trinomial.

(d) $5y^2 - \dfrac{2}{x}$ is not a polynomial since the exponent on x is -1.

● ● ● **CHECK YOURSELF 3**

Which of the following are polynomials? Classify the polynomials as monomial, binomial, or trinomial.

(a) $5x^2 - 6x$ **(b)** $8x^5$ **(c)** $5x^3 - 3xy + 7y^2$ **(d)** $9x - \dfrac{3}{x}$

It is also useful to classify polynomials by their **degree.**

> The *degree* of a monomial is the sum of the exponents of the variable factors.

● Example 4

Determining the Degree of a Monomial

(a) $5x^2$ has degree 2.

(b) $7n^5$ has degree 5.

(c) $4a^2b^4$ has degree 6. (The sum of the powers, 2 and 4, is 6.)

(d) 9 has degree 0 (because $9 = 9 \cdot 1 = 9x^0$).

● ● ● **CHECK YOURSELF 4**

Give the degree of each monomial.

(a) $4x^2$ **(b)** $7x^3y^2$ **(c)** $8p^2s$ **(d)** 5

> The *degree* of a polynomial is that of the term with the highest degree.

• Example 5

Determining the Degree of a Polynomial

(a) $7x^3 - 5x^2 + 5$ has degree 3.

(b) $5y^7 - 3y^2 + 5y - 7$ has degree 7.

(c) $4a^2b^3 - 5abc^2$ has degree 5 because the sum of the variable powers in the lead term ($4a^2b^3$) is 5.

Polynomials such as those in Examples 4(a) and (b) are called **polynomials in one variable,** and they are usually written in *descending form* so that the power of the variable decreases from left to right. In that case, the coefficient of the first term is called the **leading coefficient.**

CHECK YOURSELF 5

Give the degree of each polynomial. For those polynomials in one variable, write in descending form and give the leading coefficient.

(a) $7x^4 - 5xy + 2$ **(b)** 5 **(c)** $4x^2 - 7x^3 - 8x + 5$

A **polynomial function** is a function in which the expression on the right-hand side is a polynomial expression. For example,

$$f(x) = 3x^3 + 2x^2 - x + 5$$
$$g(x) = 2x^3 - 5x - 1$$
$$P(x) = 2x^7 + 2x^6 - x^3 + 2x^4 + 5x^2 - 6$$

are all polynomial functions. Because they are functions, every x value determines a unique ordered pair.

• Example 6

Finding Ordered Pairs

Given $f(x) = 3x^3 + 2x^2 - x + 5$ and $g(x) = 2x^3 - 5x - 1$, find the following ordered pairs.

(a) $(0, f(0))$

To find $f(0)$, we substitute 0 for x in the function $f(x) = 3x^3 + 2x^2 - x + 5$.

$$f(0) = 3(0)^3 + 2(0)^2 - (0) + 5$$
$$= 0 + 0 - 0 + 5$$
$$= 5$$

Therefore, $(0, f(0)) = (0, 5)$.

(b) $(2, f(2))$

$$f(2) = 3(2)^3 + 2(2)^2 - (2) + 5$$
$$= 24 + 8 - 2 + 5$$
$$= 35$$

Therefore, $(2, f(2)) = (2, 35)$.

(c) $(-2, f(-2))$

$f(-2) = 3(-2)^3 + 2(-2)^2 - (-2) + 5$

$\qquad = -24 + 8 + 2 + 5$

$\qquad = -9$

Therefore, $(-2, f(-2)) = (-2, -9)$.

(d) $(0, g(0))$

$g(0) = 2(0)^3 - 5(0) - 1$

$\qquad = -1$

Therefore, $(0, g(0)) = (0, -1)$.

(e) $(2, g(2))$

$g(2) = 2(2)^3 - 5(2) - 1$

$\qquad = 16 - 10 - 1$

$\qquad = 5$

Therefore, $(2, g(2)) = (2, 5)$.

(f) $(-2, g(-2))$

$g(-2) = 2(-2)^3 - 5(-2) - 1$

$\qquad = -16 + 10 - 1$

$\qquad = -7$

Therefore, $(-2, g(-2)) = (-2, -7)$.

● ● ● CHECK YOURSELF 6

Given $f(x) = x^3 - 7x + 1$ and $g(x) = x^3 + 2x^2 - 3x - 4$, find the following ordered pairs.

(a) $(0, f(0))$ (b) $(2, f(2))$ (c) $(-2, f(-2))$

(d) $(0, g(0))$ (e) $(2, g(2))$ (f) $(-2, g(-2))$

Example 7 demonstrates that the sum of two polynomial functions is a polynomial function.

● Example 7

Adding Two Polynomial Functions

Given $f(x) = 3x^3 + 2x^2 - x + 5$ and $g(x) = 2x^3 - 5x - 1$, and letting $h(x) = f(x) + g(x)$, find $h(x)$.

$h(x) = f(x) + g(x)$

$\qquad = (3x^3 + 2x^2 - x + 5) + (2x^3 - 5x - 1)$ Substitute the function values.

$\qquad = (3x^3 + 2x^3) + 2x^2 + (-5x - x) + (5 - 1)$ Collect like terms.

$\qquad = 5x^3 + 2x^2 - 6x + 4$

● ● ●   **CHECK YOURSELF 7**

Given $f(x) = x^3 - 7x + 1$ and $g(x) = x^3 + 2x^2 - 3x - 4$, and letting $h(x) = f(x) + g(x)$, find $h(x)$.

Example 8 demonstrates a method by which we can check our results when we add two polynomials.

● **Example 8**

Adding Two Polynomial Functions

Given $f(x) = 3x^3 + 2x^2 - x + 5$ and $g(x) = 2x^3 - 5x - 1$, and letting $h(x) = f(x) + g(x)$, find the following.

(a) $f(2) + g(2)$

From Example 6, we know that $f(2) = 35$ and $g(2) = 5$, so

$$f(2) + g(2) = 35 + (5)$$
$$= 40$$

(b) Use the result from Example 7 to find $h(2)$

From Example 7, we have

$$h(x) = 5x^3 + 2x^2 - 6x + 4$$

Therefore,

$$h(2) = 5(2)^3 + 2(2)^2 - 6(2) + 4$$
$$= 40 + 8 - 12 + 4$$
$$= 40$$

Note that $h(2) = f(2) + g(2)$. This helps us confirm that we have correctly added the polynomials.

● ● ●   **CHECK YOURSELF 8**

Given $f(x) = x^3 - 7x + 1$ and $g(x) = x^3 + 2x^2 - 3x - 4$, and letting $h(x) = f(x) + g(x)$, complete the following.

(a) Find $f(2) + g(2)$.   **(b)** Use the result of Check Yourself 7 to find $h(2)$.
(c) Compare the results of **(a)** and **(b).**

Subtracting polynomials proceeds in a similar fashion. We view the subtraction of a quantity as adding the opposite of that quantity, which is an application of the distributive property.

Both statements are just applications of the distributive property. This shows us that the *opposite* of $a + b$ is $-a - b$ and that the *opposite* of $a - b$ is $-a + b$.

Distribution of the Negative

$$-(a + b) = -a - b$$

and

$$-(a - b) = -a + b$$

We can now go on to subtracting polynomials.

• Example 9

Subtracting Polynomial Functions

Given $f(x) = 7x^2 - 2x$ and $g(x) = 4x^2 + 5x$, and letting $h(x) = f(x) - g(x)$, find the following.

(*a*) $h(x)$

$$
\begin{aligned}
h(x) &= f(x) - g(x) \\
&= (7x^2 - 2x) - (4x^2 + 5x) \\
&= (7x^2 - 4x^2) + (-2x - 5x) \\
&= 3x^2 - 7x
\end{aligned}
$$

(*b*) $f(2) - g(2)$

$$
\begin{aligned}
f(2) - g(2) &= (7(2)^2 - 2(2)) - (4(2)^2 + 5(2)) \\
&= (28 - 4) - (16 + 10) \\
&= (24) - (26) \\
&= -2
\end{aligned}
$$

(*c*) Use the results of part (*a*) to find $h(2)$.

$$
\begin{aligned}
h(2) &= 3(2)^2 - 7(2) \\
&= 12 - 14 \\
&= -2
\end{aligned}
$$

As was the case with addition, we have found an easy way to check our work when we are subtracting polynomial functions. We have found that $(2, h(2)) = (2, -2) = (2, f(2) - g(2))$.

● ● ● CHECK YOURSELF 9

Given $f(x) = 8x^2 - x$ and $g(x) = x^2 - 2x$, and letting $h(x) = f(x) - g(x)$, find the following.

(a) $h(x)$.   **(b)** $f(2) - g(2)$.   **(c)** Use the result of part (*a*) to find $h(2)$.

Example 10 requires that we distribute the subtraction to a negative coefficient.

• Example 10

Subtracting Polynomial Functions

Given $f(x) = 5x^2 + 2x$ and $g(x) = 3x^2 - 5x$, and letting $h(x) = f(x) - g(x)$, find the following.

(a) $h(x)$

$$h(x) = f(x) - g(x)$$
$$= (5x^2 + 2x) - (3x^2 - 5x)$$
$$= (5x^2 + 2x) + (-3x^2 + 5x)$$
$$= (5x^2 - 3x^2) + (2x + 5x)$$
$$= 2x^2 + 7x$$

(b) $f(1) - g(1)$

$$f(1) - g(1) = (5(1)^2 + 2(1)) - (3(1)^2 - 5(1))$$
$$= (5 + 2) - (3 - 5)$$
$$= (7) - (-2)$$
$$= 9$$

(c) Use the result of part (a) to find $h(1)$.

$$h(1) = 2(1)^2 + 7(1)$$
$$= 2 + 7$$
$$= 9$$

We see that $(1, h(1)) = (1, 9) = (1, f(1) - g(1))$.

● ● ● **CHECK YOURSELF 10**

Given $f(x) = 3x^2 - 4x$ and $g(x) = 6x^2 - x$, and letting $h(x) = f(x) - g(x)$, find the following.

(a) $h(x)$.   **(b)** $f(1) - g(1)$.   **(c)** Use the results of part (a) to find $h(1)$.

● ● ● **CHECK YOURSELF ANSWERS**

1. (a) Term; **(b)** not a term; **(c)** term; **(d)** term.   **2. (a)** Not like terms;
(b) like terms; **(c)** not like terms; **(d)** not like terms.   **3. (a)** Binomial;
(b) monomial; **(c)** trinomial; **(d)** not a polynomial.   **4. (a)** 2; **(b)** 5; **(c)** 3; **(d)** 0.
5. (a) Degree is 4; **(b)** degree is 0; **(c)** degree is 3, $-7x^3 + 4x^2 - 8x + 5$,
lead coefficient is -7.   **6. (a)** (0, 1); **(b)** (2, -5); **(c)** (-2, 7); **(d)** (0, -4);

(e) $(2, 6)$; **(f)** $(-2, 2)$.   **7.** $2x^3 + 2x^2 - 10x - 3$.   **8. (a)** 1; **(b)** 1;
(c) $h(2) = f(2) - g(2)$.   **9. (a)** $h(x) = 7x^2 + x$; **(b)** 30; **(c)** 30.
10. (a) $h(x) = -3x^2 - 3x$; **(b)** -6; **(c)** -6.

3.2 Exercises

In Exercises 1 to 10, identify the following as a monomial, binomial, or trinomial. Give the degree of each polynomial.

1. $4x - 3$

2. -5

3. $4y^3$

4. $5x^7 - 3$

5. $4x^6 + 2x - 5$

6. $x^2 - 2x + 3$

7. $5 - 2x$

8. x^2y^3z

9. $2x^2y^3 + 3x^2y$

10. $5x^3y - 3x^2y + 7xy$

In Exercises 11 to 18, the polynomial functions $f(x)$ and $g(x)$ are given. Find the ordered pairs.

(a) $(0, f(0))$, **(b)** $(2, f(2))$, **(c)** $(-2, f(-2))$, **(d)** $(0, g(0))$, **(e)** $(2, g(2))$, **(f)** $(-2, g(-2))$

11. $f(x) = 2x^2 + 3x - 5$ and $g(x) = 4x^2 - 5x - 7$

12. $f(x) = 3x^2 + 7x - 9$ and $g(x) = 7x^2 + 8x - 9$

13. $f(x) = x^3 + 8x^2 - 4x + 10$ and $g(x) = 3x^3 + 4x^2 - 5x - 3$

14. $f(x) = -x^3 + 3x^2 - 7x + 8$ and $g(x) = -2x^3 + 4x^2 - 9x - 2$

15. $f(x) = -4x^3 - 5x^2 + 8x - 12$ and $g(x) = -x^3 + 5x^2 + 7x - 14$

16. $f(x) = 7x^3 + 5x - 9$ and $g(x) = 4x^3 + 3x^2 + 8$

17. $f(x) = 10x^3 + 5x^2 - 3x$ and $g(x) = 8x^3 + 4x^2 + 9$

18. $f(x) = 7x^2 - 5$ and $g(x) = 3x^3 + 5x^2 + 6x + 9$

In Exercises 19 to 26, $f(x)$ and $g(x)$ are given. Let $h(x) = f(x) + g(x)$. Find **(a)** $h(x)$, **(b)** $f(1) + g(1)$, and **(c)** use the results of part **(a)** to find $h(1)$.

19. $f(x) = 5x - 3$ and $g(x) = 4x + 7$ **20.** $f(x) = 7x^2 + 3x$ and $g(x) = 5x^2 - 7x$

21. $f(x) = 5x^2 + 3x$ and $g(x) = 4x + 2x^2$ **22.** $f(x) = 8x^2 - 3x + 10$ and $g(x) = 7x^2 + 2x - 12$

23. $f(x) = -3x^2 - 5x - 7$ and $g(x) = 2x^2 + 3x + 5$

24. $f(x) = 2x^3 + 5x^2 + 8$ and $g(x) = -5x^3 - 2x^2 + 7x$

25. $f(x) = -5x^2 - 3x - 15$ and $g(x) = 5x^3 - 8x - 10$

26. $f(x) = 5x^2 + 12x - 5$ and $g(x) = 3x^3 + 7x - 9$

In Exercises 27 to 34, $f(x)$ and $g(x)$ are given. Let $h(x) = f(x) - g(x)$. Find **(a)** $h(x)$, **(b)** $f(1) - g(1)$, and **(c)** use the results of part **(a)** to find $h(1)$.

27. $f(x) = 7x + 10$ and $g(x) = 5x - 3$ **28.** $f(x) = 5x - 12$ and $g(x) = 8x - 7$

29. $f(x) = 7x^2 - 3x$ and $g(x) = -5x^2 - 2x$ **30.** $f(x) = -10x^2 + 3x$ and $g(x) = -3x^2 - 3x$

31. $f(x) = 8x^2 - 5x - 7$ and $g(x) = 5x^2 - 3x$ **32.** $f(x) = 5x^3 - 2x^2 - 8x$ and $g(x) = 7x^3 - 3x^2 - 8$

33. $f(x) = 5x^2 - 5$ and $g(x) = 8x^2 - 7x$ **34.** $f(x) = 5x^2 - 7x$ and $g(x) = 9x^2 - 3$

In Exercises 35 to 46, simplify each polynomial function.

35. $f(x) = (2x - 3) + (4x + 7) - (2x - 3)$ **36.** $f(x) = (5x - 2) - (2x + 3) + (7x - 7)$

37. $f(x) = (5x^2 - 2x + 7) - (2x^2 + 3x + 1) - (4x^2 + 3x + 3)$

38. $f(x) = (8x^2 - 8x + 5) + (3x^2 + 2x - 7) - (x^2 - 6x + 4)$

39. $f(x) = (8x^2 - 3) + (5x^2 + 7x) - (7x - 8)$ **40.** $f(x) = (9x^2 + 7x) - (5x^2 - 8x) - (4x^2 + 3)$

41. $f(x) = x - [5x - (x - 3)]$ **42.** $f(x) = x^2 - [7x^2 - (x^2 + 7)]$

43. $f(x) = (2x - 3) - [x - (2x + 7)]$ **44.** $f(x) = (3x^2 + 5x) - [x^2 - (2x^2 - 3x)]$

45. $f(x) = 2x - \{3x + 2[x - 2(x - 3)]\}$

46. $f(x) = 3x - \{5x - 2[x - 3(x + 4)]\}$

Suppose that revenue is given by the polynomial $R(x)$ and cost is given by the polynomial $C(x)$. Profit $P(x)$ can then be found with the formula

$$P(x) = R(x) - C(x)$$

In Exercises 47 to 50, find the polynomial representing profit in each expression.

47. $R(x) = 100x$
$C(x) = 2000 + 50x$

48. $R(x) = 250x$
$C(x) = 5000 + 175x$

49. $R(x) = 100x + 2x^2$
$C(x) = 2000 + 50x + 5x^2$

50. $R(x) = 250x + 5x^2$
$C(x) = 5000 + 175x + 10x^2$

51. If $P(x) = -2x + 1$, find $P(a + h) - P(a)$.

52. If $P(x) = 5x$, find $\dfrac{P(a + h) - P(a)}{h}$.

53. Find the difference when $4x^2 + 2x + 1$ is subtracted from the sum of $x^2 - 2x - 3$ and $3x^2 + 5x - 7$.

54. Subtract $7x^2 + 5x - 3$ from the sum of $2x^2 - 5x + 7$ and $-9x^2 - 2x + 5$.

55. Find the difference when $7x^2 - 3x + 2$ is subtracted from the sum of $2x^2 + 5x - 4$ and $6x^2 - 7x + 9$.

56. Subtract $4x^2 + 3x - 8$ from the sum of $2x^2 + 4x + 3$ and $5x^2 - 3x - 7$.

57. Subtract $8x^2 - 2x$ from the sum of $x^2 - 5x$ and $7x^2 + 5$.

58. Find the difference when $9a^2 - 7$ is subtracted from the sum of $5a^2 - 5$ and $-2a^2 - 2$.

59. The length of a rectangle is 1 cm more than twice its width. Represent the width of the rectangle by w, and write a polynomial to express the perimeter of the rectangle in terms of w. Be sure to simplify your result.

60. One integer is 2 more than twice the first. Another is 3 less than 3 times the first. Represent the first integer by x, and then write a polynomial to express the sum of the three integers in terms of x. Be sure to simplify your result.

Let $P(x) = 2x^3 - 3x^2 + 5x - 5$ and $Q(x) = -x^2 + 2x - 2$. In Exercises 61 and 62, find each of the following.

61. $P[Q(1)]$

   Hint: First find $Q(1)$. Then evaluate $P(x)$ for that value.

62. $Q[P(1)]$

 In Exercises 63 to 66, use your calculator to approximate the value of $f(x)$ for each given x. Express your answer to the nearest integer.

$$f(x) = 12x^5 - 16x^3 + 3x^2 + 5x - 9$$

63. $f(-4)$ **64.** $f\left(\dfrac{7}{3}\right)$ **65.** $f(\sqrt{2})$ **66.** $f(\pi)$

 67. Copy the following patterns. Cut out the circle and the rectangle, and tape them together to make a right cylinder (like an empty can). Find the surface area of the cylinder.

68. Copy the following pattern. Cut it out and fold it to form a prism. Find the surface area of the prism.

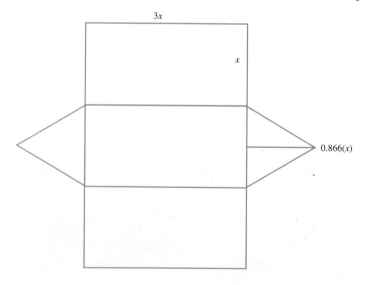

 69. The number 3078 can be written as the polynomial

$$3(10)^3 + 0(10)^2 + 7(10)^1 + 8(10)^0$$

because 10 is the *base* of the number system we commonly use. All numbers can be written as a polynomial. Interpret the following polynomials by writing them the way they would normally appear.

   (a) $7(10)^4 + 5(10)^3 + 0(10)^2 + 2(10) + 0(10)^0 = $ _____

(b) $4(10)^2 + 2(10) + 3(10)^{-1} + 2(10)^{-2} + 5(10)^{-3} =$ _____

Write these numbers as polynomials:

(c) $6525 =$ _____

(d) $99.95 =$ _____

70. In Exercise 69, the first number in the list could be written with a variable in place of the 10: $7(n)^4 + 5(n)^3 + 0(n)^2 + 2(n) + 0(n)^0$. The number could still be written as 75020, but the value of the number would be very different from 75 thousand 20 if the *base* were different. Try $n = 8$ and calculate the value of 75020_8. This is read "75020 base 8". Did you get 31248?

The number 11011 in base 10 is eleven thousand eleven. Written as a polynomial:

$$1(n)^4 + 1(n)^3 + 0(n)^2 + 1(n) + 1(n)^0$$

If $n = 2$, we have another value for 11011 but this time in *base 2*, another very widely used number system because it is used by computers. Evaluate 11011_2 in base 10.

Write the following as polynomials and then evaluate the numbers in base 10:

(a) $546302_7 =$ _____ $=$ _____ in base 10

(b) $111100111_2 =$ _____ $=$ _____ in base 10

(c) $21112_3 =$ _____ $=$ _____ in base 10

(d) $21112_5 =$ _____ $=$ _____ in base 10

You may want to find out more about base 2 numbers or the binomial number system because versions of it are widely used in computers, bar code scanners, and other electronic devices.

Answers

1. Binomial, degree is 1   **3.** Monomial, degree is 3   **5.** Trinomial, degree is 6   **7.** Binomial, degree is 1
9. Binomial, degree is 5   **11. (a)** $(0, -5)$, **(b)** $(2, 9)$, **(c)** $(-2, -3)$, **(d)** $(0, -7)$, **(e)** $(2, -1)$, **(f)** $(-2, 19)$
13. (a) $(0, 10)$, **(b)** $(2, 42)$, **(c)** $(-2, 42)$, **(d)** $(0, -3)$, **(e)** $(2, 27)$, **(f)** $(-2, -1)$ **15. (a)** $(0, -12)$,
(b) $(2, -48)$, **(c)** $(-2, -16)$, **(d)** $(0, -14)$, **(e)** $(2, 12)$, **(f)** $(-2, 0)$ **17. (a)** $(0, 0)$, **(b)** $(2, 94)$, **(c)** $(-2, -54)$,
(d) $(0, 9)$, **(e)** $(2, 89)$, **(f)** $(-2, -39)$ **19. (a)** $9x + 4$, **(b)** 13, **(c)** 13   **21. (a)** $7x^2 + 7x$, **(b)** 14, **(c)** 14
23. (a) $-x^2 - 2x - 2$, **(b)** -5, **(c)** -5 **25. (a)** $5x^3 - 5x^2 - 11x - 25$, **(b)** -36, **(c)** -36 **27. (a)** $2x + 13$,
(b) 15, **(c)** 15   **29. (a)** $12x^2 - x$, **(b)** 11, **(c)** 11   **31. (a)** $3x^2 - 2x - 7$, **(b)** -6, **(c)** -6
33. (a) $-3x^2 + 7x - 5$, **(b)** -1, **(c)** -1 **35.** $4x + 7$ **37.** $-x^2 - 8x + 3$ **39.** $13x^2 + 5$ **41.** $-3x - 3$
43. $3x + 4$ **45.** $x - 12$ **47.** $50x - 2000$ **49.** $-3x^2 + 50x - 2000$ **51.** $-2h$ **53.** $x - 11$
55. $x^2 + x + 3$ **57.** $-3x + 5$ **59.** $6w + 2$ **61.** -15 **63.** $\approx -11{,}245$ **65.** ≈ 27 **67.**
69. (a) 75,020, **(b)** 420.325, **(c)** $6(10)^3 + 5(10)^2 + 2(10) + 5(10)^0$,
(d) $9(10)^1 + 9(10)^0 + 9(10)^{-1} + 5(10)^{-2}$

3.3 Multiplying Polynomials and Special Products

3.3 OBJECTIVES

1. Evaluate $f(x) \cdot g(x)$ for a given x
2. Multiply two polynomial functions
3. Square a polynomial
4. Find the product of two binomials as a difference of squares

In Section 3.1, you saw the first exponent property and used that property to multiply monomials. Let's review.

• Example 1

Multiplying Monomials

Multiply.

Remember: $a^m a^n = a^{m+n}$

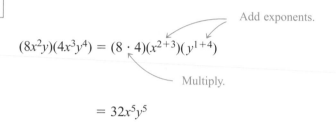

Add exponents.

$(8x^2y)(4x^3y^4) = (8 \cdot 4)(x^{2+3})(y^{1+4})$

Multiply.

Note the use of the commutative and associative properties to "regroup" and "reorder" the factors.

$$= 32x^5y^5$$

CHECK YOURSELF 1

Multiply.

(a) $(4a^3b)(9a^3b^2)$

(b) $(-5m^3n)(7mn^5)$

We now want to extend the process to multiplying polynomial functions.

• Example 2

Multiplying a Monomial and a Binomial Function

Given $f(x) = 5x^2$ and $g(x) = 3x^2 - 5x$, and letting $h(x) = f(x) \cdot g(x)$, find $h(x)$.

$h(x) = f(x) \cdot g(x)$

$\quad = 5x^2 \cdot (3x^2 - 5x)$ Apply the distributive property.

$\quad = 5x^2 \cdot 3x^2 - 5x^2 \cdot 5x$

$\quad = 15x^4 - 25x^3$

CHECK YOURSELF 2

Given $f(x) = 3x^2$ and $g(x) = 4x^2 + x$, and letting $h(x) = f(x) \cdot g(x)$, find $h(x)$.

238

We can check this result by comparing the values of $h(x)$ and of $f(x) \cdot g(x)$ for a specific value of x. This is illustrated in Example 3.

● Example 3

Multiplying a Monomial and a Binomial Function

Given $f(x) = 5x^2$ and $g(x) = 3x^2 - 5x$, and letting $h(x) = f(x) \cdot g(x)$, compare $f(1) \cdot g(1)$ with $h(1)$.

$$f(1) \cdot g(1) = 5(1)^2 \cdot (3(1)^2 - 5(1))$$
$$= 5(3 - 5)$$
$$= 5(-2)$$
$$= -10$$

From Example 2, we know that

$$h(x) = 15x^4 - 25x^3$$

So

$$h(1) = 15(1)^4 - 25(1)^3$$
$$= 15 - 25$$
$$= -10$$

Therefore, $h(1) = f(1) \cdot g(1)$.

● ● ● CHECK YOURSELF 3

Given $f(x) = 3x^2$ and $g(x) = 4x^2 + x$, and letting $h(x) = f(x) \cdot g(x)$, compare $f(1) \cdot g(1)$ with $h(1)$.

The distributive property is also used to multiply two polynomial functions. To consider the pattern, let's start with the product of two binomial functions.

● Example 4

Multiplying Binomial Functions

Given $f(x) = x + 3$ and $g(x) = 2x + 5$, and letting $h(x) = f(x) \cdot g(x)$, find the following.

(a) $h(x)$

$$
\begin{aligned}
h(x) &= f(x) \cdot g(x) \\
&= (x + 3)(2x + 5) && \text{Apply the distributive property.} \\
&= (x + 3)(2x) + (x + 3)(5) && \text{Apply the distributive property again.} \\
&= (x)(2x) + (3)(2x) + (x)(5) + (3)(5) \\
&= 2x^2 + 6x + 5x + 15 \\
&= 2x^2 + 11x + 15
\end{aligned}
$$

Notice that this ensures that each term in the first polynomial is multiplied by each term in the second polynomial.

(b) $f(1) \cdot g(1)$

$$
\begin{aligned}
f(1) \cdot g(1) &= (1 + 3)(2(1) + 5) \\
&= 4(7) \\
&= 28
\end{aligned}
$$

(c) $h(1)$

From part (a), we have $h(x) = 2x^2 + 11x + 15$, so

$$
\begin{aligned}
h(1) &= 2(1)^2 + 11(1) + 15 \\
&= 2 + 11 + 15 \\
&= 28
\end{aligned}
$$

Again, we see that $h(1) = f(1) \cdot g(1)$.

● ● ● ● **CHECK YOURSELF 4**

Given $f(x) = 3x - 2$ and $g(x) = x + 3$, and letting $h(x) = f(x) \cdot g(x)$, find the following.

(a) $h(x)$ **(b)** $f(1) \cdot g(1)$ **(c)** $h(1)$

Certain products occur frequently enough in algebra that it is worth learning special formulas for dealing with them. Consider these products of two equal binomial factors.

$a^2 \cdot 2ab + b^2$

and

$a^2 - 2ab \cdot b^2$

are called **perfect-square trinomials**.

$$
\begin{aligned}
(a + b)^2 &= (a + b)(a + b) \\
&= a^2 + 2ab + b^2 && \text{(1)} \\
(a - b)^2 &= (a - b)(a - b) \\
&= a^2 - 2ab + b^2 && \text{(2)}
\end{aligned}
$$

We can summarize these statements as follows.

Squaring a Binomial

The square of a binomial has three terms. It is the sum of (1) the square of the first term, (2) twice the product of the two terms, and (3) the square of the last term.

$(a + b)^2 = a^2 + 2ab + b^2$

and

$(a - b)^2 = a^2 - 2ab + b^2$

• Example 5

Squaring a Binomial

Find each of the following binomial squares.

Be sure to write out the expansion in detail.

(a) $(x + 5)^2 = x^2 + 2(x)(5) + 5^2$

Square of first term   Twice the product   Square of last term
of the two terms

$= x^2 + 10x + 25$

CAUTION

Be Careful! A very common mistake in squaring binomials is to forget *the middle* term!

$(y + 7)^2$

is not equal to

$y^2 + (7)^2$

The correct square is

$y^2 - 14y - 49$

$(b - 6)^2$

is not equal to

$b^2 - (6)^2$

The correct square is $b^2 - 12b = 36$. The square of a binomial is *always* a trinomial

(b) $(2a - 7)^2 = (2a)^2 - 2(2a)(7) + (-7)^2$

$= 4a^2 - 28a + 49$

● ● ● **CHECK YOURSELF 5**

Find each of the following binomial squares.

(a) $(x + 8)^2$ **(b)** $(3x - 5)^2$

Another special product involves binomials that differ only in sign. It will be extremely important in your work later in this chapter on factoring. Consider the following:

$(a + b)(a - b) = a^2 - ab + ab + b^2$

$= a^2 - b^2$

Product of Binomials Differing in Sign

$(a + b)(a - b) = a^2 - b^2$

In words, the product of two binomials that differ only in the signs of their second terms is the difference of the squares of the two terms of the binomials.

•Example 6

Finding a Special Product

Multiply.

(a) $(x - 3)(x + 3) = x^2 - (3)^2$
$$= x^2 - 9$$

CAUTION

The entire term $2x$ is squared, not just the x.

(b) $(2x - 3y)(2x + 3y) = (2x)^2 - (3y)^2$
$$= 4x^2 - 9y^2$$

(c) $(5a + 4b^2)(5a - 4b^2) = (5a)^2 - (4b^2)^2$
$$= 25a^2 - 16b^4$$

● ● ●   **CHECK YOURSELF 6**

Find each of the following products.

(a) $(y + 5)(y - 5)$ **(b)** $(2x - 3)(2x + 3)$ **(c)** $(4r + 5s^2)(4r - 5s^2)$

This format ensures that each term of one polynomial multiplies each term of the other.

When multiplying two polynomials that don't fit one of the special product patterns, there are two different ways to set up the multiplication. Example 7 will illustrate the vertical approach.

•Example 7

Multiplying Polynomials

Multiply $3x^3 - 2x^2 + 5$ and $3x + 2$.

Step 1
$$\begin{array}{r} 3x^3 - 2x^2 + \qquad 5 \\ 3x + \; 2 \\ \hline 6x^3 - 4x^2 \qquad + 10 \end{array}$$

Multiply by 2.

Step 2
$$\begin{array}{r} 3x^3 - 2x^2 + \qquad\quad 5 \\ 3x + \; 2 \\ \hline 6x^3 - 4x^2 + \qquad 10 \\ 9x^4 - 6x^3 \qquad + 15x \end{array}$$

Multiply by $3x$. Note that we align the terms in the partial product.

Step 3
$$\begin{array}{r} 3x^3 - 2x^2 + \qquad\quad 5 \\ 3x + \; 2 \\ \hline 6x^3 - 4x^2 + \qquad 10 \\ 9x^4 - 6x^3 \qquad + 15x \\ \hline 9x^4 \qquad\quad - 4x^2 + 15x + 10 \end{array}$$

Add the partial products.

● ● ● **CHECK YOURSELF 7**

Find the following product, using the vertical method.

$(4x^3 - 6x - 7)(3x - 2)$

A horizontal approach to the multiplication in Example 7 is also possible by the distributive property. As we see in Example 8, we first distribute $3x$ over the trinomial and then we distribute 2 over the trinomial.

● Example 8

Multiplying Polynomials

Multiply $(3x + 2)(3x^3 - 2x^2 + 5)$, using a horizontal format.

Again, this ensures that each term of one polynomial multiplies each term of the other.

Step 1

$(3x + 2)(3x^3 - 2x^2 + 5)$

Step 2

$= \underbrace{9x^4 - 6x^3 + 15x}_{\text{Step 1}} + \underbrace{6x^3 - 4x^2 - 10}_{\text{Step 2}}$ Combine like terms.

$= 9x^4 - 4x^2 + 15x + 10$ Write the product in desending form.

● ● ● **CHECK YOURSELF 8**

Find the product of Check Yourself 7, using a horizontal format.

Multiplication sometimes involves the product of more than two polynomials. In such cases, the associative property of multiplication allows us to choose the order of multiplication that we find easiest. Generally, we choose to start with the product of binomials. Example 9 illustrates this approach.

● Example 9

Multiplying Polynomials

Find the products.

(a) $x(x + 3)(x - 3) = x(x^2 - 9)$ Find the product $(x + 3)(x - 3)$.

$= x^3 - 9x$ Then distribute x as the last step.

(b) $2x(x + 3)(2x - 1) = 2x(2x^2 - 5x - 3)$ Find the product of the binomials.

$= 4x^3 + 10x^2 - 6x$ Then distribute $2x$.

● ● ●   **CHECK YOURSELF 9**

Find each of the following products.

(a) $m(2m + 3)(2m - 3)$ **(b)** $3a(2a + 5)(a - 3)$

● ● ●   **CHECK YOURSELF ANSWERS**

1. (a) $36a^6b^3$, **(b)** $-35m^4n^6$.   **2.** $h(x) = 12x^4 + 3x^3$.

3. $f(1) \cdot g(1) = 15 = h(1)$.   **4. (a)** $h(x) = 3x^2 + 7x - 6$, **(b)** $f(1) \cdot g(1) = 4$,

(c) $h(1) = 4$.   **5. (a)** $x^2 + 16x + 64$, **(b)** $9x^2 - 30x + 25$.   **6. (a)** $y^2 - 25$,

(b) $4x^2 - 9$, **(c)** $16r^2 - 25s^4$.   **7.** $12x^4 - 8x^3 - 18x^2 - 9x + 14$.

8. $12x^4 - 8x^3 - 18x^2 - 9x + 14$.   **9. (a)** $4m^3 - 9m$, **(b)** $6a^3 - 3a^2 - 45a$.

Pascal's Triangle. The triangle below is sometimes called "Pascal's Triangle" after Blaise Pascal, a mathematician and philosopher who lived in France in the 1600s. He is credited with developing the triangle to answer a question about odds in gambling. The same triangle shows up in Chinese manuscripts dating from the early 1300s when it was called The Old Method Chart of the Seven Multiplying Squares and gave the binomial coefficients up to the eighth power.

1. Look at the triangle. Can you find the pattern to complete the next row?

$$
\begin{array}{ccccccccccccc}
 & & & & & & 1 & & & & & & \\
 & & & & & 1 & & 1 & & & & & \\
 & & & & 1 & & 2 & & 1 & & & & \\
 & & & 1 & & 3 & & 3 & & 1 & & & \\
 & & 1 & & 4 & & 6 & & 4 & & 1 & & \\
 & 1 & & 5 & & 10 & & 10 & & 5 & & 1 & \\
1 & & 6 & & 15 & & 20 & & 15 & & 6 & & 1 \\
\end{array}
$$

— — — — — — — —

2. This triangle gives the binomial coefficients up to the eighth power. How does this work? Multiply the following and see where the multiplication and the chart agree.

$(x + 1)^0 = $ _____

$(x + 1)^1 = $ _____

$(x + 1)^2 = $ _____

$(x + 1)^3 = $ _____

$(x + 1)^4 = $ _____

$(x + 1)^5 = $ _____

How does the triangle help with these products?

3. What would be the product of $(x + 1)^6$?

4. Can you see how the triangle would help with $(2x + 5y)^3$? Explain.

3.3 Exercises

In Exercises 1 to 6, multiply each polynomial.

1. $(4x)(5y)$

2. $(-3m)(5n)$

3. $(6x^2)(-3x^3)$

4. $(5y^4)(3y^2)$

5. $(5r^2s)(6r^3s^4)$

6. $(-8a^2b^5)(-3a^3b^2)$

In Exercises 7 to 14, $f(x)$ and $g(x)$ are given. Let $h(x) = f(x) \cdot g(x)$. Find **(a)** $h(x)$, **(b)** $f(1) \cdot g(1)$, and **(c)** use the result of **(a)** to find $h(1)$.

7. $f(x) = 3x$ and $g(x) = 2x^2 - 3x$

8. $f(x) = 4x$ and $g(x) = 2x^2 - 7x$

9. $f(x) = -5x$ and $g(x) = -3x^2 - 5x + 8$

10. $f(x) = 2x^2$ and $g(x) = -7x^2 + 2x$

11. $f(x) = 4x^3$ and $g(x) = 9x^2 + 3x - 5$

12. $f(x) = 2x^3$ and $g(x) = 2x^3 - 4x$

13. $f(x) = 3x$ and $g(x) = 5x^2 - 4x$

14. $f(x) = -x^2$ and $g(x) = -7x^3 - 5x^2$

In Exercises 15 to 24, multiply each polynomial expression.

15. $(x + y)(x + 3y)$

16. $(x - 3y)(x + 5y)$

17. $(x - 2y)(x + 7y)$

18. $(x + 7y)(x - 3y)$

19. $(5x - 7y)(5x - 9y)$

20. $(3x - 5y)(7x + 2y)$

21. $(7x - 5y)(7x - 4y)$

22. $(9x + 7y)(3x - 2y)$

23. $(5x^2 - 2y)(3x + 2y^2)$

24. $(6x^2 - 5y^2)(3x^2 - 2y)$

In Exercises 25 to 38, multiply polynomial expressions using the special product formulas.

25. $(x + 5)^2$

26. $(x - 7)^2$

27. $(2x - 3)^2$

28. $(5x + 3)^2$

29. $(4x - 3y)^2$

30. $(7x - 5y)^2$

31. $(4x + 3y^2)^2$

32. $(3x^3 - 7y)^2$

33. $(x - 3y)(x + 3y)$

34. $(x + 5y)(x - 5y)$

35. $(2x - 3y)(2x + 3y)$

36. $(5x + 3y)(5x - 3y)$

37. $(4x^2 + 3y)(4x^2 - 3y)$

38. $(7x - 6y^2)(7x + 6y^2)$

In Exercises 39 to 42, multiply using the vertical format.

39. $(3x - y)(x^2 + 3xy - y^2)$

40. $(5x + y)(x^2 - 3xy + y^2)$

41. $(x - 2y)(x^2 + 2xy + 4y^2)$

42. $(x + 3y)(x^2 - 3xy + 9y^2)$

In Exercises 43 to 46, simplify each function.

43. $f(x) = x(x - 3)(x + 1)$

44. $f(x) = x(x + 4)(x - 2)$

45. $f(x) = 2x(x - 5)(x + 4)$

46. $f(x) = x^2(x - 4)(x^2 + 5)$

 47. Does $(2x + 3)^2 = 4x^2 + 9$? If not, explain why and give the correct answer.

 48. What are some advantages of using the vertical format in multiplying polynomials?

 49. What are the advantages of using the horizontal format in multiplying polynomials?

 50. Explain why the products $(x - y)^2$ and $(y - x)^2$ are the same.

51. You are given three integers such that the second integer is 3 more than the first and the third integer is 1 less than twice the first. Represent the first integer by x. Then write and simplify the polynomial that represents the product of the three integers.

52. The length of a box is 2 cm more than its width. The height is 3 cm less than twice its width. Represent the width of the box by w. Then write and simplify the polynomial that represents the volume of the box.

If the polynomial $p(x)$ represents the selling price of an object, then the polynomial $R(x)$, where $R(x) = x \cdot p(x)$, is the revenue produced by selling x objects. Use this information to solve Exercises 53 and 54.

53. If $p(x) = 100 - 0.2x$, find $R(x)$. Find $R(50)$.

54. If $p(x) = 250 - 0.5x$, find $R(x)$. Find $R(20)$.

Note that $(28)(32) = (30 - 2)(30 + 2) = 900 - 4 = 896$. In Exercises 55 to 60, use the difference-of-squares formula to find the products.

55. $(49)(51)$

56. $(27)(33)$

57. $(34)(26)$

58. $(98)(102)$

59. $(55)(65)$

60. $(56)(64)$

Answers

1. $20xy$ **3.** $-18x^5$ **5.** $30r^5s^5$ **7.** (a) $6x^3 - 9x^2$, (b) -3, (c) -3 **9.** (a) $15x^3 + 25x^2 - 40x$, (b) 0, (c) 0   **11.** (a) $36x^5 + 12x^4 - 20x^3$, (b) 28, (c) 28   **13.** (a) $15x^5 - 12x^4$, (b) 3, (c) 3   **15.** $x^2 + 4xy + 3y^2$ **17.** $x^2 + 5xy - 14y^2$ **19.** $25x^2 - 80xy + 63y^2$ **21.** $49x^2 - 63xy + 20y^2$

23. $15x^3 + 10x^2y^2 - 6xy - 4y^3$ **25.** $x^2 + 10x + 25$ **27.** $4x^2 - 12x + 9$ **29.** $16x^2 - 24xy + 9y^2$
31. $16x^2 + 24xy^2 + 9y^4$ **33.** $x^2 - 9y^2$ **35.** $4x^2 - 9y^2$ **37.** $16x^4 - 9y^2$ **39.** $3x^3 + 8x^2y - 6xy^2 + y^3$
41. $x^3 - 8y^3$ **43.** $x^3 - 2x^2 - 3x$ **45.** $2x^3 - 2x^2 - 40x$ **47.** **49.**
51. $2x^3 + 5x^2 - 3x$ **53.** $100x - 0.2x^2, 4500$ **55.** 2499
57. 884 **59.** 3575

3.4 Factoring Polynomials and Special Polynomials

3.4 OBJECTIVES

1. Remove the greatest common factor (GCF)
2. Factor by grouping
3. Factor the difference of two squares
4. Factor the sum and difference of two cubes

Recall that a prime number is any integer greater than 1 that has only itself and 1 as factors. Writing

$15 = 3 \cdot 5$

as a product of prime factors is called the **completely factored form** for 15.

In fact, we will see that factoring out the GCF is the *first* method to try in any of the factoring problems we will discuss.

When the integers 3 and 5 are multiplied, the product is 15. We call 3 and 5 the **factors** of 15.

Writing $3 \cdot 5 = 15$ indicates multiplication, but when we write $15 = 3 \cdot 5$, we say we have **factored** 15. In general, factoring is the reverse of multiplication. We can extend this idea to algebra.

From the last section on multiplying polynomials and special products, we know that

$$(2x + 3)(x - 2) = 2x^2 - x - 6$$

But what if we begin with $2x^2 - x - 6$? How do we write the polynomial as a product of other polynomials?

There are a number of methods that we can use to factor polynomials: factoring out the greatest common factor (GCF), factoring by grouping, factoring the difference of two squares, and factoring the sum or difference of two cubes.

FACTORING OUT THE GREATEST COMMON FACTOR (GCF)

The first step in factoring is always to factor out the **greatest common factor (GCF)**, if any.

> The *greatest common factor (GCF)* of a polynomial is the monomial with the highest degree and the largest numerical coefficient that is a factor of each term of the polynomial.

Once the GCF is found, we apply the distributive property to write the original polynomial as a product of the GCF and the polynomial formed by dividing each term by that GCF. Example 1 illustrates this approach.

• Example 1

Factoring Out a Monomial

Here 4 is the GCF of the numerical coefficients, and the highest common power of x is 1, so the GCF of $4x^3 - 12x$ is $4x$.

(a) Factor $4x^3 - 12x$.

Note that the numerical coefficient of the GCF is 4 and the variable factor is x (the highest power common to each term). So

$$4x^3 - 12x = 4x \cdot x^2 - 4x \cdot 3$$
$$= 4x(x^2 - 3)$$

Here 6 is the GCF of the numerical coefficients, the highest common power of a is 2, and the highest common power of b is 2.

(b) Factor $6a^3b^2 - 12a^2b^3 + 24a^4b^4$.

Here the GCF is $6a^2b^2$, and we can write

$$6a^3b^2 - 12a^2b^3 + 24a^4b^4$$
$$= 6a^2b^2 \cdot a - 6a^2b^2 \cdot 2b + 6a^2b^2 \cdot 4a^2b^2$$
$$= 6a^2b^2(a - 2b + 4a^2b^2)$$

(c) Factor $8m^4n^2 - 16m^2n^2 + 24mn^3 - 32mn^4$.

Here the GCF is $8mn^2$, and we have

$$8m^4n^2 - 16m^2n^2 + 24mn^3 - 32mn^4$$
$$= 8mn^2 \cdot m^3 - 8mn^2 \cdot 2m + 8mn^2 \cdot 3n - 8mn^2 \cdot 4n^2$$
$$= 8mn^2(m^3 - 2m + 3n - 4n^2)$$

Notice that in Example 1(b) it is also true that

$$6a^3b^2 - 12a^2b^3 + 24a^4b^4 = 3ab(2a^2b - 4ab^2 + 8a^3b^3)$$

However, this is not in *completely factored form* since we agree that this means factoring out the GCF (that monomial with the largest possible coefficient and degree). In this case, we must remove $6a^2b^2$.

• • • CHECK YOURSELF 1

Write each of the following in completely factored form.

(a) $7x^3y - 21x^2y^2 + 28xy^3$ **(b)** $15m^4n^4 - 5mn^3 + 20mn^2 - 25m^2n^2$

FACTORING BY GROUPING

A related factoring method is called **factoring by grouping.** We introduce this method in Example 2.

• Example 2

Finding a Common Factor

(*a*) Factor $3x(x + y) + 2(x + y)$.

We see that *the binomial $x + y$* is a common factor and can be removed.

Because of the commutative property, the factors can be written in either order.

$$3x(x + y) + 2(x + y)$$
$$= (x + y) \cdot 3x + (x + y) \cdot 2$$
$$= (x + y)(3x + 2)$$

(*b*) Factor $3x^2(x - y) + 6x(x - y) + 9(x - y)$.

We note that here the GCF is $3(x - y)$. Factoring as before, we have

$$3(x - y)(x^2 + 2x + 3)$$

● ● ● CHECK YOURSELF 2

Completely factor each of the polynomials.

(a) $7a(a - 2b) + 3(a - 2b)$ **(b)** $4x^2(x + y) - 8x(x + y) - 16(x + y)$

If the terms of a polynomial have no common factor (other than 1), factoring by grouping is the preferred method, as illustrated in Example 3.

• Example 3

Factoring by Grouping Terms

Suppose we want to factor the polynomial

Note that our example has *four* terms. That is the clue for trying the factoring by grouping method.

$$ax - ay + bx - by$$

As you can see, the polynomial has no common factors. However, look at what happens if we separate the polynomial into *two groups* of *two terms*.

$$ax - ay + bx - by$$
$$= \underbrace{ax - ay}_{(1)} + \underbrace{bx - by}_{(2)}$$

Now *each* group has a common factor, and we can write the polynomial as

$$a(x - y) + b(x - y)$$

In this form, we can see that $x - y$ is the GCF. Factoring out $x - y$, we get

$$a(x - y) + b(x - y) = (x - y)(a + b)$$

● ● ● **CHECK YOURSELF 3**

Use the factoring by grouping method.

$x^2 - 2xy + 3x - 6y$

Be particularly careful of your treatment of algebraic signs when applying the factoring by grouping method. Consider Example 4.

● Example 4

Factoring by Grouping Terms

Factor $2x^3 - 3x^2 - 6x + 9$.

We group the polynomial as follows.

$$\underbrace{2x^3 - 3x^2}_{(1)} \underbrace{- 6x + 9}_{(2)}$$ Remove the common factor of -3 from the second two terms.

Note that $9 = (-3)(-3)$.

$$= x^2(2x - 3) - 3(2x - 3)$$
$$= (2x - 3)(x^2 - 3)$$

● ● ● **CHECK YOURSELF 4**

Factor by grouping.

$3y^3 + 2y^2 - 6y - 4$

It may also be necessary to change the order of the terms as they are grouped. Look at Example 5.

● Example 5

Factoring by Grouping Terms

Factor $x^2 - 6yz + 2xy - 3xz$.

Grouping the terms as before, we have

$$\underbrace{x^2 - 6yz}_{(1)} + \underbrace{2xy - 3xz}_{(2)}$$

Do you see that we have accomplished nothing because there are no common factors in the first group?

We can, however, rearrange the terms to write the original polynomial as

$$\underbrace{x^2 + 2xy}_{(1)} - \underbrace{3xz - 6yz}_{(2)}$$

$$= x(x + 2y) - 3z(x + 2y)$$ We can now remove the common factor of $x + 2y$ in group (1) and group (2).

$$= (x + 2y)(x - 3z)$$

Note: It is often true that the grouping can be done in more than one way. The factored form will be the same.

● ● ●   **CHECK YOURSELF 5**

We can write the polynomial of Example 5 as

$$x^2 - 3xz + 2xy - 6yz$$

Factor, and verify that the factored form is the same in either case.

FACTORING THE DIFFERENCE OF TWO SQUARES

Another factoring method uses the **difference of two squares.**

The Difference of Two Squares

$$a^2 - b^2 = (a + b)(a - b) \tag{1}$$

The product of the sum and difference of two terms gives the *difference of two squares.*

Equation (1) is easy to apply in factoring. It is just a matter of recognizing a binomial as the difference of two squares.

CAUTION

What about the sum of two squares, such as

$$x^2 + 25$$

In general, it is *not possible* to factor (using real numbers) a sum of two squares. So

$$(x^2 + 25) \neq (x + 5)(x + 5)$$

To confirm this identity, use the FOIL method to multiply

$$(a + b)(a - b)$$

We are looking for perfect squares—the exponents must be multiples of 2 and the coefficients perfect squares—1, 4, 9, 16, and so on.

● Example 6

Factoring the Difference of Two Squares

(*a*) Factor $x^2 - 25$.

Note that our example has two terms—a clue to try factoring as the difference of two squares.

$$x^2 - 25 = (x)^2 - (5)^2$$

$$= (x + 5)(x - 5)$$

(*b*) Factor $9a^2 - 16$.

$$9a^2 - 16 = (3a)^2 - (4)^2$$

$$= (3a + 4)(3a - 4)$$

(c) Factor $25m^4 - 49n^2$.

$$25m^4 - 49n^2 = (5m^2)^2 - (7n)^2$$
$$= (5m^2 + 7n)(5m^2 - 7n)$$

● ● ● **CHECK YOURSELF 6**

Factor each of the following binomials.

(a) $y^2 - 36$ **(b)** $25m^2 - n^2$ **(c)** $16a^4 - 9b^2$

We mentioned earlier that factoring out a common factor should always be considered your first step. Then other steps become obvious. Consider Example 7.

● Example 7

Factoring the Difference of Two Squares

Factor $a^3 - 16ab^2$.

First note the common factor of a. Removing that factor, we have

$$a^3 - 16ab^2 = a(a^2 - 16b^2)$$

We now see that the binomial factor is a difference of squares, and we can continue to factor as before. So

$$a^2 - 16ab^2 = a(a + 4b)(a - 4b)$$

● ● ● **CHECK YOURSELF 7**

Factor $2x^3 - 18xy^2$.

You may also have to apply the difference of two squares method *more than once* to completely factor a polynomial.

● Example 8

Factoring the Difference of Two Squares

Factor $m^4 - 81n^4$.

$$m^4 - 81n^4 = (m^2 + 9n^2)(m^2 - 9n^2)$$

Do you see that we are not done in this case? Since $m^2 - 9n^2$ is still factorable, we can continue to factor as follows.

Note The other binomial factor, $m^2 + 9n^2$, is a *sum of two squares*, which cannot be factored further.

$$m^4 - 81n^4 = (m^2 + 9n^2)(m + 3n)(m - 3n)$$

● ● ● CHECK YOURSELF 8

Factor $x^4 - 16y^4$.

FACTORING THE SUM OR DIFFERENCE OF TWO CUBES

Two additional methods for factoring certain binomials include finding the sum or difference of two cubes.

The Sum or Difference of Two Cubes

Be sure you take the time to expand the product on the right-hand side to confirm the identity.

$$a^3 + b^3 = (a + b)(a^2 - ab + b^2) \tag{2}$$

$$a^3 - b^3 = (a - b)(a^2 + ab + b^2) \tag{3}$$

● Example 9

Factoring the Sum or Difference of Two Cubes

We are now looking for perfect cubes—the exponents must be multiples of 3 and the coefficients perfect cubes—1, 8, 27, 64, and so on.

(*a*) Factor $x^3 + 27$.

The first term is the cube of x, and the second is the cube of 3, so we can apply equation (2). Letting $a = x$ and $b = 3$, we have

$$x^3 + 27 = (x + 3)(x^2 - 3x + 9)$$

Again, looking for a *common factor* should be your first step.

(*b*) Factor $8w^3 - 27z^3$.

This is a difference of cubes, so use equation (3).

$$8w^3 - 27z^3 = (2w - 3z)[(2w)^2 + (2w)(3z) + (3z)^2]$$

$$= (2w - 3z)(4w^2 + 6wz + 9z^2)$$

Remember to write the GCF as a part of the final factored form

(*c*) Factor $5a^3b - 40b^4$.

First note the common factor of $5b$. The binomial is the difference of cubes, so use equation (3).

$$5a^3b - 40b^4 = 5b(a^3 - 8b^3)$$

$$= 5b(a - 2b)(a^2 + 2ab + 4b^2)$$

● ● ●   **CHECK YOURSELF 9**

Factor completely.

(a) $27x^3 + 8y^3$

(b) $3a^4 - 24ab^3$

In each example in this section, we factored a polynomial expression. If we are given a polynomial function to factor, there is no change in the ordered pairs represented by the function after it is factored.

● **Example 10**

Factoring a Polynomial Function

Given the function $f(x) = 9x^2 + 15x$, complete the following.

(a) Find $f(1)$.

$f(1) = 9(1)^2 + 15(1)$

$\quad = 9 + 15$

$\quad = 24$

(b) Factor $f(x)$.

$f(x) = 9x^2 + 15x$

$\quad = 3x(3x + 5)$

(c) Find $f(1)$ from the factored form of $f(x)$.

$f(1) = 3(1)(3(1) + 5)$

$\quad = 3(8)$

$\quad = 24$

● ● ●   **CHECK YOURSELF 10**

Given the function $f(x) = 16x^5 + 10x^2$, complete the following.

(a) Find $f(1)$.   **(b)** Factor $f(x)$.   **(c)** Find $f(1)$ from the factored form of $f(x)$.

● ● ●   **CHECK YOURSELF ANSWERS**

1. (a) $7xy(x^2 - 3xy + 4y^2)$; **(b)** $5mn^2(3m^3n^2 - n + 4 - 5m)$.
2. $(a - 2b)(7a + 3)$; **(b)** $4(x + y)(x^2 - 2x - 4)$.   **3.** $(x - 2y)(x + 3)$.
4. $(3y + 2)(y^2 - 2)$.   **5.** $(x - 3z)(x + 2y)$.   **6. (a)** $(y + 6)(y - 6)$;
(b) $(5m + n)(5m - n)$; **(c)** $(4a^2 + 3b)(4a^2 - 3b)$.   **7.** $2x(x + 3y)(x - 3y)$.
8. $(x^2 + 4y^2)(x + 2y)(x - 2y)$.   **9. (a)** $(3x + 2y)(9x^2 - 6xy + 4y^2)$;
(b) $3a(a - 2b)(a^2 + 2ab + 4b^2)$.   **10. (a)** 26; **(b)** $2x^2(8x^3 + 5)$; **(c)** 26.

3.4 Exercises

In Exercises 1 to 54, completely factor each polynomial.

1. $6x + 9y$

2. $7a - 21b$

3. $4x^2 - 12x$

4. $5a^2 + 25a$

5. $18m^2n + 27mn^2$

6. $24c^2d^3 - 30c^3d^2$

7. $5x^3 - 15x^2 + 25x$

8. $28r^3 - 21r^2 + 7r$

9. $12m^3n - 6mn + 18mn^2$

10. $18w^2z + 27wz - 36wz^2$

11. $4a^3b^2 - 8a^2b + 12ab^2 - 4ab$

12. $9r^3r^3 + 27r^3s^2 - 6r^2s^2 + 3rs$

13. $x(y - z) + 3(y - z)$

14. $2a(c - d) - b(c - d)$

15. $3(m - n) + 5(m - n)^2$

16. $4(r + 2s) - 3(r + 2s)^2$

17. $5x^2(x - y) - 10x(x - y) + 15(x - y)$

18. $7a^2(a + 2b) + 21a(a + 2b) - 14(a + 2b)$

19. $x^2 - 49$

20. $m^2 - 64$

21. $a^2 - 81$

22. $b^2 - 36$

23. $9p^2 - 1$

24. $4x^2 - 9$

25. $25a^2 - 16$

26. $16m^2 - 49$

27. $x^2y^2 - 25$

28. $m^2n^2 - 9$

29. $4c^2 - 25d^2$

30. $9a^2 - 49b^2$

31. $49p^2 - 64q^2$

32. $25x^2 - 36y^2$

33. $x^4 - 16y^2$

34. $a^2 - 25b^4$

35. $a^3 - 4ab^2$

36. $9p^2q - q^3$

37. $a^4 - 16b^4$

38. $81x^4 - y^4$

39. $x^3 + 64$

40. $y^3 - 8$

41. $m^3 - 125$

42. $b^3 + 27$

43. $a^3b^3 - 27$

44. $p^3q^3 - 64$

45. $8w^3 + z^3$

46. $c^3 - 27d^3$

47. $r^3 - 64s^3$

48. $125x^3 + y^3$

49. $8x^3 - 27y^3$

50. $64m^3 - 27n^3$

51. $8x^3 + y^6$

52. $m^6 - 27n^3$

53. $4x^3 - 32y^3$

54. $3a^3 + 81b^3$

In Exercises 55 to 60, factor each polynomial by grouping.

55. $ab - ac + b^2 - bc$

56. $ax + 2a + bx + 2b$

57. $6r^2 + 12rs - r - 2s$

58. $2mn - 4m^2 + 3n - 6m$ **59.** $ab^2 - 2b^2 + 3a - 6$ **60.** $r^2s^2 - 3s^2 - 2r^2 + 6$

In Exercises 61 to 64, factor each polynomial by grouping. *Hint:* Consider a rearrangement of terms.

61. $x^2 - 10y - 5xy + 2x$ **62.** $a^2 - 12b + 3ab - 4a$

63. $m^2 - 6n^3 + 2mn^2 - 3mn$ **64.** $r^2 - 3rs^2 - 12s^3 + 4rs$

In Exercises 65 to 72, factor each polynomial completely. *Hint:* Try factoring by grouping as your first step.

65. $x^3 + 3x^2 - 4x - 12$ **66.** $a^3 - 5a^2 - 9a + 45$ **67.** $8x^3 + 12x^2 - 2x - 3$

68. $18b^3 - 9b^2 - 2b + 1$ **69.** $9a^3 + 27a^2 - 4a - 12$ **70.** $4m^3 + 12m^2 - 25m - 75$

71. $x^4 - x^3y + xy^3 - y^4$ **72.** $a^4 + a^3b - 8ab^3 - 8b^4$

For each of the functions in Exercises 73 to 78, **(a)** find $f(1)$, **(b)** factor $f(x)$, and **(c)** find $f(1)$ from the factored form of $f(x)$.

73. $f(x) = 12x^5 + 21x^2$ **74.** $f(x) = -6x^3 - 10x$ **75.** $f(x) = -8x^5 + 20x$

76. $f(x) = 5x^5 - 35x^3$ **77.** $f(x) = x^5 + 3x^2$ **78.** $f(x) = 6x^6 - 16x^5$

In Exercises 79 to 82, factor. *Hint:* Consider *three* groups of *two* terms.

79. $x^3 - x^2 + 3x + x^2y - xy + 3y$ **80.** $m^3 - m^2 - 4m + 2m^2n - 2mn - 8n$

81. $a^3 - a^2b - 3a^2 + 3ab + 3a - 3b$ **82.** $r^3 + 2r^2s + r^2 + 2rs - 3r - 6s$

 83. For the monomials x^4y^2, x^8y^6, and x^9y^4, explain how you can determine the GCF by inspecting exponents.

 84. It is not possible to use the grouping method to factor $2x^3 + 6x^2 + 8x + 4$. Is it correct to conclude that the polynomial is prime? Justify your answer.

85. The area of a rectangle of length l is given by $36l - l^2$. Factor the expression and determine the width of the rectangle.

86. The area of a rectangle of width w is given by $18w^2 - 5w$. Factor the expression and determine the width of the rectangle.

87. Verify the formula for factoring the sum of two cubes by finding the product $(a + b)(a^2 - ab + b^2)$.

88. Verify the formula for factoring the difference of two cubes by finding the product $(a - b)(a^2 + ab + b^2)$.

89. What are the characteristics of a monomial that is a perfect cube?

90. Suppose you factored the polynomial $4x^2 - 16$ as follows:

$$4x^2 - 16 = (2x + 4)(2x - 4)$$

Would this be in completely factored form? If not, what would be the final form?

91. Completely factor $x^6 - y^6$ by considering the binomial as the difference of squares

$$(x^3)^2 - (y^3)^2$$

92. Completely factor $x^6 - y^6$ by considering the binomial as the difference of cubes

$$(x^2)^3 - (y^2)^3$$

93. Compare your results in Exercises 91 and 92. What can you conclude about the factors of

$$x^4 + x^2y^2 + y^4$$

94. In the drawing below, squares of length $\dfrac{y}{2}$ are cut from the corners of a sheet of cardboard with sides of length x.

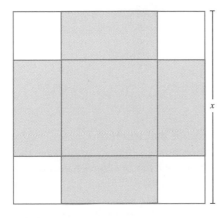

Show that the shaded area can be represented by $x^2 - y^2$.

95. Show that the shaded area in Exercise 94 can also be represented by

$$(x - y)^2 + 2y(x - y)$$

96. Factor and simplify your result in Exercise 95. Compare your result with that in Exercise 94. What factoring technique does this verify?

Answers

1. $3(2x + 3y)$ **3.** $4x(x - 3)$ **5.** $9mn(2m + 3n)$ **7.** $5x(x^2 - 3x + 5)$ **9.** $6mn(2m^2 - 1 + 3n)$

11. $4ab(a^2b - 2a + 3b - 1)$ **13.** $(y - z)(x + 3)$ **15.** $(m - n)(3 + 5m - 5n)$ **17.** $5(x - y)(x^2 - 2x + 3)$

19. $(x + 7)(x - 7)$ **21.** $(a + 9)(a - 9)$ **23.** $(3p + 1)(3p - 1)$ **25.** $(5a + 4)(5a - 4)$

27. $(xy + 5)(xy - 5)$ **29.** $(2c + 5d)(2c - 5d)$ **31.** $(7p + 8q)(7p - 8q)$ **33.** $(x^2 + 4y)(x^2 - 4y)$

35. $a(a + 2b)(a - 2b)$ **37.** $(a^2 + 4b^2)(a + 2b)(a - 2b)$ **39.** $(x + 4)(x^2 - 4x + 16)$

41. $(m - 5)(m^2 + 5m + 25)$ **43.** $(ab - 3)(a^2b^2 + 3ab + 9)$ **45.** $(2w + z)(4w^2 - 2wz + z^2)$

47. $(r - 4s)(r^2 + 4rs + 16s^2)$ **49.** $(2x - 3y)(4x^2 + 6xy + 9y^2)$ **51.** $(2x + y^2)(4x^2 - 2xy^2 + y^4)$

53. $4(x - 2y)(x^2 + 2xy + 4y^2)$ **55.** $(b - c)(a + b)$ **57.** $(r + 2s)(6r - 1)$ **59.** $(a - 2)(b^2 + 3)$

61. $(x + 2)(x - 5y)$ **63.** $(m - 3n)(m + 2n^2)$ **65.** $(x + 3)(x + 2)(x - 2)$ **67.** $(2x + 3)(2x + 1)(2x - 1)$

69. $(a + 3)(3a + 2)(3a - 2)$ **71.** $(x - y)(x + y)(x^2 - xy + y^2)$ **73.** (a) 33, (b) $3x^2(4x^3 + 7)$, (c) 33

75. (a) 12, (b) $4x(-2x^4 + 5)$, (c) 12   **77.** (a) 4, (b) $x^2(x^3 + 3)$, (c) 4   **79.** $(x + y)(x^2 - x + 3)$

81. $(a - b)(a^2 - 3a + 3)$ **83.**   **85.** $36 - l$ **87.** $a^3 + b^3$ **89.**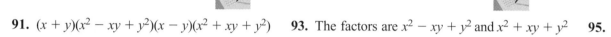

91. $(x + y)(x^2 - xy + y^2)(x - y)(x^2 + xy + y^2)$ **93.** The factors are $x^2 - xy + y^2$ and $x^2 + xy + y^2$ **95.**

Factoring Trinomials

The product of two binomials of the form

$$(\_x + \_)(\_x + \_)$$

will always be a trinomial. In your earlier mathematics classes, you used the FOIL method to find the product of two binomials. In this section, we will use the factoring by grouping method to find the binomial factors for a trinomial.

First, let's look at some factored trinomials.

● Example 1

Matching Trinomials and Their Factors

Determine which of the following are true statements.

(a) $x^2 - 2x - 8 = (x - 4)(x + 2)$

This is a true statement. Using the FOIL method, we see that

$$(x - 4)(x + 2) = x^2 + 2x - 4x - 8$$
$$= x^2 - 2x - 8$$

(b) $x^2 - 6x + 5 = (x - 2)(x - 3)$
Not a true statement, since

$$(x - 2)(x - 3) = x^2 - 3x - 2x + 6$$
$$= x^2 - 5x + 6$$

(c) $x^2 + 5x - 14 = (x - 2)(x + 7)$
True, since

$$(x - 2)(x + 7) = x^2 + 7x - 2x - 14$$
$$= x^2 + 5x - 14$$

(d) $x^2 - 8x - 15 = (x - 5)(x - 3)$
False, since

$$(x - 5)(x - 3) = x^2 - 3x - 5x + 15$$
$$= x^2 - 8x + 15$$

● ● ● **CHECK YOURSELF 1**

Determine which of the following are true statements.

(a) $2x^2 - 2x - 3 = (2x - 3)(x + 1)$ **(b)** $3x^2 + 11x - 4 = (3x - 1)(x + 4)$
(c) $2x^2 - 7x + 3 = (x - 3)(2x - 1)$

The first step in learning to factor a trinomial is to identify its coefficients. To be consistent, we first write the trinomial in standard $ax^2 + bx + c$ form, then label the three coefficients as a, b, and c.

● Example 2

Identifying the Coefficients of $ax^2 + bx + c$

Where necessary, rewrite the trinomial in $ax^2 + bx + c$ form. Then label a, b, and c.

(a) $x^2 - 3x - 18$

$a = 1$ $b = -3$ $c = -18$

(b) $x^2 - 24x + 23$

$a = 1$ $b = -24$ $c = 23$

(c) $x^2 + 8 - 11x$

First rewrite the trinomial in descending order.

$$x^2 - 11x + 8$$

Then,

$$a = 1 \qquad b = -11 \qquad c = 8.$$

CHECK YOURSELF 2

Where necessary, rewrite the trinomial in $ax^2 + bx + c$ form. Then label a, b, and c.

(a) $x^2 + 5x - 14$ **(b)** $x^2 - 18x + 17$ **(c)** $x - 6 + 2x^2$

Not all trinomials can be factored. To discover if a trinomial is factorable, we try the *ac* test.

The *ac* Test

A trinomial of the form $ax^2 + bx + c$ is factorable if (and only if) there are two numbers, m and n, such that

$$ac = mn \qquad b = m + n$$

In Example 3, we will determine whether each trinomial is factorable by finding the values of m and n.

• Example 3

Using the *ac* Test

Use the *ac* test to determine which of the following trinomials can be factored. Find the values of m and n for each trinomial that can be factored.

(a) $x^2 - 3x - 18$

First, we note that $a = 1$, $b = -3$, and $c = -18$, so $ac = 1(-18) = -18$.

Then, we look for two numbers, m and n, such that $mn = ac$ and $m + n = b$. In this case, that means

$$mn = -18 \qquad m + n = -3$$

We will look at every pair of integers with a product of -18. We then look at the sum of each pair.

mn	$m + n$
$1(-18) = -18$	$1 + -18 = -17$
$2(-9) \ = -18$	$2 + -9 \ = -7$
$3(-6) \ = -18$	$3 + -6 \ = -3$

We need look no further because we have found two integers whose ac product is -18 and $m + n$ sum is -3.

$$m = 3 \qquad n = -6$$

(b) $x^2 - 24x + 23$

   We see that $a = 1$, $b = -24$, and $c = 23$. So, $ac = 23$ and $b = -24$. Therefore,

$$mn = 23 \qquad m + n = -24$$

We now work with integer pairs, looking for two integers with a product of 23 and a sum of -24.

mn	$m + n$
$1(23) \quad\ = 23$	$1 + 23 \quad\ = 24$
$1(-23) = 23$	$-1 + -23 = -24$

We find that $m = -1$ and $n = -23$.

(c) $x^2 - 11x + 8$

   We see that $a = 1$, $b = -11$, and $c = 8$. So, $ac = 8$ and $b = -11$. Therefore,

$$mn = 8 \qquad m + n = -11$$

mn	$m + n$
$1(8) \quad\ = 8$	$1 + 8 \quad\ = 9$
$2(4) \quad\ = 8$	$2 + 4 \quad\ = 6$
$-1(-8) = 8$	$-1 + -8 = -9$
$-2(-4) = 8$	$-2 + -4 = -6$

There are no other pair of integers with a product of 8, and none has a sum of -11. The trinomial $x^2 - 11x + 8$ is not factorable.

(d) $2x^2 + 7x - 15$

   We see that $a = 2$, $b = 7$, and $c = -15$. So, $ac = -30$ and $b = 7$. Therefore,

$$mn = -30 \qquad m + n = 7$$

mn	$m + n$
$1(-30) = -30$	$1 + -30 = -29$
$2(-15) = -30$	$2 + -15 = -14$
$3(-10) = -30$	$3 + -10 = -7$
$5(-6) \quad\ = -30$	$5 + -6 \quad\ = -1$
$6(-5) \quad\ = -30$	$6 + -5 \quad\ = 1$
$10(-3) = -30$	$10 + -3 = 7$

There is no need to go further. We have found two integers with a product of -30 and a sum of 7. So $m = 10$ and $n = -3$.

   In this example, you may have noticed patterns and shortcuts that make it easier to find m and n. By all means, use those patterns. This is essential in mathematical thinking. You are taught a step-by-step process that will always work for solving a problem; this process is called an *algorithm*. It is very easy to teach a computer an algorithm. It is very difficult (some would say impossible) for a computer to have insight. Shortcuts that you discover are insights. They may be the most important part of your mathematical education.

● ● ● **CHECK YOURSELF 3**

Use the *ac* test to determine which of the following trinomials can be factored. Find the values of m and n for each trinomial that can be factored.

(a) $x^2 - 7x + 12$ **(b)** $x^2 + 5x - 14$ **(c)** $2x^2 + x - 6$ **(d)** $3x^2 - 6x + 7$

So far we have used the results of the *ac* test only to determine whether a trinomial is factorable. The results can also be used to help factor the trinomial.

● Example 4

Using the Results of the *ac* Test to Factor a Trinomial

Rewrite the middle term as the sum of two terms, then factor by grouping.

(*a*) $x^2 - 3x - 18$
We see that $a = 1$, $b = -3$, and $c = -18$, so

$$ac = -18 \qquad b = -3$$

We are looking for two numbers, m and n, where

$$mn = -18 \qquad m + n = -3$$

In Example 3, we found that the two integers were 3 and -6 because $3(-6) = -18$ and $3 + (-6) = -3$. That result is used to rewrite the middle term (here $-3x$) as the sum of two terms. We now rewrite the middle term as the sum of $3x$ and $-6x$.

$$x^2 + 3x - 6x - 18$$

Then, we factor by grouping:

$$x^2 + 3x - 6x - 18 = x(x + 3) - 6(x + 3)$$
$$= (x - 6)(x + 3)$$

(*b*) $x^2 - 24x + 23$
We use the results of Example 3(*b*), in which we found $m = -1$ and $n = -23$, to rewrite the middle term of the expression.

$$x^2 - 24x + 23 = x^2 - x - 23x + 23$$

Then, we factor by grouping:

$$x^2 - x - 23x + 23 = x(x - 1) - 23(x - 1)$$
$$= (x - 1)(x - 23)$$

(*c*) $2x^2 + 7x - 15$
From Example 3(*d*), we know that this trinomial is factorable and that $m = 10$ and $n = -3$. We use that result to rewrite the middle term of the trinomial.

$$2x^2 + 7x - 15 = 2x^2 + 10x - 3x - 15$$
$$= 2x(x + 5) - 3(x + 5)$$
$$= (x + 5)(2x - 3)$$

● ● ● ● **CHECK YOURSELF 4**

Rewrite the middle term as the sum of two terms, then factor by grouping.

(a) $x^2 - 7x + 12$.   **(b)** $x^2 + 5x - 14$.   **(c)** $2x^2 - x - 6$.   **(d)** $3x^2 - 7x - 6$.

Not all product pairs need to be tried to find m and n. A look at the sign pattern will eliminate many of the possibilities. Assuming the lead coefficient to be positive, there are four possible sign patterns.

Pattern	Example	Conclusion
1. b and c are both positive.	$2x^2 + 13x + 15$	m and n must be positive.
2. b is negative and c is positive.	$x^2 - 3x + 2$	m and n must both be negative.
3. b is positive and c is negative.	$x^2 + 5x - 14$	m and n are of opposite signs. (The value with the larger absolute value is positive.)
4. b and c are both negative.	$x^2 - 4x - 4$	m and n are of opposite signs. (The value with the larger absolute value is negative.)

TRIAL AND ERROR

Sometimes the factors of a trinomial seem obvious. At other times you might be certain that there are only a couple of possible sets of factors for a trinomial. It is perfectly acceptable to check these proposed factors to see if they work. If you find the factors in this manner, we say that you have used the **trial and error method.** The difficulty in using this method is in the name. Remember, the second part is ERROR.

FACTORING AND POLYNOMIAL FUNCTIONS

To this point we have been factoring polynomial expressions. When a function is defined by a polynomial expression, we can factor that expression without affecting any of the ordered pairs associated with the function. Factoring the expression makes it easier to find some of the ordered pairs.

In particular, we will be looking for values of x that cause $f(x)$ to be 0. We do this by using the **zero product rule.**

Zero Product Rule

If $0 = ab$, then either $a = 0$, $b = 0$, or both are zero.

Another way to say this is, if the product of two numbers is zero, then at least one of those numbers must be zero.

• Example 5

Factoring Polynomial Functions

Given the function $f(x) = 2x^2 + 7x - 15$, complete the following.

(a) Rewrite the function in factored form.
From Example 4(c) we have

$$f(x) = (x + 5)(2x - 3)$$

(b) Find the ordered pair associated with $f(0)$.

$$f(0) = (0 + 5)(0 - 3) = -15$$

The ordered pair is $(0, -15)$.

(c) Find all ordered pairs $(x, 0)$
We are looking for the x value for which $f(x) = 0$, so

$$0 = (x + 5)(2x - 3)$$

By the zero product rule, we know that either

$$(x + 5) = 0 \qquad \text{or} \qquad (2x - 3) = 0$$

which means that

$$x = -5 \qquad \text{or} \qquad x = \frac{3}{2}$$

The ordered pairs are $(-5, 0)$ and $\left(\frac{3}{2}, 0\right)$. Check the original function to see that these ordered pairs are associated with the graph of that function.

● ● ● **CHECK YOURSELF 5**

Given the function $f(x) = 2x^2 - x - 6$, complete the following.

(a) Rewrite the function in factored form. (b) Find the ordered pair associated with $f(0)$. (c) Find all ordered pairs $(x, 0)$.

● ● ● **CHECK YOURSELF ANSWERS**

1. (a) False, **(b)** true, **(c)** true. **2. (a)** $a = 1$, $b = 5$, $c = -4$; **(b)** $a = 1$, $b = -18$, $c = 17$; **(c)** $a = 2$, $b = 1$, $c = -6$. **3. (a)** Factorable, $m = -4$, $n = -3$; **(b)** factorable, $m = 7$, $n = -2$; **(c)** factorable, $m = 4$, $n = -3$; **(d)** not factorable. **4. (a)** $x^2 - 3x - 4x + 12 = (x - 3)(x - 4)$; **(b)** $x^2 + 7x - 2x - 14 = (x + 7)(x - 2)$; **(c)** $2x^2 - x - 6 = 2x^2 - 4x + 3x - 6 = (2x + 3)(x - 2)$; **(d)** $3x^2 - 7x - 6 = 3x^2 - 9x + 2x = (3x + 2)(x - 3)$.

5. (a) $f(x) = (2x + 3)(x - 2)$; **(b)** $(0, -6)$; **(c)** $\left(-\frac{3}{2}, 0\right)$ and $(2, 0)$.

3.5 Exercises

In Exercises 1 to 8, determine which are true statements.

1. $x^2 - 2x - 3 = (x + 1)(x - 3)$

2. $x^2 - 2x - 8 = (x - 2)(x + 4)$

3. $2x^2 - 5x + 4 = (2x - 1)(x - 4)$

4. $3x^2 - 13x - 10 = (3x + 2)(x - 5)$

5. $x^2 - x - 6 = (x - 5)(x + 1)$

6. $6x^2 + 7x - 3 = (3x - 1)(2x + 3)$

7. $-2x^2 + 11x - 5 = (-x + 5)(2x + 1)$

8. $-6x^2 + 13x - 6 = (2x - 3)(-3x + 2)$

In Exercises 9 to 16, where necessary, rewrite the trinomial in $ax^2 + bx + c$ form, then label a, b, and c.

9. $x^2 + 3x - 5$

10. $x^2 - 2x - 1$

11. $2x^2 + 5x + 3$

12. $-3x^2 - x + 2$

13. $x - 1 + 2x^2$

14. $4 - 5x - 3x^2$

15. $2x + 3x^2 - 5$

16. $x - x^2 + 4$

In Exercises 17 to 24, use the *ac* test to determine which trinomials can be factored. Find the values of m and n for each trinomial that can be factored.

17. $x^2 + 3x - 10$

18. $x^2 - x - 12$

19. $x^2 - 2x + 3$

20. $6x^2 - 7x + 2$

21. $2x^2 - 3x + 2$

22. $3x^2 - 10x - 8$

23. $2x^2 + 5x + 2$

24. $3x^2 + x - 2$

In Exercises 25 to 70, completely factor each polynomial expression.

25. $x^2 + 7x + 12$

26. $x^2 + 9x + 20$

27. $x^2 - 9x + 8$

28. $x^2 - 11x + 10$

29. $x^2 - 15x + 50$

30. $x^2 - 13x + 40$

31. $x^2 + 7x - 30$

32. $x^2 - 7x - 18$

33. $x^2 - 10x + 24$

34. $x^2 + 13x - 30$

35. $x^2 - 7x - 44$

36. $x^2 - 15x - 54$

37. $x^2 + 8xy + 15y^2$ **38.** $x^2 - 9xy + 20y^2$ **39.** $x^2 - 16xy + 55y^2$ **40.** $x^2 - 9xy - 22y^2$

41. $3x^2 + 11x - 20$ **42.** $2x^2 + 9x - 18$ **43.** $5x^2 + 18x - 8$ **44.** $3x^2 - 20x - 7$

45. $12x^2 + 23x + 5$ **46.** $8x^2 + 30x + 7$ **47.** $4x^2 + 20x + 25$ **48.** $9x^2 - 24x + 16$

49. $5x^2 + 19x - 30$ **50.** $3x^2 + 17x - 28$ **51.** $5x^2 + 24x - 36$ **52.** $3x^2 - 14x - 24$

53. $10x^2 - 7x - 12$ **54.** $6x^2 + 5x - 21$ **55.** $16x^2 + 40x + 25$ **56.** $18x^2 + 45x + 7$

57. $7x^2 - 17xy + 6y^2$ **58.** $5x^2 + 17xy - 12y^2$ **59.** $8x^2 - 30xy + 7y^2$ **60.** $8x^2 - 14xy - 15y^2$

61. $3x^2 - 24x + 45$ **62.** $2x^2 + 10x - 28$ **63.** $2x^2 - 26x + 72$ **64.** $3x^2 + 39x + 120$

65. $6x^3 - 31x^2 + 5x$ **66.** $8x^3 + 25x^2 + 3x$ **67.** $5x^3 + 14x^2 - 24x$ **68.** $3x^4 + 17x^3 - 28x^2$

69. $3x^3 - 15x^2y - 18xy^2$ **70.** $2x^3 - 10x^2y - 72xy^2$

In Exercises 71 to 76, for each function, **(a)** rewrite the function in factored form, **(b)** find the ordered pair associated with $f(0)$, and **(c)** find all ordered pairs $(x, 0)$.

71. $f(x) = x^2 - 2x - 3$ **72.** $f(x) = x^2 - 3x - 10$

73. $f(x) = 2x^2 + 3x - 2$ **74.** $f(x) = 3x^2 - 11x + 6$

75. $f(x) = 3x^2 + 5x - 28$ **76.** $f(x) = 10x^2 + 13x - 3$

Certain trinomials in quadratic form can be factored with similar techniques. For instance we can factor $x^4 - 5x^2 - 6$ as $(x^2 - 6)(x^2 + 1)$. In Exercises 77 to 88, apply a similar method to completely factor each polynomial.

77. $x^4 + 3x^2 + 2$ **78.** $x^4 - 7x^2 + 10$ **79.** $x^4 - 8x^2 - 33$ **80.** $x^4 + 5x^2 - 14$

81. $y^6 - 2y^3 - 15$ **82.** $x^6 + 10x^3 + 21$ **83.** $x^5 - 6x^3 - 16x$ **84.** $x^6 - 8x^4 + 15x^2$

85. $x^4 - 5x^2 - 36$ **86.** $x^4 - 5x^2 + 4$ **87.** $x^6 - 6x^3 - 16$ **88.** $x^6 - 2x^3 - 3$

In Exercises 89 to 96, determine a value of the number k so that the polynomial can be factored.

89. $x^2 + 5x + k$ **90.** $x^2 + 3x + k$ **91.** $6x^2 + x + k$ **92.** $4x^2 - x + k$

93. $x^2 + kx - 6$ **94.** $x^2 + kx - 15$ **95.** $6x^2 + kx - 3$ **96.** $2x^2 + kx - 15$

97. The product of three numbers is $x^3 + 6x^2 + 8x$. Show that the numbers are consecutive even integers. (*Hint:* Factor the expression.)

98. The product of three numbers is $x^3 + 3x^2 + 2x$. Show that the numbers are consecutive integers.

Answers

1. True   **3.** False   **5.** False   **7.** False   **9.** $a = 1, b = 3, c = -5$ **11.** $a = 2, b = 5, c = 3$
13. $a = 2, b = 1, c = -1$ **15.** $a = 3, b = 2, c = -5$ **17.** Factorable, $m = 5, n = -2$ **19.** Not
factorable   **21.** Not factorable   **23.** Factorable, $m = 4, n = 1$ **25.** $(x + 3)(x + 4)$
27. $(x - 8)(x - 1)$ **29.** $(x - 10)(x - 5)$ **31.** $(x + 10)(x - 3)$ **33.** $(x - 6)(x - 4)$
35. $(x - 11)(x + 4)$ **37.** $(x + 3y)(x + 5y)$ **39.** $(x - 11y)(x - 5y)$ **41.** $(3x - 4)(x + 5)$
43. $(5x - 2)(x + 4)$ **45.** $(3x + 5)(4x + 1)$ **47.** $(2x + 5)^2$ **49.** $(5x - 6)(x + 5)$
51. $(5x - 6)(x + 6)$ **53.** $(2x - 3)(5x + 4)$ **55.** $(4x + 5)^2$ **57.** $(7x - 3y)(x - 2y)$
59. $(4x - y)(2x - 7y)$ **61.** $3(x - 5)(x - 3)$ **63.** $2(x - 4)(x - 9)$ **65.** $x(6x - 1)(x - 5)$
67. $x(x + 4)(5x - 6)$ **69.** $3x(x - 6y)(x + y)$ **71. (a)** $(x - 3)(x + 1)$, **(b)** $(0, -3)$, **(c)** $(3, 0)$ and $(-1, 0)$

73. (a) $(2x - 1)(x + 2)$, **(b)** $(0, -2)$, **(c)** $\left(\frac{1}{2}, 0\right)$ and $(-2, 0)$ **75. (a)** $(x + 4)(3x - 7)$, **(b)** $(0, -28)$,

(c) $(-4, 0)$ and $\left(\frac{7}{3}, 0\right)$ **77.** $(x^2 + 1)(x^2 + 2)$ **79.** $(x^2 - 11)(x^2 + 3)$ **81.** $(y^3 - 5)(y^3 + 3)$
83. $x(x^2 + 2)(x^2 - 8)$ **85.** $(x + 3)(x - 3)(x^2 + 4)$ **87.** $(x - 2)(x^2 + 2x + 4)(x^3 + 2)$ **89.** 4
91. $-2, -1,$ or -5 **93.** $-5, 5, -1,$ or 1 **95.** $7, -7, -3,$ or 3 **97.** $x(x + 2)(x + 4)$

3.6 Synthetic Substitution and Division

3.6 OBJECTIVES

1. Evaluate $f(0) \div g(0)$
2. Use the notation of synthetic substitution
3. Find the remainders of polynomial division synthetically
4. Completely factor the polynomial by using synthetic division

In earlier algebra classes, you may have learned to divide polynomials by the long division method. In this book, we will use the process of **synthetic division**, which involves substitution, to find the quotient when one polynomial is divided by another.

Before we work on synthetic division, let's look at an alternative to the polynomial substitution method we discussed in Section 1.3. Let's use the notation $P(x)$ to represent a polynomial function. We might have

$$P(x) = 3x^2 + 5x - 4$$

If you were to try to find $P(2)$ by the substitution method, you would replace x with 2 and simplify the expression; thus,

$$P(2) = 3(2)^2 + 5(2) - 4$$

$$= 12 + 10 - 4$$

$$= 18$$

Let's factor an x out of the first two terms of the polynomial and look at a slightly different approach.

$$P(x) = (3x + 5)x - 4$$

Now, we substitute 2 for x and simplify.

$$P(2) = [3(2) + 5](2) - 4$$

$$= (6 + 5)(2) - 4$$

$$= (11)(2) - 4$$

$$= 22 - 4$$

$$= 18$$

This process can be duplicated using synthetic division. We begin by writing the number to be substituted (separated as shown) and the coefficients of the polynomial. The leading coefficient is rewritten below the line.

$$\underline{2|} \quad 3 \quad\quad 5 \quad\quad -4$$
$$\overline{}$$
$$3$$

We multiply the number to be substituted and the leading coefficient. The result is added to the second coefficient.

$2 \times 3 = 6$

$5 + 6 = 11$

The process is repeated until we have used all the coefficients.

$2 \times 11 = 22$

$-4 + 22 = 18$

$$\begin{array}{r|rrr} 2 & 3 & 5 & -4 \\ & & 6 & 22 \\ \hline & 3 & 11 & 18 \end{array}$$

The last number, 18, is $P(2)$.

●Example 1

Evaluating a Function by Synthetic Substitution

If $P(x) = 3x^3 - 4x^2 + 2x - 1$, then $P(-1)$.

$$\begin{array}{r|rrrr} -1 & 3 & -4 & 2 & -1 \\ & & -3 & 7 & -9 \\ \hline & 3 & -7 & 9 & -10 \end{array}$$

Therefore $P(-1) = -10$.

● ● ● CHECK YOURSELF 1

If $P(x) = 2x^3 + 3x^2 - 2x - 5$, find $P(2)$.

What does it mean when $P(k) = 0$? We know then that k is a solution of the equation $P(x) = 0$. Example 2 illustrates this concept.

●Example 2

Evaluating a Function by Synthetic Substitution

If $P(x) = x^2 + 6x + 8$, find $P(-4)$.

$$\begin{array}{r|rrr} -4 & 1 & 6 & 8 \\ & & -4 & -8 \\ \hline & 1 & 2 & 0 \end{array}$$

Since $P(-4) = 0$, we know that -4 is a solution of the equation

$x^2 + 6x + 8 = 0$

● ● ● CHECK YOURSELF 2

If $P(x) = 2x^3 + 7x^2 + 12x + 12$, find $P(-2)$.

We can also use synthetic substitution when we want to divide a polynomial and the divisor is of the form $x - k$. When we divide a polynomial by a lesser-degree polynomial, we get a quotient, which we call $Q(x)$, and a remainder, which we call r, so

Think of dividing 29 by 4.

$29 = 4 \cdot 7 + 1$

$$P(x) = (x - k) \cdot Q(x) + r$$

This must be true for every value of x, so let $x = k$ and

$$P(k) = (k - k) \cdot Q(k) + r$$
$$= 0 \cdot Q(k) + r$$
$$= r$$

We now know that the value we get by synthetic substitution is the remainder when we divide $P(x)$ by $x - k$.

• Example 3

Finding the Remainder

Find the remainder when $x^3 + 3x^2 - 8x + 1$ is divided by $x - 3$.

You could check your result by using long division. And $x - 3$ is of the form $x - k$, where $k = 3$

$$
\begin{array}{r|rrrr}
3\rfloor & 1 & 3 & -8 & 1 \\
 & & 3 & 18 & 30 \\
\hline
 & 1 & 6 & 10 & 31 \\
\end{array}
$$

So the remainder is 31.

● ● ● **CHECK YOURSELF 3**

Find the remainder when $2x^3 + 7x^2 + 12x - 8$ is divided by $x - 1$.

To complete the division, we must find the quotient. Let's look at an example of long division.

Let's divide $2x^2 + 5x - 3$ by $x + 2$.

$$
\begin{array}{r}
2x + 1 \\
x + 2 \overline{\smash{)}2x^2 + 5x - 3} \\
\underline{2x^2 + 4x } \\
x - 3 \\
\underline{x + 2} \\
-5 \\
\end{array}
$$

So,

$$\frac{2x^2 + 5x - 3}{x + 2} = 2x + 1 + \frac{-5}{x + 2}$$

Let's compare this to synthetic substitution. Find the remainder when $2x^2 + 5x - 3$ is divided by $x + 2$.

$x + 2$ is of the form $x - k$, since

$x + 2 = x - (-2)$

$k = -2$

$$
\begin{array}{r|rrr}
-2 & 2 & 5 & -3 \\
 & & -4 & -2 \\
\hline
 & 2 & 1 & -5 \\
\end{array}
$$

You will notice several similarities between the long division and the synthetic substitution. The most interesting is along the bottom row of the substitution. Note that the first two numbers are the coefficients of the quotient and that the last number is the remainder. This will always be the case when we use synthetic substitution.

● Example 4

Using Synthetic Division

Use synthetic division to divide $2x^3 - 4x^2 - 7x + 5$ by $x - 3$.

$$
\begin{array}{r|rrrr}
3 & 2 & -4 & -7 & 5 \\
 & & 6 & 6 & -3 \\
\hline
 & 2 & 2 & -1 & 2 \quad \longleftarrow \text{Remainder} \\
\end{array}
$$

Coefficients of the quotient

So $2x^3 - 4x^2 - 7x + 5 = (x - 3)(2x^2 + 2x - 1) + 2$.

● ● ●   **CHECK YOURSELF 4**

Divide $2x^3 + 7x^2 + 12x - 8$ by $x - 1$.

As was true with long division, we must be careful when one of the coefficients is 0. Use a placeholder of 0 when this occurs.

● Example 5

Using Synthetic Division

Divide $4x^4 - 3x^2 + 2x - 7$ by $x + 1$.

Zero represents the coefficient of the x^3 term.

$$
\begin{array}{r|rrrrr}
-1 & 4 & 0 & -3 & 2 & -7 \\
 & & -4 & 4 & -1 & -1 \\
\hline
 & 4 & -4 & 1 & 1 & -8 \\
\end{array}
$$

Coefficients of the quotient   Remainder

So $4x^4 - 3x^2 + 2x - 7 = (x + 1)(4x^3 - 4x^2 + x + 1) + (-8)$.

● ● ● **CHECK YOURSELF 5**

Divide $3x^4 - 3x^3 + 2$ by $x + 1$.

Synthetic division is commonly used to determine whether $x - k$ is a factor of a given polynomial.

●**Example 6**

Using Synthetic Division

Show that $x - 2$ is a factor of $2x^3 - 7x^2 + 10x - 8$.

$$
\begin{array}{r|rrrr}
2\rule{0pt}{0pt}] & 2 & -7 & 10 & -8 \\
 &   & 4 & -6 & 8 \\
\hline
 & 2 & -3 & 4 & 0
\end{array}
\quad \text{The remainder is 0.}
$$

This tells us that $2x^3 - 7x^2 + 10x - 8 = (x - 2)(2x^2 - 3x + 4)$.

● ● ● **CHECK YOURSELF 6**

Show that $x + 5$ is a factor of $x^3 + 5x^2 - 3x - 15$.

By using synthetic division in Example 7, we will completely factor third-degree polynomials.

●**Example 7**

Using Synthetic Division

(*a*) Completely factor $x^3 + 2x^2 - 9x - 18$, given that $x + 2$ is a factor.

$$
\begin{array}{r|rrrr}
-2\rule{0pt}{0pt}] & 1 & 2 & -9 & -18 \\
 &   & -2 & 0 & 18 \\
\hline
 & 1 & 0 & -9 & 0
\end{array}
$$

We now know that

$$x^3 + 2x^2 - 9x - 18 = (x + 2)(x^2 - 9)$$
$$= (x + 2)(x + 3)(x - 3)$$

(*b*) Completely factor $2x^3 - 3x^2 - 8x - 3$, given that $x + 1$ is a factor.

$$\begin{array}{r}
-1 \underline{\big|} \quad 2 \quad -3 \quad -8 \quad 3 \\
 -2 \quad 5 \quad 3 \\
\hline
2 \quad -5 \quad -3 \quad 0
\end{array}$$

So we have

$$2x^3 - 3x^2 - 8x - 3 = (x + 1)(2x^2 - 5x - 3)$$

Using the techniques in Section 3.5, we have

$$2x^2 - 5x - 3 = (2x + 1)(x - 3)$$

So

$$2x^3 - 3x^2 - 8x - 3 = (x + 1)(2x + 1)(x - 3)$$

● ● ● **CHECK YOURSELF 7**

Completely factor $2x^3 - 3x^2 - 8x + 12$, given that $x - 2$ is a factor.

● ● ● **CHECK YOURSELF ANSWERS**

1. 19.   **2.** 0.   **3.** 13.   **4.** $(x - 1)(2x^2 + 9x + 21) + 13$.
5. $(x + 1)(3x^3 - 6x^2 + 6x - 6) + 8$.   **6.** Remainder $= 0$.
7. $(x - 2)(x + 2)(2x - 3)$.

3.6 Exercises

$P(x) = 3x^3 - 2x^2 + x - 4$. In Exercises 1 to 4, use synthetic substitution to find the following.

1. $P(2)$ **2.** $P(3)$ **3.** $P(-3)$ **4.** $P(-5)$

In Exercises 5 to 10, use synthetic division to find the remainder for each division.

5. $(x^2 - 3x + 5) \div (x - 3)$ **6.** $(x^2 + 2x + 4) \div (x - 2)$ **7.** $(2x^3 + x^2 - 5x - 1) \div (x + 1)$

8. $(3x^3 - 2x^2 + x - 2) \div (x + 3)$ **9.** $(3x^3 - 4x + 1) \div (x - 3)$ **10.** $(5x^3 - 6x + 2) \div (x - 2)$

In Exercises 11 to 16, use synthetic division to find the quotient and remainder.

11. $(x^2 + 4x + 5) \div (x - 1)$ **12.** $(x^2 - 3x + 4) \div (x - 3)$ **13.** $(3x^3 + 2x^2 - 5x + 2) \div (x + 1)$

14. $(2x^3 - 5x^2 + 2x - 2) \div (x + 2)$ **15.** $(4x^3 - 3x + 2) \div (x - 3)$

16. $(3x^3 - 6x + 2) \div (x - 4)$

 17. Can the quotient and remainder for $\dfrac{x^3 - 4x^2 + 5x - 6}{x^2 - 4}$ be found by using synthetic division? If not, why not? If so, what are they?

 18. What are the advantages of using synthetic division to find the value of a polynomial for a specific number?

In Exercises 19 to 22, use synthetic division to confirm the following.

19. $x + 2$ is a factor of $x^2 - x - 6$.   **20.** $x - 4$ is a factor of $x^2 - 2x - 8$.

21. $x + 3$ is a factor of $x^3 + 27$.   **22.** $x - 2$ is a factor of $x^4 - 16$.

23. Completely factor $x^3 + 2x^2 - 11x - 12$, given that $x + 4$ is a factor.

24. Completely factor $x^3 - 4x^2 - 7x + 10$, given that $x - 5$ is a factor.

25. Completely factor $2x^3 + 3x^2 - 23x - 12$, given that $x + 4$ is a factor.

26. Completely factor $3x^3 - 4x^2 - 12x + 16$, given that $x - 2$ is a factor.

Answers

1. 14   **3.** -106 **5.** 5   **7.** 3   **9.** 70   **11.** Quotient: $x + 5$; remainder: 10
13. Quotient: $3x^2 - x - 4$; remainder: 6   **15.** Quotient: $4x^2 + 12x + 33$; remainder: 101   **17.**
19. $P(-2) = 0$ **21.** $P(-3) = 0$ **23.** $(x + 4)(x - 3)(x + 1)$ **25.** $(2x + 1)(x - 3)(x + 4)$

Summary

Positive Integer Exponents and the Scientific Calculator [3.1]

Properties of Exponents

For any nonzero real numbers a and b and integers m and n:

Product Rule

$x^5 \cdot x^7 = x^{5+7} = x^{12}$

$$a^m \cdot a^n = a^{m+n}$$

Quotient Rule

$\dfrac{x^7}{x^5} = x^{7-5} = x^2$

$$\frac{a^m}{a^n} = a^{m-n}$$

Product-Power Rule

$(x^5)^3 = x^{5 \cdot 3} = x^{15}$

$$(ab)^n = a^n b^n$$

Power Rule

$$(a^m)^n = a^{mn}$$

Adding or Subtracting Polynomials [3.2]

To add

$4x^2 + 3x + 2$

and

$3x^2 - 5x - 7$

write

$(4x^2 + 3x + 2) +$
$(3x^2 - 5x - 7)$

$= 7x^2 - 2x - 5$

To simplify:

$3x - 2[4 - (3x + 5)]$
$= 3x - 2(4 - 3x - 5)$
$= 3x - 2(-3x - 1)$
$= 3x + 6x + 2$
$= 9x + 2$

A **term** is a number or the product of a number and one or more variables, raised to a power. If terms contain exactly the same variables raised to the same powers, they are called **like terms.** A **monomial** is a term in which only whole numbers can appear as exponents. A **polynomial** may be a monomial or any finite sum (or difference) of monomials. We call a polynomial with two terms a **binomial,** one with three terms a **trinomial.**

The **degree** of a monomial is the sum of the exponents of the variable factors. The degree of a polynomial is that of the term with the highest degree.

A **polynomial function** is a function in which the expression on the right-hand side is a polynomial expression.

Adding Polynomials To add polynomials, simply combine any like terms by using the distributive property.

To subtract

$3x^2 - 2x - 5$

from

$4x^2 - 6x + 2$

write

$(4x^2 - 6x + 2) -$
$(3x^2 - 2x - 5)$
$= 4x^2 - 6x + 2 -$
$\quad 3x^2 + 2x + 5$
$= x^2 - 4x + 7$

Subtracting Polynomials To subtract polynomials, enclose the polynomial being subtracted in parentheses, preceded by a negative sign. Then remove the parentheses according to the previous rule and combine like terms.

Multiplying Polynomials and Special Products [3.3]

To multiply:

$$\begin{array}{r} 3x^2 - 3x + 5 \\ 5x - 2 \\ \hline -6x^2 + 4x - 10 \\ 15x^3 - 10x^2 + 25x \\ \hline 15x^3 - 16x^2 + 29x - 10 \end{array}$$

Multiplying Polynomials To multiply two polynomials, multiply each term of the first polynomial by each term of the second polynomial.

$(2x - 3)(2x + 3) = 4x^2 - 9$

$(2x + 5)^2 = 4x^2 + 20x + 25$

$(3a - b)^2 = 9a^2 - 6ab + b^2$

Special Products Certain special products can be found by applying the following formulas.

$(a + b)(a - b) = a^2 - b^2$ — Product of binomials differing in sign.

$(a + b)^2 = a^2 + 2ab + b^2$ — Squaring a binomial.

$(a - b)^2 = a^2 - 2ab + b^2$ — Squaring a binomial.

Factoring Polynomials and Special Polynomials [3.4]

$5x^2y^2 - 10xy^2 + 35x^2y$ has GCF $5xy$.

The **greatest common factor (GCF)** of a polynomial is the monomial with highest degree and largest numerical coefficient that is a factor of each term of the polynomial.

Factor:

$5x^2y^2 - 10xy^2 + 35x^2y$
$= 5xy(xy - 2y + 7x)$

Factor:

$16x^2 - 9y^2$
$= (4x + 3y)(4x - 3y)$

Factor:

$x^3 + 8y^3$
$= (x + 2y)(x^2 - 2xy + 4y^2)$

To Factor a Polynomial In general, you can apply the following steps.

1. If a polynomial has a GCF other than 1, factor out that *greatest common factor*.

2. If the terms of a polynomial have no common factor (other than 7), **factoring by grouping** is the preferred method.

3. If a polynomial is a binomial, try factoring it as a **difference of two squares** or as a **sum or difference of two cubes**.

Factoring Trinomials [3.4 and 3.5]

To Factor a Trinomial In general, you can apply the following steps.

1. Write the trinomial in standard $ax^2 + bx + c$ form.
2. Then, label the three coefficients a, b, and c.

Not all trinomials are factorable. To discover if a trinomial is factorable, try the *ac* test.

The *ac* Test A trinomial of the form $ax^2 + bx + c$ is factorable if (and only if) there are two numbers, m and n, such that

$$ac = mn \qquad b = m + n$$

Zero Product Rule

If $0 = ab$, then either $a = 0$, $b = 0$, or both are zero.

Synthetic Substitution and Division [3.6]

To find $P(k)$, we can use **synthetic substitution.** If $P(k) = 0$, then k is a root of the polynomial $P(x)$.

The value we get by synthetic substitution with k is the remainder when $P(x)$ is divided by $x - k$.

When we use the same process to do **synthetic division,** we find the coefficients to the quotient along the bottom row.

If $P(x) = x^2 + x - 1$, find $P(1)$.

```
1 |  1   1   -1
   |      1    2
   ------------
      1   2    1 = P(1)
```

Divide $2x^3 + 3x - 4$ by $x - 2$.

```
2 |  2   3   -4
   |      4   14
   ------------
      2   7   10
```

$Q(x) = 2x + 7;\ R = 10$

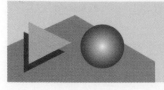

Summary Exercises

This summary exercise set is provided to give you practice with each of the objectives in the chapter. Each exercise is keyed to the appropriate chapter section. The answers are provided in the *Instructor's Manual*.

[3.1] Simplify each expression, using the properties of exponents.

1. $r^4 \cdot r^9$

2. $(-3)^2(-3)^3$

3. $(a^3b^2)(a^8b^3)$

4. $(6c^0d^4)(-3c^2d^2)$

5. $\dfrac{x^{12}}{x^5}$

6. $\dfrac{(2x-1)^8}{(2x-1)^5}$

7. $(x^2y)^3$

8. $(2c^3d^4)^3$

9. $(2a^3)^0(-3a^4)^2$

10. $\left(\dfrac{x}{y^2}\right)^2$

11. $\left(\dfrac{3m^2n^3}{p^4}\right)^3$

12. $\left(\dfrac{a^3}{b^4}\right)\left(\dfrac{b^2}{2a^2}\right)^3$

13. Given $f(x) = 4x^2 + 5x - 6$, find $f(-2)$.

14. Given $f(x) = 5x^2 - 4x + 8$, find $f(3)$.

15. Given $f(x) = -2x^3 + 4x^2 - 5x + 8$, find $f(3)$.

16. Given $f(x) = 3x^3 + 5x^2 - 2x + 6$, find $f(-1)$.

17. Given $f(x) = -2x^4 - 5x^2 + 3x - 6$, find $f(2)$.

18. Given $f(x) = -x^5 - 3x^4 + 5x^3 - x - 4$, find $f(-2)$.

[3.2] Identify each of the following polynomials as a monomial, binomial, or trinomial. Give the degree of each.

19. $7x^4 - 8x^6$

20. $5x^4$

21. 5

22. $4x^2 + 5x^4 + 6x^6$

In each of the following, the polynomial functions $f(x)$ and $g(x)$ are given. Find the following ordered pairs:
(a) $(0, f(0))$, **(b)** $(2, f(2))$, **(c)** $(0, g(0))$, and **(d)** $(2, g(2))$.

23. $f(x) = 5x^2 - 3x - 4$ and $g(x) = 3x^2 + 6x - 5$

24. $f(x) = 2x^2 + 4x + 6$ and $g(x) = 3x^2 - 4x + 3$

25. $f(x) = x^3 - 3x^2 + 7x$ and $g(x) = -2x^3 - x^2 + 3x + 1$

26. $f(x) = -x^4 - 3x^2 + 2x$ and $g(x) = -2x^4 + 3x^3 - 2x^2 + 2$

In each of the following, $f(x)$ and $g(x)$ are given. Let $h(x) = f(x) + g(x)$. Find **(a)** $h(x)$, **(b)** $f(1) + g(1)$, and **(c)** $h(1)$.

27. $f(x) = 4x^2 + 5x - 3$ and $g(x) = -2x^2 + x - 5$

28. $f(x) = -3x^3 + 2x^2 - 5$ and $g(x) = 4x^3 - 4x^2 + 5x + 6$

29. $f(x) = 2x^4 + 4x^2 + 5$ and $g(x) = x^3 - 5x^2 + 6x$

30. $f(x) = 3x^3 + 5x - 5$ and $g(x) = -2x^3 + 2x^2 + 5x$

Simplify each of the following polynomial functions.

31. $f(x) = [(3x^2 + 4x) - 3] - [2x^2 - 6x - 2]$

32. $f(x) = [2x^3 + 4x] + [-3x^3 - 2x^2 + 5] - [4x^2 - 6x - 7]$

33. $f(x) = (3x - 7) - [5x^2 - (8x + 6)] + (-6x^2 - 9)$

34. $f(x) = (-3x^2 + 6x + 9) - [(7x^2 + 5x - 8)] + [2 - (-3x^2 + 7x + 9)]$

35. $f(x) = -4x^2(-6x + 5) - [(3x^3 - 5x^2 + 6) - (x^3 + 6x^2 + 4x - 2)]$

36. $f(x) = -5x^2(3x^2 + 6x - 2) + [-2x(-4x + 5) + (-3x^2 - x - 1)]$

[3.3] In each of the following, $f(x)$ and $g(x)$ are given. Let $h(x) = f(x) \cdot g(x)$. Find **(a)** $h(x)$, **(b)** $f(1) \cdot g(1)$, and **(c)** $h(1)$.

37. $f(x) = 4x$ and $g(x) = 3x^2 - 5x$ **38.** $f(x) = -6x$ and $g(x) = -4x^2 - 5x + 7$

39. $f(x) = 3x^3$ and $g(x) = -2x^2 - 8x + 1$ **40.** $f(x) = -7x^3$ and $g(x) = 5x^2 - 4x + 10$

Multiply each polynomial expression.

41. $5x(3x^2 - 4x)$ **42.** $5y^2(2y^3 - 3y^2 + 5y)$ **43.** $(x - 2y)(x + 3y)$ **44.** $(a - 5b)(a - 6b)$

45. $(3c - 5d)(5c + 2d)$ **46.** $(4x^2 - y)(2x + 3y^2)$ **47.** $x(x - 3)(x + 2)$ **48.** $2y(2y + 3)(3y + 2)$

Multiply the following polynomial expressions using the special product formulas.

49. $(x + 8)^2$

50. $(y - 5)^2$

51. $(2a - 3b)^2$

52. $(5x + 2y)^2$

53. $(x - 4y)(x + 4y)$

54. $(2c - 3d)(2c + 3d)$

Multiply each polynomial expression using the vertical method.

55. $(2x - 3)(x^2 - 5x + 2)$

56. $(5a - b)(2a^2 - 3ab - 2b^2)$

[3.4]   Factor each of the following polynomials completely.

57. $18x^2y + 24xy^2$

58. $35a^3 - 28a^2 + 7a$

59. $18m^2n^2 - 27m^2n + 45m^2n^3$

60. $x(2x - y) + y(2x - y)$

61. $5(w - 3z) - 10(w - 3z)^2$

For each of the following functions, **(a)** find $f(1)$, **(b)** factor $f(x)$, and **(c)** find $f(1)$ from the factored form.

62. $f(x) = 3x^3 + 15x^2$

63. $f(x) = -6x^3 - 2x^2$

64. $f(x) = 12x^6 - 8x^4$

65. $f(x) = 2x^5 - 2x$

[3.4]   Factor each of the following binomials completely.

66. $x^2 - 64$

67. $25a^2 - 16$

68. $16m^2 - 49n^2$

69. $3w^3 - 12wz^2$

70. $a^4 - 16b^4$

71. $m^3 - 64$

72. $8x^3 + 1$

73. $8c^3 - 27d^3$

74. $125m^3 + 64n^3$

75. $2x^4 + 54x$

[3.5]   Use the *ac* test to determine which of the following trinomials can be factored. Find the values of m and n for each trinomial that can be factored.

76. $x^2 - x - 30$

77. $x^2 + 3x + 2$

78. $2x^2 - 11x + 12$

79. $4x^2 - 23x + 15$

Completely factor each of the following polynomial expressions.

80. $x^2 + 12x + 20$

81. $a^2 - a - 12$

82. $w^2 - 13w + 40$

83. $r^2 - 9r - 36$

84. $x^2 - 8xy - 48y^2$

85. $a^2 + 17ab + 30b^2$

86. $5x^2 + 13x - 6$

87. $2a^2 + 3a - 35$

88. $4r^2 + 20r + 21$ **89.** $6c^2 - 19c + 10$ **90.** $6m^2 - 19mn + 10n^2$ **91.** $8x^2 + 14xy - 15y^2$

92. $9x^2 - 15x - 6$ **93.** $5w^2 - 25wz + 30z^2$ **94.** $3c^3 + 18c^2 + 15c$ **95.** $2a^3 + 4a^2b - 6ab^2$

96. $x^4 + 6x^2 + 5$ **97.** $a^4 - 3a^2b^2 - 4b^4$

[3.6]   Let $P(x) = 4x^3 - 2x^2 + 3x - 7$. Use synthetic substitution to find the following.

98. $P(2)$ **99.** $P(1)$ **100.** $P(-3)$ **101.** $P(-4)$

Using synthetic substitution to find the remainder for each division.

102. $(x^2 - 2x + 7) \div (x - 2)$ **103.** $(x^2 + 3x - 5) \div (x - 1)$ **104.** $(2x^2 - x + 9) \div (x + 3)$

105. $(3x^2 + 2x - 8) \div (x + 2)$ **106.** $(3x^3 + x - 2) \div (x + 1)$ **107.** $(2x^3 + 3x - 1) \div (x - 3)$

For each of the following, use synthetic division to find the quotient and remainder.

108. $(x^2 + 3x + 4) \div (x - 1)$ **109.** $(2x^2 - x + 7) \div (x - 3)$ **110.** $(2x^2 + x - 2) \div (x + 2)$

111. $(x^2 + 3x - 5) \div (x - 4)$ **112.** $(4x^3 + 3x - 5) \div (x - 2)$ **113.** $(2x^3 + x - 3) \div (x - 3)$

114. Completely factor $x^3 + 4x^2 - 11x - 30$, given that $(x - 3)$ is a factor.

115. Completely factor $x^3 - 2x^2 - 13x - 10$, given that $(x + 2)$ is a factor.

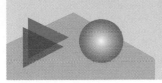

The purpose of this self-test is to help you check your progress and to review for a chapter test in class. Allow yourself about 1 hour to take the test. When you are done, check your answers in the back of the book. If you missed any answers, be sure to go back and review the appropriate sections in the chapter and the exercises that are provided.

Simplify each expression, using the properties of exponents.

1. $(3x^2y)(\ 2xy^3)$

2. $\left(\dfrac{8m^2n^5}{2p^3}\right)^2$

3. $(x^4y^5)^2$

4. $\dfrac{9c^5d^3}{18c^7d^4}$

5. $(3x^2y)^3(-2xy^2)^2$

6. $(-2xy)^3(4x^2y)^2$

7. Given $f(x) = 3x^2 - 4x + 5$, find $f(-2)$.

In the following, $f(x)$ and $g(x)$ are given. Find **(a)** $h(x) = f(x) + g(x)$, **(b)** $p(x) = f(x) - g(x)$, **(c)** $f(1) + g(1)$, **(d)** $f(1) - g(1)$, **(e)** $h(1)$, **(f)** $p(1)$.

8. $f(x) = 4x^2 - 3x + 7$ and $g(x) = 2x^2$

9. $f(x) = -3x^3 + 5x^2 - 2x - 7$ and $g(x) = -2x^2 + 7x - 2$

Simplify the following polynomial function.

10. $f(x) = 5x - \{4x + 2[x - 3(x + 2)]\}$

Multiply each of the following polynomials.

11. $(2a - 5b)(3a + 7b)$

12. $(5m - 3n)(5m + 3n)$

13. $(2a + 3b)^2$

14. $(2x - 5)(x^2 - 4x + 3)$

Factor each of the following polynomials completely.

15. $14a^2b^2 - 21a^2b + 35ab^2$

16. $x^2 - 3xy + 5x - 15y$

17. $25c^2 - 64d^2$

18. $27x^3 - 1$

19. $16a^4 + 2ab^3$

20. $x^2 - 2x - 48$

21. $10x^2 - 39x + 14$

22. $6x^3 + 3x^2 - 45x$

23. Use synthetic substitution to find the remainder if $x^2 - 3x + 8$ is divided by $x + 2$.

24. Use synthetic division to find the quotient and remainder, if $(3x^2 + 5x - 1) \div (x + 1)$.

25. Completely factor $x^3 - x^2 - 14x + 24$, given that $(x - 2)$ is a factor.

This test is provided to help you in the process of reviewing previous chapters. Answers are provided in the back of the book. If you missed any answers, be sure to go back and review the appropriate chapter sections.

1. Solve the equation $4x - 2(x + 1) = 3(5 - x) - 7$.

2. If $f(x) = 4x^3 - 5x^2 + 7x - 11$, find $f(-2)$.

3. Find the x and y intercepts and graph the equation $2x + 3y = 12$.

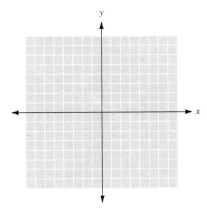

4. Find the equation of the line that passes through the point $(-1, -2)$ and is perpendicular to the line $4x + 5y = 15$.

5. Simplify the expression $(4x^3y^2)^3(-2x^2y^3)^2$.

6. If $P(x) = 4x^5 + 7x^4 - 3x^2 - 5x + 1$, find $P(-1)$ using synthetic substitution.

7. Find the domain and range of the relation $4x - 3y = 15$.

8. Use the given graph to estimate **(a)** $f(-2)$, **(b)** $f(0)$, and **(c)** $f(3)$.

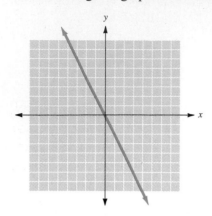

Simplify each of the following polynomial functions.

9. $f(x) = (2x^2 - 3x + 4) - (3x^2 + 2x - 5)$.

10. $f(x) = (2x + 3)(5x - 4)$.

Factor each of the following completely.

11. $3x^3 - x^2 - 2x$

12. $16x^2 - 25y^2$

13. $3x^2 - 3xy + x - y$

14. If $f(x) = -5x + 1$ and $g(x) = 8x + 6$, find **(a)** $f + g$, **(b)** $f - g$, **(c)** $f \cdot g$, and **(d)** $\dfrac{f}{g}$.

Solve the following equations.

15. $-6x - 6(2x - 9) = -3(x + 7)$

16. $|2x + 46| = 10$

17. $4(3x - 6) = -8(x - 2)$

Solve the following inequalities.

18. $4(3 + x) \geq 8$

19. $|x + 9| \leq 4$

20. $-3|x + 5| > -3$

RATIONAL EXPRESSIONS AND FUNCTIONS

INTRODUCTION

The House of Representatives comprises officials elected from congressional districts within each state. The number of representatives a state sends to the U.S. House of Representatives depends on the state's population. The total number of representatives to the House has grown from 106 in 1790 to 435 today, the maximum number established in 1930. These 435 representatives are apportioned to the 50 states on the basis of population. This apportionment is revised after every decennial (10-year) census.

If a particular state has a population, A, and its number of representatives is equal to a, then $\dfrac{A}{a}$ represents the ratio of people in the state to their total number of representatives in the U.S. House. It follows that the total population of the country, P, and its total number of representatives, r, is represented by the ratio $\dfrac{P}{r}$. If another state, with population, E, has e number of representatives, then $\dfrac{E}{e}$ should also be equal to $\dfrac{P}{r}$ and to $\dfrac{A}{a}$, if the apportionment is fair.

A comparison of these ratios for states in 1990 finds Pennsylvania with 546,880 people per representative and Arizona with 610,800—Arizona was above the national average of 571,750 people per representative for 1990, and Pennsylvania below. This is not so much a result of political backroom bargaining as it is a result of ratios that do not divide out evenly—there are remainders when the proportions are solved. Should the numbers be rounded up or down? If they are all rounded down, the total is too small, if rounded up, the total number of representatives would be more than the 435 seats in the House. So, since all the states cannot be treated equally, the question of what is fair and how to decide who gets the additional representatives has been debated in Congress since its inception.

© Susan Van Etten/The Picture Cube, Inc.

The method currently used was devised in the 1920s. Its creator, E. V. Huntington, said that if $\dfrac{A}{a}$ were larger than $\dfrac{E}{e}$, then the difference between them is not so important as the **relative difference,** which he defined as

$$\frac{\dfrac{A}{a} - \dfrac{E}{e}}{\dfrac{E}{e}}$$

Huntington said that it is only correct to assign an additional representative to state A provided

$$\frac{\dfrac{E}{e} - \dfrac{A}{a+1}}{\dfrac{A}{a+1}} < \frac{\dfrac{A}{a} - \dfrac{E}{e+1}}{\dfrac{E}{e+1}}$$

This inequality can be simplified to

$$\frac{A}{\sqrt{a(a+1)}} > \frac{E}{\sqrt{e(e+1)}}$$

This statement forms the basis upon which the "extra" representatives are apportioned.

This method has its share of critics who claim it fails to fairly apportion representatives. So far, they have not convinced enough Congressional representatives to change from Huntington's method, but with a new census occurring every 10 years, a situation may soon develop that will cause some state to challenge it.

In this chapter, we will work with rational expressions and functions. Among the skills you will develop will be those necessary to simplify the inequality mentioned above. _____ ∎

4.1 Simplifying Rational Expressions and Functions

Our work in this chapter will expand your experience with algebraic expressions to include algebraic fractions or **rational expressions.** We consider the four basic operations of addition, subtraction, multiplication, and division in the next sections. Fortunately, you will observe many parallels to your previous work with arithmetic fractions.

EVALUATING RATIONAL EXPRESSIONS

First, let's define what we mean by a rational expression. Recall that a rational number is the ratio of two integers. Similarly, a rational expression can be written as the ratio of two polynomials, in which the denominator cannot have the value 0.

The word "rational" comes from "ratio."

> A *rational expression* is the ratio of two polynomials. It can be written as
>
> $$\frac{P}{Q}$$
>
> where P and Q are polynomials and Q cannot have the value 0.

The expressions

$$\frac{x-3}{x+1} \qquad \frac{x^2+5}{x-3} \quad \text{and} \quad \frac{x^2-2x}{x^2+3x+1}$$

are all rational expressions. The restriction that the denominator of the expressions not be 0 means that certain values for the variable may have to be excluded because division by 0 is undefined.

• Example 1

Precluding Division by Zero

(*a*) For what values of x is the following expression undefined?

$$\frac{x}{x-5}$$

A fraction is undefined when its denominator is equal to 0.

Note that when $x = 5$, $\dfrac{x}{x-5}$

becomes $\dfrac{5}{5-5}$, or $\dfrac{5}{0}$.

To answer this question, we must find where the denominator is 0. Set

$$x - 5 = 0$$

or

$$x = 5$$

289

The expression $\dfrac{x}{x-5}$ is undefined for $x=5$.

(b) For what values of x is the following expression undefined?

$$\frac{3}{x+5}$$

Again, set the denominator equal to 0:

$$x + 5 = 0$$

or

$$x = -5$$

The expression $\dfrac{3}{x+5}$ is undefined for $x=-5$.

● ● ●   **CHECK YOURSELF 1**

For what values of the variable are the following expressions undefined?

(a) $\dfrac{1}{r+7}$ **(b)** $\dfrac{5}{2x-9}$

Scientific calculators are often used to evaluate rational expressions for values of the variable. The *parentheses keys* help in this process.

● Example 2

Evaluating a Rational Expression

Using a calculator, evaluate the following expressions for the given value of the variable.

(a) $\dfrac{3x}{2x-5}$ for $x=4$

Enter the expression in your calculator as follows:

$3\;\boxed{\times}\;4\;\boxed{\div}\;\boxed{(}\;\boxed{(}\;2\;\boxed{\times}\;4\;\boxed{-}\;5\;\boxed{)}\;\boxed{=}$

The display will read the value 4.

(b) $\dfrac{2x+7}{4x-11}$ for $x=4$

CAUTION

Be sure to use the parentheses keys before the 2 and after the 5.

Enter the expression as follows:

$\boxed{(}\ 2\ \boxed{\times}\ 4\ \boxed{+}\ 7\ \boxed{)}\ \boxed{\div}\ \boxed{(}\ 4\ \boxed{\times}\ 4\ \boxed{-}\ 11\ \boxed{)}\ \boxed{=}$

The display will read the value 3.

CHECK YOURSELF 2

Using a scientific calculator, evaluate each of the following.

(a) $\dfrac{5x}{3x - 2}$ for $x = 4$ **(b)** $\dfrac{2x + 9}{3x - 4}$ for $x = 3$

SIMPLIFYING RATIONAL EXPRESSIONS

Generally, we want to write rational expressions in the simplest possible form. To begin our discussion of simplifying rational expressions, let's review for a moment. As we pointed out previously, there are many parallels to your work with arithmetic fractions. Recall that

$$\frac{3}{5} = \frac{3 \cdot 2}{5 \cdot 2} = \frac{6}{10}$$

so

$$\frac{3}{5} \quad \text{and} \quad \frac{6}{10}$$

name equivalent fractions. In a similar fashion,

$$\frac{10}{15} = \frac{5 \cdot 2}{5 \cdot 3} = \frac{2}{3}$$

so

$$\frac{10}{15} \quad \text{and} \quad \frac{2}{3}$$

name equivalent fractions.

We can always multiply or divide the numerator and denominator of a fraction by the same nonzero number. The same pattern is true in algebra.

Fundamental Principle of Rational Expressions

For polynomials P, Q, and R,

$$\frac{P}{Q} = \frac{PR}{QR} \qquad \text{where } Q \neq 0 \qquad R \neq 0$$

This principle can be used in two ways. We can multiply or divide the numerator and denominator of a rational expression by the same nonzero polynomial. The result will always be an expression that is equivalent to the original one.

In simplifying arithmetic fractions, we used this principle to divide the numerator and denominator by all common factors. With arithmetic fractions, those common factors are generally easy to recognize. Given rational expressions where the numerator and denominator are polynomials, we must determine those factors as our first step. The most important tools for simplifying expressions are the factoring techniques in Chapter 3.

In fact, you will see that most of the methods in this chapter depend on factoring polynomials. Additional factoring techniques can be found in Appendix B.

• Example 3

Simplifying Rational Expressions

Simplify each rational expression. Assume denominators are not 0.

We find the common factors 4, x, and y in the numerator and denominator. We divide the numerator and denominator by the common factor $4xy$. Note that

$$\frac{4xy}{4xy} = 1$$

$$(a) \quad \frac{4x^2y}{12xy^2} = \frac{4xy \cdot x}{4xy \cdot 3y}$$

$$= \frac{x}{3y}$$

We have *divided* the numerator and denominator by the common factor $x - 2$. Again note that

$$\frac{x - 2}{x - 2} = 1$$

$$(b) \quad \frac{3x - 6}{x^2 - 4} = \frac{3(x - 2)}{(x + 2)(x - 2)} \quad \text{Factor the numerator and the denominator.}$$

We can now divide the numerator and denominator by the common factor $x - 2$:

$$\frac{3(x - 2)}{(x + 2)(x - 2)} = \frac{3}{x + 2}$$

and the rational expression is in simplest form.

C A U T I O N

Pick any value other than 0 for the variable x, and substitute. You will quickly see that

$$\frac{x + 2}{x + 3} \neq \frac{2}{3}$$

Be Careful! Given the expression

$$\frac{x + 2}{x + 3}$$

students, are often tempted to divide by variable x, as in

$$\frac{x + 2}{x + 3} \overset{?}{=} \frac{2}{3}$$

This is not a valid operation. We can only divide by common *factors,* and in the expression above the variable x is a *term* in both the numerator and the denominator. The numerator and denominator of a rational expression must be factored *before* common factors are divided out. Therefore,

$$\frac{x + 2}{x + 3}$$

is in its simplest possible form.

● ● ● **CHECK YOURSELF 3**

Simplify each expression.

(a) $\dfrac{36a^3b}{9ab^2}$

(b) $\dfrac{x^2 - 25}{4x + 20}$

The same techniques are used when trinomials need to be factored. Example 4 further illustrates the simplification of rational expressions.

● Example 4

Simplifying Rational Expressions

Simplify each rational expression.

Divide by the common factor $x + 1$, using the fact that

$$\dfrac{x + 1}{x + 1} = 1$$

where $x \neq -1$

(a) $\dfrac{5x^2 - 5}{x^2 - 4x - 5} = \dfrac{5(x - 1)(x + 1)}{(x - 5)(x + 1)}$

$$= \dfrac{5(x - 1)}{x - 5}$$

In part (c) we factor by grouping in the numerator and use the sum of cubes in the denominator. Note that

$x^3 + 2x^2 - 3x - 6$

$= x^2(x + 2) - 3(x + 2)$

$= (x + 2)(x^2 - 3)$s

(b) $\dfrac{2x^2 + x - 6}{2x^2 - x - 3} = \dfrac{(x + 2)(2x - 3)}{(x + 1)(2x - 3)}$

$$= \dfrac{x + 2}{x + 1}$$

(c) $\dfrac{x^3 + 2x^2 - 3x - 6}{x^3 + 8} = \dfrac{(x + 2)(x^2 - 3)}{(x + 2)(x^2 - 2x + 4)} = \dfrac{x^2 - 3}{x^2 - 2x + 4}$

● ● ● **CHECK YOURSELF 4**

Simplify each rational expression.

(a) $\dfrac{x^2 - 5x + 6}{3x^2 - 6x}$

(b) $\dfrac{3x^2 + 14x - 5}{3x^2 + 2x - 1}$

Simplifying certain algebraic expressions involves recognizing a particular pattern. Verify for yourself that

$$3 - 9 = -(9 - 3)$$

In general, it is true that

$$a - b = -(-a + b) = -(b - a) = -1(b - a)$$

or, by dividing both sides of the equation by $b - a$,

Note that

$$\frac{a - b}{a - b} = 1$$

but

$$\frac{a - b}{b - a} = -1$$

where $a \neq b$.

$$\frac{a - b}{b - a} = \frac{-(b - a)}{b - a} = -1$$

Example 5 makes use of this result.

● Example 5

Simplifying Rational Expressions

Simplify each rational expression.

Note:

$$\frac{x - 2}{2 - x} = -1$$

(a) $\quad \dfrac{2x - 4}{4 - x^2} = \dfrac{2\overset{-1}{\cancel{(x - 2)}}}{(2 + x)\cancel{(2 - x)}}$

$\qquad\qquad = \dfrac{2(-1)}{2 + x} = \dfrac{-2}{2 + x}$

(b) $\quad \dfrac{9 - x^2}{x^2 + 2x - 15} = \dfrac{(3 + x)\overset{-1}{\cancel{(3 - x)}}}{(x + 5)\cancel{(x - 3)}}$

$\qquad\qquad\quad = \dfrac{(3 + x)(-1)}{x + 5}$

$\qquad\qquad\quad = \dfrac{-x - 3}{x + 5}$

● ● ● **CHECK YOURSELF 5**

Simplify each rational expression.

(a) $\dfrac{5x - 20}{16 - x^2}$
(b) $\dfrac{x^2 - 6x - 27}{81 - x^2}$

The following algorithm summarizes our work with simplifying rational expressions.

Simplifying Rational Expressions

1. Completely factor both the numerator and denominator of the expression.
2. Divide the numerator and denominator by *all* common factors.
3. The resulting expression will be in simplest form (or in lowest terms).

IDENTIFYING RATIONAL FUNCTIONS

We begin with a definition.

A **rational function** is a function that is defined by a rational expression. It can be written as

$$f(x) = \frac{P}{Q}$$

where P and Q are polynomials and Q does not have a value of zero.

• Example 6

Identifying Rational Functions

Which of the following are rational functions?

(a) $f(x) = 3x^3 - 2x + 5$ This is a rational function; it could be written over the denominator 1, and 1 is a polynomial.

(b) $f(x) = \dfrac{3x^2 - 5x + 2}{2x - 1}$ This is a rational function; it is the ratio of two polynomials.

Recall from Chapter 3 that there are no square roots of variables in a polynomial.

(c) $f(x) = 3x^3 + 3\sqrt{x}$ This is not a rational function; it is not the ratio of two polynomials.

● ● ●   **CHECK YOURSELF 6**

Which of the following are rational functions?

(a) $f(x) = x^5 - 2x^4 + 3x - 1$ **(b)** $f(x) = \dfrac{x^2 - x + 7}{\sqrt{x} - 1}$

(c) $f(x) = \dfrac{3x^3 + 3x}{2x + 1}$

SIMPLIFYING RATIONAL FUNCTIONS

When we simplify a rational function, it is important that we note the x values that need to be excluded, particularly when we are trying to draw the graph of a function. The set of ordered pairs of the simplified function will be exactly the same as the set of ordered pairs of the original function. If we plug the excluded value(s) for x into the simplified expression, we get a set of ordered pairs that represent "holes" in the graph. These holes are breaks in the curve. We use an open circle to designate them on a graph.

• Example 7

Simplifying a Rational Function

Given the function

$$f(x) = \frac{x^2 + 2x + 1}{x + 1}$$

complete the following.

(a) Simplify the rational expression on the right.

$$\frac{x^2 + 2x + 1}{x + 1} = \frac{(x + 1)(x + 1)}{(x + 1)}$$

$$= (x + 1) \qquad x \neq -1$$

(b) Rewrite the function in simplified form.

$$f(x) = x + 1 \qquad x \neq -1$$

(c) Find the ordered pair associated with the hole in the graph of the original function.

Plugging -1 into the simplified function yields the ordered pair $(-1, 0)$. This represents the hole in the graph of the function

$$f(x) = \frac{x^2 + 2x + 1}{x + 1}$$

● ● ●   **CHECK YOURSELF 7**

Given the function

$$f(x) = \frac{5x^2 - 10x}{5x}$$

complete the following.

(a) Rewrite the function in simplified form.
(b) Find the ordered pair associated with the hole in the graph of the original function.

• Example 8

Graphing a Rational Function

Graph the following function.

$$f(x) = \frac{x^2 + 2x + 1}{x + 1}$$

From Example 7, we know that

$$\frac{x^2 + 2x + 1}{x + 1} = x + 1 \qquad x \neq -1$$

Therefore,

$$f(x) = x + 1 \qquad x \neq -1$$

The graph will be the graph of the line $f(x) = x + 1$, with an open circle at the point $(-1, 0)$.

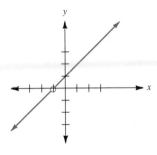

CHECK YOURSELF 8

Graph the function $f(x) = \dfrac{5x^2 - 10x}{5x}$.

CHECK YOURSELF ANSWERS

1. (a) $r = -7$; (b) $x = \dfrac{9}{2}$.   2. (a) 2; (b) 3.   3. (a) $\dfrac{4a^2}{b}$; (b) $\dfrac{x - 5}{4}$.

4. (a) $\dfrac{x - 3}{3x}$; (b) $\dfrac{x + 5}{x + 1}$.   5. (a) $\dfrac{-5}{x + 4}$; (b) $\dfrac{-x - 3}{x + 9}$.   6. (a) A rational

function, (b) not a rational function, and (c) a rational function.

7. (a) $f(x) = x - 2$, $x \neq 0$; (b) $(0, -2)$.   8.

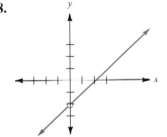

Age Ratios. Your friend has a 4-year-old cousin, Amy, who has a 9-year-old brother. The younger child is upset because not only does her brother refuse to let her play with him and his friends, he teases her because she can never catch up in age! You explain to the child that as she gets older, this age difference will not seem like such a big deal.

Write an expression for the ratio of the younger child's age to her brother's age. Then complete the following table.

Amy's Age, years	Brother's Age	Amy's Age Brother's Age
4		
5		
10		
15		
20		
25		

1. What happens to the ratio as the children grow older?
2. Draw a graph of the ratio as a function of Amy's age to show how this ratio changes.
3. Assume that Amy and her brother live to be 100 and 105 years old, respectively. What is the ratio between their ages now?
4. Draw a graph of the ratio of Amy's brother's age to Amy's age. Does this graph have anything in common with the first graph? Explain.
5. Write a short explanation for Amy (who is only 4, remember!) of your conclusions about how their age ratios change over time.

4.1   Exercises

In Exercises 1 to 12, for what values of the variable is each rational expression undefined?

1. $\dfrac{x}{x-5}$ **2.** $\dfrac{y}{y+7}$ **3.** $\dfrac{x+5}{3}$ **4.** $\dfrac{x-6}{4}$

5. $\dfrac{2x-3}{2x-1}$ **6.** $\dfrac{3x-2}{3x+1}$ **7.** $\dfrac{2x+5}{x}$ **8.** $\dfrac{3x-7}{x}$

9. $\dfrac{\;\;\;+1)}{\;\;\;\;}$ **10.** $\dfrac{x+2}{3x-7}$ **11.** $\dfrac{4-x}{x}$ **12.** $\dfrac{2x+7}{3x+\dfrac{1}{3}}$

In Exercises 13 to 16, evaluate each expression, using a calculator.

13. $\dfrac{3x}{2x - 1}$ for $x = 2$

14. $\dfrac{5x}{4x - 3}$ for $x = 2$

15. $\dfrac{2x + 3}{x + 3}$ for $x = -6$

16. $\dfrac{4x - 7}{2x - 1}$ for $x = -2$

In Exercises 17 to 48, simplify each expression. Assume the denominators are not 0.

17. $\dfrac{14}{21}$

18. $\dfrac{45}{75}$

19. $\dfrac{4x^5}{6x^2}$

20. $\dfrac{25x^6}{20x^2}$

21. $\dfrac{10x^2y^5}{25xy^2}$

22. $\dfrac{18a^2b^3}{24a^4b^3}$

23. $\dfrac{-42x^3y}{14xy^3}$

24. $\dfrac{-15x^3y^3}{-20xy^2}$

25. $\dfrac{28a^5b^3c^2}{84a^2bc^4}$

26. $\dfrac{-52p^5q^3r^2}{39p^3q^5r^2}$

27. $\dfrac{6x - 24}{x^2 - 16}$

28. $\dfrac{x^2 - 25}{3x - 15}$

29. $\dfrac{x^2 + 2x + 1}{6x + 6}$

30. $\dfrac{5y^2 - 10y}{y^2 + y - 6}$

31. $\dfrac{x^2 - 5x - 14}{x^2 - 49}$

32. $\dfrac{2m^2 + 11m - 21}{4m^2 - 9}$

33. $\dfrac{3b^2 - 14b - 5}{b - 5}$

34. $\dfrac{a^2 - 9b^2}{a^2 + 8ab + 15b^2}$

35. $\dfrac{2y^2 + 3yz - 5z^2}{2y^2 + 11yz + 15z^2}$

36. $\dfrac{6x^2 - x - 2}{3x^2 - 5x + 2}$

37. $\dfrac{x^3 - 64}{x^2 - 16}$

38. $\dfrac{r^2 - rs - 6s^2}{r^3 + 8s^3}$

39. $\dfrac{a^4 - 81}{a^2 + 5a + 6}$

40. $\dfrac{c^4 - 16}{c^2 - 3c - 10}$

41. $\dfrac{xy - 2x + 3y - 6}{x^2 + 8x + 15}$

42. $\dfrac{cd - 3c + 5d - 15}{d^2 - 7d + 12}$

43. $\dfrac{x^2 + 3x - 18}{x^3 - 3x^2 - 2x + 6}$

44. $\dfrac{y^2 + 2y - 35}{y^2 - 5y - 3y + 15}$

45. $\dfrac{2m - 10}{25 - m^2}$

46. $\dfrac{5x - 20}{16 - x^2}$

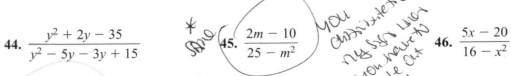

47. $\dfrac{49 - x^2}{2x^2 - 13x - 7}$

48. $\dfrac{2x^2 - 7x + 3}{9 - x^2}$

In Exercises 49 to 54, identify which functions are rational functions.

49. $f(x) = 4x^2 - 5x + 6$

50. $f(x) = \dfrac{x^3 - 2x^2 + 7}{\sqrt{x} + 2}$

51. $f(x) = \dfrac{x^2 - x - 1}{x + 2}$

52. $f(x) = \dfrac{\sqrt{x} - x + 3}{x - 2}$

53. $f(x) = 5x^2 - \sqrt[3]{x}$

54. $f(x) = \dfrac{x^2 - x + 5}{x}$

For the given functions in Exercises 55 to 60, **(a)** rewrite the function in simplified form, and **(b)** find the ordered pair associated with the hole in the graph of the original function.

55. $f(x) = \dfrac{x^2 - x - 2}{x + 1}$

56. $f(x) = \dfrac{x^2 + x - 12}{x + 4}$

57. $f(x) = \dfrac{3x^2 + 5x - 2}{x + 2}$

58. $f(x) = \dfrac{2x^2 - 7x + 5}{2x - 5}$

59. $f(x) = \dfrac{x^2 + 4x + 4}{5(x + 2)}$

60. $f(x) = \dfrac{x^2 - 6x + 9}{7(x - 3)}$

In Exercises 61 to 66, graph the rational functions. Indicate the coordinates of the hole in the graph.

61. $f(x) = \dfrac{x^2 - 2x - 8}{x + 2}$

62. $f(x) = \dfrac{x^2 + 4x - 5}{x + 5}$

63. $f(x) = \dfrac{x^2 + 4x + 3}{x + 1}$

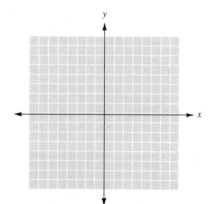
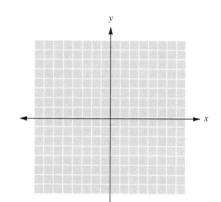
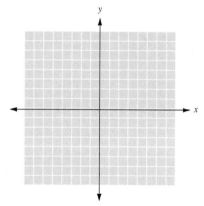

64. $f(x) = \dfrac{x^2 + 7x + 10}{x + 2}$

65. $f(x) = \dfrac{x^2 - 4x + 3}{x - 1}$

66. $f(x) = \dfrac{x^2 - 6x + 8}{x - 4}$

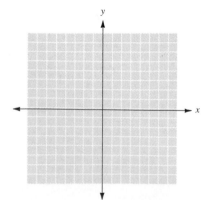
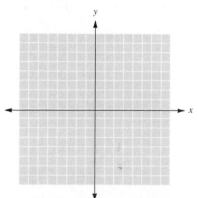
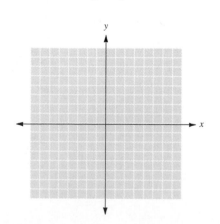

67. Explain why the following statement is false.

$$\frac{6m^2 + 2m}{2m} = 6m^2 + 1$$

68. State and explain the fundamental principle of fractions.

69. The rational expression $\dfrac{x^2 - 4}{x + 2}$ can be simplified to $x - 2$. Is this reduction true for all values of x? Explain.

70. What is meant by a rational expression in lowest terms?

In Exercises 71 to 76, simplify.

71. $\dfrac{2(x + h) - 2x}{(x + h) - x}$

72. $\dfrac{-3(x + h) - (-3x)}{(x + h) - x}$

73. $\dfrac{3(x + h) - 3 - (3x - 3)}{(x + h) - x}$

74. $\dfrac{2(x + h) + 5 - (2x + 5)}{(x + h) - x}$

75. $\dfrac{(x + h)^2 - x^2}{(x + h) - x}$

76. $\dfrac{(x + h)^3 - x^3}{(x + h) - x}$

Given $f(x) = \dfrac{P(x)}{Q(x)}$, if the graphs of $P(x)$ and $Q(x)$ intersect at $(a, 0)$, then $x - a$ is a factor of both $P(x)$ and $Q(x)$. Use a graphing calcuator to find the common factor for the expressions in Exercises 77 and 78.

77. $f(x) = \dfrac{x^2 + 4x - 5}{x^2 + 3x - 10}$

78. $f(x) = \dfrac{2x^2 + 11x - 21}{2x^2 + 15x + 7}$

Answers

1. 5   **3.** Never undefined   **5.** $\dfrac{1}{2}$ **7.** 0   **9.** -2 **11.** 0   **13.** 2   **15.** 3   **17.** $\dfrac{2}{3}$

19. $\dfrac{2x^3}{3}$ **21.** $\dfrac{2xy^3}{5}$ **23.** $\dfrac{-3x^2}{y^2}$ **25.** $\dfrac{a^3b^2}{3c^2}$ **27.** $\dfrac{6}{x + 4}$ **29.** $\dfrac{x + 1}{6}$ **31.** $\dfrac{x + 2}{x + 7}$

33. $3b + 1$ **35.** $\dfrac{y - z}{y + 3z}$ **37.** $\dfrac{x^2 + 4x + 16}{x + 4}$ **39.** $\dfrac{(a^2 + 9)(a - 3)}{a + 2}$ **41.** $\dfrac{y - 2}{x + 5}$

43. $\dfrac{x + 6}{x^2 - 2}$ **45.** $\dfrac{-2}{m + 5}$ **47.** $\dfrac{-x - 7}{2x + 1}$ **49.** Rational   **51.** Rational   **53.** Not rational

55. (a) $f(x) = x - 2$, (b) $(-1, -3)$ **57.** (a) $f(x) = 3x - 1$, (b) $(-2, -7)$ **59.** (a) $f(x) = \dfrac{x + 2}{5}$, (b) $(-2, 0)$

61.

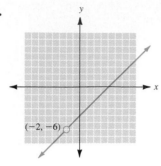

63.

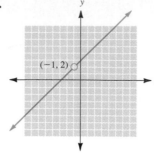

(−1, 2)

65.

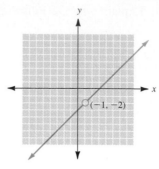

(−1, −2)

(−2, −6)

67.    **69.**   **71.** 2   **73.** 3   **75.** $2x + h$ **77.** $x + 5$

 4.2

Multiplying and Dividing Rational Expressions and Functions

4.2 OBJECTIVES

1. Multiply and divide rational expressions
2. Multiply and divide two rational functions

For all problems with rational expressions, assume denominators are not 0.

Once again, let's turn to an example from arithmetic to begin our discussion of multiplying rational expressions. Recall that to multiply two fractions, we multiply the numerators and multiply the denominators. For instance,

$$\frac{2}{5} \cdot \frac{3}{7} = \frac{2 \cdot 3}{5 \cdot 7} = \frac{6}{35}$$

In algebra, the pattern is exactly the same.

Multiplying Rational Expressions

For polynomials P, Q, R, and S,

$$\frac{P}{Q} \cdot \frac{R}{S} = \frac{PR}{QS} \qquad \text{where } Q \neq 0 \qquad S \neq 0$$

● Example 1

Multiplying Rational Expressions

Multiply.

$$\frac{2x^3}{5y^2} \cdot \frac{10y}{3x^2} = \frac{20x^3y}{15x^2y^2}$$

$$= \frac{5x^2y \cdot 4x}{5x^2y \cdot 3y} \qquad \text{Divide by the common factor } 5x^2y \text{ to simplify.}$$

$$= \frac{4x}{3y}$$

● ● ● **CHECK YOURSELF 1**

Multiply.

$$\frac{9a^2b^3}{5ab^4} \cdot \frac{20ab^2}{27ab^3}$$

The factoring methods in Chapter 3 are used to simplify rational expressions.

Generally, you will find it best to divide by any common factors before you multiply, as Example 2 illustrates.

● Example 2

Multiplying Rational Expressions

Multiply as indicated.

(a) $\dfrac{x}{x^2 - 3x} \cdot \dfrac{6x - 18}{9x}$ $\qquad$ Factor.

$$= \frac{x}{x(x - 3)} \cdot \frac{\overset{2}{6}(x - 3)}{\underset{3}{9}x} \qquad \text{Divide by the common factors of 3, } x, \text{ and } x - 3.$$

$$= \frac{2}{3x}$$

(b) $\dfrac{x^2 - y^2}{5x^2 - 5xy} \cdot \dfrac{10xy}{x^2 + 2xy + y^2}$ $\qquad$ Factor and divide by the common factors of 5, x, $x - y$, and $x + y$.

$$= \frac{(x + y)(x - y)}{\underset{1}{5}x(x - y)} \cdot \frac{\overset{2}{10}xy}{(x + y)(x + y)}$$

$$= \frac{2y}{x + y}$$

Note that

$$\frac{2-x}{x-2} = -1$$

(c) $\dfrac{4}{x^2 - 2x} \cdot \dfrac{10x - 5x^2}{8x + 24}$

$$= \frac{4}{x\cancel{(x-2)}} \cdot \frac{5x\cancel{(2-x)}^{-1}}{8(x+3)}$$

$$= \frac{-5}{2(x+3)}$$

● ● ● **CHECK YOURSELF 2**

Multiply as indicated.

(a) $\dfrac{x^2 - 5x - 14}{4x^2} \cdot \dfrac{8x + 56}{x^2 - 49}$

(b) $\dfrac{x}{2x - 6} \cdot \dfrac{3x - x^2}{2}$

The following algorithm summarizes our work in multiplying rational expressions.

Multiplying Rational Expressions

STEP 1 Write each numerator and denominator in completely factored form.
STEP 2 Divide by any common factors appearing in both the numerator and denominator.
STEP 3 Multiply as needed to form the desired product.

DIVIDING RATIONAL EXPRESSIONS

In dividing rational expressions, you can again use your experience from arithmetic. Recall that

We invert the *divisor* (the second fraction) and multiply.

$$\frac{3}{5} \div \frac{2}{3} = \frac{3}{5} \cdot \frac{3}{2} = \frac{9}{10}$$

Once more, the pattern in algebra is identical.

Dividing Rational Expressions

For polynomials P, Q, R, and S,

$$\frac{P}{Q} \div \frac{R}{S} = \frac{P}{Q} \cdot \frac{S}{R} = \frac{PS}{QR}$$

where $Q \neq 0 \qquad R \neq 0 \qquad S \neq 0$

To divide rational expressions, invert the divisor and multiply as before, as Example 3 illustrates.

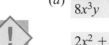

Dividing Rational Expressions

Divide as indicated.

Invert the divisor and multiply.

CAUTION

Be Careful! Invert the divisor, then factor.

(a) $\dfrac{3x^2}{8x^3y} \div \dfrac{9x^2y^2}{4y^4} = \dfrac{3x^2}{8x^3y} \cdot \dfrac{4y^4}{9x^2y^2} = \dfrac{y}{6x^3}$

(b) $\dfrac{2x^2 + 4xy}{9x - 18y} \div \dfrac{4x + 8y}{3x - 6y} = \dfrac{2x^2 + 4xy}{9x - 18y} \cdot \dfrac{3x - 6y}{4x + 8y}$

$= \dfrac{2x(x + 2y)}{9(x - 2y)} \cdot \dfrac{3(x - 2y)}{4(x + 2y)} = \dfrac{x}{6}$

(c) $\dfrac{2x^2 - x - 6}{4x^2 + 6x} \div \dfrac{x^2 - 4}{4x} = \dfrac{2x^2 - x - 6}{4x^2 + 6x} \cdot \dfrac{4x}{x^2 - 4}$

$= \dfrac{(2x + 3)(x - 2)}{2x(2x + 3)} \cdot \dfrac{4x^2}{(x + 2)(x - 2)} = \dfrac{2x}{x + 2}$

● ● ●   **CHECK YOURSELF 3**

Divide and simplify.

(a) $\dfrac{5xy}{7x^3} \div \dfrac{10y^2}{14x^3}$

(b) $\dfrac{3x - 9y}{2x + 10y} \div \dfrac{x^2 - 3xy}{4x^2 + 20xy}$

(c) $\dfrac{x^2 - 9}{x^3 - 27} \div \dfrac{x^2 - 2x - 15}{2x^2 - 10x}$

We summarize our work in dividing fractions with the following algorithm.

Dividing Rational Expressions

STEP 1 Invert the divisor (the *second* rational expression) to write the problem as one of multiplication.

STEP 2 Proceed as in the algorithm for the multiplication of rational expressions.

MULTIPLYING RATIONAL FUNCTIONS

The product of two rational functions is always a rational function. Given two rational functions, $f(x)$ and $g(x)$, we can rename the product, so

$$h(x) = f(x) \cdot g(x)$$

This will always be true for values of x for which both f and g are defined. So, for example, $h(1) = f(1) \cdot g(1)$ as long as both $f(1)$ and $g(1)$ exist.

If

$$h(x) = f(x) \cdot g(x)$$

then the set of ordered pairs

$$(x, f(x) \cdot g(x)) = (x, h(x))$$

Example 4 illustrates this concept.

• Example 4

Given the rational functions

$$f(x) = \frac{x^2 - 3x - 10}{x + 1} \quad \text{and} \quad g(x) = \frac{x^2 - 4x - 5}{x - 5}$$

find the following.

(a) $f(0) \cdot g(0)$

    If $f(0) = -10$ and $g(0) = 1$, then $f(0) \cdot g(0) = (-10)(1) = -10$.

(b) $f(5)g(5)$

    Although we can find $f(5)$, $g(5)$ is undefined. 5 is an excluded value for the range of the function. Therefore, $f(5)g(5)$ is undefined.

(c) $h(x) = f(x) \cdot g(x)$

$$h(x) = f(x) \cdot g(x)$$

$$= \frac{x^2 - 3x - 10}{x + 1} \cdot \frac{x^2 - 4x - 5}{x - 5}$$

$$= \frac{(x - 5)(x + 2)}{\cancel{(x + 1)}} \cdot \frac{\cancel{(x + 1)}(x - 5)}{\cancel{(x - 5)}}$$

$$= (x - 5)(x + 2) \qquad x \neq -1, x \neq 5$$

(d) $h(0)$

$$h(0) = (0 - 5)(0 + 2) = -10$$

(e) $h(5)$

    Although the temptation is to substitute 5 for x in part (c), notice that the function is undefined when x is -1 or 5. As was true in part (b), the function is undefined at that point.

● ● ● **CHECK YOURSELF 4**

Given the rational functions

$$f(x) = \frac{x^2 - 2x - 8}{x + 2} \qquad \text{and} \qquad g(x) = \frac{x^2 - 3x - 10}{x - 4}$$

find the following.

(a) $f(0)g(0)$ **(b)** $f(4)g(4)$ **(c)** $h(x) = f(x)g(x)$ **(d)** $h(0)$ **(e)** $h(4)$

DIVIDING RATIONAL FUNCTIONS

When we divide two rational expressions to create a third rational expression, we must be certain to exclude values for which the polynomial in the denominator is equal to zero, as Example 5 illustrates.

● Example 5

Dividing Polynomial Functions

Given the rational functions

$$f(x) = \frac{x^3 - 2x^2}{x + 2} \qquad \text{and} \qquad g(x) = \frac{x^2 - 3x + 2}{x - 4}$$

complete the following.

(a) Find $\dfrac{f(0)}{g(0)}$.

If $f(0) = 0$ and $g(0) = -\dfrac{1}{2}$, then

$$\frac{f(0)}{g(0)} = \frac{0}{-\dfrac{1}{2}} = 0$$

(b) Find $\dfrac{f(1)}{g(1)}$.

Although we can find both $f(1)$ and $g(1)$, $g(1) = 0$, so division is undefined. 1 is an excluded value for the range of the quotient.

(c) Find $h(x) = \dfrac{f(x)}{g(x)}$.

$$h(x) = \frac{f(x)}{g(x)}$$

$$= \frac{\dfrac{x^3 - 2x^2}{x + 2}}{\dfrac{x^2 - 3x + 2}{x - 4}} \qquad \text{Invert and multiply.}$$

$$= \frac{x^3 - 2x^2}{x + 2} \cdot \frac{x - 4}{x^2 - 3x + 2}$$

$$= \frac{x^2(x - 2)}{x + 2} \cdot \frac{x - 4}{(x - 1)(x - 2)}$$

$$= \frac{x^2(x - 4)}{(x + 2)(x - 1)} \qquad x \neq -2, 1, 2, 4$$

(*d*) For which values of *x* is *h*(*x*) undefined?

h(*x*) will be undefined for any value of *x* that would cause division by zero. *h*(*x*) is undefined for the values −2, 1, 2, and 4.

●●● **CHECK YOURSELF 5**

Given the rational functions

$$f(x) = \frac{x^2 - 2x + 1}{x + 3} \qquad \text{and} \qquad g(x) = \frac{x^2 - 5x + 4}{x - 2}$$

complete the following.

(**a**) Find $\dfrac{f(0)}{g(0)}$. (**b**) Find $\dfrac{f(1)}{g(1)}$. (**c**) Find $h(x) = \dfrac{f(x)}{g(x)}$.

(**d**) For which values of *x* is *h*(*x*) undefined?

───────────────────────

●●● **CHECK YOURSELF ANSWERS**

1. $\dfrac{4a}{3b^2}$. **2.** (**a**) $\dfrac{2(x + 2)}{x^2}$; (**b**) $\dfrac{-x^2}{4}$. **3.** (**a**) $\dfrac{x}{y}$; (**b**) 6; (**c**) $\dfrac{2x}{x^2 + 3x + 9}$.

4. (**a**) −10, (**b**) undefined, (**c**) $h(x) = (x - 5)(x + 2)$, (**d**) −10, and (**e**) undefined.

5. (**a**) $-\dfrac{1}{6}$, (**b**) undefined, (**c**) $h(x) = \dfrac{(x - 1)(x - 2)}{(x + 3)(x - 4)}$, and (**d**) $x \neq -3, 1, 2, 4$.

───────────────────────

Calculating Probability. Probability, or the study of chance, measures the relative likelihood of events. Suppose a situation has *N* equally likely possible outcomes, and an event includes *E* of these. Let *P* be the probability that the event will occur. Then

$$P = \frac{E}{N}.$$ From this definition, it is clear that probability will always be between 0 and 1: $0 \leq P \leq 1$. The probability of an event not happening is $1 - P$.

The probability of two separate, independent events, Event 1 and Event 2, *both* happening is the *product* of the probability of the first event times the probability of the second event: $P(E_1 \text{ and } E_2) = P(E_1) \cdot P(E_2)$. The events must be independent, which means the first event cannot have any effect on the outcome of the second.

Work with a partner to complete the following.

1. What is the probability that two people in your class have the same birthday? Surprisingly, it is pretty common in a class of 30 people to find at least two people who have the same birthday (not the same year). You can use the probability of two independent events—the probability that two people will *not* have the same birthday—to figure this out.

    Begin with one person from the class. This person can have a birthday on any day. What is the probability that the second person will *not* have the same birthday? (How many days are in the year?) Let us say that they do not have the same birthday as the first person. Now there are two days in the year when no one else in the class can have a birthday. If there are not two people with the same birthday, then the probability that the third person does not have the same birthday is. . . . If these two probabilities are multiplied, we get the probability that three people do not have the same birthday. Carry on until all 30 people have been accounted for. What is the probability that none of them have the same birthday? You may want to use the following formula.

    $$P(\text{no birthdays are the same for } n \text{ people}) = 1 - (P_2)(P_3)(P_4) \ldots (P_n)$$

    where

    $$P_i = \frac{365 - (i - 1)}{365}$$

    and is the probability that the *i*th person does not have the same birthday as anyone who came before.

2. Suppose there are 2000 lottery tickets in a large jar. You have bought one of these tickets. What is the probability that your ticket might be drawn if 5 tickets are drawn one after the other? (This is the probability of winning on the first draw, or on the second draw, or on the third draw, or on the fourth, etc.) Once, again, compute the probability that your ticket will not be drawn on each draw, and multiply to get the probability that none of the tickets will be yours. Remember that the tickets are not being returned to the jar after being drawn. Write an algebraic formula for this probability.

4.2 Exercises

In Exercises 1 to 36, multiply or divide as indicated. Express your result in simplest form.

1. $\dfrac{x^2}{3} \cdot \dfrac{6x}{x^4}$

2. $\dfrac{-y^3}{10} \cdot \dfrac{15y}{y^6}$

3. $\dfrac{a}{7a^3} \div \dfrac{a^2}{21}$

4. $\dfrac{p^5}{8} \div \dfrac{-p^2}{12p}$

5. $\dfrac{4xy^2}{15x^3} \cdot \dfrac{25xy}{16y^3}$

6. $\dfrac{3x^3y}{10xy^3} \cdot \dfrac{5xy^2}{-9xy^3}$

7. $\dfrac{8b^3}{15ab} \div \dfrac{2ab^2}{20ab^3}$

8. $\dfrac{4x^2y^2}{9x^3} \div \dfrac{-8y^2}{27xy}$

9. $\dfrac{m^3n}{2mn} \cdot \dfrac{6mn^2}{m^3n} \div \dfrac{3mn}{5m^2n}$

10. $\dfrac{4cd^2}{5cd} \cdot \dfrac{3c^3d}{2c^2d} \div \dfrac{9cd}{20cd^3}$

11. $\dfrac{5x + 15}{3x} \cdot \dfrac{9x^2}{2x + 6}$

12. $\dfrac{a^2 - 3a}{5a} \cdot \dfrac{20a^2}{3a - 9}$

13. $\dfrac{3b - 15}{6b} \div \dfrac{4b - 20}{9b^2}$

14. $\dfrac{7m^2 + 28m}{4m} \div \dfrac{5m + 20}{12m^2}$

15. $\dfrac{x^2 - 3x - 10}{5x} \cdot \dfrac{15x^2}{3x - 15}$

16. $\dfrac{y^2 - 8y}{4y} \cdot \dfrac{12y^2}{y^2 - 64}$

17. $\dfrac{c^2 + 2c - 8}{6c} \div \dfrac{5c + 20}{18c}$

18. $\dfrac{m^2 - 49}{5m} \div \dfrac{3m + 21}{20m^2}$

19. $\dfrac{x^2 - 2x - 8}{4x - 16} \cdot \dfrac{10x}{x^2 - 4}$

20. $\dfrac{y^2 + 7y + 10}{y^2 + 5y} \cdot \dfrac{2y}{y^2 - 4}$

21. $\dfrac{d^2 - 3d - 18}{16d - 96} \div \dfrac{d^2 - 9}{20d}$

22. $\dfrac{b^2 + 6b + 8}{b^2 + 4b} \div \dfrac{b^2 - 4}{2b}$

23. $\dfrac{2x^2 - x - 3}{3x^2 + 7x + 4} \cdot \dfrac{3x^2 - 11x - 20}{4x^2 - 9}$

24. $\dfrac{4p^2 - 1}{2p^2 - 9p - 5} \cdot \dfrac{3p^2 - 13p - 10}{9p^2 - 4}$

25. $\dfrac{a^2 - 9}{2a^2 - 6a} \div \dfrac{2a^2 + 5a - 3}{4a^2 - 1}$

26. $\dfrac{2x^2 - 5x - 7}{4x^2 - 9} \div \dfrac{5x^2 + 5x}{2x^2 + 3x}$

27. $\dfrac{2w - 6}{w^2 + 2w} \cdot \dfrac{3w}{3 - w}$

28. $\dfrac{3y - 15}{y^2 + 3y} \cdot \dfrac{4y}{5 - y}$

29. $\dfrac{a - 7}{2a + 6} \div \dfrac{21 - 3a}{a^2 + 3a}$

30. $\dfrac{x - 4}{x^2 + 2x} \div \dfrac{16 - 4x}{3x + 6}$

31. $\dfrac{x^2 - 9y^2}{2x^2 - xy - 15y^2} \cdot \dfrac{4x + 10y}{x^2 + 3xy}$

32. $\dfrac{2a^2 - 7ab - 15b^2}{2ab - 10b^2} \cdot \dfrac{2a^2 - 3ab}{4a^2 - 9b^2}$

33. $\dfrac{3m^2 - 5mn + 2n^2}{9m^2 - 4n^2} \div \dfrac{m^3 - m^2n}{9m^2 + 6mn}$

34. $\dfrac{2x^2y - 5xy^2}{4x^2 - 25y^2} \div \dfrac{4x^2 + 20xy}{2x^2 + 15xy + 25y^2}$

35. $\dfrac{x^3 + 8}{x^2 - 4} \cdot \dfrac{5x - 10}{x^3 - 2x^2 + 4x}$

36. $\dfrac{a^3 - 27}{a^2 - 9} \div \dfrac{a^3 + 3a^2 + 9a}{3a^3 + 9a^2}$

37. Let $f(x) = \dfrac{x^2 - 3x - 4}{x + 2}$ and $g(x) = \dfrac{x^2 - 2x - 8}{x - 4}$. Find **(a)** $f(0) \cdot g(0)$, **(b)** $f(4) \cdot g(4)$, **(c)** $h(x) = f(x) \cdot g(x)$, **(d)** $h(0)$, and **(e)** $h(4)$.

38. Let $f(x) = \dfrac{x^2 - 4x + 3}{x + 5}$ and $g(x) = \dfrac{x^2 + 7x + 10}{x - 3}$. Find **(a)** $f(1) \cdot g(1)$, **(b)** $f(3) \cdot g(3)$, **(c)** $h(x) = f(x) \cdot g(x)$, **(d)** $h(1)$, and **(e)** $h(3)$.

39. Let $f(x) = \dfrac{2x^2 - 3x - 5}{x + 2}$ and $g(x) = \dfrac{3x^2 + 5x - 2}{x + 1}$. Find **(a)** $f(1) \cdot g(1)$, **(b)** $f(-2) \cdot g(-2)$, **(c)** $h(x) = f(x) \cdot g(x)$, **(d)** $h(1)$, and **(e)** $h(-2)$.

40. Let $f(x) = \dfrac{x^2 - 1}{x - 3}$ and $g(x) = \dfrac{x^2 - 9}{x - 1}$. Find **(a)** $f(2) \cdot g(2)$, **(b)** $f(3) \cdot g(3)$, **(c)** $h(x) = f(x) \cdot g(x)$, **(d)** $h(2)$, and **(e)** $h(3)$.

41. Let $f(x) = \dfrac{3x^2 + x - 2}{x - 2}$ and $g(x) = \dfrac{x^2 - 4x - 5}{x + 4}$. Find **(a)** $\dfrac{f(0)}{g(0)}$, **(b)** $\dfrac{f(1)}{g(1)}$, **(c)** $h(x) = \dfrac{f(x)}{g(x)}$, and **(d)** the values of x for which $h(x)$ is undefined.

42. Let $f(x) = \dfrac{x^2 + x}{x - 5}$ and $g(x) = \dfrac{x^2 - x - 6}{x - 5}$. Find **(a)** $\dfrac{f(0)}{g(0)}$, **(b)** $\dfrac{f(2)}{g(2)}$, **(c)** $h(x) = \dfrac{f(x)}{g(x)}$, and **(d)** the values of x for which $h(x)$ is undefined.

The results from multiplying and dividing rational expressions can be checked by using a graphing calculator. To do this, define one expression in Y_1 and the other in Y_2. Then define the operation in Y_3 as $Y_1 \cdot Y_2$ or $Y_1 \div Y_2$. Put your simplified result in Y_4 (sorry, you still must simplify algebraically). Deselect the graphs for Y_1 and Y_2. If you have correctly simplified the expression, the graphs of Y_3 and Y_4 will be identical. Use this technique in Exercises 43 to 46.

43. $\dfrac{x^3 - 3x^2 + 2x - 6}{x^2 - 9} \cdot \dfrac{5x^2 + 15x}{20x}$

44. $\dfrac{3a^3 + a^2 - 9a - 3}{15a^2 + 5a} \cdot \dfrac{3a^2 + 9}{a^4 - 9}$

45. $\dfrac{x^4 - 16}{x^2 + x - 6} \div (x^3 + 4x)$

46. $\dfrac{w^3 + 27}{w^2 + 2w - 3} \div (w^3 - 3w^2 + 9w)$

Answers

1. $\dfrac{2}{x}$ **3.** $\dfrac{3}{a^4}$ **5.** $\dfrac{5}{12x}$ **7.** $\dfrac{16h^3}{3a}$ **9.** $5mn$ **11.** $\dfrac{15x}{2}$ **13.** $\dfrac{9b}{8}$ **15.** $x^2 + 2x$

17. $\dfrac{3(c - 2)}{5}$ **19.** $\dfrac{5x}{2(x - 2)}$ **21.** $\dfrac{5d}{4(d - 3)}$ **23.** $\dfrac{x - 5}{2x + 3}$ **25.** $\dfrac{2a - 1}{2a}$ **27.** $\dfrac{-6}{w + 2}$

29. $\dfrac{-a}{6}$ **31.** $\dfrac{2}{x}$ **33.** $\dfrac{3}{m}$ **35.** $\dfrac{5}{x}$ **37. (a)** -4, **(b)** undefined, **(c)** $(x + 1)(x - 4)$, **(d)** -4,

(e) undefined. **39.** (a) -6, (b) undefined, (c) $(2x - 5)(3x + 1)$, (d) -6, (e) undefined. **41.** (a) $-\dfrac{4}{5}$,

(b) $\dfrac{5}{4}$, (c) $\dfrac{(3x - 2)(x + 4)}{(x - 2)(x - 5)}$, (d) $2, -4, -1, 5$ **43.** $\dfrac{x^2 + 2}{4}$ **45.** $\dfrac{x + 2}{x(x + 3)}$

4.3 Adding and Subtracting Rational Expressions and Functions

4.3 OBJECTIVES

1. Add and subtract rational expressions
2. Add and subtract rational functions

ADDING AND SUBTRACTING RATIONAL EXPRESSIONS

Recall that adding or subtracting two arithmetic fractions with the same denominator is straightforward. The same is true in algebra. To add or subtract two rational expressions with the same denominator, we add or subtract their numerators and then write that sum or difference over the common denominator.

> **Adding or Subtracting Rational Expressions**
>
> $$\frac{P}{R} + \frac{Q}{R} = \frac{P + Q}{R}$$
>
> and
>
> $$\frac{P}{R} - \frac{Q}{R} = \frac{P - Q}{R}$$
>
> where $R \neq 0$.

●Example 1

Adding and Subtracting Rational Expressions

Perform the indicated operations.

Since we have common denominators, we simply perform the indicated operations on the numerators.

$$\frac{3}{2a^2} - \frac{1}{2a^2} + \frac{5}{2a^2} = \frac{3 - 1 + 5}{2a^2}$$

$$= \frac{7}{2a^2}$$

● ● ● CHECK YOURSELF 1

Perform the indicated operations.

$$\frac{5}{3y^2} + \frac{4}{3y^2} - \frac{7}{3y^2}$$

The sum or difference of rational expressions should always be expressed in simplest form. Consider Example 2.

● Example 2

Adding and Subtracting Rational Expressions

Add or subtract as indicated.

(a) $\dfrac{5x}{x^2-9} + \dfrac{15}{x^2-9}$ Add the numerators.

$$= \frac{5x+15}{x^2-9}$$

$$= \frac{5(x+3)}{(x-3)(x+3)} = \frac{5}{x-3}$$ Factor and divide by the common factor.

(b) $\dfrac{3x+y}{2x} - \dfrac{x-3y}{2x} = \dfrac{(3x+y)-(x-3y)}{2x}$ Be sure to *enclose the second numerator* in parentheses.

$$= \frac{3x+y-x+3y}{2x}$$ Remove the parentheses by *changing each sign*.

$$= \frac{2x+4y}{2x} = \frac{2(x+2y)}{2x}$$ Factor and divide by the common factor of 2.

$$= \frac{x+2y}{x}$$

● ● ● CHECK YOURSELF 2

Perform the indicated operations.

(a) $\dfrac{6a}{a^2-2a-8} + \dfrac{12}{a^2-2a-8}$ **(b)** $\dfrac{5x-y}{3y} - \dfrac{2x-4y}{3y}$

Now, what if our rational expressions *do not* have common denominators? In that case, we must use the least common denominator (LCD). The **least common denominator** is the simplest polynomial that is divisible by each of the individual denominators. Each expression in the desired sum or difference is then "built up" to an equivalent expression having that LCD as a denominator. We can then add or subtract as before.

By **inspection,** we mean you look at the denominators and find that the LCD is obvious (as in Example 2).

Again, we see the key role that factoring plays in the process of working with rational expressions.

Although in many cases we can find the LCD by inspection, we can state an algorithm for finding the LCD that is similar to the one used in arithmetic.

Finding the Least Common Denominator

STEP 1 Write each of the denominators in completely factored form.
STEP 2 Write the LCD as the product of each prime factor to the highest power to which it appears in the factored form of any individual denominators.

Example 3 illustrates the procedure.

• Example 3

Finding the LCD for Two Rational Expressions

Find the LCD for each of the following pairs of rational expressions.

(a) $\dfrac{3}{4x^2}$ and $\dfrac{5}{6xy}$

Factor the denominators.

You may very well be able to find this LCD by inspecting the numerical coefficients and the variable factors.

$4x^2 = 2^2 \cdot x^2$

$6y = 2 \cdot 3 \cdot x \cdot y$

The LCD must have the factors

$2^2 \cdot 3 \cdot x^2 \cdot y$

and so $12x^2y$ is the desired LCD.

(b) $\dfrac{7}{x-3}$ and $\dfrac{2}{x+5}$

Here, neither denominator can be factored. The LCD must have the factors $x-3$ and $x+5$. So the LCD is

It is generally best to leave the LCD in this factored form.

$(x-3)(x+5)$

● ● ● **CHECK YOURSELF 3**

Find the LCD for the following pairs of rational expressions.

(a) $\dfrac{3}{8a^3}$ and $\dfrac{5}{6a^2}$ **(b)** $\dfrac{4}{x+7}$ and $\dfrac{3}{x-5}$

Let's see how factoring techniques are applied in Example 4.

• Example 4

Finding the LCD for Two Rational Expressions

Find the LCD for the following pairs of rational expressions.

(a) $\dfrac{2}{x^2 - x - 6}$ and $\dfrac{1}{x^2 - 9}$

Factoring, we have

$$x^2 - x - 6 = (x + 2)(x - 3)$$

and

$$x^2 - 9 = (x + 3)(x - 3)$$

The LCD of the given expressions is then

The LCD most contain *each* of the factors appearing in the original denominators.

$$(x + 2)(x - 3)(x + 3)$$

(b) $\dfrac{5}{x^2 - 4x + 4}$ and $\dfrac{3}{x^2 + 2x - 8}$

Again, we factor:

$$x^2 - 4x + 4 = (x - 2)^2$$
$$x^2 + 2x - 8 = (x - 2)(x + 4)$$

The LCD must contain $(x - 2)^2$ as a factor since $x - 2$ appears *twice* as a factor in the first denominator.

The LCD is then

$$(x - 2)^2(x + 4)$$

● ● ● **CHECK YOURSELF 4**

Find the LCD for the following pairs of rational expressions.

(a) $\dfrac{3}{x^2 - 2x - 15}$ and $\dfrac{5}{x^2 - 25}$

(b) $\dfrac{5}{y^2 + 6y + 9}$ and $\dfrac{3}{y^2 - y - 12}$

Let's look at Example 5, in which the concept of the LCD is applied in adding or subtracting rational expressions.

• Example 5

Adding and Subtracting Rational Expressions

Add or subtract as indicated.

(a) $\dfrac{5}{4xy} + \dfrac{3}{2x^2}$

The LCD for $2x^2$ and $4xy$ is $4x^2y$. We rewrite each of the rational expressions with the LCD as a denominator.

Note that in each case we are multiplying by 1: $\dfrac{x}{x}$ in the first fraction and $\dfrac{2y}{2y}$ in the second fraction, which is why the resulting fractions are equivalent to the original ones.

$$\dfrac{5}{4xy} + \dfrac{3}{2x^2} = \dfrac{5 \cdot x}{4xy \cdot x} + \dfrac{3 \cdot 2y}{2x^2 \cdot 2y}$$

$$= \dfrac{5x}{4x^2y} + \dfrac{6y}{4x^2y} = \dfrac{5x + 6y}{4x^2y}$$

Multiply the first rational expression by $\dfrac{x}{x}$ and the second by $\dfrac{2y}{2y}$ to form the LCD of $4x^2y$.

(b) $\dfrac{3}{a-3} - \dfrac{2}{a}$

The LCD for a and $a - 3$ is $a(a - 3)$. We rewrite each of the rational expressions with that LCD as a denominator.

$$\dfrac{3}{a-3} - \dfrac{2}{a}$$

$$= \dfrac{3a}{a(a-3)} - \dfrac{2(a-3)}{a(a-3)}$$

Subtract the numerators.

$$= \dfrac{3a - 2(a-3)}{a(a-3)}$$

Remove the parentheses, and combine like terms.

$$= \dfrac{3a - 2a + 6}{a(a-3)} = \dfrac{a+6}{a(a-3)}$$

●●● **CHECK YOURSELF 5**

Perform the indicated operations.

(a) $\dfrac{3}{2ab} + \dfrac{4}{5b^2}$

(b) $\dfrac{5}{y+2} - \dfrac{3}{y}$

Let's proceed to Example 6, in which factoring will be required in forming the LCD.

• Example 6

Adding and Subtracting Rational Expressions

Add or subtract as indicated.

(a) $\dfrac{-5}{x^2 - 3x - 4} + \dfrac{8}{x^2 - 16}$

We first factor the two denominators.

$$x^2 - 3x - 4 = (x + 1)(x - 4)$$
$$x^2 - 16 = (x + 4)(x - 4)$$

We see that the LCD must be

$$(x + 1)(x + 4)(x - 4)$$

Again, rewriting the original expressions with factored denominators gives

We use the facts that

$\dfrac{x + 4}{x + 4} = 1$ and

$\dfrac{x + 1}{x + 1} = 1$

$$\dfrac{-5}{(x + 1)(x - 4)} + \dfrac{8}{(x - 4)(x + 4)}$$

$$= \dfrac{-5(x + 4)}{(x + 1)(x - 4)(x + 4)} + \dfrac{8(x + 1)}{(x - 4)(x + 4)(x + 1)}$$

$$= \dfrac{-5(x + 4) + 8(x + 1)}{(x + 1)(x - 4)(x + 4)} \qquad \text{Now add the numerators.}$$

$$= \dfrac{-5x - 20 + 8x + 8}{(x + 1)(x - 4)(x + 4)} \qquad \text{Combine like terms in the numerator.}$$

$$= \dfrac{3x - 12}{(x + 1)(x - 4)(x + 4)} \qquad \text{Factor.}$$

$$= \dfrac{3(x - 4)}{(x + 1)(x - 4)(x + 4)} \qquad \text{Divide by the common factor } x - 4.$$

$$= \dfrac{3}{(x + 1)(x + 4)}$$

(b) $\dfrac{5}{x^2 - 5x + 6} - \dfrac{3}{4x - 12}$

Again, factor the denominators.

$$x^2 - 5x + 6 = (x - 2)(x - 3)$$
$$4x - 12 = 4(x - 3)$$

The LCD is $4(x - 2)(x - 3)$, and proceeding as before, we have

$$\dfrac{5}{(x - 2)(x - 3)} - \dfrac{3}{4(x - 3)}$$

$$= \dfrac{5 \cdot 4}{4(x - 2)(x - 3)} - \dfrac{3(x - 2)}{4(x - 2)(x - 3)}$$

$$= \dfrac{20 - 3(x - 2)}{4(x - 2)(x - 3)}$$

$$= \dfrac{20 - 3x + 6}{4(x - 2)(x - 3)} = \dfrac{-3x + 26}{4(x - 2)(x - 3)} \qquad \text{Simplify the numerator and combine like terms.}$$

● ● ● **CHECK YOURSELF 6**

Add or subtract as indicated.

(a) $\dfrac{-4}{x^2 - 4} + \dfrac{7}{x^2 - 3x - 10}$ **(b)** $\dfrac{5}{3x - 9} - \dfrac{2}{x^2 - 9}$

Example 7 looks slightly different from those you have seen thus far, but the reasoning involved in performing the subtraction is exactly the same.

● Example 7

Subtracting Rational Expressions

Subtract.

$$3 - \dfrac{5}{2x - 1}$$

To perform the subtraction, remember that 3 is equivalent to the fraction $\dfrac{3}{1}$, so

$$3 - \dfrac{5}{2x - 1} = \dfrac{3}{1} - \dfrac{5}{2x - 1}$$

The LCD for 1 and $2x - 1$ is just $2x - 1$. We now rewrite the first expression with that denominator.

$$3 - \dfrac{5}{2x - 1} = \dfrac{3(2x - 1)}{2x - 1} - \dfrac{5}{2x - 1}$$

$$= \dfrac{3(2x - 1) - 5}{2x - 1} = \dfrac{6x - 8}{2x - 1}$$

● ● ● **CHECK YOURSELF 7**

Subtract.

$$\dfrac{4}{3x + 1} - 3$$

Example 8 uses an observation from Section 4.1. Recall that

$$a - b = -(b - a)$$
$$= -1(b - a)$$

Let's see how this is used in adding rational expressions.

• Example 8

Adding and Subtracting Rational Expressions

Add.

$$\frac{x^2}{x - 5} + \frac{3x + 10}{5 - x}$$

Your first thought might be to use a denominator of $(x - 5)(5 - x)$. However, we can simplify our work considerably if we multiply the numerator and denominator of the second fraction by -1 to find a common denominator.

Use

$$\frac{-1}{-1} = 1$$

Note that

$(-1)(5 - x) = x - 5$

The fractions now have a common denominator, and we can add as before.

$$\frac{x^2}{x - 5} + \frac{3x + 10}{5 - x}$$

$$= \frac{x^2}{x - 5} + \frac{(-1)(3x + 10)}{(-1)(5 - x)}$$

$$= \frac{x^2}{x - 5} + \frac{-3x - 10}{x - 5}$$

$$= \frac{x^2 - 3x - 10}{x - 5}$$

$$= \frac{(x + 2)(x - 5)}{x - 5}$$

$$= x + 2$$

● ● ● ● **CHECK YOURSELF 8**

Add.

$$\frac{x^2}{x - 7} + \frac{10x - 21}{7 - x}$$

ADDING RATIONAL FUNCTIONS

The sum of two rational functions is always a rational function. Given two rational functions, $f(x)$ and $g(x)$, we can rename the sum, so $h(x) = f(x) + g(x)$. This will always be true for values of x for which both f and g are defined. So, for example, $h(-2) = f(-2) + g(-2)$, so long as both $f(-2)$ and $g(-2)$ exist.

If

$h(x) = f(x) + g(x)$

then the set of ordered pairs

$(x, f(x) + g(x)) = (x, h(x))$

Example 9 illustrates this approach.

• Example 9

Adding Two Rational Functions

Given

$$f(x) = \frac{3x}{x+5} \quad \text{and} \quad g(x) = \frac{x}{x-4}$$

complete the following.

(a) Find $f(1) + g(1)$.

If $f(1) = \frac{1}{2}$ and $g(1) = -\frac{1}{3}$, then

$$f(1) + g(1) = \frac{1}{2} + \left(-\frac{1}{3}\right)$$

$$= \frac{3}{6} + \left(-\frac{2}{6}\right)$$

$$= \frac{1}{6}$$

(b) Find $h(x) = f(x) + g(x)$.

$$h(x) = f(x) + g(x)$$

$$= \frac{3x}{x+5} + \frac{x}{x-4}$$

$$= \frac{3x(x-4) + x(x+5)}{(x+5)(x-4)}$$

$$= \frac{3x^2 - 12x + x^2 + 5x}{(x+5)(x-4)}$$

$$= \frac{4x^2 - 7x}{(x+5)(x-4)} \quad x \neq -5, 4$$

(c) Find the ordered pair $(1, h(1))$.

$$h(1) = \frac{-3}{-18} = \frac{1}{6}$$

The ordered pair is $\left(1, \frac{1}{6}\right)$.

● ● ● CHECK YOURSELF 9

Given

$$f(x) = \frac{x}{2x-5} \quad \text{and} \quad g(x) = \frac{2x}{3x-1}$$

complete the following.

(a) Find $f(1) + g(1)$.

(b) Find $h(x) = f(x) + g(x)$.

(c) Find the ordered pair $(1, h(1))$.

SUBTRACTING RATIONAL FUNCTIONS

When subtracting rational functions, one must take particular care with the signs in the numerator of the expression being subtracted.

● Example 10

Subtracting Rational Functions

Given

$$f(x) = \frac{3x}{x + 5} \quad \text{and} \quad g(x) = \frac{x - 2}{x - 4}$$

complete the following.

(a) Find $f(1) - g(1)$.

If $f(1) = \dfrac{1}{2}$ and $g(1) = \dfrac{1}{3}$, then

$$f(x) + g(x) = \frac{1}{2} - \left(\frac{1}{3}\right)$$

$$= \frac{3}{6} - \left(\frac{2}{6}\right)$$

$$= \frac{1}{6}$$

(b) Find $h(x) = f(x) - g(x)$.

$$h(x) = \frac{3x}{x + 5} - \frac{x - 2}{x - 4}$$

$$= \frac{3x(x - 4) - (x - 2)(x + 5)}{(x + 5)(x - 4)}$$

$$= \frac{(3x^2 - 12x) - (x^2 + 3x - 10)}{(x + 5)(x - 4)}$$

$$= \frac{2x^2 - 15x + 10}{(x + 5)(x - 4)} \qquad x \neq -5, 4$$

(c) Find the ordered pair $(1, h(1))$

$$h(1) = \frac{-3}{-18} = \frac{1}{6}$$

The ordered pair is $\left(1, \frac{1}{6}\right)$

● ● ● CHECK YOURSELF 10

Given

$$f(x) = \frac{x}{2x - 5} \quad \text{and} \quad g(x) = \frac{2x - 1}{3x - 1}$$

complete the following.

(a) Find $f(1) - g(1)$.   **(b)** Find $h(x) = f(x) - g(x)$.
(c) Find the ordered pair $(1, h(1))$.

● ● ● CHECK YOURSELF ANSWERS

1. $\frac{2}{3y^2}$.   **2. (a)** $\frac{6}{a - 4}$; **(b)** $\frac{x + y}{y}$.   **3. (a)** $24a^3$; **(b)** $(x + 7)(x - 5)$.

4. (a) $(x - 5)(x + 5)(x + 3)$; **(b)** $(y + 3)^2(y - 4)$.   **5. (a)** $\frac{8a + 15b}{10ab^2}$;

(b) $\frac{2y - 6}{y(y + 2)}$.   **6. (a)** $\frac{3}{(x - 2)(x - 5)}$; **(b)** $\frac{5x + 9}{3(x + 3)(x - 3)}$.   **7.** $\frac{-9x + 1}{3x + 1}$.

8. $x - 3$.   **9. (a)** $\frac{2}{3}$; **(b)** $h(x) = \frac{7x^2 - 11x}{(2x - 5)(3x - 1)}$; **(c)** $\left(1, \frac{2}{3}\right)$.   **10. (a)** $\frac{-5}{6}$;

(b) $h(x) = \frac{-x^2 + 11x - 5}{(2x - 5)(3x - 1)}$; **(c)** $\left(1, -\frac{5}{6}\right)$.

Probability and Pari-Mutual Betting. In most gambling games, payoffs are determined by the **odds**. At horse and dog tracks, the odds (D) are a ratio that is calculated by taking into account the total amount wagered (A), the amount wagered on a particular animal (a), and the government share, called the take-out (f). The ratio is then rounded down to a comparison of integers like 99 to 1, 3 to 1, or 5 to 2. Below is the formula that tracks use to find odds.

$$(D) = \frac{A(1 - f)}{a} - 1$$

Work with a partner to complete the following.

1. Assume that the government takes 10%, and simplify the expression for D. Use this formula to compute the odds on each horse if a total of $10,000 were bet on all the horses and the amounts were distributed as shown in the table.

Horse	Total Amount Wagered on This Horse to Win	Odds: Amount Paid on Each Dollar Bet If Horse Wins
1	$5000	
2	$1000	
3	$2000	
4	$1500	
5	$ 500	

2. Odds can be used as a guide in determining the chance that a given horse will win. The probability of a horse winning is related to many variables, such as track condition, how the horse is feeling, and weather. However, the odds do reflect the consensus opinion of racing fans and can be used to give some idea of the probability.

The relationship between odds and probability is given by the equations

$$P(\text{win}) = \frac{1}{D + 1}$$

and $P(\text{loss}) = 1 - P(\text{win})$

or $P(\text{loss}) = 1 - \dfrac{1}{D + 1}$

Solve this equation for D, the odds against the horse winning. Do the probabilities for each horse winning all add up to 1? Should they add to 1?

4.3 Exercises

In Exercises 1 to 36, perform the indicated operations. Express your results in simplest form.

1. $\dfrac{7}{2x^2} + \dfrac{5}{2x^2}$

2. $\dfrac{11}{3b^3} - \dfrac{2}{3b^3}$

3. $\dfrac{5}{3a + 7} + \dfrac{2}{3a + 7}$

4. $\dfrac{6}{5x + 3} - \dfrac{3}{5x + 3}$

5. $\dfrac{2x}{x - 3} - \dfrac{6}{x - 3}$

6. $\dfrac{7w}{w + 3} + \dfrac{21}{w + 3}$

7. $\dfrac{y^2}{2y + 8} + \dfrac{3y - 4}{2y + 8}$

8. $\dfrac{x^2}{4x - 12} - \dfrac{9}{4x - 12}$

9. $\dfrac{4m - 7}{m - 5} - \dfrac{2m + 3}{m - 5}$

10. $\dfrac{3b - 8}{b - 6} + \dfrac{b - 16}{b - 6}$

11. $\dfrac{x - 7}{x^2 - x - 6} + \dfrac{2x - 2}{x^2 - x - 6}$

12. $\dfrac{5x - 12}{x^2 - 8x + 15} - \dfrac{3x - 2}{x^2 - 8x + 15}$

13. $\dfrac{5}{3x} + \dfrac{3}{2x}$

14. $\dfrac{4}{5w} - \dfrac{3}{4w}$

15. $\dfrac{6}{a} + \dfrac{3}{a^2}$

16. $\dfrac{3}{p} - \dfrac{7}{p^2}$

17. $\dfrac{2}{m} - \dfrac{2}{n}$

18. $\dfrac{3}{x} + \dfrac{3}{y}$

19. $\dfrac{3}{4b^2} - \dfrac{5}{3b^3}$

20. $\dfrac{4}{5x^3} - \dfrac{3}{2x^2}$

21. $\dfrac{2}{a} - \dfrac{1}{a-2}$

22. $\dfrac{4}{c} + \dfrac{3}{c+1}$

23. $\dfrac{2}{x+1} + \dfrac{3}{x+2}$

24. $\dfrac{4}{y-1} + \dfrac{2}{y+3}$

25. $\dfrac{5}{y-3} - \dfrac{1}{y+1}$

26. $\dfrac{4}{x+5} - \dfrac{3}{x-1}$

27. $\dfrac{2w}{w-7} + \dfrac{w}{w-2}$

28. $\dfrac{3n}{n+5} + \dfrac{n}{n-4}$

29. $\dfrac{3x}{3x-2} - \dfrac{2x}{2x+1}$

30. $\dfrac{5c}{5c-1} + \dfrac{2c}{2c-3}$

31. $\dfrac{6}{m-7} + \dfrac{2}{7-m}$

32. $\dfrac{5}{a-5} - \dfrac{3}{5-a}$

33. $\dfrac{3}{x^2-16} + \dfrac{2}{x-4}$

34. $\dfrac{5}{y^2+5y+6} + \dfrac{2}{y+2}$

35. $\dfrac{4m}{m^2-3m+2} - \dfrac{1}{m-2}$

36. $\dfrac{x}{x^2-1} - \dfrac{2}{x-1}$

 As we saw in Section 4.2 exercises, the graphing calculator can be used to check our work. In Exercises 37 to 42, enter the first rational expression in Y_1 and the second in Y_2. In Y_3, you will enter either $Y_1 + Y_2$ or $Y_1 - Y_2$. Enter your algebraically simplified rational expression in Y_4. The graphs of Y_3 and Y_4 will be identical if you have correctly simplified the expression.

37. $\dfrac{6y}{y^2-8y+15} + \dfrac{9}{y-3}$

38. $\dfrac{8a}{a^2-8a+12} + \dfrac{4}{a-2}$

39. $\dfrac{6x}{x^2-10x+24} - \dfrac{18}{x-6}$

40. $\dfrac{21p}{p^2-3p-10} - \dfrac{15}{p-5}$

41. $\dfrac{2}{z^2-4} + \dfrac{3}{z^2+2z-8}$

42. $\dfrac{5}{x^2-3x-10} + \dfrac{2}{x^2-25}$

In Exercises 43 to 46, find **(a)** $f(1) + g(1)$, **(b)** $h(x) = f(x) + g(x)$, and **(c)** the ordered pair $(1, h(1))$.

43. $f(x) = \dfrac{3x}{x+1}$ and $g(x) = \dfrac{2x}{x-3}$

44. $f(x) = \dfrac{4x}{x-4}$ and $g(x) = \dfrac{x+4}{x+1}$

45. $f(x) = \dfrac{x}{x+1}$ and $g(x) = \dfrac{1}{x^2 + 2x + 1}$

46. $f(x) = \dfrac{x+2}{x-4}$ and $g(x) = \dfrac{x+3}{x+4}$

In Exercises 47 to 50, find **(a)** $f(1) - g(1)$, **(b)** $h(x) = f(x) - g(x)$, and **(c)** the ordered pair $(1, h(1))$.

47. $f(x) = \dfrac{x+5}{x-5}$ and $g(x) = \dfrac{x-5}{x+5}$

48. $f(x) = \dfrac{2x}{x-4}$ and $g(x) = \dfrac{3x}{x+7}$

49. $f(x) = \dfrac{x+9}{4x-36}$ and $g(x) = \dfrac{x-9}{x^2 - 18x + 81}$

50. $f(x) = \dfrac{4x+1}{x+5}$ and $g(x) = -\dfrac{2}{x}$

In Exercises 51 to 60, evaluate each expression at the given variable value(s).

51. $\dfrac{5x+5}{x^2 + 3x + 2} - \dfrac{x-3}{x^2 + 5x + 6}, \; x = -4$

52. $\dfrac{y-3}{y^2 - 6y + 8} + \dfrac{2y-6}{y^2 - 4}, \; y = 3$

53. $\dfrac{2m+2n}{m^2 - n^2} + \dfrac{m-2n}{m^2 + 2mn + n^2}, \; m = 3, n = 2$

54. $\dfrac{w-3z}{w^2 - 2wz + z^2} - \dfrac{w+2z}{w^2 - z^2}, \; w = 2, z = 1$

55. $\dfrac{1}{a-3} - \dfrac{1}{a+3} + \dfrac{2a}{a^2 - 9}, \; a = 4$

56. $\dfrac{1}{m+1} + \dfrac{1}{m-3} - \dfrac{4}{m^2 - 2m - 3}, \; m = -2$

57. $\dfrac{3w^2 + 16w - 8}{w^2 + 2w - 8} + \dfrac{w}{w+4} - \dfrac{w-1}{w-2}, \; w = 3$

58. $\dfrac{4x^2 - 7x - 45}{x^2 - 6x + 5} - \dfrac{x+2}{x-1} - \dfrac{x}{x-5}, \; x = -3$

59. $\dfrac{a^2 - 9}{2a^2 - 5a - 3} \cdot \left(\dfrac{1}{a-2} + \dfrac{1}{a+3} \right), \; a = -3$

60. $\dfrac{m^2 - 2mn + n^2}{m^2 + 2mn - 3n^2} \cdot \left(\dfrac{2}{m-n} - \dfrac{1}{m+n} \right), \; m = 4, n = -3$

Answers

1. $\dfrac{6}{x^2}$ **3.** $\dfrac{7}{3a+7}$ **5.** 2   **7.** $\dfrac{y-1}{2}$ **9.** 2   **11.** $\dfrac{3}{x+2}$ **13.** $\dfrac{19}{6x}$ **15.** $\dfrac{3(2a+1)}{a^2}$

17. $\dfrac{2(n-m)}{mn}$ **19.** $\dfrac{9b+20}{12b^3}$ **21.** $\dfrac{a-4}{a(a-2)}$ **23.** $\dfrac{5x+7}{(x+1)(x+2)}$ **25.** $\dfrac{4(y+2)}{(y-3)(y+1)}$

27. $\dfrac{w(3w-11)}{(w-7)(w-2)}$ **29.** $\dfrac{7x}{(3x-2)(2x+1)}$ **31.** $\dfrac{4}{m-7}$ **33.** $\dfrac{2x+11}{(x+4)(x-4)}$

35. $\dfrac{3m+1}{(m-1)(m-2)}$ **37.** $\dfrac{15}{y-5}$ **39.** $\dfrac{-12}{x-4}$ **41.** $\dfrac{5z+14}{(z+2)(z-2)(z+4)}$ **43. (a)** $\dfrac{1}{2}$,

(b) $\dfrac{5x^2 - 7x}{(x+1)(x-3)}$, **(c)** $\left(1, \dfrac{1}{2}\right)$ **45. (a)** $\dfrac{3}{4}$, **(b)** $\dfrac{x^2 + x + 1}{(x+1)^2}$,   **(c)** $\left(1, \dfrac{3}{4}\right)$ **47. (a)** $-\dfrac{5}{6}$,

(b) $\dfrac{20x}{(x-5)(x+5)}$, **(c)** $\left(1, \dfrac{5}{6}\right)$ **49. (a)** $-\dfrac{3}{16}$,   **(b)** $\dfrac{x+5}{4(x-9)}$, **(c)** $\left(1, \dfrac{-3}{16}\right)$ **51.** 1   **53.** $\dfrac{49}{25}$

55. 2   **57.** 8   **59.** Undefined

 **4.4** ## Simplifying Complex Fractions

4.4 OBJECTIVES

1. Use the fundamental principle to simplify complex fractions
2. Use division to simplify complex fractions

Our work in this section deals with two methods for simplifying complex fractions. We begin with a definition. A **complex fraction** is a fraction that has a fraction in its numerator or denominator (or both). Some examples are

$$\frac{\dfrac{5}{6}}{\dfrac{3}{4}} \qquad \frac{\dfrac{4}{x}}{\dfrac{3}{x+1}} \qquad \text{and} \qquad \frac{1+\dfrac{1}{x}}{1-\dfrac{1}{x}}$$

Two methods can be used to simplify complex fractions. Method 1 involves the fundamental principle, and Method 2 involves inverting and multiplying.

METHOD 1 FOR SIMPLIFYING COMPLEX FRACTIONS

Fundamental principle:

$$\frac{P}{Q} = \frac{PR}{QR}$$

where $Q \neq 0$ and $R \neq 0$.

Recall that by the *fundamental principle* we can always multiply the numerator and denominator of a fraction by the same nonzero quantity. In simplifying a complex fraction, we multiply the numerator and denominator by the LCD of all fractions that appear within the complex fraction.

Here the denominators are 5 and 10, so we can write

Again, we are multiplying by $\dfrac{10}{10}$ or 1.

$$\frac{\dfrac{3}{5}}{\dfrac{7}{10}} = \frac{\dfrac{3}{5}\cdot 10}{\dfrac{7}{10}\cdot 10} = \frac{6}{7}$$

METHOD 2 FOR SIMPLIFYING COMPLEX FRACTIONS

Our second approach interprets the complex fraction as indicating division and applies our earlier work in dividing fractions in which we *invert and multiply*.

$$\frac{\dfrac{3}{5}}{\dfrac{7}{10}} = \frac{3}{5} \div \frac{7}{10} = \frac{3}{5} \cdot \frac{10}{7} = \frac{6}{7} \qquad \text{Invert and multiply.}$$

Which method is better? The answer depends on the expression you are trying to simplify. Both approaches are effective, and you should be familiar with both. With practice you will be able to tell which method may be easier to use in a particular situation.

Let's look at the same two methods applied to the simplification of an algebraic complex fraction.

• Example 1

Simplifying Complex Fractions

Simplify.

$$\frac{1 + \dfrac{2x}{y}}{2 - \dfrac{x}{y}}$$

Method 1 The LCD of 1, $\dfrac{2x}{y}$, 2, and $\dfrac{x}{y}$ is y. So we multiply the numerator and denominator by y.

$$\frac{1 + \dfrac{2x}{y}}{2 - \dfrac{x}{y}} = \frac{\left(1 + \dfrac{2x}{y}\right) \cdot y}{\left(2 - \dfrac{x}{y}\right) \cdot y}$$ Distribute y over the numerator and denominator.

$$= \frac{1 \cdot y + \dfrac{2x}{y} \cdot y}{2 \cdot y - \dfrac{x}{y} \cdot y}$$ Simplify.

$$= \frac{y + 2x}{2y - x}$$

Method 2 In this approach, we must *first work separately* in the numerator and denominator to form single fractions.

Make sure you understand the steps in forming a single fraction in the numerator and denominator.

$$\frac{1 + \dfrac{2x}{y}}{2 - \dfrac{x}{y}} = \frac{\dfrac{y}{y} + \dfrac{2x}{y}}{\dfrac{2y}{y} - \dfrac{x}{y}} = \frac{\dfrac{y + 2x}{y}}{\dfrac{2y - x}{y}}$$

$$= \frac{y + 2x}{y} \cdot \frac{y}{2y - x}$$ Invert the divisor and multiply.

$$= \frac{y + 2x}{2y - x}$$

● ● ●   CHECK YOURSELF 1

Simplify.

$$\frac{\dfrac{x}{y} - 1}{\dfrac{2x}{y} + 2}$$

Again, simplifying a complex fraction means writing an equivalent simple fraction in lowest terms, as Example 2 illustrates.

• Example 2

Simplifying Complex Fractions

Simplify.

$$\frac{1 - \dfrac{2y}{x} + \dfrac{y^2}{x^2}}{1 - \dfrac{y^2}{x^2}}$$

We choose the first method of simplification in this case. The LCD of all the fractions that appear is x^2. So we multiply the numerator and denominator by x^2.

$$\frac{1 - \dfrac{2y}{x} + \dfrac{y^2}{x^2}}{1 - \dfrac{y^2}{x^2}} = \frac{\left(1 - \dfrac{2y}{x} + \dfrac{y^2}{x^2}\right) \cdot x^2}{\left(1 - \dfrac{y^2}{x^2}\right) \cdot x^2}$$

Distribute x^2 over the numerator and denominator, and simplify.

$$= \frac{x^2 - 2xy + y^2}{x^2 - y^2}$$

Factor the numerator and denominator.

$$= \frac{(x - y)(x - y)}{(x + y)(x - y)} = \frac{x - y}{x + y}$$

Divide by the common factor $x - y$.

● ● ● **CHECK YOURSELF 2**

Simplify.

$$\frac{1 + \dfrac{5}{x} + \dfrac{6}{x^2}}{1 - \dfrac{9}{x^2}}$$

In Example 3, we will illustrate the second method of simplification for purposes of comparison.

• Example 3

Simplifying Complex Fractions

Simplify.

$$\frac{1 - \dfrac{1}{x+2}}{x - \dfrac{2}{x-1}}$$

Again, take time to make sure you understand how the numerator and denominator are rewritten as single fractions.

Note: Method 2 is probably the more efficient in this case. The LCD of the denominators would be $(x+2)(x-1)$, leading to a somewhat more complicated process if method 1 were used.

$$\frac{1 - \dfrac{1}{x+2}}{x - \dfrac{2}{x-1}} = \frac{\dfrac{x+2}{x+2} - \dfrac{1}{x+2}}{\dfrac{x(x-1)}{x-1} - \dfrac{2}{x-1}} = \frac{\dfrac{x+1}{x+2}}{\dfrac{x^2-x-2}{x-1}}$$

$$= \frac{x+1}{x+2} \cdot \frac{x-1}{x^2-x-2}$$

$$= \frac{x+1}{x+2} \cdot \frac{x-1}{(x-2)(x+1)}$$

$$= \frac{x-1}{(x+2)(x-2)}$$

● ● ● CHECK YOURSELF 3

Simplify.

$$\frac{2 + \dfrac{5}{x-3}}{x - \dfrac{1}{2x+1}}$$

The following algorithm summarizes our work with complex fractions.

Simplifying Complex Fractions

METHOD 1

1. Multiply the numerator and denominator of the complex fraction by the LCD of all the fractions that appear within the numerator and denominator.
2. Simplify the resulting rational expression, writing the expression in lowest terms.

METHOD 2

1. Write the numerator and denominator of the complex fraction as single fractions, if necessary.
2. Invert the denominator and multiply as before, writing the result in lowest terms.

● ● ● CHECK YOURSELF ANSWERS

1. $\dfrac{x-y}{2x+2y}$.   2. $\dfrac{x+2}{x-3}$.   3. $\dfrac{2x+1}{(x-3)(x+1)}$.

4.4 Exercises

In Exercises 1 to 39, simplify each complex fraction.

1. $\dfrac{\dfrac{2}{3}}{\dfrac{6}{8}}$

2. $\dfrac{\dfrac{5}{6}}{\dfrac{10}{15}}$

3. $\dfrac{\dfrac{2}{3} + \dfrac{1}{2}}{\dfrac{3}{4} - \dfrac{1}{3}}$

4. $\dfrac{\dfrac{3}{4} + \dfrac{1}{2}}{\dfrac{7}{8} - \dfrac{1}{4}}$

5. $\dfrac{2 + \dfrac{1}{3}}{3 - \dfrac{1}{5}}$

6. $\dfrac{1 + \dfrac{3}{4}}{2 - \dfrac{1}{8}}$

7. $\dfrac{\dfrac{x}{8}}{\dfrac{x^2}{4}}$

8. $\dfrac{\dfrac{a^2}{10}}{\dfrac{a^3}{15}}$

9. $\dfrac{\dfrac{3}{m}}{\dfrac{6}{m^2}}$

10. $\dfrac{\dfrac{15}{x^2}}{\dfrac{20}{x^3}}$

11. $\dfrac{\dfrac{y + 1}{y}}{\dfrac{y - 1}{2y}}$

12. $\dfrac{\dfrac{x + 3}{4x}}{\dfrac{x - 3}{2x}}$

13. $\dfrac{\dfrac{a + 2b}{3a}}{\dfrac{a^2 + 2ab}{9b}}$

14. $\dfrac{\dfrac{m - 3n}{4m}}{\dfrac{m^2 - 3mn}{8n}}$

15. $\dfrac{\dfrac{x - 2}{x^2 - 9}}{\dfrac{x^2 - 4}{x^2 + 3x}}$

16. $\dfrac{\dfrac{x + 5}{x^2 - 6x}}{\dfrac{x^2 - 25}{x^2 - 36}}$

17. $\dfrac{2 - \dfrac{1}{x}}{2 + \dfrac{1}{x}}$

18. $\dfrac{3 + \dfrac{1}{b}}{3 - \dfrac{1}{b}}$

19. $\dfrac{\dfrac{1}{x} - \dfrac{1}{y}}{\dfrac{1}{xy}}$

20. $\dfrac{\dfrac{1}{ab}}{\dfrac{1}{a} + \dfrac{1}{b}}$

21. $\dfrac{\dfrac{x^2}{y^2} - 1}{\dfrac{x}{y} + 1}$

330

22. $\dfrac{\dfrac{m}{n} + 2}{\dfrac{m^2}{n^2} - 4}$

23. $\dfrac{1 + \dfrac{3}{a} - \dfrac{4}{a^2}}{1 + \dfrac{2}{a} - \dfrac{3}{a^2}}$

24. $\dfrac{1 - \dfrac{2}{x} - \dfrac{8}{x^2}}{1 - \dfrac{1}{x} - \dfrac{6}{x^2}}$

25. $\dfrac{\dfrac{x^2}{y} + 2x + y}{\dfrac{1}{y^2} - \dfrac{1}{x^2}}$

26. $\dfrac{\dfrac{a}{b} + 1 - \dfrac{2b}{a}}{\dfrac{1}{b^2} - \dfrac{4}{a^2}}$

27. $\dfrac{1 + \dfrac{1}{x - 1}}{1 - \dfrac{1}{x - 1}}$

28. $\dfrac{2 - \dfrac{1}{m - 2}}{2 + \dfrac{1}{m - 2}}$

29. $\dfrac{1 - \dfrac{1}{y - 1}}{y - \dfrac{8}{y + 2}}$

30. $\dfrac{1 + \dfrac{1}{x + 2}}{x - \dfrac{18}{x - 3}}$

31. $\dfrac{\dfrac{1}{x - 3} + \dfrac{1}{x + 3}}{\dfrac{1}{x - 3} - \dfrac{1}{x + 3}}$

32. $\dfrac{\dfrac{2}{m - 2} + \dfrac{1}{m - 3}}{\dfrac{2}{m - 2} - \dfrac{1}{m - 3}}$

33. $\dfrac{\dfrac{x}{x + 1} + \dfrac{1}{x - 1}}{\dfrac{x}{x - 1} - \dfrac{1}{x + 1}}$

34. $\dfrac{\dfrac{y}{y - 4} + \dfrac{1}{y + 2}}{\dfrac{4}{y - 4} - \dfrac{1}{y + 2}}$

35. $\dfrac{\dfrac{a + 1}{a - 1} - \dfrac{a - 1}{a + 1}}{\dfrac{a + 1}{a - 1} + \dfrac{a - 1}{a + 1}}$

36. $\dfrac{\dfrac{x + 2}{x - 2} - \dfrac{x - 2}{x + 2}}{\dfrac{x + 2}{x - 2} + \dfrac{x - 2}{x + 2}}$

37. $1 + \dfrac{1}{1 + \dfrac{1}{x}}$

38. $1 + \dfrac{1}{1 - \dfrac{1}{y}}$

39. $1 + \dfrac{1}{1 + \dfrac{1}{1 + \dfrac{1}{x}}}$

 40. Extend the "continued fraction" patterns in Exercises 37 and 39 to write the next complex fraction.

 41. Outline the two different methods used to simplify a complex fraction. What are the advantages of each method?

 42. Can the expression $\dfrac{x^{-1} + y^{-1}}{x^{-2} + y^{-2}}$ be written as $\dfrac{x^2 + y^2}{x + y}$? If not, what is the correct simplified form?

43. Simplify the complex fraction in Exercise 42.

44. Write and simplify a complex fraction that is the reciprocal of $x + \dfrac{6}{x-1}$.

45. Let $f(x) = \dfrac{3}{x}$. Write and simplify a complex fraction whose numerator is $f(3 + h) - f(3)$ and whose denominator is h.

46. Write and simplify a complex fraction that is the arithmetic mean of $\dfrac{1}{x}$ and $\dfrac{1}{x-1}$.

47. Compare your results in Exercises 37, 41, and 42. Could you have predicted the result?

Suppose you drive at 40 mi/h from city A to city B. You then return along the same route from city B to city A at 50 mi/h. What is your average rate for the round trip? Your obvious guess would be 45 mi/h, but you are in for a surprise.

Suppose that the cities are 200 mi apart. Your time from city A to city B is the distance divided by the rate, or

$$\frac{200 \text{ mi}}{40 \text{ mi/h}} = 5 \text{ h}$$

Similarly, your time from city B to city A is

$$\frac{200 \text{ mi}}{50 \text{ mi/h}} = 4 \text{ h}$$

The total time is then 9 h, and now using *rate equals distance divided by time,* we have

$$\frac{400 \text{ mi}}{9 \text{ h}} = \frac{400}{9} \text{ mi/h} = 44\frac{4}{9} \text{ mi/h}$$

Note that the rate for the round trip is independent of the distance involved. For instance, try the same computations above if cities A and B are 400 mi apart.

The answer to the problem above is the complex fraction

$$R = \frac{2}{\dfrac{1}{R_1} + \dfrac{1}{R_2}}$$

where R_1 = rate going
   R_2 = rate returning
   R = rate for round trip

Use this information to solve Exercises 48 to 51.

48. Verify that if $R_1 = 40$ mi/h and $R_2 = 50$ mi/h, then $R = 44\dfrac{4}{9}$ mi/h, by simplifying the complex fraction *after* substituting those values.

49. Simplify the given complex fraction first. *Then* substitute 40 for R_1 and 50 for R_2 to calculate R.

50. Repeat Exercise 48, where $R_1 = 50$ mi/h and $R_2 = 60$ mi/h.

51. Use the procedure in Exercise 49 with the above values for R_1 and R_2.

52. Show that the inequality regarding relative difference in apportionment given at the beginning of the chapter

$$\frac{\dfrac{E}{e} - \dfrac{A}{a+1}}{\dfrac{A}{a+1}} < \frac{\dfrac{A}{a} - \dfrac{E}{e+1}}{\dfrac{E}{e+1}}$$

can be simplified to

$$\frac{A}{\sqrt{a(a+1)}} > \frac{E}{\sqrt{e(e+1)}}.$$

Here, A and E represent the populations of two states of the United States, and a and e are the number of representatives each of these two states have in the U. S. House of Representatives.

53. Mathematicians have shown that there are situations in which the method for apportionment described in the chapter's introduction does not work, and a state may not even get its basic quota of representatives. They give the table below of a hypothetical seven states and their populations as an example.

State	Population	Exact Quota	Number of Reps.
A	325	1.625	2
B	788	3.940	4
C	548	2.740	3
D	562	2.810	3
E	4,263	21.315	21
F	3,219	16.095	15
G	295	1.475	2
Total	10,000	50	50

In this case, the total population of all states is 10,000, and there are 50 representatives in all, so there should be no more than 10,000/50 or 200 people per representative. The quotas are found by dividing the population by 200. Whether a state, A, should get an additional representative before another state, E, should get one is decided in this method by using the simplified inequality below. If the ratio

$$\frac{A}{\sqrt{a(a+1)}} > \frac{E}{\sqrt{e(e+1)}}$$

is true, then A gets an extra representative before E does.

(a) If you go through the process of comparing the inequality above for each pair of states, state F loses a representative to state G. Do you see how this happens? Will state F complain?

(b) Alexander Hamilton, one of the signers of the Constitution, proposed that the extra representative positions be given one at a time to states with the largest remainder until all the "extra" positions were filled. How would this affect the table? Do you agree or disagree?

54. In Italy in the 1500s, Pietro Antonio Cataldi expressed square roots as infinite, continued fractions. It is not a difficult process to follow. For instance, if you want the square root of 5, then let

$$x + 1 = \sqrt{5}$$

Squaring both sides gives

$$(x + 1)^2 = 5 \quad \text{or} \quad x^2 + 2x + 1 = 5$$

which can be written

$$x(x + 2) = 4$$

$$x = \frac{4}{x + 2}$$

One can continue replacing x with $\dfrac{4}{x + 2}$:

$$x = \cfrac{4}{2 + \cfrac{4}{2 + \cfrac{4}{2 + \cfrac{4}{2 + \dots}}}}$$

to obtain

$$\sqrt{5} - 1$$

(a) Evaluate this complex fraction and see how close it is to the square root of 5. What should you put where the ellipses (. . .) are? Try a number you feel is close to $\sqrt{5}$. How far would you have to go to get the square root correct to the nearest hundredth?

(b) Develop an infinite complex fraction for $\sqrt{10}$.

Answers

1. $\dfrac{8}{9}$ **3.** $\dfrac{14}{5}$ **5.** $\dfrac{5}{6}$ **7.** $\dfrac{1}{2x}$ **9.** $\dfrac{m}{2}$ **11.** $\dfrac{2(y + 1)}{y - 1}$ **13.** $\dfrac{3b}{a^2}$ **15.** $\dfrac{x}{(x + 2)(x - 3)}$

17. $\dfrac{2x - 1}{2x + 1}$ **19.** $y - x$ **21.** $\dfrac{x - y}{y}$ **23.** $\dfrac{a + 4}{a + 3}$ **25.** $\dfrac{x^2 y(x + y)}{x - y}$ **27.** $\dfrac{x}{x - 2}$

29. $\dfrac{y + 2}{(y - 1)(y + 4)}$ **31.** $\dfrac{x}{3}$ **33.** 1 **35.** $\dfrac{2a}{a^2 + 1}$ **37.** $\dfrac{2x + 1}{x + 1}$ **39.** ✎   **41.** $\dfrac{3x + 2}{2x + 1}$

43. $\dfrac{5x + 3}{3x + 2}$ **45.** $\dfrac{-1}{3 + h}$ **47.** ✎   **49.** $44\dfrac{4}{9}$ mi/h   **51.** $54\dfrac{6}{11}$ mi/h   **53.** ✎

 **4.5 Rational Equations and Inequalities in One Variable**

4.5 OBJECTIVES

1. Solve rational equations in one variable algebraically
2. Solve literal equations involving a rational expression
3. Find the zeros of a rational function
4. Solve rational inequalities in one variable algebraically

RATIONAL EQUATIONS

Applications of your work in algebra will often result in equations involving rational expressions. Our objective in this section is to develop methods to find the solution of such equations.

The usual solution technique is to multiply both sides of the equation by the lowest common denominator (LCD) of all the rational expressions appearing in the equation. The resulting equation will be cleared of fractions, and we can then proceed to solve the equation as before. Example 1 illustrates the process.

• Example 1

Clearing Equations of Fractions

Solve.

$$\frac{2x}{3} + \frac{x}{5} = 13$$

The LCD for 3 and 5 is 15. Multiplying both sides of the equation by 15, we have

$$15\left(\frac{2x}{3} + \frac{x}{5}\right) = 15 \cdot 13 \qquad \text{Distribute 15 on the left.}$$

$$15 \cdot \frac{2x}{3} + 15 \cdot \frac{x}{5} = 15 \cdot 13$$

$$10x + 3x = 195 \qquad \text{Simplify. The equation is now cleared of fractions.}$$

$$13x = 195$$

$$x = 15$$

To check, substitute 15 in the original equation.

$$\frac{2 \cdot 15}{3} + \frac{15}{5} \stackrel{?}{=} 13$$

$$10 + 3 \stackrel{?}{=} 13$$

$$13 = 13 \qquad \text{A true statement.}$$

So 15 is the solution for the equation.

C A U T I O N

Be Careful! A common mistake is to confuse an *equation* such as

$$\frac{2x}{3} + \frac{x}{5} = 13$$

and an *expression* such as

$$\frac{2x}{3} + \frac{x}{5}$$

Let's compare.

Equation: $\dfrac{2x}{3} + \dfrac{x}{5} = 13$

Here we want to *solve the equation for x,* as in Example 1. We multiply both sides by the LCD to clear fractions and proceed as before.

Expression: $\dfrac{2x}{3} + \dfrac{x}{5}$

Here we want to find *a third fraction* that is equivalent to the given expression. We write each fraction as an equivalent fraction with the LCD as a common denominator.

$$\frac{2x}{3} + \frac{x}{5} = \frac{2x \cdot 5}{3 \cdot 5} + \frac{x \cdot 3}{5 \cdot 3}$$

$$= \frac{10x}{15} + \frac{3x}{15} = \frac{10x + 3x}{15}$$

$$= \frac{13x}{15}$$

● ● ● **CHECK YOURSELF 1**

Solve.

$$\frac{3x}{2} - \frac{x}{3} = 7$$

The process is similar when variables are in the denominators. Consider Example 2.

● Example 2

Solving an Equation Involving Rational Expressions

Solve.

We assume that *x* cannot have the value 0. Do you see why?

$$\frac{7}{4x} - \frac{3}{x^2} = \frac{1}{2x^2}$$

The LCD of $4x$, x^2, and $2x^2$ is $4x^2$. So, multiplying both sides by $4x^2$, we have

$$4x^2\left(\frac{7}{4x} - \frac{3}{x^2}\right) = 4x^2 \cdot \frac{1}{2x^2} \qquad \text{Distribute } 4x^2 \text{ on the left side.}$$

$$4x^2 \cdot \frac{7}{4x} - 4x^2 \cdot \frac{3}{x^2} = 4x^2 \cdot \frac{1}{2x^2} \qquad \text{Simplify.}$$

$$7x - 12 = 2$$

$$7x = 14$$

$$x = 2$$

We leave the check of the solution, $x = 2$, to you. Be sure to return to the original equation and substitute 2 for x.

CHECK YOURSELF 2

Solve.

$$\frac{5}{2x} - \frac{4}{x^2} = \frac{7}{2x^2}$$

Example 3 illustrates the same solution process when there are binomials in the denominators.

• Example 3

Solving an Equation Involving Rational Expressions

Solve.

Here we assume that x *cannot* have the value -2 or 3.

$$\frac{4}{x + 2} + 3 = \frac{3x}{x - 3}$$

The LCD is $(x + 2)(x - 3)$. Multiplying by that LCD, we have

Note that multiplying *each term* by the LCD is the same as multiplying both sides of the equation by the LCD.

$$(x + 2)(x - 3)\left(\frac{4}{x + 2}\right) + (x + 2)(x - 3)(3) = (x + 2)(x - 3)\left(\frac{3x}{x - 3}\right)$$

Or, simplifying each term, we have

$$4(x - 3) + 3(x + 2)(x - 3) = 3x(x + 2)$$

We now clear the parentheses and proceed as before.

$$4x - 12 + 3x^2 - 3x - 18 = 3x^2 + 6x$$

$$3x^2 + x - 30 = 3x^2 + 6x$$

$$x - 30 = 6x$$

$$-5x = 30$$

$$x = -6$$

Again, we leave the check of this solution to you.

● ● ● **CHECK YOURSELF 3**

Solve.

$$\frac{5}{x-4} + 2 = \frac{2x}{x-3}$$

Factoring plays an important role in solving equations containing rational expressions.

● Example 4

Solving an Equation Involving Rational Expressions

Solve.

$$\frac{3}{x-3} - \frac{7}{x+3} = \frac{2}{x^2-9}$$

In factored form, the denominator on the right side is $(x-3)(x+3)$, which forms the LCD, and we multiply each term by that LCD.

$$(x-3)(x+3)\left(\frac{3}{x-3}\right) - (x-3)(x+3)\left(\frac{7}{x+3}\right) = (x-3)(x+3)\left[\frac{2}{(x-3)(x+3)}\right]$$

Again, simplifying each term on the right and left sides, we have

$$3(x+3) - 7(x-3) = 2$$
$$3x + 9 - 7x + 21 = 2$$
$$-4x = -28$$
$$x = 7$$

Be sure to check this result by substitution in the original equation.

● ● ● **CHECK YOURSELF 4**

Solve $\dfrac{4}{x-4} - \dfrac{3}{x+1} = \dfrac{5}{x^2-3x-4}$.

Whenever we multiply both sides of an equation by an expression containing a variable, there is the possibility that a proposed solution may make that multiplier 0. As we pointed out earlier, multiplying by 0 does not give an equivalent equation, and therefore verifying solutions by substitution serves not only as a check of our work but also as a check for extraneous solutions. Consider Example 5.

•Example 5

Solving an Equation Involving Rational Expressions

Solve.

Note that we must assume that $x \neq 2$.

$$\frac{x}{x-2} - 7 = \frac{2}{x-2}$$

The LCD is $x - 2$, and multiplying, we have

Note that each of the three terms gets multiplied by $(x - 2)$.

$$\left(\frac{x}{x-2}\right)(x-2) - 7(x-2) = \left(\frac{2}{x-2}\right)(x-2)$$

Simplifying yields

$$x - 7(x - 2) = 2$$
$$x - 7x + 14 = 2$$
$$-6x = -12$$
$$x = 2$$

CAUTION

Because division by 0 is undefined, we conclude that 2 is *not a solution* for the original equation. It is an extraneous solution. The original equation has no solution.

To check this result, by substituting 2 for x, we have

$$\frac{2}{2-2} - 7 \stackrel{?}{=} \frac{2}{2-2}$$

$$\frac{2}{0} - 7 \stackrel{?}{=} \frac{2}{0}$$

● ● ● **CHECK YOURSELF 5**

Solve $\dfrac{x-3}{x-4} = 4 + \dfrac{1}{x-4}$.

Equations involving rational expressions may also lead to quadratic equations, as illustrated in Example 6.

•Example 6

Solving an Equation Involving Rational Expressions

Solve.

Assume $x \neq 3$ and $x \neq 4$.

$$\frac{x}{x-4} = \frac{15}{x-3} - \frac{2x}{x^2 - 7x + 12}$$

After factoring the trinomial denominator on the right, the LCD of $x - 3$, $x - 4$, and $x^2 - 7x + 12$ is $(x - 3)(x - 4)$. Multiplying by that LCD, we have

$$(x - 3)(x - 4)\left(\frac{x}{x - 4}\right) = (x - 3)(x - 4)\left(\frac{15}{x - 3}\right) - (x - 3)(x - 4)\left[\frac{2x}{(x - 3)(x - 4)}\right]$$

Simplifying yields

$x(x - 3) = 15(x - 4) - 2x$ Remove the parentheses.

$x^2 - 3x = 15x - 60 - 2x$ Write in standard form and factor.

$x^2 - 16x + 60 = 0$

$(x - 6)(x - 10) = 0$

So

$x = 6$ or $x = 10$

Verify that 6 and 10 are both solutions for the original equation.

● ● ● **CHECK YOURSELF 6**

Solve $\dfrac{3x}{x + 2} - \dfrac{2}{x + 3} = \dfrac{36}{x^2 + 5x + 6}$.

The following algorithm summarizes our work in solving equations containing rational expressions.

> **Solving Literal Equations Containing Rational Expressions**
>
> STEP 1 Clear the equation of fractions by multiplying both sides of the equation by the LCD of all the fractions that appear.
> STEP 2 Solve the equation resulting from step 1.
> STEP 3 Check all solutions by substitution in the original equation.

The method in this section may also be used to solve certain literal equations for a specified variable. Consider Example 7.

• Example 7

Solving a Literal Equation

A parallel electric circuit. The symbol for a resistor is

—⋎⋎⋎— .

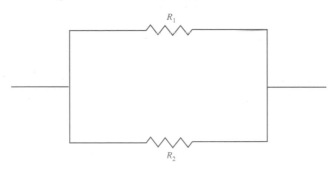

Recall that the numbers 1 and 2 are *subscripts*. We read R_1 as "R sub 1" and R_2 as "R sub 2."

If two resistors with resistances R_1 and R_2 are connected in parallel, the combined resistance R can be found from

$$\frac{1}{R} = \frac{1}{R_1} + \frac{1}{R_2}$$

Solve the formula for R.

First, the LCD is RR_1R_2, and we multiply:

$$RR_1R_2 \cdot \frac{1}{R} = RR_1R_2 \cdot \frac{1}{R_1} + RR_1R_2 \cdot \frac{1}{R_2}$$

Simplifying yields

$R_1R_2 = RR_2 + RR_1$ Factor out R on the right.

$R_1R_2 = R(R_2 + R_1)$ Divide by $R_2 + R_1$ to isolate R.

Symmetric property of equality.

$$\frac{R_1R_2}{R_2 + R_1} = R \qquad \text{or} \qquad R = \frac{R_1R_2}{R_1 + R_2}$$

● ● ●   **CHECK YOURSELF 7**

Solve for D_1.

Note: This formula involves the focal length of a convex lens.

$$\frac{1}{F} = \frac{1}{D_1} + \frac{1}{D_2}$$

The techniques we have just discussed can also be used to find the zeros of rational functions. Remember that a zero of a function is a value of x for which $f(x) = 0$.

• Example 8

Finding the Zeros of a Function

Find the zeros of

$$f(x) = \frac{1}{x} - \frac{3}{7x} - \frac{4}{21}$$

Set the function equal to 0, and solve the resulting equation for x.

$$f(x) = \frac{1}{x} - \frac{3}{7x} - \frac{4}{21} = 0$$

The LCD for x, $7x$, and 21 is $21x$. Multiplying both sides by $21x$, we have

$$21x\left(\frac{1}{x} - \frac{3}{7x} - \frac{4}{21}\right) = 21x \cdot 0$$

$$21 - 9 - 4x = 0 \qquad \text{Distribute } 21x \text{ on the left side.}$$

$$12 - 4x = 0 \qquad \text{Simplify.}$$

$$12 = 4x$$

$$3 = x$$

So 3 is the value of x for which $f(x) = 0$, that is, 3 is a zero of $f(x)$.

● ● ●   **CHECK YOURSELF 8**

Find the zeros of the function.

$$f(x) = \frac{5x + 2}{x - 6} - \frac{11}{4}$$

RATIONAL INEQUALITIES

To solve inequalities involving rational expressions, we need some properties of division over the real numbers. Recall that

1. The quotient of two positive numbers is always positive.
2. The quotient of two negative numbers is always positive.
3. The quotient of a positive number and a negative number is always negative.

As with quadratic inequalities, we solve rational inequalities by using sign graphs, as Example 9 illustrates.

• Example 9

The graph of

$$y = \frac{x - 3}{x + 2} \text{ is}$$

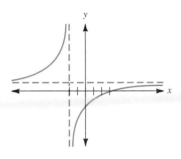

For what values of x is y less than 0?

Solving a Rational Inequality

Solve.

$$\frac{x - 3}{x + 2} < 0$$

The inequality states that the quotient of $x - 3$ and $x + 2$ must be negative (less than 0). This means that the numerator and denominator must have opposite signs.

We start by finding the critical points. These are points where either the numerator or denominator is 0. In this case, the critical points are 3 and -2.

The solution depends on determining whether the numerator and denominator are positive or negative. To visualize the process, start with a number line and label it as shown below.

Examining the sign of the numerator and denominator, we see that

For any x less than -2, the quotient is positive (quotient of two negatives).

For any x between -2 and 3, the quotient is negative (quotient of a negative and a positive).

For any x greater than 3, the quotient is positive (quotient of two positives).

We return to the original inequality

$$\frac{x - 3}{x + 2} < 0$$

This inequality is true only when the quotient is negative, that is, when x is between -2 and 3. This solution can be written as $\{x \mid -2 < x < 3\}$ and represented on a graph as follows.

● ● ● CHECK YOURSELF 9

Solve and graph the solution set.

$$\frac{x - 4}{x + 2} > 0$$

As with quadratic inequalities, the solution process illustrated in Example 9 is valid only when the rational expression is isolated on one side of the inequality and is related to 0. If this is not the case, we must write an equivalent inequality as the first step, as Example 10 illustrates.

• Example 10

Solving a Rational Inequality

Solve.

$$\frac{2x - 3}{x + 1} \geq 1$$

Since the rational expression is not related to 0, we use the following procedure.

We have subtracted 1 from both sides.

$$\frac{2x - 3}{x + 1} - 1 \geq 0 \qquad \text{Form a common denominator on the left side.}$$

$$\frac{2x - 3}{x + 1} - \frac{x + 1}{x + 1} \geq 0 \qquad \text{Combine the expressions on the left side.}$$

$$\frac{2x - 3 - (x + 1)}{x + 1} \geq 0 \qquad \text{Simplify.}$$

$$\frac{x - 4}{x + 1} \geq 0$$

We can now proceed as before since the rational expression is related to 0. The critical points are 4 and -1, and the sign graph is formed as shown below.

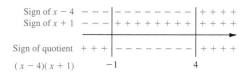

We want a *positive* quotient.

From the sign graph the solution is

$$\{x | x < -1 \text{ or } x \geq 4\}$$

The graph is shown below.

Note that 4 is included, but -1 cannot be included in the solution set. Why?

● ● ● **CHECK YOURSELF 10**

Solve and graph the solution set.

$$\frac{2x - 3}{x - 2} \leq 1$$

● ● ● **CHECK YOURSELF ANSWERS**

1. 6.   **2.** 3.   **3.** 9.   **4.** -11.   **5.** No solution.   **6.** $-5, \dfrac{8}{3}$.

7. $\dfrac{FD_2}{D_2 - F}$. **8.** $-\dfrac{74}{9}$. **9.** $\{x \mid x < -2 \text{ or } x > 4\}$ ← ○———————○ → -2 4

10. $\{x \mid 1 \le x < 2\}$ ← ●———○ → 1 2

4.5 Exercises

In Exercises 1 to 8, decide whether each of the following is an expression or an equation. If it is an equation, find a solution. If it is an expression, write it as a single fraction.

1. $\dfrac{x}{2} - \dfrac{x}{3} = 6$

2. $\dfrac{x}{4} - \dfrac{x}{7} = 3$

3. $\dfrac{x}{2} - \dfrac{x}{5}$

4. $\dfrac{x}{6} - \dfrac{x}{8}$

5. $\dfrac{3x + 1}{4} = x - 1$

6. $\dfrac{3x - 1}{2} - \dfrac{x}{5} - \dfrac{x + 3}{4}$

7. $\dfrac{x}{4} = \dfrac{x}{12} + \dfrac{1}{2}$

8. $\dfrac{2x - 1}{3} + \dfrac{x}{2}$

In Exercises 9 to 50, solve each equation.

9. $\dfrac{x}{3} + \dfrac{3}{2} = \dfrac{x}{6} + \dfrac{7}{3}$

10. $\dfrac{x}{10} - \dfrac{1}{5} = \dfrac{x}{5} + \dfrac{1}{2}$

11. $\dfrac{4}{x} + \dfrac{3}{4} = \dfrac{10}{x}$

12. $\dfrac{3}{x} = \dfrac{5}{3} - \dfrac{7}{x}$

13. $\dfrac{5}{4x} - \dfrac{1}{2} = \dfrac{1}{2x}$

14. $\dfrac{7}{6x} - \dfrac{1}{3} = \dfrac{1}{2x}$

15. $\dfrac{3}{x + 4} = \dfrac{2}{x + 3}$

16. $\dfrac{5}{x - 2} = \dfrac{4}{x - 1}$

17. $\dfrac{9}{x} + 2 = \dfrac{2x}{x + 3}$

18. $\dfrac{6}{x} - 3 = \dfrac{3x}{x + 1}$

19. $\dfrac{3}{x + 2} - \dfrac{5}{x} = \dfrac{13}{x + 2}$

20. $\dfrac{7}{x} - \dfrac{2}{x - 3} = \dfrac{6}{x}$

21. $\dfrac{3}{2} + \dfrac{2}{2x - 4} = \dfrac{1}{x - 2}$

22. $\dfrac{2}{x - 1} + \dfrac{5}{2x - 2} = \dfrac{3}{4}$

23. $\dfrac{x}{3x+12} + \dfrac{x-1}{x+4} = \dfrac{5}{3}$

24. $\dfrac{x}{4x-12} - \dfrac{x-4}{x-3} = \dfrac{1}{8}$

25. $\dfrac{x-1}{x+3} - \dfrac{x-3}{x} = \dfrac{3}{x^2+3x}$

26. $\dfrac{x+1}{x-2} - \dfrac{x+3}{x} = \dfrac{6}{x^2-2x}$

27. $\dfrac{1}{x-2} - \dfrac{2}{x+2} = \dfrac{2}{x^2-4}$

28. $\dfrac{1}{x+4} + \dfrac{1}{x-4} = \dfrac{12}{x^2-16}$

29. $\dfrac{7}{x+5} - \dfrac{1}{x-5} = \dfrac{x}{x^2-25}$

30. $\dfrac{2}{x-2} = \dfrac{3}{x+2} + \dfrac{x}{x^2-4}$

31. $\dfrac{11}{x+2} - \dfrac{5}{x^2-x-6} = \dfrac{1}{x-3}$

32. $\dfrac{5}{x-4} = \dfrac{1}{x+2} - \dfrac{2}{x^2-2x-8}$

33. $\dfrac{5}{x-2} - \dfrac{3}{x+3} = \dfrac{24}{x^2+x-6}$

34. $\dfrac{3}{x+1} - \dfrac{5}{x+6} = \dfrac{2}{x^2+7x+6}$

35. $\dfrac{x}{x-3} - 2 = \dfrac{3}{x-3}$

36. $\dfrac{x}{x-5} + 2 = \dfrac{5}{x-5}$

37. $\dfrac{2}{x^2-3x} - \dfrac{1}{x^2+2x} = \dfrac{2}{x^2-x-6}$

38. $\dfrac{2}{x^2-x} - \dfrac{4}{x^2+5x-6} = \dfrac{3}{x^2+6x}$

39. $\dfrac{2}{x^2-4x+3} - \dfrac{3}{x^2-9} = \dfrac{2}{x^2+2x-3}$

40. $\dfrac{2}{x^2-4} - \dfrac{1}{x^2+x-2} = \dfrac{3}{x^2-3x+2}$

41. $2 - \dfrac{6}{x^2} = \dfrac{1}{x}$

42. $3 - \dfrac{7}{x} - \dfrac{6}{x^2} = 0$

43. $1 - \dfrac{7}{x-2} + \dfrac{12}{(x-2)^2} = 0$

44. $1 + \dfrac{3}{x+1} = \dfrac{10}{(x+1)^2}$

45. $1 + \dfrac{3}{x^2-9} = \dfrac{10}{x+3}$

46. $3 - \dfrac{7}{x^2-x-6} = \dfrac{5}{x-3}$

47. $\dfrac{2x}{x-3} + \dfrac{2}{x-5} = \dfrac{3x}{x^2-8x+15}$

48. $\dfrac{x}{x-4} = \dfrac{5x}{x^2-x-12} - \dfrac{3}{x+3}$

49. $\dfrac{2x}{x+2} = \dfrac{5}{x^2-x-6} - \dfrac{1}{x-3}$

50. $\dfrac{3x}{x-1} = \dfrac{2}{x-2} - \dfrac{2}{x^2-3x+2}$

In Exercises 51 to 58, solve each equation for the indicated variable.

51. $\dfrac{1}{x} = \dfrac{1}{a} - \dfrac{1}{b}$ for x

52. $\dfrac{1}{x} = \dfrac{1}{a} + \dfrac{1}{b}$ for a

53. $\dfrac{1}{R} = \dfrac{1}{R_1} + \dfrac{1}{R_2}$ for R_1

54. $\dfrac{1}{F} = \dfrac{1}{D_1} + \dfrac{1}{D_2}$ for D_2

55. $y = \dfrac{x+1}{x-1}$ for x

56. $y = \dfrac{x-3}{x-2}$ for x

57. $t = \dfrac{A-P}{Pr}$ for P

58. $I = \dfrac{nE}{R+nr}$ for n

In Exercises 59 to 72, solve each inequality, and graph the solution set.

59. $\dfrac{x-2}{x+1} < 0$

60. $\dfrac{x+3}{x-2} > 0$

61. $\dfrac{x-4}{x-2} > 0$

62. $\dfrac{x+6}{x-3} < 0$

63. $\dfrac{x-5}{x+3} \le 0$

64. $\dfrac{x+3}{x-2} \le 0$

65. $\dfrac{2x-1}{x+3} \ge 0$

66. $\dfrac{3x-2}{x-4} \le 0$

67. $\dfrac{x}{x-3} + \dfrac{2}{x-3} \le 0$

68. $\dfrac{x}{x+5} - \dfrac{3}{x+5} > 0$

69. $\dfrac{x}{x+3} \le \dfrac{4}{x+3}$

70. $\dfrac{x}{x-5} \ge \dfrac{2}{x-5}$

71. $\dfrac{2x-5}{x-2} > 1$

72. $\dfrac{2x+3}{x+4} \ge 1$

In Exercises 73 to 80, find the zeros of each function.

73. $f(x) = \dfrac{x}{10} - \dfrac{12}{5}$

74. $f(x) = \dfrac{4x}{3} - \dfrac{x}{6}$

75. $f(x) = \dfrac{12}{x+5} - \dfrac{5}{x}$

76. $f(x) = \dfrac{1}{x-2} - \dfrac{3}{x}$

77. $f(x) = \dfrac{1}{x-3} + \dfrac{2}{x} - \dfrac{5}{3x}$

78. $f(x) = \dfrac{2}{x} - \dfrac{1}{x+1} - \dfrac{3}{x^2+x}$

79. $f(x) = 1 + \dfrac{39}{x^2} - \dfrac{16}{x}$

80. $f(x) = x - \dfrac{72}{x} + 1$

81. What special considerations must be made when an equation contains rational expressions with variables in the denominator?

82. In solving the inequality $\dfrac{x}{x-1} > 5$, is it incorrect to find the solution by multiplying both sides by $x - 1$? Why? What technique should be used?

83. A cellular phone base station is located at point A and has a transmitting and receiving radius of 10 miles. There is another station 20 miles due east of A. This station is active over an area with a radius of 12 miles. There is a road from point A that goes straight to a town at point C (see figure below).

(a) How many miles is it from A to C?
(b) As you travel this road, are you ever out of range of a cellular phone station? If so, for how many miles along the road does a cellular phone not work?
(c) By how many miles must the radius of the station at A or at C be increased to cover this gap?
(d) Write a letter to the Zephyr Cellular Phone Co. explaining where the gap in service is and how big it is and what they have to do to correct it. Explain how you were able to find all this using algebra.

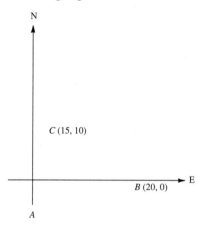

Answers

1. Equation, 36   **3.** Expression, $\dfrac{3x}{10}$ **5.** Equation, 5   **7.** Equation, 3   **9.** 5   **11.** 8

13. $\dfrac{3}{2}$ **15.** -1 **17.** $-\dfrac{9}{5}$ **19.** 2   **21.** No solution   **23.** -23 **25.** 6   **27.** 4   **29.** 8

31. 4   **33.** $\dfrac{3}{2}$ **35.** No solution   **37.** 7   **39.** 5   **41.** 2, $-\dfrac{3}{2}$ **43.** 5, 6   **45.** 4, 6

47. $-\dfrac{1}{2}$, 6   **49.** $-\dfrac{1}{2}$ **51.** $\dfrac{ab}{b-a}$ **53.** $\dfrac{RR_2}{R_2-R}$ **55.** $\dfrac{y+1}{y-1}$ **57.** $\dfrac{A}{1+rt}$ **59.** $-1 < x < 2$

61. $x < 2$ or $x < 4$ **63.** $-3 < x \le 5$ **65.** $x < -3$ or $x \ge \dfrac{1}{2}$ **67.** $-2 \le x < 3$ **69.** $-3 < x \le 4$

71. $x < 2$ or $x > 3$ **73.** 24   **75.** $\dfrac{25}{7}$ **77.** $\dfrac{3}{4}$ **79.** 3, 13   **81.**   **83.**

Summary

Simplifying Rational Expressions and Functions [4.1]

$\dfrac{x^2 - 5x}{x - 3}$ is a rational expression. The variable x cannot have the value 3.

Rational expressions have the form

$$\frac{P}{Q}$$ where P and Q are polynomials and Q cannot have the value 0.

Fundamental Principle of Rational Expressions

This uses the fact that

$$\frac{R}{R} = 1$$

where $R \neq 0$.

For polynomials P, Q, and R,

$$\frac{P}{Q} = \frac{PR}{QR}$$ where $Q \neq 0$ and $R \neq 0$

This principle can be used in two ways. We can multiply or divide the numerator and denominator of a rational expression by the same nonzero polynomial.

Simplifying Rational Expressions

$$\frac{x^2 - 4}{x^2 - 2x - 8}$$

$$= \frac{(x - 2)(x + 2)}{(x - 4)(x + 2)}$$

$$= \frac{x - 2}{x - 4}$$

To simplify a rational expression, use the following algorithm.

1. Completely factor both the numerator and denominator of the expression.
2. Divide the numerator and denominator by *all* common factors.
3. The resulting expression will be in simplest form (or in lowest terms).

Identifying Rational Functions

A rational function is a function that is defined by a rational expression. It can be written as

$$f(x) = \frac{P}{Q}$$ where P and Q are polynomials $Q \neq 0$

Simplifying Rational Functions

When we simplify a rational function, it is important that we note the x values that need to be excluded, particularly when we are trying to draw the graph of a function. The set of ordered pairs of the simplified function will be exactly the same as the set of ordered pairs of the original function. If we plug the excluded value(s) for x into the simplified expression, we get a set of ordered pairs that represent "holes" in the graph. These holes are breaks in the curve. We use an open circle to designate them on a graph.

Multiplying and Dividing Rational Expressions and Functions [4.2]

Multiplying Rational Expressions

For polynomials P, Q, R, and S,

$$\frac{2x-6}{x^2-9} \cdot \frac{x^2+3x}{6x+24}$$

$$= \frac{2(x-3)}{(x-3)(x+3)} \cdot \frac{x(x+3)}{6(x+4)}$$

$$= \frac{x}{3(x+4)}$$

$$\frac{P}{Q} \cdot \frac{R}{S} = \frac{PR}{QS} \qquad \text{where } Q \neq 0 \qquad S \neq 0$$

In practice, we apply the following algorithm to multiply two rational expressions.

1. Write each numerator and denominator in completely factored form.
2. Divide by any common factors appearing in both the numerator and denominator.
3. Multiply as needed to form the desired product.

Dividing Rational Expressions

For polynomials P, Q, R, and S,

$$\frac{5y}{2y-8} \div \frac{10y^2}{y^2-y-12}$$

$$= \frac{5y}{2y-8} \cdot \frac{y^2-y-12}{10y^2}$$

$$= \frac{5y}{2(y-4)} \cdot \frac{(y-4)(y+3)}{10y^2}$$

$$= \frac{y+3}{4y}$$

$$\frac{P}{Q} \div \frac{R}{S} = \frac{P}{Q} \cdot \frac{S}{R} = \frac{PS}{QR} \qquad \text{where } Q \neq 0 \qquad R \neq 0 \qquad S \neq 0$$

To divide two rational expressions, you can apply the following algorithm.

1. Invert the divisor (the *second* rational expression) to write the problem as one of multiplication.
2. Proceed as in the algorithm for the multiplication of rational expressions.

Multiplying Rational Functions

The product of two rational functions is always a rational function. If

$$h(x) = f(x) \cdot g(x)$$

then the set of ordered pairs

$$(x, f(x) \cdot g(x)) = (x, h(x))$$

Dividing Rational Functions

As in dividing rational expressions, invert the divisor and multiply as before.

Adding and Subtracting Rational Expressions and Functions [4.3]

Adding and Subtracting Rational Expressions

To add or subtract rational expressions with the same denominator, add or subtract their numerators and then write that sum over the common denominator. The result should be written in lowest terms.

In symbols,

$$\frac{5w}{w^2 - 16} - \frac{20}{w^2 - 16}$$

$$= \frac{5w - 20}{w^2 - 16}$$

$$\frac{P}{R} + \frac{Q}{R} = \frac{P + Q}{R}$$

and

$$= \frac{5(w - 4)}{(w + 4)(w - 4)}$$

$$\frac{P}{R} - \frac{Q}{R} = \frac{P - Q}{R}$$

$$= \frac{5}{w + 4}$$

where $R \neq 0$.

Least Common Denominator

To find the LCD for

$$\frac{2}{x^2 + 2x + 1} \quad \text{and} \quad \frac{3}{x^2 + x}$$

write

$$x^2 + 2x + 1 = (x + 1)(x + 1)$$

$$x^2 - x = x(x + 1)$$

The LCD is

$$x(x + 1)(x + 1)$$

$$\frac{2}{(x + 1)^2} - \frac{3}{x(x + 1)}$$

$$= \frac{2 \cdot x}{(x + 1)^2 x}$$

$$= \frac{3(x + 1)}{x(x + 1)(x + 1)}$$

$$= \frac{2x - 3(x + 1)}{x(x + 1)(x + 1)}$$

$$= \frac{-x - 3}{x(x + 1)(x + 1)}$$

The **least common denominator (LCD)** of a group of rational expressions is the simplest polynomial that is divisible by each of the individual denominators of the rational expressions. To find the LCD, use the following algorithm.

1. Write each of the denominators in completely factored form.
2. Write the LCD as the product of each prime factor, to the highest power to which it appears in the factored form of any individual denominators.

Now to add or subtract rational expressions with different denominators, we first find the LCD by the procedure outlined above. We then rewrite each of the rational expressions with that LCD as a common denominator. Then we can add or subtract as before.

Adding and Subtracting Rational Functions

The sum of two rational functions is always a rational function. If

$$h(x) = f(x) + g(x)$$

then the set of ordered pairs

$$(x, f(x) + g(x)) = (x, h(x))$$

When subtracting rational functions, take particular care with the signs in the numerator of the expression being subtracted.

Simplifying Complex Fractions [4.4]

Complex fractions are fractions that have a fraction in their numerator or denominator (or both).

There are two commonly used methods for simplifying complex fractions: methods 1 and 2.

Simplify
$$\frac{1 - \dfrac{2}{x}}{1 - \dfrac{4}{x^2}}$$

Method 1:

$$\frac{\left(1 - \dfrac{2}{x}\right)x^2}{\left(1 - \dfrac{4}{x^2}\right)x^2}$$

$$= \frac{x^2 - 2x}{x^2 - 4} = \frac{x(x - 2)}{(x + 2)(x - 2)}$$

$$= \frac{x}{x + 2}$$

Method 2:

$$\frac{\dfrac{x - 2}{x}}{\dfrac{x^2 - 4}{x^2}}$$

$$= \frac{x - 2}{x} \cdot \frac{x^2}{x^2 - 4}$$

$$= \frac{x - 2}{x} \cdot \frac{x^2}{(x + 2)(x - 2)}$$

$$= \frac{x}{x + 2}$$

Solve

$$\frac{3}{x - 3} - \frac{2}{x + 2} = \frac{19}{x^2 - x - 6}$$

Multiply by the LCD
$(x - 3)(x + 2)$:

$$3(x + 2) - 2(x - 3) = 19$$
$$3x + 6 - 2x + 6 = 19$$
$$x = 7$$

Check:

$$\frac{3}{4} - \frac{2}{9} = \frac{19}{36}$$

$$\frac{19}{36} = \frac{19}{36}$$

Method 1

1. Multiply the numerator and denominator of the complex fraction by the LCD of all the fractions that appear within the numerator and denominator.
2. Simplify the resulting rational expression, writing the result in lowest terms.

Method 2

1. Write the numerator and denominator of the complex fraction as single fractions, if necessary.
2. Invert the denominator and multiply as before, writing the result in lowest terms.

Rational Equations and Inequalities in One Variable [4.5]

To solve an equation involving rational expressions, you should apply the following algorithm.

1. Clear the equation of fractions by multiplying both sides of the equation by the LCD of all the fractions that appear.

2. Solve the equation resulting from step 1.

3. Check all solutions by substitution in the original equation.

Solving Rational Inequalities

To solve

$$\frac{x+2}{x-1} \geq 0$$

Critical points are -2 and 1.

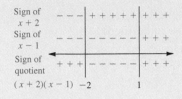

The solution set is

$$\{x \mid x \leq -2 \text{ or } x > 1\}$$

Inequalities containing rational expressions are solved in a similar fashion to quadratic inequalities. The rational expression must first be related to 0. Critical points in this case occur when the numerator is equal to 0 or when the denominator is equal to 0. A sign graph is used to indicate where the quotient is negative or positive, and the solution set is determined.

Summary Exercises

This summary exercise set is provided to give you practice with each of the objectives in the chapter. Each exercise is keyed to the appropriate chapter section. The answers are provided in the *Instructor's Manual*.

[4.1] For what value of the variable will each of the following rational expressions be undefined?

1. $\dfrac{x}{2}$

2. $\dfrac{3}{y}$

3. $\dfrac{2}{x-5}$

4. $\dfrac{3x}{2x-5}$

[4.1] Simplify each of the following rational expressions.

5. $\dfrac{18x^5}{24x^3}$

6. $\dfrac{15m^3n}{-5mn^2}$

7. $\dfrac{8y-64}{y-8}$

8. $\dfrac{5x-20}{x^2-16}$

9. $\dfrac{9-x^2}{x^2+2x-15}$

10. $\dfrac{3w^2+8w-35}{2w^2+13w+15}$

11. $\dfrac{6a^2-ab-b^2}{9a^2-b^2}$

12. $\dfrac{6w-3z}{8w^3-z^3}$

Graph the following rational functions. Indicate the coordinates of the hole in the graph.

13. $f(x) = \dfrac{x^2-3x-4}{x+1}$

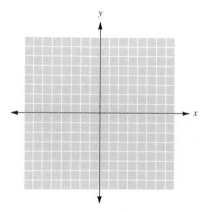

14. $f(x) = \dfrac{x^2+x-6}{x-2}$

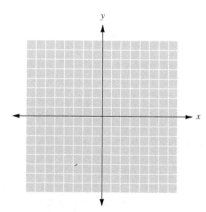

[4.2] Multiply or divide as indicated. Express your results in simplest form.

15. $\dfrac{x^5}{24} \cdot \dfrac{20}{x^3}$

16. $\dfrac{a^3b}{4ab^2} \div \dfrac{ab}{12ab^2}$

17. $\dfrac{6y-18}{9y} \cdot \dfrac{10}{5y-15}$

18. $\dfrac{m^2 - 3m}{m^2 - 5m + 6} \cdot \dfrac{m^2 - 4}{m^2 + 7m + 10}$

19. $\dfrac{a^2 - 2a}{a^2 - 4} \div \dfrac{2a^2}{3a + 6}$

20. $\dfrac{r^2 + 2rs}{r^3 - r^2 s} \div \dfrac{5r + 10s}{r^2 - s + s^2}$

21. $\dfrac{x^2 - 2xy - 3y^2}{x^2 - xy - 2y^2} \cdot \dfrac{x^2 - 4y^2}{x^2 - 8xy + 15y^2}$

22. $\dfrac{w^3 + 3w^2 + 2w + 6}{w^4 - 4} \div (w^3 + 27)$

23. Let $f(x) = \dfrac{x^2 - 16}{x - 5}$ and $g(x) = \dfrac{x^2 - 25}{x + 4}$. Find **(a)** $f(3) \cdot g(3)$, **(b)** $h(x) = f(x) \cdot g(x)$, **(c)** $h(3)$.

24. Let $f(x) = \dfrac{2x^2 - 5x - 3}{x - 4}$ and $g(x) = \dfrac{x^2 - 3x - 4}{2x^2 + 5x + 2}$. Find **(a)** $f(3) \cdot g(3)$, **(b)** $h(x) = f(x) \cdot g(x)$, **(c)** $h(3)$.

[4.3] Perform the indicated operations. Express your results in simplified form.

25. $\dfrac{5x + 7}{x + 4} - \dfrac{2x - 5}{x - 4}$

26. $\dfrac{3}{4x^2} + \dfrac{5}{6x}$

27. $\dfrac{2}{x - 5} - \dfrac{1}{x}$

28. $\dfrac{2}{y + 5} + \dfrac{3}{y + 4}$

29. $\dfrac{2}{3m - 3} - \dfrac{5}{2m - 2}$

30. $\dfrac{7}{x - 3} - \dfrac{5}{3 - x}$

31. $\dfrac{5}{4x + 4} + \dfrac{5}{2x - 2}$

32. $\dfrac{2a}{a^2 - 9a + 20} + \dfrac{8}{a - 4}$

33. $\dfrac{2}{s - 1} - \dfrac{6s}{s^2 + s - 2}$

34. $\dfrac{4}{x^2 - 9} - \dfrac{3}{x^2 - 4x + 3}$

35. $\dfrac{x^2 - 14x - 8}{x^2 - 2x - 8} + \dfrac{2x}{x - 4} - \dfrac{3}{x + 2}$

36. $\dfrac{w^2 + 2wz + z^2}{w^2 - wz - 2z^2} \cdot \left(\dfrac{3}{w + z} - \dfrac{1}{w - z} \right)$

37. Let $f(x) = \dfrac{2x}{x - 2}$ and $g(x) = \dfrac{x}{x - 3}$. Find **(a)** $f(4) + g(4)$, **(b)** $h(x) = f(x) + g(x)$, **(c)** the ordered pair $(4, h(4))$.

38. Let $f(x) = \dfrac{x + 2}{x - 2}$ and $g(x) = \dfrac{x + 1}{x - 7}$. Find **(a)** $f(3) + g(3)$, **(b)** $h(x) = f(x) + g(x)$, **(c)** the ordered pair $(3, h(3))$.

[4.4]   Simplify each of the following complex fractions.

39. $\dfrac{\frac{x^2}{12}}{\frac{x^3}{8}}$

40. $\dfrac{\frac{y-1}{y^2-4}}{\frac{y^2-1}{y^2-y-2}}$

41. $\dfrac{1+\frac{a}{b}}{1-\frac{a}{b}}$

42. $\dfrac{2-\frac{x}{y}}{4-\frac{x^2}{y^2}}$

43. $\dfrac{\frac{1}{r}-\frac{1}{s}}{\frac{1}{r^2}-\frac{1}{s^2}}$

44. $\dfrac{1-\frac{1}{x+2}}{1+\frac{1}{x+2}}$

45. $\dfrac{1-\frac{2}{x-1}}{x+\frac{3}{x-4}}$

46. $\dfrac{\frac{w}{w+1}-\frac{1}{w-1}}{\frac{w}{w-1}+\frac{1}{w+1}}$

47. $\dfrac{1}{1-\dfrac{1}{1-\frac{1}{y-1}}}$

48. $1-\dfrac{1}{1+\dfrac{1}{1-\frac{1}{x}}}$

49. $\dfrac{1-\frac{1}{x-1}}{x-\frac{8}{x+2}}$

50. $\dfrac{1}{1-\dfrac{1}{1+\frac{1}{y+1}}}$

[4.5]   Solve each of the following equations.

51. $\dfrac{1}{2x}+\dfrac{1}{3x}=\dfrac{1}{6}$

52. $\dfrac{5}{2x^2}-\dfrac{1}{4x}=\dfrac{1}{x}$

53. $\dfrac{x}{x-2}+1=\dfrac{x+4}{x-2}$

54. $\dfrac{2x-1}{x-3}-\dfrac{5}{x-3}=1$

55. $\dfrac{2}{3x+1}=\dfrac{1}{x+2}$

56. $\dfrac{5}{x+1}+\dfrac{1}{x-2}=\dfrac{7}{x+1}$

57. $\dfrac{4}{x-1}-\dfrac{5}{3x-7}=\dfrac{3}{x-1}$

58. $\dfrac{7}{x}-\dfrac{1}{x-3}=\dfrac{9}{x^2-3x}$

59. $\dfrac{2}{x-3}-\dfrac{11}{x^2-9}=\dfrac{3}{x+3}$

60. $\dfrac{5}{x+3}+\dfrac{1}{x-5}=1$

61. $\dfrac{2}{x-4}=\dfrac{x}{x-2}-\dfrac{x+4}{x^2-6x+8}$

62. $\dfrac{x}{x-5}=\dfrac{3x}{x^2-7x+10}+\dfrac{8}{x-2}$

Solve each of the following inequalities.

63. $\dfrac{x - 2}{x + 1} < 0$

64. $\dfrac{x + 3}{x - 2} > 0$

65. $\dfrac{x - 4}{x - 2} > 0$

66. $\dfrac{x + 6}{x + 3} < 0$

67. $\dfrac{x - 5}{x + 3} \le 0$

68. $\dfrac{x + 3}{x - 2} \le 0$

69. $\dfrac{2x - 1}{x + 3} \ge 0$

70. $\dfrac{3x - 2}{x - 4} \le 0$

71. $\dfrac{x}{x - 3} + \dfrac{2}{x - 3} \le 0$

72. $\dfrac{x}{x + 5} - \dfrac{3}{x + 5} > 0$

73. $\dfrac{x}{x + 3} \le \dfrac{4}{x + 3}$

74. $\dfrac{x}{x - 5} \ge \dfrac{2}{x - 5}$

75. $\dfrac{2x - 5}{x - 2} > 1$

76. $\dfrac{2x + 3}{x + 4} \ge 1$

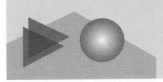

The purpose of this self-test is to help you check your progress and to review for a chapter test in class. Allow yourself about 1 hour to take the test. When you are done, check your answers in the back of the book. If you missed any answers, be sure to go back and review the appropriate sections in the chapter and the exercises that are provided.

Simplify each of the following rational expressions.

1. $\dfrac{-21x^5y^3}{28xy^5}$

2. $\dfrac{3w^2 + w - 2}{3w^2 - 8w + 4}$

3. $\dfrac{x^3 + 2x^2 - 3x}{x^3 - 3x^2 + 2x}$

4. Graph the following.

$$f(x) = \dfrac{x^2 - 5x + 4}{x - 4}$$

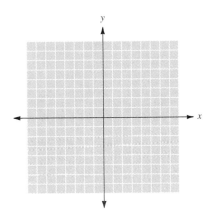

Multiply or divide as indicated.

5. $\dfrac{3ab^2}{5ab^3} \cdot \dfrac{20a^2b}{21b}$

6. $\dfrac{m^2 - 3m}{m^2 - 9} \div \dfrac{4m}{m^2 - m - 12}$

7. $\dfrac{x^2 - 3x}{5x^2} \cdot \dfrac{10x}{x^2 - 4x + 3}$

8. $\dfrac{x^2 + 3xy}{2x^3 - x^2y} \div \dfrac{x^2 + 6xy + 9y^2}{4x^2 - y^2}$

9. $\dfrac{9x^2 - 9x - 4}{6x^2 - 11x + 3} \cdot \dfrac{15 - 10x}{3x - 4}$

Add or subtract as indicated.

10. $\dfrac{5}{x - 2} - \dfrac{1}{x}$

11. $\dfrac{2}{x + 3} + \dfrac{12}{x^2 - 9}$

12. $\dfrac{6x}{x^2 - x - 2} - \dfrac{2}{x + 1}$

13. $\dfrac{3}{x^2 - 3x - 4} + \dfrac{5}{x^2 - 16}$

Simplify each of the following complex fractions.

14. $\dfrac{3 - \dfrac{x}{y}}{9 - \dfrac{x^2}{y^2}}$

15. $\dfrac{1 - \dfrac{10}{z + 3}}{2 - \dfrac{12}{z - 1}}$

16. $\dfrac{\dfrac{1}{x} + \dfrac{1}{y}}{x^2 - y^2}$

Solve the following equations.

17. $\dfrac{x}{x + 3} + 1 = \dfrac{3x - 6}{x + 3}$

18. $\dfrac{2x}{x + 1} = \dfrac{3}{x - 2} + \dfrac{1}{x^2 - x - 2}$

Solve the following and graph the solution set.

19. $\dfrac{x + 4}{x - 3} \le 0$ $\longleftrightarrow$

20. $\dfrac{x + 7}{x + 3} \ge 3$ $\longleftrightarrow$

This test is provided to help you in the process of reviewing the previous chapters. Answers are provided in the back of the book. If you missed any answers, be sure to go back and review the appropriate chapter section.

1. Solve the equation $5x - 3(2x + 6) = 4 - (3x - 2)$.

2. If $f(x) = 5x^4 - 3x^2 + 7x - 9$, find $f(-1)$.

3. Find the equation of the line that is parallel to the line $6x + 7y = 42$ and has a y intercept of -3.

4. Find the x and y intercepts of the equation $7x - 6y = -42$.

Simplify each of the following polynomial functions.

5. $3x - 2(x - (3x - 1)) + 6x(x - 2) = f(x)$

6. $x(2x - 1)(x + 3) = f(x)$

7. Find the domain and range of the relation $7x - 14 = 0$.

8. If $f(x) = 5x^3 - 3x^2 + 7x - 1$, find $f(2)$ using synthetic substitution.

Factor each of the following completely.

9. $6x^3 + 7x^2 - 3x$

10. $16x^{16} - 9y^8$

Simplify each of the following rational expressions.

11. $\dfrac{5}{x - 1} - \dfrac{2x + 6}{x^2 + 2x - 3}$

12. $\dfrac{x + 1}{x^2 - 5x - 6} \div \dfrac{x^2 - 1}{x - 6}$

13. $\dfrac{1 - \dfrac{3}{x + 3}}{\dfrac{1}{x^2 - 9}}$

14. If $f(x) = \dfrac{1}{x - 5}$ and $g(x) = 8x + 6$, find **(a)** $f + g$, **(b)** $\dfrac{f}{g}$, **(c)** domain of $\dfrac{f}{g}$.

Solve the following equations.

15. $7x + (x - 10) = -12(x - 5)$

16. $|-9x - 6| = 2$

17. $-4(7x + 6) = 8(5x + 12)$

Solve the following inequalities.

18. $-4(-2x - 7) > -6x$

19. $|5x - 4| < 3$

20. $-6|2x + 6| \leq -12$

RADICALS AND EXPONENTS

INTRODUCTION

As we have all experienced firsthand, consumer goods increase in price from year to year. This increase is usually measured by the Consumer Price Index (CPI), which measures a change in the prices of such everyday goods and services as energy, food, shelter, apparel, transportation, medical care, and utilities. The percent change in the CPI is a reflection of the purchasing power of the dollar and indicates the rate of inflation.

Since many labor contracts and government benefits programs such as Social Security increase or decrease along with the CPI, the method used to calculate this index is hotly debated by economists and statisticians. Beginning in April 1997, the Bureau of Labor Statistics began releasing an experimental CPI that uses a **geometric mean formula.** This new method may more accurately reflect the true cost-of-living increase or decrease because it takes into consideration that consumers' buying habits change as prices fluctuate. For instance, consumers may switch from romaine lettuce to iceberg lettuce or spinach if the price of romaine lettuce is too high.

To compute the CPI, prices of individual items are averaged together to produce relative price changes or indexes for 9,108 item-area categories. The number 9,108 is found by multiplying 207 items by 44 geographic areas from around the United States. For example, the cost to consumers of lettuce and spinach in one city is reflected in the table on the next page.

© Bob Daemmrich Photos/The Image Works

Item	Cost Last Month	Cost This Month	Relative Price Change: $\dfrac{\textit{Cost this month}}{\textit{Cost last month}}$
1 pound of iceberg lettuce	$0.50	$0.62	$\dfrac{0.62}{0.50}$
1 pound of romaine lettuce	$0.75	$0.80	$\dfrac{0.80}{0.75}$
1 pound of spinach	$0.70	$0.78	$\dfrac{0.78}{0.70}$

The new formula computes the relative price change, R, for salad greens the following way:

$$R = \left(\frac{0.62}{0.50} \cdot \frac{0.80}{0.75} \cdot \frac{0.78}{0.70} \right)^{\frac{1}{3}} = \sqrt[3]{\frac{0.62}{0.50}} \cdot \sqrt[3]{\frac{0.80}{0.75}} \cdot \sqrt[3]{\frac{0.78}{0.70}}$$

$$\approx 1.14 \text{ or a } 14\% \text{ increase}$$

From December 1990 through February 1997, the Bureau of Labor Statistics computed an inflation rate of 16.2%, which is equivalent to an annual growth rate of 2.46%. The old method, which is also still being used for computing the index, computed an annual growth rate for this same period of 2.80%. In this chapter, we will work with the exponents and radicals used to compute these rates. ——————■

Zero and Negative Integer Exponents and Scientific Notation

5.1 OBJECTIVES

1. Define the zero exponent
2. Simplify expressions with negative exponents
3. Write numbers in scientific notation
4. Solve an application of scientific notation

EXPONENT OF ZERO

In Section 3.1, all the exponents we looked at were positive integers. In this section, we look at the meaning of zero and negative integer exponents. First, let's look at an application of the quotient rule that will yield a zero exponent.

Recall that, in the quotient rule, to divide expressions with the same base, keep the base and subtract the exponents.

$$\frac{a^m}{a^n} = a^{m-n}$$

Now, suppose that we allow m to equal n. We then have

$$\frac{a^m}{a^m} = a^{m-m} = a^0 \qquad (1)$$

But we know that it is also true that

$$\frac{a^m}{a^m} = 1 \qquad (2)$$

Comparing equations (1) and (2), we see that the following definition is reasonable.

We must have $a \neq 0$. The form 0^0 is called **indeterminate** and is considered in later mathematics classes.

The Zero Exponent

For any real number a where $a \neq 0$,

$$a^0 = 1$$

● Example 1

The Zero Exponent

Use the above definition to simplify each expression.

Note that in $6x^0$ the exponent 0 applies *only* to x.

(a) $17^0 = 1$ (b) $(a^3b^2)^0 = 1$

(c) $6x^0 = 6 \cdot 1 = 6$ (d) $-3y^0 = -3$

● ● ●   **CHECK YOURSELF 1**

Simplify each expression.

(a) 25^0 **(b)** $(m^4n^2)^0$ **(c)** $8s^0$ **(d)** $-7t^0$

Recall that, in the product rule, to multiply expressions with the same base, keep the base and add the exponents.

$$a^m \cdot a^n = a^{m+n}$$

Now, what if we allow one of the exponents to be negative and apply the product rule? Suppose, for instance, that $m = 3$ and $m = -3$. Then

$$a^m \cdot a^n = a^3 \cdot a^{-3} = a^{3+(-3)}$$
$$= a^0 = 1$$

so

$$a^3 \cdot a^{-3} = 1$$

Dividing both sides by a^3, we get

$$a^{-3} = \frac{1}{a^3}$$

John Wallis (1616–1702), an English mathematician, was the first to fully discuss the meaning of 0, negative, and rational exponents (which we discuss in Section 5.3).

Negative Integer Exponents

For any nonzero real number a and whole number n,

$$a^{-n} = \frac{1}{a^n}$$

and a^{-n} is the **multiplicative inverse** of a^n.

Example 2 illustrates this definition.

● Example 2

Using Properties of Exponents

From this point on, to *simplify* will mean to write the expression with *positive exponents only.*

Also, we will restrict all variables so that they represent nonzero real numbers.

Simplify the following expressions.

(a) $y^{-5} = \dfrac{1}{y^5}$

(b) $4^{-2} = \dfrac{1}{4^2} = \dfrac{1}{16}$

(c) $(-3)^3 = \dfrac{1}{(-3)^3} = \dfrac{1}{-27} = -\dfrac{1}{27}$

(d) $\left(\dfrac{2}{3}\right)^{-3} = \dfrac{1}{\left(\dfrac{2}{3}\right)^3} = \dfrac{1}{\dfrac{8}{27}} = \dfrac{27}{8}$

● ● ● **CHECK YOURSELF 2**

Simplify each of the following expressions.

(a) a^{-10} **(b)** 2^{-4} **(c)** $(-4)^{-2}$ **(d)** $\left(\dfrac{5}{2}\right)^{-2}$

Example 3 illustrates the case where coefficients are involved in an expression with negative exponents. As will be clear, some caution must be used.

● Example 3

Using Properties of Exponents

Simplify each of the following expressions.

$(a)\ \ 2x^{-3} = 2 \cdot \dfrac{1}{x^3} = \dfrac{2}{x^3}$

CAUTION

The exponent -3 applies only to the variable x, and *not* to the coefficient 2.

The expressions

$4w^{-2}$ and $(4w)^{-2}$

are *not* the same. Do you see why?

$(b)\ \ 4w^{-2} = 4 \cdot \dfrac{1}{w^2} = \dfrac{4}{w^2}$

$(c)\ \ (4w)^{-2} = \dfrac{1}{(4w)^2} = \dfrac{1}{16w^2}$

● ● ● **CHECK YOURSELF 3**

Simplify each of the following expressions.

(a) $3w^{-4}$ **(b)** $10x^{-5}$ **(c)** $(2y)^{-4}$ **(d)** $-5t^{-2}$

Suppose that a variable with a negative exponent appears in the denominator of an expression. Our previous definition can be used to write a complex fraction that can then be simplified. For instance,

$$\frac{1}{a^{-2}} = \frac{1}{\dfrac{1}{a^2}} = 1 \cdot \frac{a^2}{1} = a^2 \ \longleftarrow \text{Positive exponent in numerator}$$

Negative exponent in denominator

To divide, we invert and multiply.

To avoid the intermediate steps, we can write that, in general,

> For any nonzero real number a and integer n,
>
> $$\frac{1}{a^{-n}} = a^n$$

• Example 4

Using Properties of Exponents

Simplify each of the following expressions.

(a) $\dfrac{1}{y^{-3}} = y^3$

(b) $\dfrac{1}{2^{-5}} = 2^5 = 32$

(c) $\dfrac{3}{4x^{-2}} = \dfrac{3x^2}{4}$ The exponent -2 applies only to x, not to 4

(d) $\dfrac{a^{-3}}{b^{-4}} = \dfrac{b^4}{a^3}$

● ● ● **CHECK YOURSELF 4**

Simplify each of the following expressions.

(a) $\dfrac{1}{x^{-4}}$ (b) $\dfrac{1}{3^{-3}}$ (c) $\dfrac{2}{3a^{-2}}$ (d) $\dfrac{c^{-5}}{d^{-7}}$

To review these properties, return to Section 3.1.

The product and quotient rules for exponents apply to expressions that involve any integral exponent—positive, negative, or 0. Example 5 illustrates this concept.

• Example 5

Using Properties of Exponents

Simplify each of the following expressions, and write the result, using positive exponents only.

(a) $x^3 \cdot x^{-7} = x^{3+(-7)}$ Add the exponents by the product rule.

$$= x^{-4} = \frac{1}{x^4}$$

(b) $\dfrac{m^{-5}}{m^{-3}} = m^{-5-(-3)} = m^{-5+3}$ Subtract the exponents by the quotient rule.

$$= m^{-2} = \dfrac{1}{m^2}$$

(c) $\dfrac{x^5 x^{-3}}{x^{-7}} = \dfrac{x^{5+(-3)}}{x^{-7}} = \dfrac{x^2}{x^{-7}} = x^{2-(-7)} = x^9$ We apply first the product rule and then the quotient rule.

Note that m^{-5} in the numerator becomes m^5 in the denominator, and m^{-3} in the denominator becomes m^3 in the numerator. We then simplify as before.

In simplifying expressions involving negative exponents, there are often alternate approaches. For instance, in Example 5(b), we could have made use of our earlier work to write

$$\dfrac{m^{-5}}{m^{-3}} = \dfrac{m^3}{m^5} = m^{3-5} = m^{-2} = \dfrac{1}{m^2}$$

● ● ● ● **CHECK YOURSELF 5**

Simplify each of the following expressions.

(a) $x^9 \cdot x^{-5}$ (b) $\dfrac{y^{-7}}{y^{-3}}$ (c) $\dfrac{a^{-3} a^2}{a^{-5}}$

The properties of exponents can be extended to include negative exponents. One of these properties, the Quotient-Power Rule, is particularly useful when rational expressions are raised to a negative power. Let's look at the rule and apply it to negative exponents.

Quotient-Power Rule

$$\left(\dfrac{a}{b}\right)^n = \dfrac{a^n}{b^n}$$

Raising Quotients to a Negative Power

$$\left(\dfrac{a}{b}\right)^{-n} = \dfrac{a^{-n}}{b^{-n}} = \dfrac{b^n}{a^n} = \left(\dfrac{b}{a}\right)^n \qquad a \neq 0, b \neq 0$$

● Example 6

Extending the Properties of Exponents

Simplify each expression.

(a) $\left(\dfrac{s^3}{t^2}\right)^{-2} = \dfrac{s^{-6}}{t^{-4}} = \dfrac{t^4}{s^6}$

(b) $\left(\dfrac{m^2}{n^{-2}}\right)^{-3} = \dfrac{m^{-6}}{n^6} = \dfrac{1}{n^6 m^6}$

● ● ● **CHECK YOURSELF 6**

Simplify each expression.

(a) $\left(\dfrac{3t^2}{s^3}\right)^3$
(b) $\left(\dfrac{x^5}{y^{-2}}\right)^{-3}$

● Example 7

Using Properties of Exponents

Simplify each of the following expressions.

(a) $\left(\dfrac{3}{q^5}\right)^{-2} = \left(\dfrac{q^5}{3}\right)^2$

$\qquad = \dfrac{q^{10}}{9}$

(b) $\left(\dfrac{x^3}{y^4}\right)^{-3} = \left(\dfrac{y^4}{x^3}\right)^3$

$\qquad = \dfrac{(y^4)^3}{(x^3)^3} = \dfrac{y^{12}}{x^9}$

● ● ● **CHECK YOURSELF 7**

Simplify each of the following expressions.

(a) $\left(\dfrac{r^4}{5}\right)^{-2}$
(b) $\left(\dfrac{a^4}{b^3}\right)^{-3}$

As you might expect, more complicated expressions require the use of more than one of the properties, for simplification. Example 8 illustrates such cases.

● Example 8

Using Properties of Exponents

Simplify each of the following expressions.

(a) $\dfrac{(a^2)^{-3}(a^3)^4}{(a^{-3})^3} = \dfrac{a^{-6} \cdot a^{12}}{a^{-9}}$
　　　Apply the power rule to each factor.

$\qquad = \dfrac{a^{-6+12}}{a^{-9}} = \dfrac{a^6}{a^{-9}}$
　　　Apply the product rule.

$\qquad = a^{6-(-9)} = a^{6+9} = a^{15}$
　　　Apply the quotient rule.

It may help to separate the problem into three fractions, one for the coefficients and one for each of the variables.

(b) $\dfrac{8x^{-2}y^{-5}}{12x^{-4}y^3} = \dfrac{8}{12} \cdot \dfrac{x^{-2}}{x^{-4}} \cdot \dfrac{y^{-5}}{y^3}$

$= \dfrac{2}{3}x^{-2-(-4)} \cdot y^{-5-3}$

$= \dfrac{2}{3}x^2 \cdot y^{-8} = \dfrac{2x^2}{3y^8}$

(c) $\left(\dfrac{pr^3s^{-5}}{p^3r^{-3}s^{-2}}\right)^{-2} = (p^{1-3}r^{3-(-3)}s^{-5-(-2)})^{-2}$

$= (p^{-2}r^6s^{-3})^{-2}$ Apply the power rule inside the parentheses.

$= (p^{-2})^{-2}(r^6)^{-2}(s^{-3})^{-2}$ Apply the rule for a product to a power.

$= p^4r^{-12}s^6 = \dfrac{p^4s^6}{r^{12}}$ Apply the power rule.

CAUTION

Be Careful! Another possible first step (and generally an efficient one) is to rewrite an expression by using our earlier definitions.

$a^{-n} = \dfrac{1}{a^n}$ and $\dfrac{1}{a^{-n}} = a^n$

For instance, in Example 8(b), we would *correctly* write

$\dfrac{8x^{-2}y^{-5}}{12x^{-4}y^3} = \dfrac{8x^4}{12x^2y^3y^5}$

A *common error* is to write

$\dfrac{8x^{-2}y^{-5}}{12x^{-4}y^3} = \dfrac{12x^4}{8x^2y^3y^5}$ This is *not* correct.

The coefficients should not have been moved along with the factors in x. Keep in mind that the negative exponents apply *only* to the variables. The coefficients remain *where they were* in the original expression when the expression is rewritten by using this approach.

● ● ● **CHECK YOURSELF 8**

Simplify each of the following expressions.

(a) $\dfrac{(x^5)^{-2}(x^2)^3}{(x^{-4})^3}$ (b) $\dfrac{12a^{-3}b^{-2}}{16a^{-2}b^3}$ (c) $\left(\dfrac{xy^{-3}z^{-5}}{x^{-4}y^{-2}z^3}\right)^{-3}$

You may have noticed that throughout this section we frequently used 10 as a base in the examples. You will find that experience useful as we discuss scientific notation.

We begin the discussion with a calculator exercise. On most calculators, if you multiply 2.3 times 1000, the display will read

2300

Multiply by 1000 a second time. Now you will see

2300000.

Multiplying by 1000 a third time will result in the display

This must equal
2,300,000,000.

2.3   09

And multiplying by 1000 again yields

2.3   12

Consider the following table:

$$2.3 = 2.3 \times 10^0$$
$$23 = 2.3 \times 10^1$$
$$230 = 2.3 \times 10^2$$
$$2300 = 2.3 \times 10^3$$
$$23{,}000 = 2.3 \times 10^4$$
$$230{,}000 = 2.3 \times 10^5$$

Can you see what is happening? This is the way calculators display very large numbers: The number on the left is always between 1 and 10, and the number on the right indicates the number of places the decimal point must be moved to the right to put the answer in standard (or decimal) form.

This notation is used frequently in science. It is not uncommon in scientific applications of algebra to find yourself working with very large or very small numbers. Even in the time of Archimedes (287–212 BC), the study of such numbers was not unusual. Archimedes estimated that the universe was 23,000,000,000,000,000 m in diameter, which is the approximate distance light travels in $2\frac{1}{2}$ years. By comparison, Polaris (the North Star) is 680 light-years from the earth. Example 10 will discuss the idea of light-years.

In scientific notation, his estimate for the diameter of the universe would be

$$2.3 \times 10^{16}\ \text{m}$$

In general, we can define scientific notation as follows.

Scientific Notation

Any number written in the form

$$a \times 10^n$$

where $1 \le a < 10$ and n is an integer, is written in scientific notation.

• Example 9

Using Scientific Notation

Write each of the following numbers in scientific notation.

Note the pattern for writing a number in scientific notation.

(a) $120{,}000. = 1.2 \times 10^5$

    5 places   The power is 5

(b) $88{,}000{,}000. = 8.8 \times 10^7$

    7 places   The power is 7

The exponent on 10 shows the *number of places* we must move the decimal point so that the multiplier will be a number between 1 and 10. A positive exponent tells us to move right, while a negative exponent indicates to move left.

Note: To convert back to standard or decimal form, the process is simply reversed.

(c) $520,000,000. = 5.2 \times 10^8$

8 places

(d) $4,000,000,000. = 4 \times 10^9$

9 places

(e) $0.0005 = 5 \times 10^{-4}$

4 places   If the decimal point is to be moved to the left, the exponent will
                            be negative.

(f) $0.0000000081 = 8.1 \times 10^{-9}$

9 places

●●● **CHECK YOURSELF 9**

Write in scientific notation.

(a) $212,000,000,000,000,000$ **(b)** 0.00079
(c) $5,600,000$ **(d)** 0.0000007

●**Example 10**

An Application of Scientific Notation

(a) Light travels at a speed of 3.05×10^8 meters per second (m/s). There are approximately 3.15×10^7 s in a year. How far does light travel in a year?
    We multiply the distance traveled in 1 s by the number of seconds in a year. This yields

Note that $9.6075 \times 10^{15} = 10 \times 10^{15} = 10^{16}$

$(3.05 \times 10^8)(3.15 \times 10^7) = (3.05 \cdot 3.15)(10^8 \cdot 10^7)$ Multiply the coefficients,
                                                                               and add the exponents.

$$= 9.6075 \times 10^{15}$$

For our purposes we round the distance light travels in 1 year to 10^{16} m. This unit is called a **light-year,** and it is used to measure astronomical distances.

(b) The distance from earth to the star Spica (in Virgo) is 2.2×10^{18} m. How many light-years is Spica from earth?

We divide the distance (in meters) by the number of meters in 1 light-year.

$$\frac{2.2 \times 10^{18}}{10^{16}} = 2.2 \times 10^{18-16}$$

$$= 2.2 \times 10^2 = 220 \text{ light-years}$$

●●● **CHECK YOURSELF 10**

The farthest object that can be seen with the unaided eye is the Andromeda galaxy. This galaxy is 2.3×10^{22} m from earth. What is this distance in light-years?

● ● ● **CHECK YOURSELF ANSWERS**

1. (a) 1, (b) 1, (c) 8, (d) -7. **2.** (a) $\dfrac{1}{a^{10}}$, (b) $\dfrac{1}{16}$, (c) $\dfrac{1}{16}$, (d) $\dfrac{4}{25}$.

3. (a) $\dfrac{3}{w^4}$, (b) $\dfrac{10}{x^5}$, (c) $\dfrac{1}{16y^4}$, (d) $-\dfrac{5}{t^2}$. **4.** (a) x^4, (b) 27, (c) $\dfrac{2}{3}a^2$, (d) $\dfrac{d^7}{c^5}$.

5. (a) x^4, (b) $\dfrac{1}{y^4}$, (c) a^4. **6.** (a) $\dfrac{m^{12}}{n^4}$, (b) $\dfrac{27t^6}{s^9}$, (c) $x^{15}y^6$, (d) $\dfrac{p^8q^4}{9}$.

7. (a) $\dfrac{25}{r^8}$, (b) $\dfrac{b^9}{a^{12}}$. **8.** (a) x^8, (b) $\dfrac{3}{4ab^5}$, (c) $\dfrac{y^3z^{24}}{x^{15}}$. **9.** (a) 2.12×10^{17},

(b) 7.9×10^{-4}, (c) 5.6×10^6, (d) 7×10^{-7}. **10.** 2,300,000 light-years.

5.1 Exercises

In Exercises 1 to 22, simplify each expression.

1. x^{-5}

2. 3^{-3}

3. 5^{-2}

4. x^{-8}

5. $(-5)^{-2}$

6. $(-3)^{-3}$

7. $(-2)^{-3}$

8. $(-2)^{-4}$

9. $\left(\dfrac{2}{3}\right)^{-3}$

10. $\left(\dfrac{3}{4}\right)^{-2}$

11. $3x^{-2}$

12. $4x^{-3}$

13. $-5x^{-4}$

14. $(-2x)^{-4}$

15. $(-3x)^{-2}$

16. $-5x^{-2}$

17. $\dfrac{1}{x^{-3}}$

18. $\dfrac{1}{x^{-5}}$

19. $\dfrac{2}{5x^{-3}}$

20. $\dfrac{3}{4x^{-4}}$

21. $\dfrac{x^{-3}}{y^{-4}}$

22. $\dfrac{x^{-5}}{y^{-3}}$

In Exercises 23 to 32, use the properties of exponents to simplify expressions.

23. $x^5 \cdot x^{-3}$

24. $y^{-4} \cdot y^5$

25. $a^{-9} \cdot a^6$

26. $w^{-5} \cdot x^3$

27. $z^{-2} \cdot z^{-8}$

28. $b^{-7} \cdot b^{-1}$

29. $a^{-5} \cdot a^5$

30. $x^{-4} \cdot x^4$

31. $\dfrac{x^{-5}}{x^{-2}}$

32. $\dfrac{x^{-3}}{x^{-6}}$

In Exercises 33 to 58, use the properties of exponents to simplify the following.

33. $(x^5)^3$

34. $(w^4)^6$

35. $(2x^{-3})(x^2)^4$

36. $(p^4)(3p^3)^2$

37. $(3a^{-4})(a^3)(a^2)$

38. $(5y^{-2})(2y)(y^5)$

39. $(x^4y)(x^2)^3(y^3)^0$

40. $(r^4)^2(r^2s)(s^3)^2$

41. $(ab^2c)(a^4)^4(b^2)^3(c^3)^4$

42. $(p^2qr^2)(p^2)(q^3)^2(r^2)^0$

43. $(x^5)^{-3}$

44. $(x^{-2})^{-3}$

45. $(b^{-4})^{-2}$

46. $(a^0b^{-4})^3$

47. $(x^5y^{-3})^2$

48. $(p^{-2}q^3)^{-2}$

49. $(x^{-4}y^{-2})^{-3}$

50. $(3x^{-2}y^{-2})^3$

51. $(2x^{-3}y^0)^{-5}$

52. $\dfrac{a^{-6}}{b^{-4}}$

53. $\dfrac{x^{-2}}{y^{-4}}$

54. $\left(\dfrac{x^{-3}}{y^2}\right)^{-3}$

55. $\dfrac{x^{-4}}{y^{-2}}$

56. $\dfrac{(3x^{-4})^2(2x^2)}{x^6}$

57. $(4x^{-2})^2(3x^{-4})$

58. $(5x^{-4})^{-4}(2x^3)^{-5}$

In Exercises 59 to 90, simplify each expression.

59. $(2x^5)^4(x^3)^2$

60. $(3x^2)^3(x^2)^4(x^2)$

61. $(2x^{-3})^3(3x^3)^2$

62. $(x^2y^3)^4(xy^3)^0$

63. $(xy^5z)^4(xyz^2)^8(x^6yz)^5$

64. $(x^2y^2z^2)^0(xy^2z)^2(x^3yz^2)$

65. $(3x^{-2})(5x^2)^2$

66. $(2a^3)^2(a^0)^5$

67. $(2w^3)^4(3w^{-5})^2$

68. $(3x^3)^2(2x^4)^5$

69. $\dfrac{3x^6}{2y^9} \cdot \dfrac{y^5}{x^3}$

70. $\dfrac{x^8}{y^6} \cdot \dfrac{2y^9}{x^3}$

71. $(-7x^2y)(-3x^5y^6)^4$

72. $\left(\dfrac{2w^5z^3}{3x^3y^9}\right)\left(\dfrac{x^5y^4}{w^4z^0}\right)^2$

73. $(2x^2y^{-3})(3x^{-4}y^{-2})$

74. $(-5a^{-2}b^{-4})(2a^5b^0)$

75. $\dfrac{(x^{-3})(y^2)}{y^{-3}}$

76. $\dfrac{6x^3y^{-4}}{24x^{-2}y^{-2}}$

77. $\dfrac{15x^{-3}y^2z^{-4}}{20x^{-4}y^{-3}z^2}$

78. $\dfrac{24x^{-5}y^{-3}z^2}{36x^{-2}y^3z^{-2}}$

79. $\dfrac{x^{-5}y^{-7}}{x^0y^{-4}}$

80. $\left(\dfrac{xy^3z^{-4}}{x^{-3}y^{-2}z^2}\right)^{-2}$

81. $\dfrac{x^{-2}y^2}{x^3y^{-2}} \cdot \dfrac{x^{-4}y^2}{x^{-2}y^{-2}}$

82. $\left(\dfrac{x^{-3}y^3}{x^{-4}y^2}\right)^3 \cdot \left(\dfrac{x^{-2}y^{-2}}{xy^4}\right)^{-1}$

83. $x^{2n} \cdot x^{3n}$

84. $x^{n+1} \cdot x^{3n}$

85. $\dfrac{x^{n+3}}{x^{n+1}}$

86. $\dfrac{x^{n-4}}{x^{n-1}}$

87. $(y^n)^{3n}$

88. $(x^{n+1})^n$

89. $\dfrac{x^{2n} \cdot x^{n+2}}{x^{3n}}$

90. $\dfrac{x^n \cdot x^{3n+5}}{x^{4n}}$

In Exercises 91 to 94, express each number in scientific notation.

91. The distance from the earth to the sun: 93,000,000 mi.

92. The diameter of a grain of sand: 0.000021 m.

93. The diameter of the sun: 130,000,000,000 cm.

94. The number of molecules in 22.4 L of a gas:

  602,000,000,000,000,000,000,000 (Avogadro's number)

95. The mass of the sun is approximately 1.98×10^{30} kg. If this were written in standard or decimal form, how many 0s would follow the digit 8?

96. Archimedes estimated the universe to be 2.3×10^{19} millimeters (mm) in diameter. If this number were written in standard or decimal form, how many 0s would follow the digit 3?

In Exercises 97 to 100, write each expression in standard notation.

97. 8×10^{-3}

98. 7.5×10^{-6}

99. 2.8×10^{-5}

100. 5.21×10^{-4}

In Exercises 101 to 104, write each of the following in scientific notation.

101. 0.0005

102. 0.000003

103. 0.00037

104. 0.000051

In Exercises 105 to 108, compute the expressions using scientific notation, and write your answer in that form.

105. $(4 \times 10^{-3})(2 \times 10^{-5})$

106. $(1.5 \times 10^{-6})(4 \times 10^2)$

107. $\dfrac{9 \times 10^3}{3 \times 10^{-2}}$

108. $\dfrac{7.5 \times 10^{-4}}{1.5 \times 10^2}$

In Exercises 109 to 114, perform the indicated calculations. Write your result in scientific notation.

109. $(2 \times 10^5)(4 \times 10^4)$

110. $(2.5 \times 10^7)(3 \times 10^5)$

111. $\dfrac{6 \times 10^9}{3 \times 10^7}$

112. $\dfrac{4.5 \times 10^{12}}{1.5 \times 10^7}$

113 $\dfrac{(3.3 \times 10^{15})(6 \times 10^{15})}{(1.1 \times 10^8)(3 \times 10^6)}$

114. $\dfrac{(6 \times 10^{12})(3.2 \times 10^8)}{(1.6 \times 10^7)(3 \times 10^2)}$

115. Megrez, the nearest of the Big Dipper stars, is 6.6×10^{17} m from earth. Approximately how long does it take light, traveling at 10^{16} m/year, to travel from Megrez to earth?

116. Alkaid, the most distant star in the Big Dipper, is 2.1×10^{18} m from earth. Approximately how long does it take light to travel from Alkaid to earth?

117. The number of liters of water on earth is 15,500 followed by 19 zeros. Write this number in scientific notation. Then use the number of liters of water on earth to find out how much water is available for each person on earth. The population of earth is 5.3 billion.

118. If there are 5.3×10^9 people on earth and there is enough freshwater to provide each person with 8.79×10^5 L, how much freshwater is on earth?

119. The United States uses an average of 2.6×10^6 L of water per person each year. The United States has 2.5×10^8 people. How many liters of water does the United States use each year?

 120. Can $(a + b)^{-1}$ be written as $\dfrac{1}{a} + \dfrac{1}{b}$ by using the properties of exponents? If not, why not? Explain.

 121. Write a short description of the difference between $(-4)^{-2}$, -4^{-3}, $(-4)^3$, and -4^3. Are any of these equal?

 122. If $n > 0$, which of the following expressions are negative?

$(-n)^{-3}, \ -n^{-3}, \ n^{-3}, \ (-n)^{-3}, \ (-n)^3, \ -n^3$

If $n < 0$, which of these expressions are negative? Explain what effect a negative in the exponent has on the sign of the result when an exponential expression is simplified.

 123. Take the Best Offer You are offered a 28-day job in which you have a choice of two different pay arrangements. Plan 1 offers a flat \$4,000,000 at the end of the 28th day on the job. Plan 2 offers 1¢ the first day, 2¢ the second day, 4¢ the third day, and so on, with the amount doubling each day. Make a table to decide which offer is the best. Write a formula for the amount you make on the nth day and a formula for the total after n days. Which job should you take? Why?

Answers

1. $\dfrac{1}{x^5}$ **3.** $\dfrac{1}{25}$ **5.** $\dfrac{1}{25}$ **7.** $-\dfrac{1}{8}$ **9.** $\dfrac{27}{8}$ **11.** $\dfrac{3}{x^2}$ **13.** $-\dfrac{5}{x^4}$ **15.** $\dfrac{1}{9x^2}$ **17.** x^3

19. $\dfrac{2x^3}{5}$ **21.** $\dfrac{y^4}{x^3}$ **23.** x^2 **25.** $\dfrac{1}{a^3}$ **27.** $\dfrac{1}{z^{10}}$ **29.** 1   **31.** $\dfrac{1}{x^3}$ **33.** x^{15} **35.** $2x^5$

37. $3a$ **39.** $x^{10}y$ **41.** $a^{17}b^8c^{13}$ **43.** $\dfrac{1}{x^{15}}$ **45.** b^8 **47.** $\dfrac{x^{10}}{y^6}$ **49.** $x^{12}y^6$ **51.** $\dfrac{x^{15}}{32}$

53. $\dfrac{y^4}{x^2}$ **55.** $\dfrac{y^2}{x^4}$ **57.** $\dfrac{48}{x^8}$ **59.** $16x^{26}$ **61.** $\dfrac{72}{x^3}$ **63.** $x^{42}y^{33}z^{25}$ **65.** $75x^2$ **67.** $144w^2$

69. $\dfrac{3x^3}{2y^4}$ **71.** $-567x^{22}y^{25}$ **73.** $\dfrac{6}{x^2y^5}$ **75.** $\dfrac{y^5}{x^3}$ **77.** $\dfrac{3xy^5}{4z^6}$ **79.** $\dfrac{1}{x^5y^3}$ **81.** $\dfrac{y^8}{x^7}$ **83.** x^{5n}

85. x^2 **87.** y^{3n^2} **89.** x^2 **91.** 9.3×10^7 **93.** 1.3×10^{11} **95.** 28   **97.** 0.008

99. 0.000028   **101.** 5×10^{-4} **103.** 3.7×10^{-4} **105.** 8×10^{-8} **107.** 3×10^5

109. 8×10^9 **111.** 2×10^2 **113.** 6×10^{16} **115.** 66 years   **117.** 1.55×10^{23}, 2.9×10^{13}

119. 6.5×10^{14} **121.**   **123.**

5.2 Evaluating Radical Expressions

5.2 OBJECTIVES

1. Evaluate radicals
2. Simplify radical expressions
3. Multiply radical expressions
4. Approximate radical values with a scientific calculator

In Chapter 3, we discussed the properties of integer exponents. In this and the next sections, we extend those properties. To achieve that objective, we first develop a notion that "reverses" the power process.

A statement such as

$$x^2 = 9$$

is read as "x squared equals nine."

In this section, we are concerned with the relationship between the base x and the number 9. Equivalently, we can say that "x is the square root of 9."

We know from experience that x must equal 3 (since $3^2 = 9$) or -3 [since $(-3)^2 = 9$]. We see that 9 has two square roots, 3 and -3. In fact, every positive number has two square roots, one positive and one negative. In general, if $x^2 = a$, we say that x is the *square root* of a.

We also know that $3^3 = 27$, and we similarly call 3 the *cube root* of 27. Here 3 is the *only* real number with that property. Every real number (positive or negative) has exactly one real cube root. This brings us to the following definition.

> Given
>
> $x^n = a$
>
> we say x is the **nth root of a.**

Recall that the symbol $\sqrt{}$ is called the *radical*. $\sqrt{a}$ is used to designate the *principal (positive) square root* of a. For example,

$$\sqrt{9} = 3$$

indicates that 3 is the principal square root of 9. In some applications, we will want to indicate the negative square root; to do so we write

$$-\sqrt{9} = -3$$

If both square roots need to be indicated, we write

$$\pm\sqrt{9} = \pm 3$$

The symbol $\sqrt[n]{a}$ is used to designate the principal nth root of a. Recall that we call n the *index* of the radical, and the expression a is called the *radicand.* Anytime the index is an even number, the radicand must be positive to produce a real value. Example 1 illustrates this approach.

● Example 1

Evaluating Radicals

Evaluate each radical.

(*a*) $\sqrt[3]{64}$ is the cube root of 64. It has a value of 4 since $4 \cdot 4 \cdot 4 = 64$.

(*b*) $\sqrt[4]{16}$ is the principal fourth root of 16. It has a value of 2 since $2^4 = 16$.

(*c*) $\sqrt[3]{-27} = -3$.

(*d*) $\sqrt[4]{-16}$ is not defined as a real value. No real number, taken to an even power, will produce a negative result.

(*e*) $\sqrt{0} = 0$.

CHECK YOURSELF 1

Evaluate each radical.

(a) $\sqrt[3]{125}$ **(b)** $\sqrt[6]{64}$ **(c)** $\sqrt[3]{-1000}$ **(d)** $\sqrt[4]{-2500}$

For an expression to be written in simplest radical form, several conditions must be satisfied. The first is that no factor in the radicand can be raised to a power equal to or greater than the index.

The Product Theorem

$$\sqrt[n]{ab} = \sqrt[n]{a} \cdot \sqrt[n]{b}$$

● Example 2

Simplify each radical expression. Assume all variables are positive.

(*a*) $\sqrt{32} = \sqrt{2^5} = \sqrt{2^2 \cdot 2^2 \cdot 2} = \sqrt{2^2} \cdot \sqrt{2^2} \cdot \sqrt{2} = 2 \cdot 2 \cdot \sqrt{2} = 4\sqrt{2}$

(b) $\sqrt{75a^3b} = \sqrt{25a^2 \cdot 3ab} = \sqrt{5^2a^2 \cdot 3ab} = 5a\sqrt{3ab}$

(c) $\sqrt[3]{8x^3y \cdot 2xy^2} = \sqrt[3]{8x^3y^3 \cdot 2x} = \sqrt[3]{2^3x^3y^3} \cdot \sqrt[3]{2x} = 2xy\sqrt[3]{2x}$

●●● **CHECK YOURSELF 2**

Write in simplest radical form.

(a) $\sqrt{50}$ (b) $\sqrt{32a^3}$ (c) $\sqrt[3]{-16x^4y^5}$

We can use the product theorem to multiply expressions containing radicals. The product of two radical expressions is the radical of the product of the radicands. The result should always be expressed in simplest radical form.

● Example 3

Multiplying Radicals

Multiply.

(a) $\sqrt{5} \cdot \sqrt{3} = \sqrt{15}$

(b) $\sqrt{3ab} \cdot \sqrt{6a^2b^3} = \sqrt{18a^3b^4} = 3ab^2\sqrt{2a}$

●●● **CHECK YOURSELF 3**

Multiply.

(a) $\sqrt{2x} \cdot \sqrt{6}$ (b) $\sqrt{3a^2b} \cdot \sqrt{12ab^3}$

A second theorem allows us to rewrite expressions that have fractions inside the radical.

THE Quotient Theorem

$$\sqrt[n]{\frac{a}{b}} = \frac{\sqrt[n]{a}}{\sqrt[n]{b}}$$

● Example 4

Removing Fractions From the Radicand

Simplify the following expression.

$$\sqrt{\frac{3a}{16a^2}} = \frac{\sqrt{3a}}{\sqrt{16a^2}} = \frac{\sqrt{3a}}{4a}$$

● ● ● **CHECK YOURSELF 4**

Simplify the following expression.

$$\sqrt{\dfrac{5x}{36y^2}}$$

Whenever the radicand is a real number, we can use a scientific calculator to approximate the value of a radical. This is particularly useful when we are trying to graph an equation or solve an application problem.

● Example 5

Approximating the Value of a Radical Expression

For each expression, use a calculator to approximate its value. Express your answer to the nearest tenth.

(a) $\sqrt{7x^3}$, $x = 8$

$\sqrt{7(8)^3} \approx 59.9$

(b) $\sqrt{3x} + 4\sqrt{2x^3} - \sqrt{7x}$, $x = 3$

$\sqrt{3(3)} + 4\sqrt{2(3)^3} - \sqrt{7(3)} \approx 27.8$

If your calculator has a MATH menu, you will probably find the command for the nth root in that menu.

(c) $\sqrt[5]{7x} + \sqrt[3]{15x^5} - \sqrt[4]{127x}$, $x = 5$

$\sqrt[5]{7(5)} + \sqrt[3]{15(5)^5} - \sqrt[4]{127(5)} \approx 34.7$

● ● ● **CHECK YOURSELF 5**

For each expression, use a calculator to approximate the value at the given x.

(a) $\sqrt{15x^5}$, $x = 2$ **(b)** $\sqrt{7x^3} + 3\sqrt{2x} - \sqrt{17x}$, $x = 5$

(c) $\sqrt[3]{9x} - \sqrt[5]{7x^2} - \sqrt[3]{12x^4}$, $x = 4$

● ● ● **CHECK YOURSELF ANSWERS**

1. (a) 5, **(b)** 2, **(c)** -10, **(d)** no real solution.   **2. (a)** $5\sqrt{2}$, **(b)** $4a\sqrt{2a}$,

(c) $-2xy\sqrt[3]{2xy^2}$.   **3. (a)** $2\sqrt{3x}$, **(b)** $6ab^2\sqrt{a}$.   **4.** $\dfrac{\sqrt{5x}}{6y}$.   **5. (a)** ≈ 21.9;

(b) ≈ 29.8, **(c)** ≈ -8.7.

The Average or Mean of Two or More Numbers. If b is the mean of two numbers, a and c, where $a < c$, then, according to the Pythagoreans, a group of individuals who studied mathematics, music, and mysticism about 500 BC, there are 10 different ways to define this mean. Two of the most well-known are the geometric and the arithmetic mean, defined by the Pythagoreans as follows.

Arithmetic mean: $\dfrac{b - a}{c - b} = \dfrac{a}{a}$

Geometric mean: $\dfrac{b - a}{c - b} = \dfrac{a}{b}$

Work with a partner to complete the following.

1. Solve each equation for b to find the formula for computing each mean. The arithmetic mean is commonly called the *average,* although this is not the only way to define "average."
2. Try computing the geometric and arithmetic mean of several pairs of numbers. When is the geometric mean greater than the arithmetic mean? When are they equal?
3. In a country in economic crisis, inflation usually soars. If one year the price of basic food items doubled, then the next year they tripled, and the third year they dropped in price by half, what was the *average* amount that prices were multiplied by each year? Compute the arithmetic and geometric mean of the increases each year. Which is a better measure of the average in this case? Explain.

5.2 Exercises

In Exercises 1 to 26, evaluate each radical.

1. $\sqrt{49}$

2. $\sqrt{36}$

3. $-\sqrt{36}$

4. $-\sqrt{81}$

5. $\sqrt{-49}$

6. $\sqrt{-25}$

7. $\sqrt[3]{64}$

8. $\sqrt[3]{-64}$

9. $-\sqrt[3]{216}$

10. $-\sqrt[3]{-216}$

11. $\sqrt[4]{81}$

12. $\sqrt[4]{-81}$

13. $\sqrt[5]{32}$

14. $\sqrt[5]{-32}$

15. $-\sqrt[5]{32}$

16. $-\sqrt[5]{-32}$

17. $\sqrt{\dfrac{4}{9}}$

18. $\sqrt{\dfrac{9}{25}}$

19. $\sqrt[3]{\dfrac{8}{27}}$

20. $\sqrt[3]{\dfrac{27}{64}}$

21. $\sqrt{6^2}$

22. $\sqrt[3]{r^3}$

23. $\sqrt{(-9)^2}$

24. $\sqrt{(-5)^2}$

25. $\sqrt[3]{(-5)^3}$

26. $\sqrt[3]{5^3}$

In Exercises 27 to 58, simplify each radical expression.

27. $\sqrt{12}$

28. $\sqrt{24}$

29. $-\sqrt{108}$

30. $-\sqrt{96}$

31. $\sqrt{32}$

32. $\sqrt{250}$

33. $\sqrt[3]{-48}$

34. $\sqrt[3]{-54}$

35. $\sqrt[4]{96}$

36. $\sqrt[4]{243}$

37. $\sqrt{63x^4}$

38. $\sqrt{54w^4}$

39. $\sqrt{75a^5}$

40. $\sqrt{98m^3}$

41. $\sqrt{80x^2y^3}$

42. $\sqrt{108p^5q^2}$

43. $\sqrt[3]{250x^8}$

44. $\sqrt[3]{128r^6s^2}$

45. $\sqrt[3]{56x^6y^5z^4}$

46. $-\sqrt[3]{250a^4b^{15}c^9}$

47. $\sqrt[4]{162y^{12}}$

48. $\sqrt[4]{243a^{15}}$

49. $\sqrt[5]{64w}$

50. $\sqrt[5]{96a^5b^{12}}$

51. $\sqrt{\dfrac{5}{16}}$

52. $\sqrt{\dfrac{11}{36}}$

53. $\sqrt{\dfrac{x^4}{25}}$

54. $\sqrt{\dfrac{a^6}{49}}$

55. $\sqrt{\dfrac{5}{9y^4}}$

56. $\sqrt{\dfrac{7}{25x^2}}$

57. $\sqrt[3]{\dfrac{3}{64}}$

58. $\sqrt[3]{\dfrac{4x^2}{27}}$

In Exercises 59 to 70, combine and simplify each expression.

59. $\sqrt{7} \cdot \sqrt{6}$

60. $\sqrt{3} \cdot \sqrt{10}$

61. $\sqrt{3} \cdot \sqrt{7} \cdot \sqrt{2}$

62. $\sqrt{5} \cdot \sqrt{7} \cdot \sqrt{3}$

63. $\sqrt[3]{4} \cdot \sqrt[3]{9}$

64. $\sqrt[3]{5} \cdot \sqrt[3]{7}$

65. $\sqrt{3} \cdot \sqrt{12}$

66. $\sqrt{5} \cdot \sqrt{20}$

67. $\sqrt{8x^2} \cdot \sqrt{4x}$

68. $\sqrt{12w^3} \cdot \sqrt{6w}$

69. $\sqrt[3]{25x^2} \cdot \sqrt[3]{10x^2}$

70. $\sqrt[3]{18r^2s^2} \cdot \sqrt[3]{9r^2s}$

In Exercises 71 to 90, use a calculator to evaluate the expressions. Round each answer to the nearest tenth.

71. $\sqrt{15}$

72. $\sqrt{29}$

73. $\sqrt{213}$

74. $\sqrt{156}$

75. $\sqrt{-15}$

76. $\sqrt{-79}$

77. $\sqrt{83}$

78. $\sqrt{97}$

79. $-\sqrt{15}$

80. $\sqrt{-29}$

81. $\sqrt{\dfrac{23}{5}}$

82. $\sqrt{\dfrac{39}{8}}$

83. $\sqrt{\dfrac{1124}{15}}$

84. $\sqrt{\dfrac{896}{7}}$

85. $\sqrt[4]{\dfrac{236}{10}}$

86. $\sqrt[4]{\dfrac{715}{11}}$

87. $\sqrt[5]{\dfrac{2110}{85}}$

88. $\sqrt[5]{\dfrac{1376}{19}}$

89. $\sqrt[8]{\dfrac{6432}{38}}$

90. $\sqrt[7]{\dfrac{4123}{31}}$

In Exercises 91 to 96, label the following as True or False.

91. $\sqrt{16x^{16}} = 4x^4$

92. $\sqrt{(x-4)^2} = x - 4$

93. $\sqrt{16x^{-4}y^{-4}}$ is a real number

94. $\sqrt{x^2 + y^2} = x + y$

95. $\dfrac{\sqrt{x^2 - 25}}{x - 5} = \sqrt{x + 5}$

96. $\sqrt{2} + \sqrt{6} = \sqrt{8}$

97. Is there any prime number whose square root is an integer? Explain your answer.

98. Explain the difference between the conjugate of a binomial and the opposite of a binomial. To illustrate, use $4 - \sqrt{7}$.

99. Determine two consecutive integers whose square roots are also consecutive integers.

100. Determine the missing binomial in the following: $(\sqrt{3} - 2)(\qquad) = -1$.

Answers

1. 7 **3.** −6 **5.** Not a real number **7.** 4 **9.** −6 **11.** 3 **13.** 2 **15.** −2 **17.** $\dfrac{2}{3}$

19. $\dfrac{2}{3}$ **21.** 6 **23.** 9 **25.** −5 **27.** $2\sqrt{3}$ **29.** $-6\sqrt{3}$ **31.** $4\sqrt{2}$ **33.** $-2\sqrt[3]{6}$ **35.** $2\sqrt[4]{6}$

37. $3x^2\sqrt{7}$ **39.** $5a^2\sqrt{3a}$ **41.** $4xy\sqrt{5y}$ **43.** $5x^2\sqrt[3]{2x^2}$ **45.** $2x^2yz\sqrt[3]{7y^2z}$ **47.** $3y^3\sqrt[4]{2}$

49. $2\sqrt[5]{2w}$ **51.** $\dfrac{\sqrt{5}}{4}$ **53.** $\dfrac{x^2}{5}$ **55.** $\dfrac{\sqrt{5}}{3y^2}$ **57.** $\dfrac{\sqrt[3]{3}}{4}$ **59.** $\sqrt{42}$ **61.** $\sqrt{42}$ **63.** $\sqrt[3]{36}$

65. 6 **67.** $4x\sqrt{2x}$ **69.** $5x\sqrt[3]{2x}$ **71.** 3.9 **73.** 14.6 **75.** Not a real number **77.** 9.1

79. −3.9 **81.** 2.1 **83.** 8.7 **85.** 2.2 **87.** 1.9 **89.** 1.9 **91.** False **93.** True **95.** False

97. 1 **99.** 0, 1

5.3 Rational Exponents

We will see later in this chapter that the property $(x^m)^n = x^{mn}$ holds for rational numbers m and n.

In Section 5.2, we discussed the radical notation, along with the concept of roots. In this section, we use that concept to develop a new notation, using exponents that provide an alternate way of writing these roots.

That new notation involves **rational numbers as exponents.** To start the development, we extend all the previous properties of exponents to include rational exponents.

Given that extension, suppose that

$$a = 4^{1/2} \tag{1}$$

Squaring both sides of the equation yields

$$a^2 = (4^{1/2})^2$$

or

$$a^2 = 4^{(1/2)(2)}$$

$$a^2 = 4^1$$

$$a^2 = 4 \tag{2}$$

From equation (2) we see that a is the number whose square is 4; that is, a is the principal square root of 4. Using our earlier notation, we can write

$$a = \sqrt{4}$$

But from (1)

$$a = 4^{1/2}$$

and to be consistent, we must have

$4^{1/2}$ indicates the *principal square root* of 4.

$$4^{1/2} = \sqrt{4}$$

This argument can be repeated for any exponent of the form $\dfrac{1}{n}$, so it seems reasonable to make the following definition.

> If a is any real number and n is a positive integer ($n > 1$), then
>
> $$a^{1/n} = \sqrt[n]{a}$$
>
> We restrict a so that a is nonnegative when n is even. In words, $a^{1/n}$ indicates the principal nth root of a.

Example 1 illustrates the use of rational exponents to represent roots.

• Example 1

Writing Expressions in Radical Form

Write each expression in radical form and then simplify.

$27^{1/3}$ is the *cube root* of 27.

(a) $25^{1/2} = \sqrt{25} = 5$

(b) $27^{1/3} = \sqrt[3]{27} = 3$

(c) $-36^{1/2} = -\sqrt{36} = -6$

(d) $(-36)^{1/2} = \sqrt{-36}$ is not a real number.

$32^{1/5}$ is the *fifth root* of 32.

(e) $32^{1/5} = \sqrt[5]{32} = 2$

● ● ● ●   **CHECK YOURSELF 1**

Write each expression in radical form and simplify.

(a) $8^{1/3}$ **(b)** $-64^{1/2}$ **(c)** $81^{1/4}$

We are now ready to extend our exponent notation to allow *any* rational exponent, again assuming that our previous exponent properties must still be valid. Note that

This is because

$$\frac{m}{n} = (m)\left(\frac{1}{n}\right) = \left(\frac{1}{n}\right)(m)$$

$$a^{m/n} = (a^{1/n})^m = (a^m)^{1/n}$$

From our earlier work, we know that $a^{1/n} = \sqrt[n]{a}$, and combining this with the above observation, we offer the following definition for $a^{m/n}$.

The two radical forms for $a^{m/n}$ are equivalent, and the choice of which form to use generally depends on whether we are evaluating numerical expressions or rewriting expressions containing variables in radical form.

> For any real number a and positive integers m and n with $n > 1$,
>
> $$a^{m/n} = (\sqrt[n]{a})^m = \sqrt[n]{a^m}$$

This new extension of our rational exponent notation is applied in Example 2.

• Example 2

Simplifying Expressions with Rational Exponents

Simplify each expression.

(a) $9^{3/2} = (\sqrt{9})^3$

$\qquad = 3^3 = 27$

(b) $\left(\frac{16}{81}\right)^{3/4} = \left(\sqrt[4]{\frac{16}{81}}\right)^3$

$\qquad = \left(\frac{2}{3}\right)^3 = \frac{8}{27}$

(c) $(-8)^{2/3} = (\sqrt[3]{-8})^2$

$\qquad\quad = (-2)^2 = 4$

This illustrates why we use $(\sqrt[n]{a})^m$ for $a^{m/n}$ when evaluating numerical expressions. The numbers involved will be smaller and easier to work with.

In (a) we could also have evaluated the expression as

$9^{3/2} = \sqrt{9^3} - \sqrt{729}$

$\qquad = 27$

● ● ● **CHECK YOURSELF 2**

Simplify each expression.

(a) $16^{3/4}$ 　　　　　 **(b)** $\left(\dfrac{8}{27}\right)^{2/3}$ 　　　　　 **(c)** $(-32)^{3/5}$

Now we want to extend our rational exponent notation. Using the definition of negative exponents, we can write

$a^{-m/n} = \dfrac{1}{a^{m/n}}$

Example 3 illustrates the use of negative rational exponents.

● Example 3

Simplifying Expressions with Rational Exponents

Simplify each expression.

(a) $16^{-1/2} = \dfrac{1}{16^{1/2}} = \dfrac{1}{4}$

(b) $27^{-2/3} = \dfrac{1}{27^{2/3}} = \dfrac{1}{(\sqrt[3]{27})^2} = \dfrac{1}{3^2} = \dfrac{1}{9}$

● ● ● **CHECK YOURSELF 3**

Simplify each expression.

(a) $16^{-1/4}$ 　　　　　　　　　　 **(b)** $81^{-3/4}$

Calculators can be used to evaluate expressions that contain rational exponents by using the $\boxed{y^x}$ key and the parentheses keys.

•Example 4

Estimating Powers Using a Calculator

Using a graphing calculator, evaluate each of the following. Round all answers to three decimal places.

(a) $45^{2/5}$

If you are using a scientific calculator, try using the $\boxed{y^x}$ key in place of the $\boxed{\wedge}$ key.

Enter 45 and press the $\boxed{\wedge}$ key. Then use the following keystrokes:

$$\boxed{(}\, \boxed{2}\, \boxed{\div}\, \boxed{5}\, \boxed{)}$$

Press $\boxed{=}$, and the display will read 4.584426407. Rounded to three decimal places, the result is 4.584.

(b) $38^{-2/3}$

Enter 38 and press the $\boxed{\wedge}$ key. Then use the following keystrokes:

$$\boxed{(}\, \boxed{(-)}\, \boxed{2}\, \boxed{\div}\, \boxed{3}\, \boxed{)}$$

The $\boxed{(-)}$ key changes the sign of the exponent to minus.

Press $\boxed{=}$, and the display will read 0.088473037. Rounded to three decimal places, the result is 0.088.

● ● ●   **CHECK YOURSELF 4**

Evaluate each of the following by using a scientific calculator. Round each answer to three decimal places.

(a) $23^{3/5}$ **(b)** $18^{-4/7}$

As we mentioned earlier in this section, we assume that all our previous exponent properties will continue to hold for rational exponents. Those properties are restated here.

> **Properties of Exponents**
>
> For any nonzero real numbers a and b and rational numbers m and n,
>
> 1. Product rule $a^m \cdot a^n = a^{m+n}$
> 2. Quotient rule $\dfrac{a^m}{a^n} = a^{m-n}$
> 3. Power rule $(a^m)^n = a^{mn}$
> 4. Product-power rule $(ab)^m = a^m b^m$
> 5. Quotient-power rule $\left(\dfrac{a}{b}\right)^m = \dfrac{a^m}{b^m}$
>
> We restrict a and b to being nonnegative real numbers when m or n indicates an even root.

Example 5 illustrates the use of our extended properties to simplify expressions involving rational exponents. Here, we assume that all variables represent positive real numbers.

● Example 5

Simplifying Expressions

Simplify each expression.

Product rule—add the exponents.

(*a*) $x^{2/3} \cdot x^{1/2} = x^{2/3+1/2}$

$$= x^{4/6+3/6} = x^{7/6}$$

Quotient rule—subtract the exponents.

(*b*) $\dfrac{w^{3/4}}{w^{1/2}} = w^{3/4-1/2}$

$$= w^{3/4-2/4} = w^{1/4}$$

Power rule—multiply the exponents.

(*c*) $(a^{2/3})^{3/4} = a^{(2/3)(3/4)}$

$$= a^{1/2}$$

CHECK YOURSELF 5

Simplify each expression.

(a) $z^{3/4} \cdot z^{1/2}$ **(b)** $\dfrac{x^{5/6}}{x^{1/3}}$ **(c)** $(b^{5/6})^{2/5}$

As you would expect from your previous experience with exponents, simplifying expressions often involves using several exponent properties.

● Example 6

Simplifying Expressions

Simplify each expression.

(*a*) $(x^{2/3} \cdot y^{5/6})^{3/2}$

$$= (x^{2/3})^{3/2} \cdot (y^{5/6})^{3/2} \qquad \text{Product rule.}$$

$$= x^{(2/3)(3/2)} \cdot (y^{5/6)(3/2)} = xy^{5/4} \qquad \text{Power rule.}$$

(*b*) $\left(\dfrac{r^{-1/2}}{s^{1/3}}\right)^6 = \dfrac{(r^{-1/2})^6}{(s^{1/3})^6} \qquad \text{Quotient-power rule.}$

$$= \dfrac{r^{-3}}{s^2} = \dfrac{1}{r^3 s^2} \qquad \text{Power rule.}$$

(c) $\left(\dfrac{4a^{-2/3} \cdot b^2}{a^{1/3} \cdot b^{-4}}\right)^{1/2} = \left(\dfrac{4b^2 \cdot b^4}{a^{1/3} \cdot a^{2/3}}\right)^{1/2} = \left(\dfrac{4b^6}{a}\right)^{1/2}$ We simplify inside the parentheses as the first step.

$$= \dfrac{(4b^6)^{1/2}}{a^{1/2}} = \dfrac{4^{1/2}(b^6)^{1/2}}{a^{1/2}}$$

$$= \dfrac{2b^3}{a^{1/2}}$$

● ● ● **CHECK YOURSELF 6**

Simplify each expression.

(a) $(a^{3/4} \cdot b^{1/2})^{2/3}$ **(b)** $\left(\dfrac{w^{1/2}}{z^{-1/4}}\right)^4$ **(c)** $\left(\dfrac{8x^{-3/4}y}{x^{1/4} \cdot y^{-5}}\right)^{1/3}$

We can also use the relationships between rational exponents and radicals to write expressions involving rational exponents as radicals and vice versa.

● **Example 7**

Writing Expressions in Radical Form

Write each expression in radical form.

Note: Here we use $a^{m/n} = \sqrt[n]{a^m}$, which is generally the preferred form in this situation.

(a) $a^{3/5} = \sqrt[5]{a^3}$

(b) $(mn)^{3/4} = \sqrt[4]{(mn)^3}$
$\qquad\qquad = \sqrt[4]{m^3 n^3}$

Note that the exponent applies *only* to the variable y.

(c) $2y^{5/6} = 2\sqrt[6]{y^5}$

Now the exponent applies to $2y$ because of the parentheses.

(d) $(2y)^{5/6} = \sqrt[6]{(2y)^5}$
$\qquad\qquad = \sqrt[6]{32y^5}$

● ● ● **CHECK YOURSELF 7**

Write each expression in radical form.

(a) $(ab)^{2/3}$ **(b)** $3x^{3/4}$ **(c)** $(3x)^{3/4}$

● **Example 8**

Writing Expressions in Exponential Form

Using rational exponents, write each expression and simplify.

(a) $\sqrt[3]{5x} = (5x)^{1/3}$

(b) $\sqrt{9a^2b^4} = (9a^2b^4)^{1/2}$

$\qquad = 9^{1/2}(a^2)^{1/2}(b^4)^{1/2} = 3ab^2$

(c) $\sqrt[4]{16w^{12}z^8} = (16w^{12}z^8)^{1/4}$

$\qquad = 16^{1/4}(w^{12})^{1/4}(z^8)^{1/4} = 2w^3z^2$

● ● ● CHECK YOURSELF 8

Using rational exponents, write each expression and simplify.

(a) $\sqrt{7a}$ **(b)** $\sqrt[3]{27p^6q^9}$ **(c)** $\sqrt[4]{81x^8y^{16}}$

● ● ● CHECK YOURSELF ANSWERS

1. (a) 2, **(b)** -8, **(c)** 3.   **2. (a)** 8, **(b)** $\dfrac{4}{9}$, **(c)** -8.   **3. (a)** $\dfrac{1}{2}$, **(b)** $\dfrac{1}{27}$.

4. (a) 6.562, **(b)** 0.192.   **5. (a)** $z^{5/4}$, **(b)** $x^{1/2}$, **(c)** $b^{1/3}$.   **6. (a)** $a^{1/2}b^{1/3}$, **(b)** w^2z,

(c) $\dfrac{2y^2}{x^{1/3}}$.   **7. (a)** $\sqrt[3]{a^2b^2}$, **(b)** $3\sqrt[4]{x^3}$, **(c)** $\sqrt[4]{27x^3}$.   **8. (a)** $(7a)^{1/2}$, **(b)** $3p^2q^3$,

(c) $3x^2y^4$.

5.3 Exercises

In Exercises 1 to 12, use the definition of $a^{1/n}$ to evaluate each expression.

1. $36^{1/2}$ **2.** $100^{1/2}$ **3.** $-25^{1/2}$ **4.** $(-64)^{1/2}$

5. $(-49)^{1/2}$ **6.** $-49^{1/2}$ **7.** $27^{1/3}$ **8.** $(-64)^{1/3}$

9. $81^{1/4}$ **10.** $-32^{1/5}$ **11.** $\left(\dfrac{4}{9}\right)^{1/2}$ **12.** $\left(\dfrac{27}{8}\right)^{1/3}$

In Exercises 13 to 22, use the definition of $a^{m/n}$ to evaluate each expression.

13. $27^{2/3}$ **14.** $16^{3/2}$ **15.** $(-8)^{4/3}$ **16.** $125^{2/3}$

17. $32^{2/5}$

18. $-81^{3/4}$

19. $81^{3/2}$

20. $(-243)^{3/5}$

21. $\left(\dfrac{8}{27}\right)^{2/3}$

22. $\left(\dfrac{9}{4}\right)^{3/2}$

In Exercises 23 to 32, use the definition of $a^{-m/n}$ to evaluate the following expression. Use your calculator to check each answer.

23. $25^{-1/2}$

24. $27^{-1/3}$

25. $81^{-1/4}$

26. $121^{-1/2}$

27. $9^{-3/2}$

28. $16^{-3/4}$

29. $64^{-5/6}$

30. $16^{-3/2}$

31. $\left(\dfrac{4}{25}\right)^{-1/2}$

32. $\left(\dfrac{27}{8}\right)^{-2/3}$

In Exercises 33 to 76, use the properties of exponents to simplify each expression. Assume all variables represent positive real numbers.

33. $x^{1/2} \cdot x^{1/2}$

34. $a^{2/3} \cdot a^{1/3}$

35. $y^{3/5} \cdot y^{1/5}$

36. $m^{1/4} \cdot m^{5/4}$

37. $b^{2/3} \cdot b^{3/2}$

38. $p^{5/6} \cdot p^{2/3}$

39. $\dfrac{x^{2/3}}{x^{1/3}}$

40. $\dfrac{a^{5/6}}{a^{1/6}}$

41. $\dfrac{s^{7/5}}{s^{2/5}}$

42. $\dfrac{z^{9/2}}{z^{3/2}}$

43. $\dfrac{w^{5/4}}{w^{1/2}}$

44. $\dfrac{b^{7/6}}{b^{2/3}}$

45. $(x^{3/4})^{4/3}$

46. $(y^{4/3})^{3/4}$

47. $(a^{2/5})^{3/2}$

48. $(p^{3/4})^{2/3}$

49. $(y^{-3/4})^{8}$

50. $(w^{-2/3})^{6}$

51. $(a^{2/3} \cdot b^{3/2})^{6}$

52. $(p^{3/4} \cdot q^{5/2})^{4}$

53. $(2x^{1/5} \cdot y^{3/5})^{5}$

54. $(3m^{3/4} \cdot n^{5/4})^{4}$

55. $(s^{3/4} \cdot t^{1/4})^{4/3}$

56. $(x^{5/2} \cdot y^{5/7})^{2/5}$

57. $(8p^{3/2} \cdot q^{5/2})^{2/3}$

58. $(16a^{1/3} \cdot b^{2/3})^{3/4}$

59. $(x^{3/5} \cdot y^{3/4} \cdot z^{3/2})^{2/3}$

60. $(p^{5/6} \cdot q^{2/3} \cdot r^{5/3})^{3/5}$

61. $\dfrac{a^{5/6} \cdot b^{3/4}}{a^{1/3} \cdot b^{1/2}}$

62. $\dfrac{x^{2/3} \cdot y^{3/4}}{x^{1/2} \cdot y^{1/2}}$

63. $\dfrac{(r^{-1} \cdot s^{1/2})^{3}}{r \cdot s^{-1/2}}$

64. $\dfrac{(w^{-2} \cdot z^{-1/4})^{6}}{w^{-8}z^{1/2}}$

65. $\left(\dfrac{x^{12}}{y^{8}}\right)^{1/4}$

66. $\left(\dfrac{p^{9}}{q^{6}}\right)^{1/3}$

67. $\left(\dfrac{m^{-1/4}}{n^{1/2}}\right)^{4}$

68. $\left(\dfrac{r^{1/5}}{s^{-1/2}}\right)^{10}$

69. $\left(\dfrac{r^{-1/2} \cdot s^{3/4}}{t^{1/4}}\right)^4$

70. $\left(\dfrac{a^{1/3} \cdot b^{-1/6}}{c^{-1/6}}\right)^6$

71. $\left(\dfrac{8x^3 \cdot y^{-6}}{z^{-9}}\right)^{1/3}$

72. $\left(\dfrac{16p^{-4} \cdot q^6}{r^2}\right)^{-1/2}$

73. $\left(\dfrac{16m^{-3/5} \cdot n^2}{m^{1/5} \cdot n^{-2}}\right)^{1/4}$

74. $\left(\dfrac{27x^{5/6} \cdot y^{-4/3}}{x^{-7/6} \cdot y^{5/3}}\right)^{1/3}$

75. $\left(\dfrac{x^{3/2} \cdot y^{1/2}}{z^2}\right)^{1/2}\left(\dfrac{x^{3/4} \cdot y^{3/2}}{z^{-3}}\right)^{1/3}$

76. $\left(\dfrac{p^{1/2} \cdot q^{4/3}}{r^{-4}}\right)^{3/4}\left(\dfrac{p^{15/8} \cdot q^{-3}}{r^6}\right)^{1/3}$

In Exercises 77 to 84, write each expression in radical form. Do not simplify.

77. $a^{3/4}$

78. $m^{5/6}$

79. $2x^{2/3}$

80. $3m^{2/3}$

81. $3x^{2/5}$

82. $2y^{-3/4}$

83. $(3x)^{2/5}$

84. $(2y)^{-3/4}$

In Exercises 85 to 88, write each expression using rational exponents, and simplify where necessary.

85. $\sqrt{7a}$

86. $\sqrt{25w^4}$

87. $\sqrt[3]{8m^6n^9}$

88. $\sqrt[5]{32r^{10}s^{15}}$

 In Exercises 89 to 92, evaluate each expression, using a calculator. Round each answer to three decimal places.

89. $46^{3/5}$

90. $23^{2/7}$

91. $12^{-2/5}$

92. $36^{-3/4}$

 93. Describe the difference between x^{-2} and $x^{1/2}$.

 94. Some rational exponents, like $\dfrac{1}{2}$, can easily be rewritten as terminating decimals (0.5). Others, like $\dfrac{1}{3}$, cannot. What is it that determines which rational numbers can be rewritten as terminating decimals?

In Exercises 95 to 104, apply the appropriate multiplication patterns. Then simplify your result.

95. $a^{1/2}(a^{3/2} + a^{3/4})$

96. $2x^{1/4}(3x^{3/4} - 5x^{-1/4})$

97. $(a^{1/2} + 2)(a^{1/2} - 2)$

98. $(w^{1/3} - 3)(w^{1/3} + 3)$

99. $(m^{1/2} + n^{1/2})(m^{1/2} - n^{1/2})$

100. $(x^{1/3} + y^{1/3})(x^{1/3} - y^{1/3})$

101. $(x^{1/2} + 2)^2$

102. $(a^{1/3} - 3)^2$

103. $(r^{1/2} + s^{1/2})^2$

104. $(p^{1/2} - q^{1/2})^2$

As is suggested by several of the preceding exercises, certain expressions containing rational exponents are factorable. For instance, to factor $x^{2/3} - x^{1/3} - 6$, let $u = x^{1/3}$. Note that $x^{2/3} = (x^{1/3})^2 = u^2$.

Substituting, we have $u^2 - u - 6$, and factoring yields $(u - 3)(u + 2)$ or $(x^{1/3} - 3)(x^{1/3} + 2)$.

In Exercises 105 to 110, use this technique to factor each expression.

105. $x^{2/3} + 4x^{1/3} + 3$ **106.** $y^{2/5} - 2y^{1/5} - 8$ **107.** $a^{4/5} - 7a^{2/5} + 12$

108. $w^{4/3} + 3w^{2/3} - 10$ **109.** $x^{4/3} - 4$ **110.** $x^{2/5} - 16$

In Exercises 111 to 120, perform the indicated operations. Assume that n represents a positive integer and that the denominators are not zero.

111. $x^{3n} \cdot x^{2n}$ **112.** $p^{1-n} \cdot p^{n+3}$ **113.** $(y^2)^{2n}$ **114.** $(a^{3n})^3$

115. $\dfrac{r^{n+2}}{r^n}$ **116.** $\dfrac{w^n}{w^{n-3}}$ **117.** $(a^3 \cdot b^2)^{2n}$ **118.** $(c^4 \cdot d^2)^{3m}$

119. $\left(\dfrac{x^{n+2}}{x^n}\right)^{1/2}$ **120.** $\left(\dfrac{b^n}{b^{n-3}}\right)^{1/3}$

In Exercises 121 to 124, write each expression in exponent form, simplify, and give the result as a single radical.

121. $\sqrt{\sqrt{x}}$ **122.** $\sqrt[3]{\sqrt{a}}$ **123.** $\sqrt[4]{\sqrt{y}}$ **124.** $\sqrt{\sqrt[3]{\sqrt{w}}}$

In Exercises 125 to 130, simplify each expression. Write your answer in scientific notation.

125. $(4 \times 10^8)^{1/2}$ **126.** $(8 \times 10^6)^{1/3}$ **127.** $(16 \times 10^{-12})^{1/4}$

128. $(9 \times 10^{-4})^{1/2}$ **129.** $(16 \times 10^{-8})^{1/2}$ **130.** $(16 \times 10^{-8})^{3/4}$

131. While investigating rainfall runoff in a region of semiarid farmland, a researcher encounters the following expression:

$$t = C\left(\frac{L}{xy^2}\right)^{1/3}$$

Evaluate t when $C = 20$, $L = 600$, $x = 3$, and $y = 5$.

132. The average velocity of water in an open irrigation ditch is given by the formula

$$V = \frac{1.5x^{2/3}y^{1/2}}{z}$$

Evaluate V when $x = 27$, $y = 16$, and $z = 12$.

133. Use the properties of exponents to decide what x should be to make each statement true. Explain your choices regarding which properties of exponents you decide to use.

(a) $(a^{2/3})^x = a$

(b) $(a^{5/6})^x = \dfrac{1}{a}$

(c) $a^{2x} \cdot a^{3/2} = 1$

(d) $(\sqrt{a^{2/3}})^x = a$

134. The geometric mean is used to measure average inflation rates or interest rates (see the group activity following Section 5.2). If prices increased by 15% over 5 years, then the average *annual* rate of inflation is obtained by taking the 5th root of 1.15:

$$(1.15)^{1/5} = 1.0283 \qquad \text{or} \qquad \sim 2.8\%$$

The 1 is added to 0.15 because we are taking the original price and adding 15% of that price. We could write that as

$P + 0.15P$

Factoring, we get

$$\begin{aligned} P + 0.15P &= P(1 + 0.15) \\ &= P(1.15) \end{aligned}$$

In the introduction to this chapter, the following statement was made: "From December 1990 through February 1997, the Bureau of Labor Statistics computed an inflation rate of 16.2%, which is equivalent to an annual growth rate of 2.46%." From December 1990 through February 1997 is 14 months. To what exponent was 1.162 raised to obtain this average annual growth rate?

Answers

1. 6   **3.** -5 **5.** Not a real number   **7.** 3   **9.** 3   **11.** $\dfrac{2}{3}$ **13.** 9   **15.** 16   **17.** 4

19. 729   **21.** $\dfrac{4}{9}$ **23.** $\dfrac{1}{5}$ **25.** $\dfrac{1}{3}$ **27.** $\dfrac{1}{27}$ **29.** $\dfrac{1}{32}$ **31.** $\dfrac{5}{2}$ **33.** x **35.** $y^{4/5}$ **37.** $b^{13/6}$

39. $x^{1/3}$ **41.** s **43.** $w^{3/4}$ **45.** x **47.** $a^{3/5}$ **49.** $\dfrac{1}{y^6}$ **51.** a^4b^9 **53.** $32xy^3$ **55.** $st^{1/3}$

57. $4pq^{5/3}$ **59.** $x^{2/5}y^{1/2}z$ **61.** $a^{1/2}b^{1/4}$ **63.** $\dfrac{s^2}{r^4}$ **65.** $\dfrac{x^3}{y^2}$ **67.** $\dfrac{1}{mn^2}$ **69.** $\dfrac{s^3}{r^2t}$ **71.** $\dfrac{2xz^3}{y^2}$

73. $\dfrac{2n}{m^{1/5}}$ **75.** $xy^{3/4}$ **77.** $\sqrt[4]{a^3}$ **79.** $2\sqrt[3]{x^2}$ **81.** $3\sqrt[5]{x^2}$ **83.** $\sqrt[5]{9x^2}$ **85.** $(7a)^{1/2}$ **87.** $2m^2n^3$

89. 9.946   **91.** 0.370   **93.**   **95.** $a^2 + a^{5/4}$ **97.** $a - 4$ **99.** $m - n$ **101.** $x + 4x^{1/2} + 4$

103. $r + 2r^{1/2}s^{1/2} + s$ **105.** $(x^{1/3} + 1)(x^{1/3} + 3)$ **107.** $(a^{2/5} - 3)(a^{2/5} - 4)$ **109.** $(x^{2/3} - 2)(x^{2/3} + 2)$

111. x^{5n} **113.** y^{4n} **115.** r^2 **117.** $a^{6n}b^{4n}$ **119.** x **121.** $\sqrt[4]{x}$ **123.** $\sqrt[8]{y}$ **125.** 2×10^4

127. 2×10^{-3} **129.** 4×10^{-4} **131.** 40   **133.**

5.4 Complex Numbers

5.4 OBJECTIVES

1. Define a complex number
2. Add and subtract complex numbers
3. Multiply and divide complex numbers

$i = \sqrt{-1}$ is called the **imaginary unit.**

Radicals such as

$$\sqrt{-4} \quad \text{and} \quad \sqrt{-49}$$

are *not* real numbers since no real number squared produces a negative number. Our work in this section will extend our number system to include these **imaginary numbers,** which will allow us to consider radicals such as $\sqrt{-4}$.

First we offer a definition.

The number i is defined as

$$i = \sqrt{-1}$$

Note that this means that

$$i^2 = -1$$

This definition of the number i gives us an alternate means of indicating the square root of a negative number.

When a is a positive real number,

$$\sqrt{-a} = \sqrt{a}\,i \quad \text{or} \quad i\sqrt{a}$$

• Example 1

Using the Number *i*

Write each expression as a multiple of i.

We simplify $\sqrt{8}$ as $2\sqrt{2}$. Note that we write *in front of* the radical to make it clear that i is *not part* of the radicand.

(a) $\sqrt{-4} = \sqrt{4}\,i = 2i$

(b) $-\sqrt{-9} = -\sqrt{9}\,i = -3i$

(c) $\sqrt{-8} = \sqrt{8}\,i = 2\sqrt{2}\,i$ or $2i\sqrt{2}$

(d) $\sqrt{-7} = \sqrt{7}\,i$ or $i\sqrt{7}$

 CHECK YOURSELF 1

Write each radical as a multiple of i.

(a) $\sqrt{-25}$

(b) $\sqrt{-24}$

We are now ready to define complex numbers in terms of the number i.

> A **complex number** is any number that can be written in the form
>
> $a + bi$
>
> where a and b are real numbers and
>
> $i = \sqrt{-1}$

The term "imaginary number" was introduced by René Descartes in 1637. Euler used i to indicate $\sqrt{-1}$ in 1748, but it was not until 1832 that Gauss used the term "complex number."

The first application of these numbers was made by Charles Steinmetz (1865–1923) in explaining the behavior of electric circuits.

Also, $5i$ is called a **pure imaginary** number.

The real numbers can be considered a subset of the set of complex numbers.

The form $a + bi$ is called the **standard form** of a complex number. We call a the **real part** of the complex number and b the **imaginary part.** Some examples follow.

$3 + 7i$ is an example of a complex number with real part 3 and imaginary part 7.

$5i$ is also a complex number since it can be written as $0 + 5i$.

-3 is a complex number since it can be written as $-3 + 0i$.

The basic operations of addition and subtraction on complex numbers are defined here.

Adding and Subtracting Complex Numbers

For the complex numbers $a + bi$ and $c + di$,

$(a + bi) + (c + di) = (a + c) + (b + d)i$

$(a + bi) - (c + di) = (a - c) + (b - d)i$

In words, we add or subtract the real parts and the imaginary parts of the complex numbers.

Example 2 illustrates the use of these definitions.

● Example 2

Adding and Subtracting Complex Numbers

Perform the indicated operations.

(a) $(5 + 3i) + (6 - 7i) = (5 + 6) + (3 - 7)i$
$$= 11 - 4i$$

(b) $5 + (7 - 5i) = (5 + 7) + (-5i)$
$$= 12 - 5i$$

(c) $(8 - 2i) - (3 - 4i) = (8 - 3) + [-2 - (-4)]i$
$$= 5 + 2i$$

● ● ● **CHECK YOURSELF 2**

Perform the indicated operations.

(a) $(4 - 7i) + (3 - 2i)$ **(b)** $-7 + (-2 + 3i)$ **(c)** $(-4 + 3i) - (-2 - i)$

Since complex numbers are binomial in form, the product of two complex numbers is found by applying our earlier multiplication pattern for binomials, as Example 3 illustrates.

●**Example 3**

Multiplying Complex Numbers

Multiply.

$(2 + 3i)(3 - 4i)$

$= 2(3 - 4i) + 3i(3 - 4i)$

$= 2(3) + 2(-4i) + (3i)(3) + (3i)(-4i)$

We can replace i^2 with -1 because of the definition of i, and we usually do so because of the resulting simplification.

$= 6 - 8i + 9i - 12i^2$

$= 6 + i - 12(-1)$

$= 6 + i + 12 = 18 + i$

● ● ● **CHECK YOURSELF 3**

Multiply $(2 - 5i)(3 - 2i)$.

We now consider the basic operation of multiplication on complex numbers.

> **Multiplying Complex Numbers**
>
> For the complex numbers $a + bi$ and $c + di$,
>
> $(a + bi)(c + di) = ac + adi + bci + bdi^2$
>
> $\qquad\qquad\qquad = ac + adi + bci - bd$
>
> $\qquad\qquad\qquad = (ac - bd) + (ad + bc)i$

This formula for the general product of two complex numbers can be memorized. However, you will find it much easier to get used to the multiplication pattern as it is applied to complex numbers than to memorize this formula.

There is one particular product form that will seem very familiar. We call $a + bi$ and $a - bi$ **complex conjugates.** For instance,

$3 + 2i$ and $3 - 2i$

are complex conjugates.

Consider the product

$$(3 + 2i)(3 - 2i) = 3^2 - (2i)^2$$
$$= 9 - 4i^2 = 9 - 4(-1)$$
$$= 9 + 4 = 13$$

The product of $3 + 2i$ and $3 - 2i$ is a real number. In general, we can write the product of two complex conjugates as

$$(a + bi)(a - bi) = a^2 + b^2$$

The fact that this product is always a real number will be very useful when we consider the division of complex numbers later in this section.

● Example 4

Multiplying Complex Numbers

Multiply.

We could get the same result by applying the formula above with $a = 7$ and $b = 4$.

$$(7 - 4i)(7 + 4i) = 7^2 - (4i)^2$$
$$= 7^2 - 4^2(-1)$$
$$= 7^2 + 4^2$$
$$= 49 + 16 = 65$$

● ● ●   **CHECK YOURSELF 4**

Multiply $(5 + 3i)(5 - 3i)$.

We are now ready to discuss the division of complex numbers. Generally, we find the quotient by multiplying the numerator and denominator by the conjugate of the denominator, as Example 5 illustrates.

•Example 5

Dividing Complex Numbers

Divide.

Think of $3i$ as $0 + 3i$ and of its conjugate as $0 - 3i$, or $-3i$.

(a) $\dfrac{6 + 9i}{3i}$

$$\dfrac{6 + 9i}{3i} = \dfrac{(6 + 9i)(-3i)}{(3i)(-3i)}$$

The conjugate of $3i$ is $-3i$, and so we multiply the numerator and denominator by $-3i$.

$$= \dfrac{-18i - 27i^2}{-9i^2}$$

Note: Multiplying the numerator and denominator in the original expression by i would yield the same result. Try it yourself.

$$= \dfrac{-18i - 27(-1)}{(-9)(-1)}$$

$$= \dfrac{27 - 18i}{9} = 3 - 2i$$

We multiply $\dfrac{3 - 2i}{3 - 2i}$, or 1.

(b) $\dfrac{3 - i}{3 + 2i} = \dfrac{(3 - i)(3 - 2i)}{(3 + 2i)(3 - 2i)}$

$$= \dfrac{9 - 6i - 3i + 2i^2}{9 - 4i^2}$$

$$= \dfrac{9 - 9i - 2}{9 + 4}$$

To write a complex number in standard form, we separate the real component from the imaginary.

$$= \dfrac{7 - 9i}{13} = \dfrac{7}{13} - \dfrac{9}{13}i$$

(c) $\dfrac{2 + i}{4 - 5i} = \dfrac{(2 + i)(4 + 5i)}{(4 - 5i)(4 + 5i)}$

$$= \dfrac{8 + 4i + 10i + 5i^2}{16 - 25i^2}$$

$$= \dfrac{8 + 14i - 5}{16 + 25}$$

$$= \dfrac{3 + 14i}{41} = \dfrac{3}{41} + \dfrac{14}{41}i$$

● ● ● **CHECK YOURSELF 5**

Divide.

(a) $\dfrac{5 + i}{5 - 3i}$

(b) $\dfrac{4 + 10i}{2i}$

We conclude this section with the diagram on page 401, which summarizes the structure of the system of complex numbers.

Complex numbers $(a + bi)$

Real numbers
$(b = 0)$

Pure imaginary
numbers $(a = 0, b \neq 0)$

Rational
numbers

Irrational
numbers

● ● ● ● **CHECK YOURSELF ANSWERS**

1. (a) $5i$, **(b)** $2i\sqrt{6}$.　**2. (a)** $7 - 9i$, **(b)** $-9 + 3i$, **(c)** $-2 + 4i$.

3. $-4 - 19i$.　**4.** 34.　**5. (a)** $\dfrac{11}{17} + \dfrac{10}{17}i$, **(b)** $5 - 2i$.

5.4 Exercises

In Exercises 1 to 10, write each root as a multiple of i. Simplify your results where necessary.

1. $\sqrt{-16}$　　　　**2.** $\sqrt{-36}$　　　　**3.** $-\sqrt{-64}$　　　　**4.** $-\sqrt{-25}$

5. $\sqrt{-21}$　　　　**6.** $\sqrt{-19}$　　　　**7.** $\sqrt{-12}$　　　　**8.** $\sqrt{-24}$

9. $-\sqrt{-108}$　　**10.** $-\sqrt{-192}$

In Exercises 11 to 26, perform the indicated operations.

11. $(3 + i) + (5 + 2i)$　　**12.** $(2 + 3i) + (4 + 5i)$　　**13.** $(3 - 2i) + (-2 + 7i)$

14. $(-5 - 3i) + (-2 + 7i)$　　**15.** $(5 + 4i) + (3 + 2i)$　　**16.** $(7 + 6i) - (3 + 5i)$

17. $(8 - 5i) - (3 + 2i)$　　**18.** $(7 - 3i) + (-2 + 5i)$　　**19.** $(5 + i) + (2 + 3i) + 7i$

20. $(3 - 2i) + (2 + 3i) + 7i$　　**21.** $(2 + 3i) - (3 - 5i) + (4 + 3i)$　　**22.** $(5 - 7i) + (7 + 3i) - (2 - 7i)$

23. $(7 + 3i) - [(3 + i) - (2 - 5i)]$　**24.** $(8 - 2i) - [(4 + 3i) - (-2 + i)]$　**25.** $(5 + 3i) + (-5 - 3i)$

26. $(8 - 7i) + (-8 + 7i)$

In Exercises 27 to 42, find each product. Write your answer in standard form.

27. $3i(3 + 5i)$ **28.** $2i(7 + 3i)$ **29.** $4i(3 - 7i)$ **30.** $2i(6 + 3i)$

31. $-2i(4 - 3i)$ **32.** $-5i(2 - 7i)$ **33.** $6i\left(\dfrac{2}{3} + \dfrac{5}{6}i\right)$ **34.** $4i\left(\dfrac{1}{2} + \dfrac{3}{4}i\right)$

35. $(3 + 2i)(2 + 3i)$ **36.** $(5 - 2i)(3 - i)$ **37.** $(4 - 3i)(2 + 5i)$

38. $(7 + 2i)(3 - 2i)$ **39.** $(-2 - 3i)(-3 + 4i)$ **40.** $(-5 - i)(-3 - 4i)$

41. $(5 - 2i)^2$ **42.** $(3 + 7i)^2$

In Exercises 43 to 50, write the conjugate of each complex number. Then find the product of the given number and the conjugate.

43. $3 - 2i$ **44.** $5 + 2i$ **45.** $2 + 3i$ **46.** $7 - i$

47. $-3 - 2i$ **48.** $-5 - 7i$ **49.** $5i$ **50.** $-3i$

In Exercises 51 to 62, find each quotient, and write your answer in standard form.

51. $\dfrac{3 + 2i}{i}$ **52.** $\dfrac{5 - 3i}{-i}$ **53.** $\dfrac{6 - 4i}{2i}$ **54.** $\dfrac{8 + 12i}{-4i}$

55. $\dfrac{3}{2 + 5i}$ **56.** $\dfrac{5}{2 - 3i}$ **57.** $\dfrac{13}{2 + 3i}$ **58.** $\dfrac{-17}{3 + 5i}$

59. $\dfrac{2 + 3i}{4 + 3i}$ **60.** $\dfrac{4 - 2i}{5 - 3i}$ **61.** $\dfrac{3 - 4i}{3 + 4i}$ **62.** $\dfrac{7 + 2i}{7 - 2i}$

 63. The first application of complex numbers was suggested by the Norwegian surveyor Caspar Wessel in 1797. He found that complex numbers could be used to represent distance and direction on a two-dimensional grid. Why would a surveyor care about such a thing?

64. To what sets of numbers does 1 belong?

In this section, we defined $\sqrt{-4} = \sqrt{4}\,i = 2i$ in the process of expressing the square root of a negative number as a multiple of i.

Particular care must be taken with products where two negative radicands are involved. For instance,

$$\sqrt{-3} \cdot \sqrt{-12} = (i\sqrt{3})(i\sqrt{12})$$
$$= i^2\sqrt{36} = (-1)\sqrt{36} = -6$$

is correct. However, if we try to apply the product property for radicals, we have

$$\sqrt{-3} \cdot \sqrt{-12} \stackrel{?}{=} \sqrt{(-3)(-12)} = \sqrt{36} = 6$$

which is *not* correct. The property $\sqrt{a} \cdot \sqrt{b} = \sqrt{ab}$ is not applicable in the case where a and b are both negative. Radicals such as $\sqrt{-a}$ must be written in the standard form $i\sqrt{a}$ *before* multiplying, in order to use the rules for real valued radicals.

In Exercises 65 to 72, find each product.

65. $\sqrt{-5} \cdot \sqrt{-7}$

66. $\sqrt{-3} \cdot \sqrt{-10}$

67. $\sqrt{-2} \cdot \sqrt{-18}$

68. $\sqrt{-4} \cdot \sqrt{-25}$

69. $\sqrt{-6} \cdot \sqrt{-15}$

70. $\sqrt{-5} \cdot \sqrt{-30}$

71. $\sqrt{-10} \cdot \sqrt{-10}$

72. $\sqrt{-11} \cdot \sqrt{-11}$

Since $i^2 = 1$, the positive integral powers of i form an interesting pattern. Consider the following.

$i = i$ $i^5 = i^4 \cdot i = 1 \cdot i = i$

$i^2 = -1$ $i^6 = i^4 \cdot i^2 = 1(-1) = -1$

$i^3 = i^2 \cdot i = (-1)i = -i$ $i^7 = i^4 \cdot i^3 = 1(-1) = -i$

$i^4 = i^2 \cdot i^2 = (-1)(-1) = 1$ $i^8 = i^4 \cdot i^4 = 1 \cdot 1 = 1$

Given the pattern above, do you see that any power of i will simplify to i, -1, $-i$, or 1? The easiest approach to simplifying higher powers of i is to write that power in terms of i^4 (because $1^4 = 1$). As an example,

$$i^{18} = i^{16} \cdot i^2 = (i^4)^4 \cdot i^2 = 1^4(-1) = -1$$

In Exercises 73 to 80, use these comments to simplify each power of i.

73. i^{10}

74. i^9

75. i^{20}

76. i^{15}

77. i^{38}

78. i^{40}

79. i^{51}

80. i^{61}

Answers

1. $4i$ **3.** $-8i$ **5.** $i\sqrt{21}$ **7.** $2i\sqrt{3}$ **9.** $-6i\sqrt{3}$ **11.** $8 + 3i$ **13.** $1 + 5i$ **15.** $2 + 2i$

17. $5 - 7i$ **19.** $7 + 11i$ **21.** $3 + 11i$ **23.** $6 - 3i$ **25.** $0 + 0i$ **27.** $-15 + 9i$ **29.** $28 + 12i$

31. $-6 - 8i$ **33.** $-5 + 4i$ **35.** $13i$ **37.** $23 + 14i$ **39.** $18 + i$ **41.** $21 - 20i$ **43.** $3 + 2i, 13$

45. $2 - 3i, 13$ **47.** $-3 + 2i, 13$ **49.** $-5i, 25$ **51.** $2 - 3i$ **53.** $-2 - 3i$ **55.** $\dfrac{6}{29} - \dfrac{15}{29}i$

57. $2 - 3i$ **59.** $\dfrac{17}{25} + \dfrac{6}{25}i$ **61.** $-\dfrac{7}{25} - \dfrac{24}{25}i$ **63.**   **65.** $-\sqrt{35}$ **67.** -6 **69.** $-3\sqrt{10}$

71. -10 **73.** -1 **75.** 1 **77.** -1 **79.** $-i$

Summary

Zero and Negative Integer Exponents and Scientific Notation [5.1]

Zero Exponent

$5^0 = 1$

For any real number a where $a \neq 0$,

$x^{-3} = \dfrac{1}{x^3}$

$$a^0 = 1$$

$2^{-4} = \dfrac{1}{2^4} = \dfrac{1}{16}$

Negative Integer Exponents

$2y^{-5} = \dfrac{2}{y^5}$

For any nonzero real number a and whole number n,

$(2y)^{-5} = \dfrac{1}{(2y)^5} = \dfrac{1}{32y^5}$

$$a^{-n} = \dfrac{1}{a^n}$$

$\dfrac{1}{x^{-4}} = x^4$

and a^{-n} is the **multiplicative inverse** of a^n.

$\dfrac{1}{3^{-3}} = 3^3 = 27$

Raising Quotients to a Negative Power

$$\left(\dfrac{a}{b}\right)^{-n} = \dfrac{a^{-n}}{b^{-n}} = \dfrac{b^n}{a^n} = \left(\dfrac{b}{a}\right) \qquad a \neq 0 \qquad b \neq 0$$

$38{,}000{,}000. = 3.8 \times 10^7$
$\qquad$ 7 places

Scientific Notation

Scientific notation is a useful way of expressing very large or very small numbers through the use of powers of 10. Any number written in the form

$0.0025 = 2.5 \times 10^{-3}$
$\qquad$ 3 places

$$a \times 10^n$$

where $1 \leq a < 10$ and n is an integer, is said to be written in scientific notation.

Evaluating Radical Expressions [5.2]

$\sqrt{3x} \cdot \sqrt{6x^2} = \sqrt{18x^3}$

Multiplying and Dividing Radical Expressions

$\qquad = \sqrt{9x^2 \cdot 2x}$

To multiply two monomial expressions, we use the product theorem

$\qquad = \sqrt{9x^2} \cdot \sqrt{2x}$

$\qquad = 3x\sqrt{2x}$

$$\sqrt[n]{ab} = \sqrt[n]{a} \cdot \sqrt[n]{b}$$

$\sqrt{2}(5 + \sqrt{8})$

$= \sqrt{2} \cdot 5 + \sqrt{2} \cdot \sqrt{8}$

and simplify the product.

$= 5\sqrt{2} + 4$

$\qquad$ The quotient theorem allows us to rewrite expressions that have fractions inside the radical.

$\dfrac{5}{\sqrt{8}} = \dfrac{\sqrt{5} \cdot \sqrt{2}}{\sqrt{8} \cdot \sqrt{2}} = \dfrac{5\sqrt{2}}{\sqrt{16}}$

$$\sqrt[n]{\dfrac{a}{b}} = \dfrac{\sqrt[n]{a}}{\sqrt[n]{b}}$$

$\qquad = \dfrac{5\sqrt{2}}{4}$

Rational Exponents [5.3]

Rational exponents are an alternate way of indicating roots. We use the following definition.

$36^{1/2} = \sqrt{36} = 6$

$-27^{1/3} = -\sqrt[3]{27} = -3$

$243^{1/5} = \sqrt[5]{243} = 3$

$25^{-1/2} = \dfrac{1}{\sqrt{25}} = \dfrac{1}{5}$

> If a is any real number and n is a positive integer ($n > 1$),
>
> $$a^{1/n} = \sqrt[n]{a}$$

We restrict a so that a is nonnegative when n is even.

We also define the following.

$27^{2/3} = (\sqrt[3]{27})^2$

$\quad = 3^2 = 9$

$(a^4 b^8)^{3/4} = \sqrt[4]{(a^4 b^8)^3}$

$\quad = \sqrt[4]{a^{12} b^{24}} = a^3 b^6$

> For any real number a and positive integers m and n, with $n > 1$, then
>
> $$a^{m/n} = (\sqrt[n]{a})^m = \sqrt[n]{a^m}$$

Properties of Exponents

The following five properties for exponents continue to hold for rational exponents.

Product Rule

$x^5 \cdot x^7 = x^{5+7} = x^{12}$

$$a^m \cdot a^n = a^{m+n}$$

Quotient Rule

$\dfrac{x^7}{x^5} = x^{7-5} = x^2$

$$\dfrac{a^m}{a^n} = a^{m-n}$$

Power Rule

$(x^5)^3 = x^{5 \cdot 3} = x^{15}$

$$(a^m)^n = a^{m \cdot n}$$

Product-Power Rule

$(2xy)^3 = 2^3 x^3 y^3$

$\quad = 8x^3 y^3$

$$(ab)^m = a^m b^m$$

Quotient-Power Rule

$\left(\dfrac{x^2}{3}\right)^2 = \dfrac{(x^2)^2}{3^2}$

$\quad = \dfrac{x^4}{9}$

$$\left(\dfrac{a}{b}\right)^m = \dfrac{a^m}{b^m}$$

Complex Numbers [5.4]

The number i is defined as

$$i = \sqrt{1}$$

Note that this means that

$$i^2 = -1$$

A **complex number** is any number that can be written in the form

$$a + bi$$

where a and b are real numbers and

$$i = \sqrt{-1}$$

Addition and Subtraction

For the complex numbers $a + bi$ and $c + di$,

$(2 + 3i) + (-3 - 5i)$

$= (2 - 3) + (3 - 5)i$

$= -1 - 2i$

$$(a + bi) + (c + di) = (a + c) + (b + d)i$$

and

$(5 - 2i) - (3 - 4i)$

$= (5 - 3) + (-2 - (-4))i$

$= 2 + 2i$

$$(a + bi) - (c + di) = (a - c) + (b - d)i$$

Multiplication

For the complex numbers $a + bi$ and $c + di$,

$(2 + 5i)(3 - 4i)$

$= 6 - 8i + 15i - 20i^2$

$= 6 + 7i - 20(-1)$

$= 26 + 7i$

$$(a + bi)(c + di) = (ac - bd) + (ad + bc)i$$

Note: It is generally easier to use the FOIL multiplication pattern and the definition of i, rather than to apply the above formula.

$\dfrac{3 + 2i}{3 - 2i} = \dfrac{(3 + 2i)(3 + 2i)}{(3 - 2i)(3 + 2i)}$

$= \dfrac{9 + 6i + 6i + 4i^2}{9 - 4i^2}$

$= \dfrac{9 + 12i + 4(-1)}{9 - 4(-1)}$

$= \dfrac{5 + 12i}{13} = \dfrac{5}{3} + \dfrac{12}{13}i$

Division

To divide two complex numbers, we multiply the numerator and denominator by the complex conjugate of the denominator and write the result in standard form.

Summary Exercises

This summary exercise set is provided to give you practice with each of the objectives in the chapter. Each exercise is keyed to the appropriate chapter section. The answers are provided in the *Instructor's Manual*.

[5.1]   Simplify each expression, using the properties of exponents.

1. $4x^{-5}$

2. $(2w)^{-3}$

3. $\dfrac{3}{m^{-4}}$

4. $\dfrac{a^{-5}}{b^{-4}}$

5. $y^{-5} \cdot y^2$

6. $\dfrac{w^{-7}}{w^{-3}}$

7. $(m^{-6})^{-2}$

8. $(m^3 n^{-5})^{-2}$

9. $\left(\dfrac{a^{-4}}{b^{-2}}\right)^3$

10. $\left(\dfrac{r^{-5}}{s^4}\right)^{-2}$

11. $(5w^{-2})^2(2w^{-2})$

12. $(5a^2 b^{-3})(2a^{-2} b^{-6})$

13. $\dfrac{7a^{-4} b^4}{28a^{-3} b^{-3}}$

14. $\left(\dfrac{m^{-3} n^{-3}}{m^{-4} n^4}\right)^3$

15. $\left(\dfrac{x^{-4} y^{-3} z^2}{x^{-3} y^2 z^{-4}}\right)^{-2}$

[5.2]   Evaluate each of the following roots over the set of real numbers.

16. $\sqrt{121}$

17. $-\sqrt{64}$

18. $\sqrt{-81}$

19. $\sqrt[3]{64}$

20. $\sqrt[3]{-64}$

21. $\sqrt[4]{81}$

22. $\sqrt{\dfrac{9}{16}}$

23. $\sqrt[3]{-\dfrac{8}{27}}$

24. $\sqrt{8^2}$

Simplify each of the following expressions. Assume that all variables represent positive real numbers for all subsequent exercises in this exercise set.

25. $\sqrt{4x^2}$

26. $\sqrt{a^4}$

27. $\sqrt{36y^2}$

28. $\sqrt{49w^4 z^6}$

29. $\sqrt[3]{x^9}$

30. $\sqrt[3]{-27b^6}$

31. $\sqrt[3]{8r^3 s^9}$

32. $\sqrt[4]{16x^4 y^8}$

33. $\sqrt[5]{32p^5 q^{15}}$

34. $\sqrt{45}$

35. $-\sqrt{75}$

36. $\sqrt{60x^2}$

37. $\sqrt{108a^3}$

38. $\sqrt[3]{32}$

39. $\sqrt[3]{-80w^4 z^3}$

40. $\sqrt{\dfrac{11}{81}}$

41. $\sqrt{\dfrac{7}{36}}$

42. $\sqrt{\dfrac{y^4}{49}}$

43. $\sqrt{\dfrac{2x}{9}}$

44. $\sqrt{\dfrac{5}{16x^2}}$

45. $\sqrt[3]{\dfrac{5a^2}{27}}$

Rationalize the denominator, and write each of the following expressions in simplified form.

46. $\sqrt{\dfrac{3}{7}}$

47. $\dfrac{\sqrt{12}}{\sqrt{x}}$

48. $\dfrac{\sqrt{10a}}{\sqrt{5b}}$

49. $\sqrt[3]{\dfrac{3}{a^2}}$

50. $\dfrac{2}{\sqrt[3]{3x}}$

51. $\dfrac{\sqrt[3]{x^2}}{\sqrt[3]{y^5}}$

Multiply and simplify each of the following expressions.

52. $\sqrt{3x} \cdot \sqrt{7y}$

53. $\sqrt{6x^2} \cdot \sqrt{18}$

54. $\sqrt[3]{4a^2b} \cdot \sqrt[3]{ab^2}$

55. $\sqrt{5}(\sqrt{3} + 2)$

56. $\sqrt{6}(\sqrt{8} - \sqrt{2})$

57. $\sqrt{a}(\sqrt{5a} + \sqrt{125a})$

58. $(\sqrt{3} + 5)(\sqrt{3} - 7)$

59. $(\sqrt{7} - \sqrt{2})(\sqrt{7} + \sqrt{3})$

60. $(\sqrt{5} - 2)(\sqrt{5} + 2)$

61. $(\sqrt{7} - \sqrt{3})(\sqrt{7} + \sqrt{3})$

62. $(2 + \sqrt{3})^2$

63. $(\sqrt{5} - \sqrt{2})^2$

Use a calculator to evaluate the following. Round each answer to the nearest tenth.

64. $\sqrt{17}$

65. $\sqrt{59}$

66. $\sqrt[3]{45}$

67. $\sqrt{\dfrac{37}{11}}$

68. $\sqrt{37}$

69. $\sqrt{156}$

70. $\sqrt[3]{278}$

71. $\sqrt{\dfrac{457}{56}}$

72. $\sqrt{315}$

73. $\sqrt{-75}$

74. $\sqrt[3]{69}$

75. $\sqrt[4]{\dfrac{567}{36}}$

76. $\sqrt{288}$

77. $\sqrt{-36}$

78. $\sqrt[3]{-159}$

79. $\sqrt[5]{\dfrac{529}{52}}$

[5.3]   Use the properties of exponents to simplify each of the following expressions.

80. $x^{3/2} \cdot x^{5/2}$

81. $b^{2/3} \cdot b^{3/2}$

82. $\dfrac{r^{8/5}}{r^{3/5}}$

83. $\dfrac{a^{5/4}}{a^{1/2}}$

84. $(x^{3/5})^{2/3}$

85. $(y^{-4/3})^6$

86. $(x^{4/5}y^{3/2})^{10}$

87. $(16x^{1/3} \cdot y^{2/3})^{3/4}$

88. $\left(\dfrac{x^{-2}y^{-1/6}}{x^{-4}y}\right)^3$

89. $\left(\dfrac{27y^3z^{-6}}{x^{-3}}\right)^{1/3}$

Write each of the following expressions in radical form.

90. $x^{3/4}$ **91.** $(w^2z)^{2/5}$ **92.** $3a^{2/3}$ **93.** $(3a)^{2/3}$

Write each of the following expressions, using rational exponents, and simplify where necessary.

94. $\sqrt[5]{7x}$ **95.** $\sqrt{16w^4}$ **96.** $\sqrt[3]{27p^3q^9}$ **97.** $\sqrt[4]{16a^8b^{16}}$

[5.4]   Write each of the following roots as a multiple of i. Simplify your result.

98. $\sqrt{-49}$ **99.** $\sqrt{-13}$ **100.** $-\sqrt{-60}$

Perform the indicated operations.

101. $(2 + 3i) + (3 - 5i)$ **102.** $(7 - 3i) + (-3 - 2i)$

103. $(5 - 3i) - (2 + 5i)$ **104.** $(-4 + 2i) - (-1 - 3i)$

Find each of the following products.

105. $4i(7 - 2i)$ **106.** $(5 - 2i)(3 + 4i)$ **107.** $(3 - 4i)^2$ **108.** $(2 - 3i)(2 + 3i)$

Find each of the following quotients, and write your answer in standard form.

109. $\dfrac{5 - 15i}{5i}$ **110.** $\dfrac{10}{3 - 4i}$ **111.** $\dfrac{3 - 2i}{3 + 2i}$ **112.** $\dfrac{5 + 10i}{2 + i}$

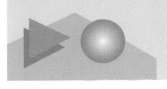

The purpose of this self-test is to help you check your progress and to review for a chapter test in class. Allow yourself about 1 hour to take the test. When you are done, check your answers in the back of the book. If you missed any answers, be sure to go back and review the appropriate sections in the chapter and the exercises that are provided.

Simplify each expression. Assume that all variables represent positive real numbers in all subsequent problems.

1. $(x^4 y^{-5})^2$

2. $\dfrac{9c^{-5}d^3}{18c^{-7}d^{-4}}$

Write the following numbers in scientific notation.

3. 4,230,000,000

4. 0.000025

Write each of the following expressions in simplified form.

5. $\sqrt{49a^4}$

6. $\sqrt[3]{-27w^6 z^9}$

7. $\dfrac{7x}{\sqrt{64y^2}}$

8. $\sqrt{\dfrac{5x}{8y}}$

9. $\sqrt{7x^3}\,\sqrt{2x^4}$

10. $\sqrt[3]{4x^5}\,\sqrt[3]{8x^6}$

Use a calculator to evaluate the following. Round each answer to the nearest tenth.

11. $\sqrt{43}$

12. $\sqrt[3]{\dfrac{73}{27}}$

Use the properties of exponents to simplify each expression.

13. $(16x^4)^{-3/2}$

14. $(27m^{3/2}n^{-6})^{2/3}$

15. $\left(\dfrac{16r^{-1/3}s^{5/3}}{rs^{-7/3}}\right)^{3/4}$

Write the expression in radical form and simplify.

16. $(a^7 b^3)^{2/5}$

Write the expression, using rational exponents. Then simplify.

17. $\sqrt[3]{125p^9 q^6}$

Perform the indicated operations.

18. $(-2 + 3i) - (-5 - 7i)$

19. $(5 - 3i)(-4 + 2i)$

20. $\dfrac{10 - 20i}{3 - i}$

411

This test is provided to help you in the process of reviewing the previous chapters. Answers are provided in the back of the book. If you missed any answers, be sure to go back and review the appropriate chapter section.

1. Solve the equation $7x - 6(x - 1) = 2(5 + x) + 11$.

2. If $f(x) = 3x^6 - 4x^3 + 9x^2 - 11$, find $f(-1)$.

3. Find the equation of the line that has a y intercept of -6 and is parallel to the line $6x - 4y = 18$.

4. Solve the equation $|3x - 5| = 4$.

Simplify each of the following polynomial functions.

5. $f(x) = 5x^2 - 8x + 11 - (-3x^2 - 2x + 8) - (-2x^2 - 4x + 3)$

6. $f(x) = (5x + 3)(2x - 9)$

Factor each of the following completely.

7. $2x^3 + x^2 - 3x$

8. $9x^4 - 36y^4$

9. $4x^2 + 8xy - 5x - 10y$

Simplify each of the following rational expressions.

10. $\dfrac{2x^2 + 13x + 15}{6x^2 + 7x - 3}$

11. $\dfrac{3}{x - 5} - \dfrac{2}{x - 1}$

12. $\dfrac{a^2 - 4a}{a^2 - 6a + 8} \cdot \dfrac{a^2 - 4}{2a^2}$

13. $\dfrac{a^2 - 9}{a^2 - a - 12} - \dfrac{a^2 - a - 6}{a^2 - 2a - 8}$

Simplify each of the following radical expressions.

14. $\sqrt{3x^3y}\,\sqrt{4x^5y^6}$

15. $(\sqrt{3} - 5)(\sqrt{2} + 3)$

Graph each equation.

16. $y = 3x - 5$

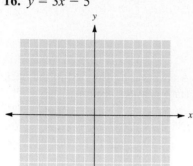

17. $x = -5$

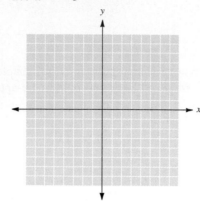

18. $2x - 3y = 12$

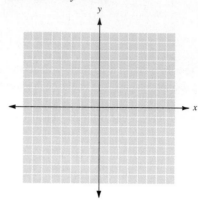

Find each of the following products.

19. $3i(6 - 5i)$

20. $(5 + 2i)(3 - 4i)$

GRAPHS OF SPECIAL FUNCTIONS AND CONIC SECTIONS

INTRODUCTION

Running power lines from a power plant, wind farm, or hydroelectric plant to a city is a very costly enterprise. Land must be cleared, towers built, and conducting wires strung from tower to tower across miles of countryside. Typical construction designs run from about 300- to 1200-foot spans, with towers about 75 to 200 feet high. Of course, if a lot of towers are needed, and many of them must be tall, the construction costs skyrocket.

Power line construction carries a unique set of problems. Towers must be built tall enough and close enough to keep the conducting lines well above the ground. The sag of these wires (how much they droop from the towers) is a function of the weight of the conductor, the span-length, and the tension in the wires. The amount of this sag, measured in feet, is approximated by the following formula:

$$\text{Sag} = \frac{wS^2}{8T}$$

where w = weight of wires in pounds/foot

S = span length in feet

T = tension in wires measured in pounds

If the weight of the conducting wires is 2074 pounds per 1000 feet, one of the largest sizes, and the tension on the wires is 5,000 pounds, the relationship between the sag and the span is given by the following equation:

$$\text{Sag} = \frac{2.074S^2}{8(5000)}$$

The following graph shows how the sag increases as the distance between the towers increases.

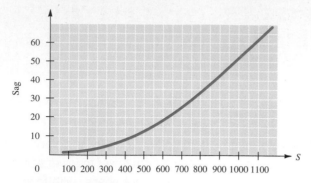

There would be a different graph for every combination of wire weight and tension. This particular graph represents a wire weight of 2.074 pounds per foot and a 5000-pound tension on the conductor.

In this graph, when the towers are 600 feet apart, the conducting wires will sag 18.7 feet; if the towers are 1000 feet apart, the wires will sag 51.8 feet. Which will be the best for a particular area? The building planners must know what kind of clearance is needed over the terrain in order to decide which costs less: very tall towers spaced far apart or shorter towers placed closer together.

The actual curve of the power lines is called a **catenary curve.** The curve that we use to approximate the sag is a **parabola,** one of the conic sections we will study in this chapter. _____ ■

6.1 Solving Quadratic Equations by Factoring

6.1 OBJECTIVES

1. Solve quadratic equations by factoring
2. Find the zeros of a quadratic function

This is a quadratic equation in one variable, here x. You can recognize such a quadratic equation by the fact that the highest power of the variable x is the second power.

The factoring techniques you learned in previous algebra classes provide us with tools for solving equations that can be written in the form

$$ax^2 + bx + c = 0 \qquad a \neq 0$$

where a, b, and c are constants.

An equation written in the form $ax^2 + bx + c = 0$ is called a **quadratic equation in standard form.** Using factoring to solve quadratic equations requires the **zero-product principle,** which says that if the product of two factors is 0, then one or both of the factors must be equal to 0. In symbols:

Zero-Product Principle

If $a \cdot b = 0$, then $a = 0$ or $b = 0$ or $a = b = 0$

Let's see how the principle is applied to solving quadratic equations.

● Example 1

Solving Equations by Factoring

To use the zero-product principle, 0 must be on one side of the equation.

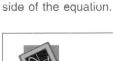

Graph the function

$$y = x^2 - 3x - 18$$

on your graphing calculator. The solutions to the equation $0 = x^2 - 3x - 18$ will be those values on the curve at which $y = 0$. Those are the points at which the graph intercepts the x axis.

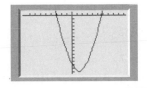

Solve.

$$x^2 - 3x - 18 = 0$$

Factoring on the left, we have

$$(x - 6)(x + 3) = 0$$

By the zero-product principle, we know that one or both of the factors must be zero. We can then write

$$x - 6 = 0 \qquad \text{or} \qquad x + 3 = 0$$

Solving each equation gives

$$x = 6 \qquad \text{or} \qquad x = -3$$

The two solutions are 6 and -3 and the solution set is written as $\{x | x = -3, 6\}$ or simply $\{-3, 6\}$.

The solutions are sometimes called the **zeros,** or **roots,** of the equation. They represent the point where the graph of the equation crosses the x axis.

Quadratic equations can be checked in the same way as linear equations were checked: by substitution. For instance, if $x = 5$, we have

$$5^2 - 2 \cdot 5 - 15 \stackrel{?}{=} 0$$

$$25 - 10 - 15 \stackrel{?}{=} 0$$

$$0 = 0$$

which is a true statement. We leave it to you to check the solution of -3.

CHECK YOURSELF 1

Solve $x^2 - 9x + 20 = 0$.

Other factoring techniques are also used in solving quadratic equations. Example 2 illustrates this concept.

●Example 2

Solving Equations by Factoring

CAUTION

A *common mistake* is to forget the statement $x = 0$ when you are solving equations of this type. Be sure to include the *two* solutions.

The graph of
$y + x^2 - 9$ is

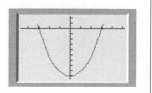

(*a*) Solve $x^2 - 5x = 0$.
Again, factor the left side of the equation and apply the zero-product principle.

$$x(x - 5) = 0$$

Now

$$x = 0 \qquad \text{or} \qquad x - 5 = 0$$
$$x = 5$$

The two solutions are 0 and 5 or $\{0, 5\}$.

(*b*) Solve $x^2 - 9 = 0$.
Factoring yields

$$(x + 3)(x - 3) = 0$$
$$x + 3 = 0 \qquad \text{or} \qquad x - 3 = 0$$
$$x = -3 \qquad\qquad x = 3 \qquad \text{or} \qquad \{\pm 3\}$$

CHECK YOURSELF 2

Solve by factoring.

(a) $x^2 + 8x = 0$ **(b)** $x^2 - 16 = 0$

Example 3 illustrates a crucial point. Our solution technique depends on the zero-product principle, which means that the product of factors *must be equal to 0*. The importance of this is shown now.

CAUTION

Consider the equation

$x(2x - 1) = 3$

Students are sometimes tempted to write

$x = 3$ or $2x - 1 = 3$

That is *not correct*. Instead, subtract 3 from both sides of the equation *as the first step* to write

$x^2 - 2x - 3 = 0$

in standard form. Only now can you factor and proceed as before.

• Example 3

Solving Equations by Factoring

Solve $2x^2 - x = 3$.

The first step in the solution is to write the equation in standard form (that is, when one side of the equation is 0). So start by adding -3 to both sides of the equation. Then,

$2x^2 - x - 3 = 0$ Make sure all terms are on one side of the equation. The other side will be 0.

You can now factor and solve by using the zero-product principle.

$(2x - 3)(x + 1) = 0$

$2x - 3 = 0$ or $x + 1 = 0$

$2x = 3$ $x = -1$

$x = \dfrac{3}{2}$ or $\left\{\dfrac{3}{2}, -1\right\}$

CHECK YOURSELF 3

Solve $3x^2 = 5x + 2$.

In all the previous examples, the quadratic equations had two distinct real number solutions. That may or may not always be the case, as we shall see.

• Example 4

Solving Equations by Factoring

Note that the graph of $y = x^2 - 6x - 9$ touches the x axis at only the point $(3, 0)$.

Solve $x^2 - 6x + 9 = 0$.

Factoring, we have

$(x - 3)(x - 3) = 0$

and

$x - 3 = 0$ or $x - 3 = 0$

$x = 3$ $x = 3$ or $\{3\}$

Even though a quadratic equation will always have two solutions, they may not always be real numbers. More about this in the next section.

A quadratic (or second-degree) equation always has *two* solutions. When an equation such as this one has two solutions that are the same number, we call 3 the **repeated** (or **double**) **solution** of the equation.

CHECK YOURSELF 4

Solve $x^2 + 6x + 9 = 0$.

Always examine the quadratic member of an equation for common factors. It will make your work much easier, as Example 5 illustrates.

• Example 5

Solving Equations by Factoring

Solve $3x^2 - 3x - 60 = 0$.

First, note the common factor 3 in the quadratic member of the equation. Factoring out the 3, we have

$$3(x^2 - x - 20) = 0$$

Now divide both sides of the equation by 3:

Note the advantage of dividing both members by 3. The coefficients in the quadratic member become smaller, and that member is much easier to factor.

$$\frac{3(x^2 - x - 20)}{3} = \frac{0}{3}$$

or

$$x^2 - x - 20 = 0$$

We can now factor and solve as before:

$$(x - 5)(x + 4) = 0$$

$$x - 5 = 0 \quad \text{or} \quad x + 4 = 0$$

$$x = 5 \qquad\qquad x = -4 \quad \text{or} \quad \{5, -4\}$$

● ● ●   **CHECK YOURSELF 5**

Solve $2x^2 - 10x - 48 = 0$.

In Chapter 1, we introduced the concept of a function and expressed the equation of a line in function form. Another type of function is called a quadratic function.

> A **quadratic function** is a function that can be written in the form
>
> $f(x) = ax^2 + bx + c$
>
> where *a, b,* and *c* are real numbers and $a \neq 0$.

For example, $f(x) = 3x^2 - 2x - 1$ and $g(x) = x^2 - 2$ are quadratic functions. In working with functions, we often want to find the values of x for which $f(x) = 0$. As in quadratic equations, these values are called the **zeros of the function.** They represent the points where the graph of the function crosses the x axis. To find the zeros of a quadratic function, a quadratic equation must be solved.

•Example 6

Finding the Zeros of a Function

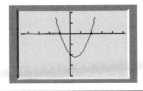

The graph of $f(x) = x^2 - x - 2$ intercepts the x axis at the points $(2, 0)$ and $(-1, 0)$, so these are the *zeros* of the equation.

Find the zeros of $f(x) = x^2 - x - 2$.

To find the zeros of $f(x) = x^2 - x - 2$, we must solve the quadratic equation $f(x) = 0$.

$$f(x) = 0$$
$$x^2 - x - 2 = 0$$
$$(x - 2)(x + 1) = 0$$
$$x - 2 = 0 \quad \text{or} \quad x + 1 = 0$$
$$x = 2 \qquad\qquad x = -1 \quad \text{or} \quad \{-1, 2\}$$

The zeros of the function are -1 and 2.

● ● ● CHECK YOURSELF 6

Find the zeros of $f(x) = 2x^2 - x - 3$.

● ● ● CHECK YOURSELF ANSWERS

1. $\{4, 5\}$.   **2. (a)** $\{0, -8\}$; **(b)** $\{4, -4\}$.   **3.** $\left\{-\dfrac{1}{3}, 2\right\}$.   **4.** $\{-3\}$.

5. $\{-3, 8\}$.   **6.** $\left\{-1, \dfrac{3}{2}\right\}$.

6.1 Exercises

In Exercises 1 to 46, solve the quadratic equations by factoring.

1. $x^2 + 4x + 3 = 0$

2. $x^2 - 5x + 4 = 0$

3. $x^2 - 2x - 15 = 0$

4. $x^2 + 4x - 32 = 0$

5. $x^2 - 11x + 30 = 0$

6. $x^2 + 14x + 48 = 0$

7. $x^2 - 4x - 21 = 0$

8. $x^2 + 5x \quad 36 = 0$

9. $x^2 - 5x = 50$

10. $x^2 + 14x = -33$

11. $x^2 = 2x + 35$

12. $x^2 = 6x + 27$

13. $x^2 - 8x = 0$

14. $x^2 + 7x = 0$

15. $x^2 + 10x = 0$

16. $x^2 - 9x = 0$

17. $x^2 = 5x$

18. $4x = x^2$

19. $x^2 - 25 = 0$

20. $x^2 - 49 = 0$

21. $x^2 = 64$

22. $x^2 = 36$

23. $4x^2 + 12x + 9 = 0$

24. $9x^2 - 30x + 25 = 0$

25. $2x^2 - 17x + 36 = 0$

26. $5x^2 + 17x - 12 = 0$

27. $3x^2 - x = 4$

28. $6x^2 = 13x - 6$

29. $6x^2 = 7x - 2$

30. $4x^2 - 3 = x$

31. $2m^2 = 12m + 54$

32. $5x^2 - 55x = 60$

33. $4x^2 - 24x = 0$

34. $6x^2 - 9x = 0$

35. $5x^2 = 15x$

36. $7x^2 = -49x$

37. $x(x - 2) = 15$

38. $x(x + 3) = 28$

39. $x(2x - 3) = 9$

40. $x(3x + 1) = 52$

41. $2x(3x + 1) = 28$

42. $3x(2x - 1) = 30$

43. $(x - 3)(x - 1) = 15$

44. $(x + 4)(x + 1) = 18$

45. $(2x + 1)(x - 4) = 11$

46. $(3x - 5)(x + 2) = 14$

In Exercises 47 to 50, find the zeros of the functions.

47. $f(x) = 3x^2 - 24x + 36$

48. $f(x) = 2x^2 - 6x - 56$

49. $f(x) = 4x^2 + 16x - 20$

50. $f(x) = 3x^2 - 33x + 54$

51. Explain the differences between solving the equations $3(x - 2)(x + 5) = 0$ and $3x(x - 2)(x + 5) = 0$.

52. How can a graphing calculator be used to determine the zeros of a quadratic function?

In Exercises 53 to 56, write an equation that has the following solutions. *Hint:* Write the binomial factors and then the quadratic member of the equation.

53. $\{2, -3\}$

54. $\{0, 5\}$

55. $\{6, 2\}$

56. $\{-4, 4\}$

The zero-product rule can be extended to three or more factors. If $a \cdot b \cdot c = 0$, then at least one of these factors is 0. In Exercises 57 to 60, use this information to solve the equations.

57. $x^3 - 3x^2 - 10x = 0$

58. $x^3 + 8x^2 + 15x = 0$

59. $x^3 - 9x = 0$

60. $x^3 = 16x$

In Exercises 61 to 64, extend the ideas in the previous exercises to find solutions for the following equations. *Hint:* Apply factoring by grouping.

61. $x^3 + x^2 - 4x - 4 = 0$

62. $x^3 - 5x^2 - x + 5 = 0$

63. $x^4 - 10x^2 + 9 = 0$

64. $x^4 - 5x^2 + 4 = 0$

The net productivity of a forested wetland as related to the amount of water moving through the wetland can be expressed by a quadratic equation. In Exercises 65 to 68, if y represents the amount of wood produced, in grams per square meter, and x represents the amount of water present, in centimeters, determine where the productivity is zero in each wetland represented by the equations.

65. $y = -3x^2 + 300x$

66. $y = -4x^2 + 500x$

67. $y = -6x^2 + 792x$

68. $y = -7x^2 + 1022x$

69. Break-Even Analysis The manager of a bicycle shop knows that the cost of selling x bicycles is $C = 20x + 60$ and the revenue from selling x bicycles is $R = x^2 - 8x$. Find the break-even value of x.

70. Break-Even Analysis A company that produces computer games has found that its operating cost in dollars is $C = 40x + 150$ and its revenue in dollars is $R = 65x - x^2$. For what value(s) of x will the company break even?

Graphing Calculator

Use your calculator to graph $f(x)$.

71. $f(x) = 3x^2 - 24x + 36$. Note the x values at which the graph crosses the x axis. Compare your answer to the solution for Exercise 47.

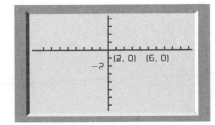

72. $f(x) = 2x^2 - 6x - 56$. Note the x values at which the graph crosses the x axis. Compare your answer to the solution for Exercise 48.

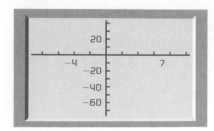

73. Work with another student and use your calculator to solve this exercise. Find an equation for each graph given. There could be more than one equation for some of the graphs. Remember the connection between the x intercepts and the zeros.

(a)

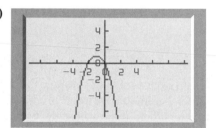

(b)

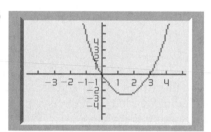

(c)

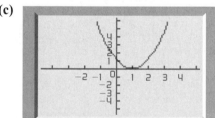

(d)

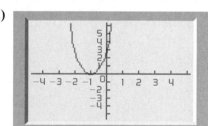

(e)

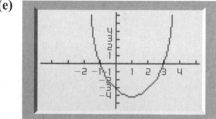

(f)

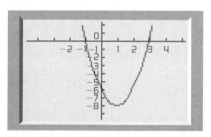

(g)

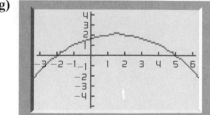

(h)

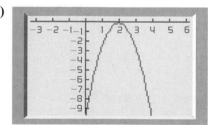

Answers

1. $\{-3, -1\}$ **3.** $\{-3, 5\}$ **5.** $\{5, 6\}$ **7.** $\{-3, 7\}$ **9.** $\{-5, 10\}$ **11.** $\{-5, 7\}$ **13.** $\{0, 8\}$

15. $\{0, -10\}$ **17.** $\{0, 5\}$ **19.** $\{-5, 5\}$ **21.** $\{-8, 8\}$ **23.** $\left\{-\dfrac{3}{2}\right\}$ **25.** $\left\{4, \dfrac{9}{2}\right\}$ **27.** $\left\{-1, \dfrac{4}{3}\right\}$

29. $\left\{\dfrac{1}{2}, \dfrac{2}{3}\right\}$ **31.** $\{-3, 9\}$ **33.** $\{0, 6\}$ **35.** $\{0, 3\}$ **37.** $\{-3, 5\}$ **39.** $\left\{-\dfrac{3}{2}, 3\right\}$ **41.** $\left\{-\dfrac{7}{3}, 2\right\}$

43. $\{-2, 6\}$ **45.** $\left\{-\dfrac{3}{2}, 5\right\}$ **47.** $\{2, 6\}$ **49.** $\{-5, 1\}$ **51.**   **53.** $x^2 + x - 6 = 0$

55. $x^2 - 8x + 12 = 0$ **57.** $\{-2, 0, 5\}$ **59.** $\{-3, 0, 3\}$ **61.** $\{-2, -1, 2\}$ **63.** $\{-3, -1, 1, 3\}$

65. $\{0\ \text{cm}, 100\ \text{cm}\}$ **67.** $\{0\ \text{cm}, 132\ \text{cm}\}$ **69.** $\{30\}$

71. 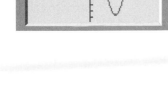   2, 6 are the zeros of Exercise 47.   **73.**

Solving Quadratic Equations by the Square Root Method and by Completing the Square

6.2

6.2 OBJECTIVES

1. Solve quadratic equations by using the square root method
2. Solve quadratic equations by completing the square

In Section 6.1, we solved quadratic equations by factoring and using the zero-product principal. However, not all equations are factorable by that method. In this section, we will learn another simple technique that can be used to solve any quadratic equation called the **square root method.** Let's begin by solving a special type of equation by factoring.

● **Example 1**

Solving Equations by Factoring

Solve the quadratic equation $x^2 = 16$ by factoring.
We write the equation in standard form:

$$x^2 - 16 = 0$$

Here, we factor the quadratic member of the equation as a difference of squares.

Factoring, we have

$$(x + 4)(x - 4) = 0$$

Finally, the solutions are

$$x = -4 \qquad \text{or} \qquad x = 4 \qquad \text{or} \qquad \{\pm 4\}$$

● ● ●   **CHECK YOURSELF 1**

Solve each of the following quadratic equations.

(a) $5x^2 = 180$ **(b)** $x^2 = 25$

THE SQUARE ROOT METHOD

The equation in Example 1 could have been solved in an alternative fashion. We could have used what is called the **square root method.** Again, given the equation

$$x^2 = 16$$

we can write the equivalent statement

$$x = \sqrt{16} \quad \text{or} \quad x = -\sqrt{16}$$

This yields the solutions

Note: Be sure to include *both* the positive and the negative square roots when you use the square root method.

$$x = 4 \quad \text{or} \quad x = -4 \quad \text{or} \quad \{\pm 4\}$$

This discussion leads us to the following general result.

> **Square Root Property**
>
> If $x^2 = k$, where k is a complex number, then
>
> $$x = \sqrt{k} \quad \text{or} \quad x = -\sqrt{k}$$

Example 2 further illustrates the use of this property.

● Example 2

Using the Square Root Method

Solve each equation by using the square root method.

(a) $x^2 = 9$

By the square root property,

$$x = \sqrt{9} \quad \text{or} \quad x = -\sqrt{9}$$
$$= 3 \quad\quad\quad\quad = -3 \quad \text{or} \quad \{\pm 3\}$$

If a calculator were used, $\sqrt{17} = 4.123$ (rounded to three decimal places).

(b) $x^2 - 17 = 0$

Add 17 to both sides of the equation.

$$x^2 = 17 \quad \text{or} \quad \{\pm\sqrt{17}\} \quad \text{or} \quad \{-\sqrt{17}, \sqrt{17}\}$$

(c) $2x^2 - 3 = 0$

$$2x^2 = 3$$

$$x^2 = \frac{3}{2}$$

$$x = \pm\sqrt{\frac{3}{2}}$$

$$x = \pm\frac{\sqrt{6}}{2} \quad \text{or} \quad \left\{\pm\frac{\sqrt{6}}{2}\right\}$$

In Example 2(*d*) we see that complex-number solutions may result.

(d) $x^2 + 1 = 0$

$$x^2 = -1$$

$$x = \pm\sqrt{-1}$$

$$x = \pm i \quad \text{or} \quad \{\pm i\}$$

● ● ●   **CHECK YOURSELF 2**

Solve each equation.

(a) $x^2 = 5$ **(b)** $x^2 - 2 = 0$ **(c)** $3x^2 - 8 = 0$ **(d)** $x^2 + 9 = 0$

We can also use the approach in Example 2 to solve an equation of the form

$$(x + 3)^2 = 16$$

As before, by the square root property we have

$x + 3 = \pm 4$ Subtract 3 from both sides of the equation.

Solving for x yields

$$x = -3 \pm 4$$

which means that there are two solutions:

$$x = -3 + 4 \quad \text{or} \quad x = -3 - 4$$

$$= 1 \qquad\qquad = -7 \quad \text{or} \quad \{1, -7\}$$

● **Example 3**

Using the Square Root Method

Use the square root method to solve each equation.

(a) $(x - 5)^2 - 5 = 0$

$$(x - 5)^2 = 5$$

The two solutions $5 + \sqrt{5}$ and $5 - \sqrt{5}$ are abbreviated as $5 \pm \sqrt{5}$.

$$x - 5 = \pm\sqrt{5}$$

$$x = 5 \pm \sqrt{5} \quad \text{or} \quad \{5 \pm \sqrt{5}\}$$

(b) $3(y + 1)^2 - 2 = 0$

$$3(y + 1)^2 = 2$$

$$(y + 1)^2 = \frac{2}{3}$$

We have solved for y and rationalized the denominator.

$$y + 1 = \pm\sqrt{\frac{2}{3}}$$

$$\sqrt{\frac{2}{3}} = \frac{\sqrt{2}}{\sqrt{3}} = \frac{\sqrt{2}\cdot\sqrt{3}}{\sqrt{3}\cdot\sqrt{3}} = \frac{\sqrt{6}}{3}$$

$$y = -1 \pm \frac{\sqrt{6}}{3}$$

Then we combine the terms on the right, using the common denominator of 3.

$$= \frac{-3 \pm \sqrt{6}}{3} \qquad \text{or} \qquad \left\{ \frac{-3 \pm \sqrt{6}}{3} \right\}$$

● ● ● **CHECK YOURSELF 3**

Using the square root method, solve each equation.

(a) $(x - 2)^2 - 3 = 0$ **(b)** $2(x - 1)^2 = 1$

COMPLETING THE SQUARE

As we stated earlier, not all quadratic equations can be solved directly by factoring or using the square root method. We must extend our techniques.

The square root method is useful in this process because any quadratic equation can be written in the form

If $(x + h)^2 = k$, then

$$(x + h)^2 = k$$

$x + h = \pm\sqrt{k}$

and

which yields the solution

$x = -h \pm \sqrt{k}$

$$x = -h \pm \sqrt{k}$$

The process of changing an equation in standard form

$$ax^2 + bx + c = 0$$

to the form

$$(x + h)^2 = k$$

is called the method of **completing the square,** and it is based on the relationship between the middle term and the last term of any perfect-square trinomial.

Let's look at three perfect-square trinomials to see whether we can detect a pattern:

$$x^2 + 4x + 4 = (x + 2)^2 \tag{1}$$

$$x^2 - 6x + 9 = (x - 3)^2 \tag{2}$$

$$x^2 + 8x + 16 = (x + 4)^2 \tag{3}$$

Note that this relationship is true *only* if the leading, or x^2, coefficient is 1. That will be important later.

Note that in each case the last (or constant) term is the square of one-half of the coefficient of x in the middle (or linear) term. For example, in equation (2),

$$x^2 - 6x + 9 = (x - 3)^2$$

$\dfrac{1}{2}$ of this coefficient is -3, and $(-3)^2 = 9$, the constant.

Verify this relationship for yourself in equation (3). To summarize, in perfect-square trinomials, the constant is always the square of one-half the coefficient of x.

We are now ready to use the above observation in the solution of quadratic equations by completing the square. Consider Example 4.

• Example 4

Completing the Square to Solve an Equation

Solve $x^2 + 8x - 7 = 0$ by completing the square.

First, we rewrite the equation with the constant on the *right-hand side:*

$\dfrac{1}{2} \cdot 8 = 4$ and $4^2 = 16$

$$x^2 + 8x = 7$$

Remember that if $(x + h)^2 = k$, then $x = -h \pm \sqrt{k}$.

Our objective is to have a perfect-square trinomial on the left-hand side. We know that we must add the square of one-half of the x coefficient to complete the square. In this case, that value is 16, so now we add 16 to each side of the equation.

$$x^2 + 8x + 16 = 7 + 16$$

When you graph the related function, $y = x^2 + 8x - 7$, you will note that the x values for the x intercepts are just below 1 and just above -9. Be certain that you see how these points relate to the exact solutions, $-4 + \sqrt{23}$ and $-4 - \sqrt{23}$.

Factor the perfect-square trinomial on the left, and combine like terms on the right to yield

$$(x + 4)^2 = 23$$

Now the square root property yields

$$x + 4 = \pm\sqrt{23}$$

Subtracting 4 from both sides of the equation gives

$$x - -4 \pm \sqrt{23} \qquad \text{or} \qquad \{-4 \pm \sqrt{23}\}$$

● ● ● **CHECK YOURSELF 4**

Solve $x^2 - 6x - 2 - 0$ by completing the square.

• Example 5

Completing the Square to Solve an Equation

Solve $x^2 + 5x - 3 = 0$ by completing the square.

$x^2 + 5x - 3 = 0$ Add 3 to both sides.

$x^2 + 5x = 3$ Make the left-hand side a perfect square.

Add the square of one-half of the x coefficient to both sides of the equation. Note that

$$\frac{1}{2} \cdot 5 = \frac{5}{2}$$

$$x^2 + 5x + \left(\frac{5}{2}\right)^2 = 3 + \left(\frac{5}{2}\right)^2$$

$$\left(x + \frac{5}{2}\right)^2 = \frac{37}{4}$$ Take the square root of both sides.

$$x + \frac{5}{2} = \pm\frac{\sqrt{37}}{2}$$ Solve for x.

$$x = \frac{-5 \pm \sqrt{37}}{2} \quad \text{or} \quad \left\{\frac{-5 \pm \sqrt{37}}{2}\right\}$$

● ● ● **CHECK YOURSELF 5**

Solve $x^2 + 3x - 7 = 0$ by completing the square.

Some equations have nonreal complex solutions, as Example 6 illustrates.

• Example 6

Note that the graph of $y = x^2 + 4x + 13$ does not intercept the x axis.

Completing the Square to Solve an Equation

Solve $x^2 + 4x + 13 = 0$ by completing the square.

$x^2 + 4x + 13 = 0$ Subtract 13 from both sides.

$x^2 + 4x = -13$ Add $\left[\frac{1}{2}(4)\right]^2$ to both sides.

$x^2 + 4x + 4 = -13 + 4$ Factor the left-hand side.

$(x + 2)^2 = -9$ Take the square root of both sides.

$x + 2 = \pm\sqrt{-9}$ Simplify the radical

$x + 2 = \pm\sqrt{9}i$

$x + 2 = \pm 3i$

$x = -2 \pm 3i \quad \text{or} \quad \{-2 \pm 3i\}$

● ● ● **CHECK YOURSELF 6**

Solve $x^2 + 10x + 41 = 0$.

Example 7 illustrates a situation in which the leading coefficient of the quadratic member is not equal to 1. As you will see, an extra step will be required.

• Example 7

CAUTION

Before you can complete the square on the left, the coefficient of x^2 must be equal to 1. Otherwise, we must *divide* both sides of the equation by that coefficient

Completing the Square to Solve an Equation

Solve $3x^2 + 6x - 7 = 0$ by completing the square.

$$3x^2 + 6x - 7 = 0 \qquad \text{Add 7 to both sides.}$$

$$3x^2 + 6x = 7 \qquad \text{Divide both sides by 3.}$$

$$x^2 + 2x = \frac{7}{3} \qquad \text{Now, complete the square on the left.}$$

$$x^2 + 2x + 1 = \frac{7}{3} + 1 \qquad \text{The left side is now a perfect square.}$$

$$(x + 1)^2 = \frac{10}{3}$$

$$x + 1 = \pm \sqrt{\frac{10}{3}}$$

$$x = 1 \pm \sqrt{\frac{10}{3}}$$

We have rationalized the denominator and combined the terms on the right side.

$$= \frac{-3 \pm \sqrt{30}}{3}$$

● ● ● CHECK YOURSELF 7

Solve $2x^2 - 8x + 3 = 0$ by completing the square.

The following algorithm summarizes our work in this section with solving quadratic equations by completing the square.

Completing the Square

STEP 1   Isolate the constant on the right side of the equation.
STEP 2   Divide both sides of the equation by the coefficient of the x^2 term if that coefficient is not equal to 1.
STEP 3   Add the square of one-half of the coefficient of the linear term to both sides of the equation. This will give a perfect-square trinomial on the left side of the equation.
STEP 4   Write the left side of the equation as the square of a binomial, and simplify on the right side.
STEP 5   Use the square root property, and then solve the resulting linear equations.

● ● ● **CHECK YOURSELF ANSWERS**

1. (a) $-6, 6$; (b) $-5, 5$.   **2.** (a) $\sqrt{5}, -\sqrt{5}$; (b) $\sqrt{2}, -\sqrt{2}$; (c) $\dfrac{2\sqrt{6}}{3}$,

$-\dfrac{2\sqrt{6}}{3}$; and (d) $3i, -3i$.   **3.** (a) $2 \pm \sqrt{3}$; (b) $\dfrac{2 \pm \sqrt{2}}{2}$.   **4.** $3 \pm \sqrt{11}$.

5. $\dfrac{-3 \pm \sqrt{37}}{2}$.   **6.** $-5 \pm 4i$.   **7.** $\dfrac{4 \pm \sqrt{10}}{2}$.

6.2 Exercises

In Exercises 1 to 8, solve by factoring or completing the square.

1. $x^2 + 6x + 5 = 0$

2. $x^2 + 5x + 6 = 0$

3. $z^2 - 2z - 35 = 0$

4. $q^2 - 5q - 24 = 0$

5. $2x^2 - 5x - 3 = 0$

6. $3x^2 + 10x - 8 = 0$

7. $6y^2 - y - 2 = 0$

8. $10z^2 + 3z - 1 = 0$

In Exercises 9 to 20, use the square root method to find solutions for the equations.

9. $x^2 = 36$

10. $x^2 = 144$

11. $y^2 = 7$

12. $p^2 = 18$

13. $2x^2 - 12 = 0$

14. $3x^2 - 66 = 0$

15. $2t^2 + 12 = 4$

16. $3u^2 - 5 = -32$

17. $(x + 1)^2 = 12$

18. $(2x - 3)^2 = 5$

19. $(2z + 1)^2 - 3 = 0$

20. $(3p - 4)^2 + 9 = 0$

In Exercises 21 to 32, find the constant that must be added to each binomial expression to form a perfect-square trinomial.

21. $x^2 + 12x$

22. $r^2 - 14r$

23. $y^2 - 8y$

24. $w^2 + 16w$

25. $x^2 - 3x$

26. $z^2 + 5z$

27. $n^2 + n$

28. $x^2 - x$

29. $x^2 + \dfrac{1}{2}x$

30. $x^2 - \dfrac{1}{3}x$

31. $x^2 + 2ax$

32. $y^2 - 4ay$

In Exercises 33 to 54, solve each equation by completing the square.

33. $x^2 + 12x - 2 = 0$

34. $x^2 - 14x - 7 = 0$

35. $y^2 - 2y = 8$

36. $z^2 + 4z - 72 = 0$

37. $x^2 - 2x - 5 = 0$

38. $x^2 - 2x = 3$

39. $x^2 + 10x + 13 = 0$

40. $x^2 + 3x - 17 = 0$

41. $z^2 - 5z - 7 = 0$

42. $q^2 - 8q + 20 = 0$

43. $m^2 - m - 3 = 0$

44. $y^2 + y - 5 = 0$

45. $x^2 + \dfrac{1}{2}x = 1$

46. $x^2 - \dfrac{1}{3}x = 2$

47. $2x^2 + 2x - 1 = 0$

48. $3x^2 - 3x - 1$

49. $3x^2 - 6x = 2$

50. $4x^2 + 8x - 1 = 0$

51. $3x^2 - 2x + 12 = 0$

52. $7y^2 - 2y + 3 = 0$

53. $x^2 + 8x + 20 = 0$

54. $x^2 - 2x + 10 = 0$

 55. Why must the leading coefficient of the quadratic member be set equal to 1 before using the technique of completing the square?

 56. What relationship exists between the solution of a quadratic equation and the graph of a quadratic function?

In Exercises 57 to 62, find the constant that must be added to each binomial to form a perfect-square trinomial. Let x be the variable; other letters represent constants.

57. $x^2 + 2ax$

58. $x^2 + 2abx$

59. $x^2 + 3ax$

60. $x^2 + abx$

61. $a^2x^2 + 2ax$

62. $a^2x^2 + 4abx$

In Exercises 63 and 64, solve each equation by completing the square.

63. $x^2 + 2ax = 4$

64. $x^2 + 2ax - 8 = 0$

In Exercises 65 to 68, use your graphing utility to find the graph. Approximate the x intercepts for each graph. (You may have to adjust the viewing window to see both intercepts.)

65. $y = x^2 + 12x - 2$

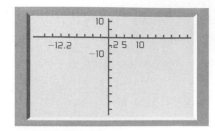

66. $y = x^2 - 14x - 7$

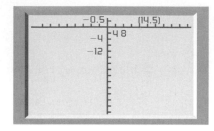

67. $y = x^2 - 2x - 8$

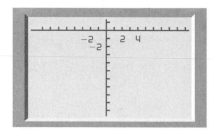

68. $y = x^2 + 4x - 72$

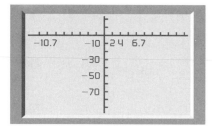

Answers

1. $\{-5, -1\}$ **3.** $\{-5, 7\}$ **5.** $\left\{-\dfrac{1}{2}, 3\right\}$ **7.** $\left\{-\dfrac{1}{2}, \dfrac{2}{3}\right\}$ **9.** $\{\pm 6\}$ **11.** $\{-\sqrt{7}, \sqrt{7}\}$

13. $\{-\sqrt{6}, \sqrt{6}\}$ **15.** $\{\pm 2i\}$ **17.** $\{-1 \pm 2\sqrt{3}\}$ **19.** $\left\{-\dfrac{1 \pm \sqrt{3}}{2}\right\}$ **21.** 36 **23.** 16 **25.** $\dfrac{9}{4}$

27. $\dfrac{1}{4}$ **29.** $\dfrac{1}{16}$ **31.** a^2 **33.** $\{-6 \pm \sqrt{38}\}$ **35.** $\{-2, 4\}$ **37.** $\{1 \pm \sqrt{6}\}$ **39.** $\{-5 \pm 2\sqrt{3}\}$

41. $\left\{\dfrac{5 \pm \sqrt{53}}{2}\right\}$ **43.** $\left\{\dfrac{1 \pm \sqrt{13}}{2}\right\}$ **45.** $\left\{\dfrac{-1 \pm \sqrt{17}}{4}\right\}$ **47.** $\left\{\dfrac{-1 \pm \sqrt{3}}{2}\right\}$ **49.** $\left\{\dfrac{3 \pm \sqrt{15}}{3}\right\}$

51. $\left\{\dfrac{1 \pm i\sqrt{35}}{3}\right\}$ **53.** $\{-4 \pm 2i\}$ **55.** 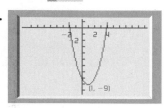 **57.** a^2 **59.** $\dfrac{9}{4}a^2$ **61.** 1 **63.** $\{-a + \sqrt{4 + a^2}\}$

65.

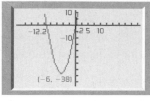

$-12.2, 0.2$, the same as Exercise 33

67.

$-2, 4$, the same as Exercise 35

6.3 The Quadratic Formula

6.3 OBJECTIVES

1. Solve quadratic equations by using the quadratic formula
2. Determine the nature of the solutions of a quadratic equations by using the discriminant
3. Use the Pythagorean theorem to solve a geometric application

Every quadratic equation can be solved by using the quadratic formula. In this section, we will first describe how the quadratic formula is derived, then we will examine its use. Recall that a quadratic equation is any equation that can be written in the form

$$ax^2 + bx + c = 0 \qquad \text{where } a \neq 0$$

Deriving the Quadratic Formula

STEP 1 Isolate the constant on the right side of the equation.

$$ax^2 + bx = -c$$

STEP 2 Divide both sides by the coefficient of the x^2 term.

$$x^2 + \frac{b}{a}x = -\frac{c}{a}$$

STEP 3 Add the square of one-half the x coefficient to both sides.

$$x^2 + \frac{b}{a}x + \frac{b^2}{4a^2} = -\frac{c}{a} + \frac{b^2}{4a^2}$$

STEP 4 Factor the left side as a perfect-square binomial. Then apply the square root property.

$$\left(x + \frac{b}{2a}\right)^2 = \frac{-4ac + b^2}{4a^2}$$

$$x + \frac{b}{2a} = \sqrt{\frac{b^2 - 4ac}{4a^2}}$$

STEP 5 Solve the resulting linear equations.

$$x = -\frac{b}{2a} \pm \frac{\sqrt{b^2 - 4ac}}{2a}$$

STEP 6 Simplify.

$$= \frac{-b \pm \sqrt{b^2 - 4ac}}{2a}$$

We now use the result derived above to state the **quadratic formula**, a formula that allows us to find the solutions for any quadratic equation.

The Quadratic Formula

Given any quadratic equation in the form

$$ax^2 + bx + c = 0 \qquad \text{where } a \neq 0$$

the two solutions to the equation are found using the formula

$$x = \frac{-b \pm \sqrt{b^2 - 4ac}}{2a}$$

Our first example uses an equation in standard form.

435

● Example 1

Using the Quadratic Formula

Note that the equation is in standard form.

Solve, using the quadratic formula.

$$6x^2 - 7x - 3 = 0$$

First, we determine the values for *a, b,* and *c.* Here,

$$a = 6 \qquad b = -7 \qquad c = -3$$

Substituting those values into the quadratic formula, we have

Since $b^2 - 4ac = 121$ is a perfect square, the two solutions in this case are rational numbers.

$$x = \frac{-(-7) \pm \sqrt{(-7)^2 - 4(6)(-3)}}{2(6)}$$

Simplifying inside the radical gives us

$$x = \frac{7 \pm \sqrt{121}}{12}$$

$$= \frac{7 \pm 11}{12}$$

Compare these solutions to the graph of $y = 6x^2 - 7x - 3$

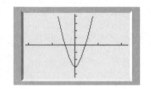

This gives us the solutions

$$x = \frac{3}{2} \qquad \text{or} \qquad x = -\frac{1}{3} \qquad \text{or} \qquad \left\{ \frac{3}{2}, -\frac{1}{3} \right\}$$

Note that since the solutions for the equation of this example are rational, the original equation could have been solved by our earlier method of factoring.

● ● ● **CHECK YOURSELF 1**

Solve, using the quadratic formula.

$$3x^2 + 2x - 8 = 0$$

To use the quadratic formula, we often must write the equation in standard form. Example 2 illustrates this approach.

● Example 2

Using the Quadratic Formula

The equation *must be in standard form* to determine *a, b,* and *c.*

Solve by using the quadratic formula.

$$9x^2 = 12x - 4$$

First, we must write the equation in standard form.

$$9x^2 - 12x + 4 = 0$$

Second, we find the values of a, b, and c. Here,

$$a = 9 \qquad b = -12 \qquad c = 4$$

Substituting these values into the quadratic formula, we find

$$x = \frac{-(-12) \pm \sqrt{(-12)^2 - 4(9)(4)}}{2(9)}$$

$$= \frac{12 + \sqrt{0}}{18}$$

and simplifying yields

$$x = \frac{2}{3} \qquad \text{or} \qquad \left\{ \frac{2}{3} \right\}$$

The graph of
$y = 9x^2 - 12x + 4$
intercepts the x axis
only at the
point $\left(\frac{2}{3}, 0 \right)$.

● ● ● **CHECK YOURSELF 2**

Use the quadratic formula to solve the equation.

$$4x^2 - 4x = -1$$

Thus far our examples and exercises have led to rational solutions. That is not always the case, as Example 3 illustrates.

● Example 3

Using the Quadratic Formula

Using the quadratic formula, solve

$$x^2 - 3x = 5$$

Once again, to use the quadratic formula, we write the equation in standard form.

$$x^2 - 3x - 5 = 0$$

We now determine values for a, b, and c and substitute.

$$x = \frac{-(-3) \pm \sqrt{(-3)^2 - 4(1)(-5)}}{2(1)}$$

Simplifying as before, we have

$$x = \frac{3 \pm \sqrt{29}}{2} \qquad \text{or} \qquad \left\{ \frac{3 \pm \sqrt{29}}{2} \right\}$$

● ● ● **CHECK YOURSELF 3**

Using the quadratic equation, solve $2x^2 = x + 7$.

Example 4 requires some special care in simplifying the solution.

● Example 4

Using the Quadratic Formula

Using the quadratic formula, solve

$$3x^2 - 6x + 2 = 0$$

Here, we have $a = 3$, $b = -6$, and $c = 2$. Substituting gives

$$x = \frac{-(-6) \pm \sqrt{(-6)^2 - 4(3)(2)}}{2(3)}$$

$$= \frac{6 \pm \sqrt{12}}{6}$$ We now look for the largest perfect-square factor of 12, the radicand.

Simplifying, we note that $\sqrt{12}$ is equal to $\sqrt{4 \cdot 3}$, or $2\sqrt{3}$. We can then write the solutions as

$$x = \frac{6 \pm 2\sqrt{3}}{6} = \frac{2(3 \pm \sqrt{3})}{6} = \frac{3 \pm \sqrt{3}}{3}$$

CAUTION

Students are sometimes tempted to reduce this result to

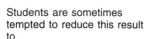

This is *not a valid step.* We must divide *each of the terms* in the numerator by 2 when simplifying the expression.

● ● ● **CHECK YOURSELF 4**

Solve by using the quadratic formula.

$$x^2 - 4x = 6$$

Let's examine a case in which the solutions are nonreal complex numbers.

● Example 5

Using the Quadratic Formula

Solve by using the quadratic formula.

$$x^2 - 2x = -2$$

Rewriting in standard form, we have

$$x^2 - 2x + 2 = 0$$

The solutions will be complex any time $\sqrt{b^2 - 4ac}$ is negative.

The graph of $y = x^2 - 2x + 2$ does not intercept the x axis, so there are no real solutions.

Labeling the coefficients, we find that

$$a = 1 \qquad b = -2 \qquad c = 2$$

Applying the quadratic formula, we have

$$x = \frac{2 \pm \sqrt{-4}}{2}$$

and noting that $\sqrt{-4}$ is $2i$, we can simplify to

$$x = 1 \pm i \qquad \text{or} \qquad \{1 \pm i\}$$

● ● ● **CHECK YOURSELF 5**

Solve by using the quadratic formula.

$$x^2 - 4x + 6 = 0$$

In attempting to solve a quadratic equation, you should first try the factoring method. If this method does not work, you can apply the quadratic formula or the square root method to find the solution. The following algorithm outlines the steps.

Solving a Quadratic Equation by Using the Quadratic Formula

STEP 1 Write the equation in standard form (one side is equal to 0).

$$ax^2 + bx + c = 0$$

STEP 2 Determine the values for a, b, and c.
STEP 3 Substitute those values into the quadratic formula.

$$x = \frac{-b \pm \sqrt{b^2 - 4ac}}{2a}$$

STEP 4 Simplify.

Although not necessarily distinct or real, every second-degree equation has two solutions.

Given a quadratic equation, the radicand $b^2 - 4ac$ determines the number of real solutions. Because of this, we call the result of substituting a, b, and c into that part of the quadratic formula the discriminant. Because the discriminant is a real number, there are three possibilities, known as the **trichotomy property.**

Graphically, we can see the number of real solutions as the number of times the related quadratic function intercepts the x axis.

The Trichotomy Property

If $b^2 - 4ac$ $\begin{cases} < 0 & \text{there are } \textit{no real solutions,} \text{ but two complex solutions.} \\ = 0 & \text{there is } \textit{one real solution} \text{ (a double solution).} \\ > 0 & \text{there are } \textit{two distinct real solutions.} \end{cases}$

• Example 6

Analyzing the Discriminant

How many real solutions are there for each of the following quadratic equations?

(a) $x^2 + 7x - 15 = 0$

The discriminant $[49 - 4(1)(-15)]$ is 109. This indicates that there are two real solutions.

We could find two complex solutions by using the quadratic formula.

(b) $3x^2 - 5x + 7 = 0$

The discriminant is negative. There are no real solutions.

(c) $9x^2 - 12x + 4 = 0$

The discriminant is 0. There is exactly one real solution (a double solution).

● ● ● **CHECK YOURSELF 6**

How many real solutions are there for each of the following quadratic equations?

(a) $2x^2 - 3x + 2 = 0$ (b) $3x^2 + x - 11 = 0$

(c) $4x^2 - 4x + 1 = 0$ (d) $x^2 = -5x - 7$

Frequently, as in Examples 3 and 4, the solutions of a quadratic equation involve square roots. When we are solving algebraic equations, it is generally best to leave solutions in this form. However, if an equation resulting from an application has been solved by the use of the quadratic formula, we will often estimate the root and sometimes accept only positive solutions. Consider the following two applications involving thrown balls that can be solved by using the quadratic formula.

• Example 7

Solving a Thrown-Ball Application

If a ball is thrown upward from the ground, the equation to find the height h of such a ball thrown with an initial velocity of 80 ft/s is

Here h measures the height above the ground, in feet, t seconds (s) after the ball is thrown upward.

$$h(t) = 80t - 16t^2$$

Find the time it takes the ball to reach a height of 48 ft.

First we substitute 48 for h, and then we rewrite the equation in standard form.

Note that the result of dividing by 16

$$\frac{0}{16} = 0$$

is 0 on the right.

$$16t^2 - 80t + 48 = 0$$

To simplify the computation, we divide both sides of the equation by the common factor, 16. This yields

$$t^2 - 5t + 3 = 0$$

We solve for t as before, using the quadratic equation, with the result

$$t = \frac{5 \pm \sqrt{13}}{2}$$

There are two solutions because the ball reaches the height *twice,* once on the way up and once on the way down.

This gives us two solutions, $\frac{5 + \sqrt{13}}{2}$ and $\frac{5 - \sqrt{13}}{2}$. But, because we have specified units of time, we generally estimate the answer to the nearest tenth or hundredth of a second.

In this case, estimating to the nearest tenth of a second gives solutions of 0.7 and 4.3 s.

CHECK YOURSELF 7

The equation to find the height h of a ball thrown with an initial velocity of 64 ft/s is

$$h(t) = 64t - 16t^2$$

Find the time it takes the ball to reach a height of 32 ft.

• Example 8

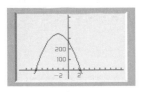

The graph of $h(t) = 240 - 64t - 16t^2$ shows the height, h, at any time t.

Solving a Thrown-Ball Application

The height, h, of a ball thrown downward from the top of a 240-ft building with an initial velocity of 64 ft/s is given by

$$h(t) = 240 - 64t - 16t^2$$

At what time will the ball reach a height of 176 ft?

Let $h(t) = 176$, and write the equation in standard form.

$$176 = 240 - 64t - 16t^2$$

$$0 = 64 - 64t - 16t^2$$

$$16t^2 + 64t - 64 = 0$$

or

$$t^2 + 4t - 4 = 0$$

Again, we divide both sides of the equation by 16 to simplify the computation.

Applying the quadratic formula with $a = 1$, $b = 4$, and $c = -4$ yields

$$t = -2 \pm 2\sqrt{2}$$

The ball has a height of 64 ft at approximately 0.8 s.

Estimating these solutions, we have $t = -4.8$ and $t = 0.8$ s, but of these two values only the *positive value* makes any sense. (To accept the negative solution would be to say that the ball reached the specified height before it was thrown.)

● ● ●   **CHECK YOURSELF 8**

The height h of a ball thrown upward from the top of a 96-ft building with an initial velocity of 16 ft/s is given by

$$h(t) = 96 + 16t - 16t^2$$

When will the ball have a height of 32 ft? (Estimate your answer to the nearest tenth of a second.)

Another geometric result that generates quadratic equations in applications is the **Pythagorean theorem.** You may recall from earlier algebra courses that the theorem gives an important relationship between the lengths of the sides of a right triangle (a triangle with a 90° angle).

The Pythagorean Theorem

In any right triangle, the square of the longest side (the hypotenuse) is equal to the sum of the squares of the two shorter sides (the legs).

$$c^2 = a^2 + b^2$$

Hypotenuse
c

a

Legs

b

In Example 9, the solution of the quadratic equation contains radicals. Substituting a pair of solutions such as $\dfrac{3 \pm \sqrt{5}}{2}$ is a very difficult process. As in our thrown-ball applications, the emphasis is on checking the "reasonableness" of the answer.

●**Example 9**

A Triangular Application

One leg of a right triangle is 4 cm longer than the other leg. The length of the hypotenuse of the triangle is 12 cm. Find the length of the two legs.

As in any geometric problem, a sketch of the information will help us visualize.

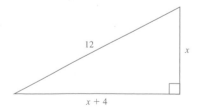

12

x

$x + 4$

We assign variable x to the shorter leg and $x + 4$ to the other leg.

Remember: The sum of the squares of the legs of the triangle is equal to the square of the hypotenuse.

Now we apply the Pythagorean theorem to write an equation for the solution.

$$x^2 + (x + 4)^2 = (12)^2$$
$$x^2 + x^2 + 8x + 16 = 144$$

or

$$2x^2 + 8x - 128 = 0$$

Dividing both sides by 2, we have the equivalent equation

Dividing both sides of a quadratic equation by a common factor is always a prudent step. It simplifies your work with the quadratic formula.

$$x^2 + 4x - 64 = 0$$

Using the quadratic formula, we get

$$x = -2 + 2\sqrt{17} \qquad \text{or} \qquad x = -2 - 2\sqrt{17}$$

Now, we check our answers for reasonableness. We can reject $-2 - 2\sqrt{17}$ (do you see why?), but we should still check the reasonableness of the value $-2 + 2\sqrt{17}$. We could substitute $-2 + 2\sqrt{17}$ into the original equation, but it seems more prudent to simply check that it "makes sense" as a solution. Remembering that $\sqrt{16} = 4$, we estimate $-2 + 2\sqrt{17}$ as

$\sqrt{17}$ is just slightly *more* than $\sqrt{16}$. or 4.

$$-2 + 2(4) = 6$$

Our equation in step 3,

$$x^2 + (x + 4)^2 = (12)^2$$

where x equals 6, becomes

$$36 + 100 = 144$$

This indicates that our answer is at least reasonable.

● ● ● **CHECK YOURSELF 9**

One leg of a right triangle is 2 cm longer than the other. The hypotenuse is 1 cm less than twice the length of the shorter leg. Find the length of each side of the triangle.

● ● ● **CHECK YOURSELF ANSWERS**

1. $\left\{-2, \dfrac{4}{3}\right\}$.　2. $\left\{\dfrac{1}{2}\right\}$.　3. $\left\{\dfrac{1 \pm \sqrt{57}}{4}\right\}$.　4. $\{2 \pm \sqrt{10}\}$.

5. $\{2 \pm i\sqrt{2}\}$.　6. **(a)** None, **(b)** two, **(c)** one, **(d)** none.　7. 0.6 and 3.4 s.

8. 2.6 s.　9. Approximately 4.3, 6.3, and 7.7 cm.

6.3 Exercises

In Exercises 1 to 8, solve each quadratic equation by first factoring and then using the quadratic formula.

1. $x^2 - 5x - 14 = 0$

2. $x^2 + 7x - 18 = 0$

3. $t^2 + 8t - 65 = 0$

4. $q^2 + 3q - 130 = 0$

5. $5x^2 + 4x - 1 = 0$

6. $3x^2 + 2x - 1 = 0$

7. $16t^2 - 24t + 9 = 0$

8. $6m^2 - 23m + 10 = 0$

In Exercises 9 to 20, solve each quadratic equation by **(a)** completing the square and **(b)** using the quatratic formula.

9. $x^2 - 2x - 5 = 0$

10. $x^2 + 6x - 1 = 0$

11. $x^2 + 3x - 27 = 0$

12. $t^2 + 4t - 7 = 0$

13. $2x^2 - 6x - 3 = 0$

14. $2x^2 - 6x + 1 = 0$

15. $2q^2 - 4q + 1 = 0$

16. $4r^2 - 2r + 1 = 0$

17. $3x^2 - x - 2 = 0$

18. $2x^2 - 8x + 3 = 0$

19. $2y^2 - y - 5 = 0$

20. $3m^2 + 2m - 1 = 0$

In Exercises 21 to 42, solve each equation by using the quadratic formula.

21. $x^2 - 4x + 3 = 0$

22. $x^2 - 7x + 3 = 0$

23. $p^2 - 8p + 16 = 0$

24. $u^2 + 7u - 30 = 0$

25. $2x^2 - 2x - 3 = 0$

26. $2x^2 - 3x - 7 = 0$

27. $-3s^2 + 2s - 1 = 0$

28. $5t^2 - 2t - 2 = 0$

Hint: Clear each of the following equations of fractions by grouping symbols first.

29. $2x^2 - \dfrac{1}{2}x - 5 = 0$

30. $3x^2 + \dfrac{1}{3}x - 3 = 0$

31. $5t^2 - 2t - \dfrac{2}{3} = 0$

32. $3y^2 + 2y + \dfrac{3}{4} = 0$

33. $(x - 2)(x + 3) = 4$

34. $(x + 1)(x - 8) = 3$

35. $(t + 1)(2t - 4) - 7 = 0$

36. $(2w + 1)(3w - 2) = 1$

37. $3x - 5 = \dfrac{1}{x}$

38. $x + 3 = \dfrac{1}{x}$

39. $2t - \dfrac{3}{t} = 3$

40. $4p - \dfrac{1}{p} = 6$

41. $\dfrac{5}{y^2} + \dfrac{2}{y} - 1 = 0$

42. $\dfrac{6}{x^2} - \dfrac{2}{x} = 1$

In Exercises 43 to 50, for each quadratic equation, find the value of the discriminant and give the number of real solutions.

43. $2x^2 - 5x = 0$

44. $3x^2 + 8x = 0$

45. $m^2 - 8m + 16 = 0$

46. $4p^2 + 12p + 9 = 0$

47. $3x^2 - 7x + 1 = 0$

48. $2x^2 - x + 5 = 0$

49. $2w^2 - 5w + 11 = 0$

50. $7q^2 - 3q + 1 = 0$

In Exercises 51 to 62, find all the solutions of each quadratic equation. Use any applicable method.

51. $x^2 - 8x + 16 = 0$

52. $4x^2 + 12x + 9 = 0$

53. $3t^2 - 7t + 1 = 0$

54. $2z^2 - z + 5 = 0$

55. $5y^2 - 2y = 0$

56. $7z^2 - 6z - 2 = 0$

57. $(x - 1)(2x + 7) = -6$

58. $4x^2 - 3 = 0$

59. $x^2 + 9 = 0$

60. $(4x - 5)(x + 2) = 1$

61. $x - 3 - \dfrac{10}{x} = 0$

62. $1 + \dfrac{2}{x} + \dfrac{2}{x^2} = 0$

The equation

$$h(t) = 112t - 16t^2$$

is the equation for the height of an arrow, shot upward from the ground with an initial velocity of 112 ft/s, where t is the time, in seconds, after the arrow leaves the ground. Use this information to solve Exercises 63 and 64. Your answers should be expressed to the nearest tenth of a second.

63. Find the time it takes for the arrow to reach a height of 112 ft.

64. Find the time it takes for the arrow to reach a height of 144 ft.

The equation

$$h(t) - 320 - 32t - 16t^2$$

is the equation for the height of a ball, thrown downward from the top of a 320-ft building with an initial velocity of 32 ft/s, where t is the time, to the nearest tenth of a second, after the ball is thrown down from the top of the building. Use this information to solve Exercises 65 and 66.

65. Find the time it takes for the ball to reach a height of 240 ft.

66. Find the time it takes for the ball to reach a height of 96 ft.

67. Number Problem The product of two consecutive integers is 72. What are the two integers?

68. Number Problem The sum of the squares of two consecutive whole numbers is 61. Find the two whole numbers.

69. Rectangles The width of a rectangle is 3 ft less than its length. If the area of the rectangle is 70 ft², what are the dimensions of the rectangle?

70. Rectangles The length of a rectangle is 5 cm more than its width. If the area of the rectangle is 84 cm², find the dimensions.

71. Rectangles The length of a rectangle is 2 cm more than 3 times its width. If the area of the rectangle is 85 cm², find the dimensions of the rectangle.

72. Rectangles If the length of a rectangle is 3 ft less than twice its width, and the area of the rectangle is 54 ft², what are the dimensions of the rectangle?

73. Triangles One leg of a right triangle is twice the length of the other. The hypotenuse is 6 m long. Find the length of each leg.

74. Triangles One leg of a right triangle is 2 ft longer than the shorter side. If the length of the hypotenuse is 14 ft, how long is each leg?

75. Triangles One leg of a right triangle is 1 in. shorter than the other leg. The hypotenuse is 3 in. longer than the shorter side. Find the length of each side.

76. Triangles The hypotenuse of a given right triangle is 5 cm longer than the shorter leg. The length of the shorter leg is 2 cm less than that of the longer leg. Find the length of the three sides.

77. Triangles The sum of the lengths of the two legs of a right triangle is 25 m. The hypotenuse is 22 m long. Find the length of the two legs.

78. Triangles The sum of the lengths of one side of a right triangle and the hypotenuse is 15 cm. The other leg is 5 cm shorter than the hypotenuse. Find the length of each side.

79. **Thrown Ball** If a ball is thrown vertically upward from the ground, its height, h, after t seconds is given by

$$h(t) = 64t - 16t^2$$

(a) How long does it take the ball to return to the ground? [*Hint:* Let $h(t) = 0$.]

(b) How long does it take the ball to reach a height of 48 ft on the way up?

80. **Thrown Ball** If a ball is thrown vertically upward from the ground, its height, h, after t seconds is given by

$$h(t) = 96t - 16t^2$$

(a) How long does it take the ball to return to the ground?

(b) How long does it take the ball to pass through a height of 128 ft on the way back down to the ground?

81. **Cost** Suppose that the cost $C(x)$, in dollars, of producing x chairs is given by

$$C(x) = 2400 - 40x + 2x^2$$

How many chairs can be produced for $5400?

82. **Profit** Suppose that the profit $T(x)$, in dollars, of producing and selling x appliances is given by

$$T(x) = -3x^2 + 240x - 1800$$

How many appliances must be produced and sold to achieve a profit of $3000?

If a ball is thrown upward from the roof of a building 70 m tall with an initial velocity of 15 m/s, its approximate height, h, after t seconds is given by

$$h(t) = 70 + 15t - 5t^2$$

Note: The difference between this equation and the one we used in Example 8 has to do with the units used. When we used feet, the t^2 coefficient was -16 (from the fact that the acceleration due to gravity is approximately 32 ft/s^2). When we use meters as the height, the t^2 coefficient is -5 (that same acceleration becomes approximately 10 m/s^2). Use this information to solve Exercises 83 and 84.

83. **Thrown Ball** How long does it take the ball to fall back to the ground?

84. **Thrown Ball** When will the ball reach a height of 80 m?

Changing the initial velocity to 25 m/s will only change the t coefficient. Our new equation becomes

$$h(t) = 70 + 25t - 5t^2$$

85. **Thrown Ball** How long will it take the ball to return to the thrower?

86. Thrown Ball When will the ball reach a height of 85 m?

The only part of the height equation that we have not discussed is the constant. You have probably noticed that the constant is always equal to the initial height of the ball (70 m in our previous problems). Now, let's have *you* develop an equation.

A ball is thrown upward from the roof of a 100-m building with an initial velocity of 20 m/s. Use this information to solve Exercises 87 to 90.

87. Thrown Ball Find the equation for the height, *h,* of the ball after *t* seconds.

88. Thrown Ball How long will it take the ball to fall back to the ground?

89. Thrown Ball When will the ball reach a height of 75 m?

90. Thrown Ball Will the ball ever reach a height of 125 m? (*Hint:* Check the discriminant.)

A ball is thrown upward from the roof of a 100-ft building with an initial velocity of 20 ft/s. Use this information to solve Exercises 91 to 94.

91. Thrown Ball Find the height, *h,* of the ball after *t* seconds.

92. Thrown Ball How long will it take the ball to fall back to the ground?

93. Thrown Ball When will the ball reach a height of 80 ft?

 94. Thrown Ball Will the ball ever reach a height of 120 ft? Explain.

95. Profit A small manufacturer's weekly profit in dollars is given by

$P(x) = -3x^2 + 270x$

Find the number of items x that must be produced to realize a profit of $5100.

96. Profit Suppose the profit in dollars is given by

$P(x) = -2x^2 + 240x$

Now how many items must be sold to realize a profit of $5100?

97. Equilibrium Price The demand equation for a certain computer chip is given by

$$D = -2p + 14$$

The supply equation is predicted to be

$$S = -p^2 + 16p - 2$$

Find the equilibrium price.

98. Equilibrium Price The demand equation for a certain type of print is predicted to be

$$D = -200p + 36{,}000$$

The supply equation is predicted to be

$$S = -p^2 + 400p - 24{,}000$$

Find the equilibrium price.

99. Can the solution of a quadratic equation with integer coefficients include one real and one imaginary number? Justify your answer.

100. Explain how the discriminant is used to predict the nature of the solutions of a quadratic equation.

In Exercises 101 to 108, solve each equation for x.

101. $x^2 + y^2 = z^2$

102. $2x^2y^2z^2 = 1$

103. $x^2 - 36a^2 = 0$

104. $ax^2 - 9b^2 = 0$

105. $2x^2 + 5ax - 3a^2 = 0$

106. $3x^2 - 16bx + 5b^2 = 0$

107. $2x^2 + ax - 2a^2 = 0$

108. $3x^2 - 2bx - 2b^2 = 0$

109. Given that the polynomial $x^3 - 3x^2 - 15x + 25 = 0$ has as one of its solutions $x = 5$, find the other two solutions. (*Hint:* If you divide the given polynomial by $x - 5$ the quotient will be a quadratic equation. The remaining solutions will be the solutions for *that* equation.)

110. Given that $2x^3 + 2x^2 - 5x - 2 = 0$ has as one of its solutions $x = -2$, find the other two solutions. (*Hint:* In this case, divide the original polynomial by $x + 2$.)

111. Find all the zeros of the function $f(x) = x^3 + 1$.

112. Find the zeros of the function $f(x) = x^2 + x + 1$.

113. Find all six solutions to the equation $x^6 - 1 = 0$. (*Hint:* Factor the left-hand side of the equation first as the difference of squares, then as the sum and difference of cubes.)

114. Find all six solutions to $x^6 = 64$.

Answers

1. $\{-2, 7\}$ **3.** $\{-13, 5\}$ **5.** $\left\{-1, \dfrac{1}{5}\right\}$ **7.** $\left\{\dfrac{3}{4}\right\}$ **9.** $\{1 \pm \sqrt{6}\}$ **11.** $\left\{\dfrac{-3 \pm 3\sqrt{13}}{2}\right\}$

13. $\left\{\dfrac{3 \pm \sqrt{15}}{2}\right\}$ **15.** $\left\{\dfrac{2 \pm \sqrt{2}}{2}\right\}$ **17.** $\left\{-\dfrac{2}{3}, 1\right\}$ **19.** $\left\{\dfrac{1 \pm \sqrt{41}}{4}\right\}$ **21.** $\{1, 3\}$ **23.** $\{4\}$

25. $\left\{\dfrac{1 \pm \sqrt{7}}{2}\right\}$ **27.** $\left\{\dfrac{1 \pm i\sqrt{2}}{3}\right\}$ **29.** $\left\{\dfrac{1 \pm \sqrt{161}}{8}\right\}$ **31.** $\left\{\dfrac{3 \pm \sqrt{39}}{15}\right\}$ **33.** $\left\{\dfrac{-1 \pm \sqrt{41}}{2}\right\}$

35. $\left\{\dfrac{1 \pm \sqrt{23}}{2}\right\}$ **37.** $\left\{\dfrac{5 \pm \sqrt{37}}{6}\right\}$ **39.** $\left\{\dfrac{3 \pm \sqrt{33}}{4}\right\}$ **41.** $\{1 \pm \sqrt{6}\}$ **43.** 25, two   **45.** 0, one

47. 37, two   **49.** -63, none   **51.** $\{4\}$ **53.** $\left\{\dfrac{7 \pm \sqrt{37}}{6}\right\}$ **55.** $\left\{0, \dfrac{2}{5}\right\}$ **57.** $\left\{\dfrac{-5 \pm \sqrt{33}}{4}\right\}$

59. $\{-3i, 3i\}$ **61.** $\{-2, 5\}$ **63.** $\{1.2 \text{ or } 5.8 \text{ s}\}$ **65.** 1.4 s   **67.** $-9, -8,$ or 8, 9
69. 7 by 10 ft   **71.** 5 by 17 cm   **73.** 2.7 cm, 5.4 cm   **75.** 5.5, 6.5, 8.5 in.   **77.** 3.2 m, 21.8 m
79. (a) 4 s, **(b)** 1 s   **81.** 50 chairs   **83.** 5.5 s   **85.** 2.5 s   **87.** $h(t) = 100 + 20t - 5t^2$ **89.** 5 s
91. $h(t) = 100 + 20t - 16t^2$ **93.** 1.9 s   **95.** 63 or 27   **97.** \$0.94   **99.**

101. $\{\pm\sqrt{z^2 - y^2}\}$ **103.** $\{-6a, 6a\}$ **105.** $\left\{-3a, \dfrac{a}{2}\right\}$ **107.** $\left\{\dfrac{-a \pm a\sqrt{17}}{4}\right\}$ **109.** $\{-1 \pm \sqrt{6}\}$

111. $\left\{-1, \dfrac{1 \pm i\sqrt{3}}{2}\right\}$ **113.** $\left\{-1, 1, \dfrac{1 \pm i\sqrt{3}}{2}, \dfrac{-1 \pm i\sqrt{3}}{2}\right\}$

 **6.4** **Graphing Parabolas**

6.4 OBJECTIVES

1. Find the axis of symmetry
2. Find the vertex
3. Graph a parabola
4. Solve an application involving the quadratic formula.

In Section 1.5, we dealt with the graph of a linear equation in which the graphs of all the linear equations were straight lines. Suppose that now we allow the terms in x and/or y to be quadratic; that is, we allow squares in one or both of those terms. The graphs of such equations will form a family of curves called the **conic sections.** Conic sections are curves formed when a plane cuts through, or forms a section of, a cone. The conic sections comprise four curves—the **parabola, circle, ellipse,** and **hyperbola.** Examples of how these curves are formed are shown below.

Parabola

Circle

Ellipse

Hyperbola

The inclination of the plane determines which of the sections is formed.

The names *ellipse, parabola,* and *hyperbola* are attributed to Apollonius, a third-century B.C. Greek mathematician and astronomer.

Our attention here is focused on the first of these sections, the parabola.

Consider an equation of the form

$$y = ax^2 + bx + c \qquad a \neq 0$$

This equation is quadratic in x and linear in y. Its graph will always be the curve called the parabola.

In an equation of the form

$$y = ax^2 + bx + c \qquad a \neq 0$$

the parabola opens upward or downward, as follows:

1. If $a > 0$, the parabola opens *upward.*
2. If $a < 0$, the parabola opens *downward.*

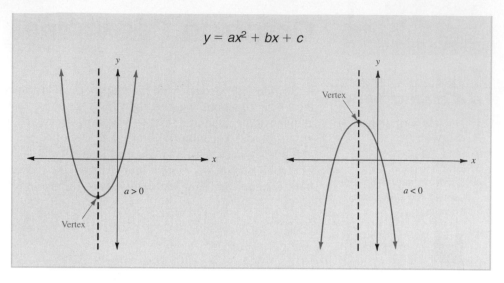

Two points regarding the parabola can be made by observation. Consider the above illustrations.

1. There is always a **minimum** (or lowest) point on the parabola if it opens upward. There is always a **maximum** (or highest) point on the parabola if it opens downward. In either case, that maximum or minimum value occurs at the **vertex** of the parabola.
2. Every parabola has an **axis of symmetry.** In the case of parabolas that open upward or downward, that axis of symmetry is a vertical line midway between any pair of symmetric points on the parabola. Also, the point where this axis of symmetry intersects the parabola is the vertex of the parabola.

The following figure summarizes these observations.

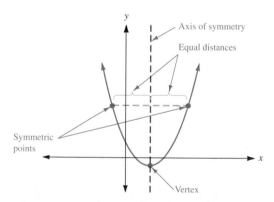

Our objective is to be able to quickly sketch a parabola. This can be done with *as few as three points* if those points are carefully chosen. For this purpose you will want to find the vertex and two symmetric points.

First, let's see how the coordinates of the vertex can be determined from the standard equation

$$y = ax^2 + bx + c \tag{1}$$

In equation (1), if $x = 0$, then $y = c$, and so $(0, c)$ gives the point where the parabola intersects the y axis (the y intercept).

Look at the sketch. To determine the coordinates of the symmetric point (x_1, c), note that it lies along the horizontal line $y = c$.

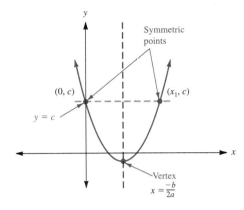

Therefore, let $y = c$ in equation (1):

$$c = ax^2 + bx + c$$

$$0 = ax^2 + bx$$

$$0 = x(ax + b)$$

and

$$x = 0 \qquad \text{or} \qquad x = -\frac{b}{a}$$

We now know that

$$(0, c) \qquad \text{and} \qquad \left(-\frac{b}{a}, c\right)$$

are the coordinates of the symmetric points shown. Since the axis of symmetry must be midway between these points, the x value along that axis is given by

$$x = \frac{0 + (-b/a)}{2} = -\frac{b}{2a} \tag{2}$$

Since the vertex for any parabola lies on the axis of symmetry, we can now state the following general result.

Vertex of a Parabola

If

$$y = ax^2 + bx + c \qquad a \neq 0$$

then the x coordinate of the vertex of the corresponding parabola is

$$x = -\frac{b}{2a}$$

Note: The y coordinate of the vertex can be found most easily by substituting the value found for x into the original equation.

In fact, we will always have *two* symmetric *x* intercepts unless the quadratic member is a perfect square, and that case can be handled by later techniques presented in this section.

We now know how to find the vertex of a parabola, and if two symmetric points can be determined, we are well on our way to the desired graph. Perhaps the simplest case is when the quadratic member of the given equation is factorable. In most cases, the two *x* intercepts will then give two symmetric points that are very easily found. Example 1 illustrates such a case.

• Example 1

Graphing a Parabola

Graph the equation

$$y = x^2 + 2x - 8$$

Sketch the information to help you solve the problem. Begin by drawing—as a dashed line—the axis of symmetry.

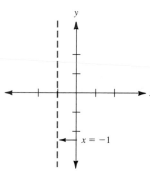

First, find the axis of symmetry. In this equation, $a = 1$, $b = 2$, and $c = -8$. We then have

$$x = -\frac{b}{2a} = \frac{-2}{2 \cdot 1} = \frac{-2}{2} = -1$$

Thus, $x = -1$ is the axis of symmetry.

Second, find the vertex. Since the vertex of the parabola lies on the axis of symmetry, let $x = -1$ in the original equation. If $x = -1$,

$$y = (-1)^2 + 2(-1) - 8 = -9$$

and $(-1, -9)$ is the vertex of the parabola.

Third, find two symmetric points. Note that the quadratic member in this case is factorable, and so setting $y = 0$ in the original equation will quickly give two symmetric points (the *x* intercepts):

At this point you can plot the vertex along the axis of symmetry.

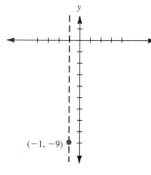

$$0 = x^2 + 2x - 8$$
$$= (x + 4)(x - 2)$$

So when $y = 0$,

$$x + 4 = 0 \qquad \text{or} \qquad x - 2 = 0$$
$$x = -4 \qquad\qquad x = 2$$

and our *x* intercepts are $(-4, 0)$ and $(2, 0)$.

Fourth, draw a smooth curve connecting the points found above, to form the parabola.

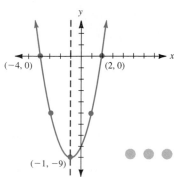

Note: You can choose to find additional pairs of symmetric points at this time if necessary. For instance, the symmetric points $(0, -8)$ and $(-2, -8)$ are easily located.

● ● ● CHECK YOURSELF 1

Graph the equation

$$y = -x^2 - 2x + 3$$

Hint: Since the coefficient of x^2 is negative, the parabola opens downward.

A similar process will work if the quadratic member of the given equation is *not* factorable. In that case, one of two things happens:

1. The *x* intercepts are irrational and therefore not particularly helpful in the graphing process.
2. The *x* intercepts do not exist.

Consider Example 2.

• Example 2

Graphing a Parabola

Graph the function

$$f(x) = x^2 - 6x + 3$$

First, find the axis of symmetry. Here $a = 1$, $b = -6$, and $c = 3$. So

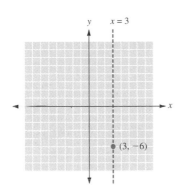

$$x = -\frac{b}{2a} = \frac{-(-6)}{2 \cdot 1} = \frac{6}{2} = 3$$

Thus, $x = 3$ is the axis of symmetry.
   Second, find the vertex. If $x = 3$,

$$f(x) = 3^2 - 6 \cdot 3 + 3 = -6$$

and $(3, -6)$ is the vertex of the desired parabola.
   Third, find two symmetric points. Here the quadratic member is not factorable, and the *x* intercepts are irrational, so we would prefer to find another pair of symmetric points.

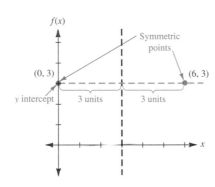

Note that $(0, 3)$ is the *y* intercept of the parabola. We found the axis of symmetry at $x = 3$ in step 1. Note that the symmetric point to $(0, 3)$ lies along the horizontal line through the *y* intercept at the same distance (3 units) from the axis of symmetry. Hence, $(6, 3)$ is our symmetric point.
   Fourth, draw a smooth curve connecting the points found above to form the parabola.

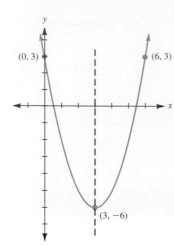

Note: An alternate method is available in step 3. Observing that 3 is the y intercept and that the symmetric point lies along the line $y = 3$, set $y = 3$ in the original equation:

$$3 = x^2 - 6x + 3$$
$$0 = x^2 - 6x$$
$$0 = x(x - 6)$$

so

$$x = 0 \quad \text{or} \quad x - 6 = 0$$
$$x = 6$$

and $(0, 3)$ and $(6, 3)$ are the desired symmetric points.

● ● ● **CHECK YOURSELF 2**

Graph the function

$$f(x) = x^2 + 4x + 5$$

Thus far the coefficient of x^2 has been 1 or -1. The following example shows the effect of different coefficients on the term in x^2.

• Example 3

Graphing a Parabola

Graph the equation

$$y = 3x^2 - 6x + 5$$

First, find the axis of symmetry.

$$x = -\frac{b}{2a} = \frac{-(-6)}{2 \cdot 3} = \frac{6}{6} = 1$$

Second, find the vertex. If $x = 1$,

$$y = 3(1)^2 - 6 \cdot 1 + 5 = 2$$

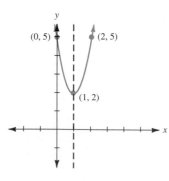

So $(1, 2)$ is the vertex.

Third, find symmetric points. Again the quadratic member is not factorable, and we use the y intercept $(0, 5)$ and its symmetric point $(2, 5)$.

Fourth, connect the points with a smooth curve to form the parabola. Compare this curve to those in previous examples. Note that the parabola is "tighter" about the axis of symmetry. That is the effect of the larger x^2 coefficient.

● ● ● CHECK YOURSELF 3

Graph the equation

$$y = \frac{1}{2}x^2 - 3x - 1$$

The following algorithm summarizes our work thus far in this section.

> **To Graph a Parabola**
>
> **STEP 1**   Find the axis of symmetry.
> **STEP 2**   Find the vertex.
> **STEP 3**   Determine two symmetric points.
> **Note:** You can use the x intercepts if the quadratic member of the given equation is factorable. Otherwise use the y intercept and its symmetric point.
> **STEP 4**   Draw a smooth curve connecting the points found above to form the parabola. You may choose to find additional pairs of symmetric points at this time.

So far we have dealt with equations of the form

$$y = ax^2 + bx + c$$

Suppose we reverse the role of x and y. We then have

$$x = ay^2 + by + c$$

which is quadratic in y but not in x. The graph of such an equation is once again a parabola, but this time the parabola is horizontally oriented.

> In an equation of the form
>
> $$x = ay^2 + by + c \qquad a \neq 0$$
>
> the parabola opens leftward or rightward, as follows:
>
> **1.** If $a > 0$, the parabola opens *rightward*.
> **2.** If $a < 0$, the parabola opens *leftward*.
>
> $$x = ay^2 + by + c$$
>
>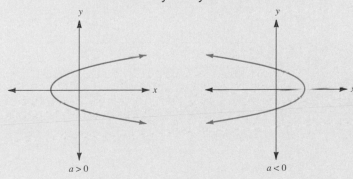
>
> $a > 0$ $a < 0$

Much of what we did earlier is easily extended to a horizontally oriented parabola. Example 4 illustrates the changes in the process.

• Example 4

Graphing a Parabola

Graph the equation

$$x = y^2 + 4y - 5$$

First, find the axis of symmetry. Now the axis of symmetry is horizontal with the equation

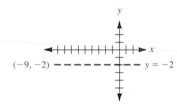

$$y = -\frac{b}{2a}$$

So

$$y = -\frac{b}{2a} = \frac{-4}{2 \cdot 1} = -2$$

Second, find the vertex. If $y = -2$,

$$x = (-2)^2 + 4(-2) - 5 = -9$$

So $(-9, -2)$ is the vertex.

Third, find two symmetric points. Here the quadratic member is factorable, so set $x = 0$ in the original equation. That gives the y intercepts:

$$0 = y^2 + 4y - 5$$
$$ = (y + 5)(y - 1)$$
$$y + 5 = 0 \qquad \text{or} \qquad y - 1 = 0$$
$$y = -5 \qquad\qquad y = 1$$

The y intercepts then are at $(0, -5)$ and $(0, 1)$.

Fourth, draw a smooth curve through the points found above, to form the parabola.

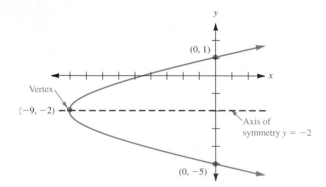

● ● ● **CHECK YOURSELF 4**

Graph the equation $x = -y^2 - 2y + 3$.

From graphs of equations of the form $y = ax^2 + bx + c$, we know that if $a > 0$, then the vertex is the lowest point on the graph (the minimum value). Also, if $a < 0$, then the vertex is the highest point on the graph (the maximum value). We can use this result to solve a variety of problems in which we want to find the maximum or minimum value of a variable. The following are just two of many typical examples.

● **Example 5**

An Application Involving a Quadratic Function

A software company sells a word processing program for personal computers. They have found that their monthly profit in dollars, P, from selling x copies of the program is approximated by

$$P(x) = -0.3x^2 + 90x - 1500$$

Find the number of copies of the program that should be sold in order to maximize the profit.

Since the relating equation is quadratic, the graph must be a parabola. Also since the coefficient of x^2 is negative, the parabola must open downward, and thus the vertex will give the maximum value for the profit, P. To find the vertex,

$$x = -\frac{b}{2a} = \frac{-90}{2(-0.3)} = \frac{-90}{-0.6} = 150$$

The maximum profit must then occur when $x = 150$, and we substitute that value into the original equation:

$$P(x) = -0.3(150)^2 + (90)(150) - 1500$$
$$= \$5250$$

The maximum profit will occur when 150 copies are sold per month, and that profit will be \$5250.

● ● ● **CHECK YOURSELF 5**

A company that sells portable radios finds that its weekly profit in dollars, P, and the number of radios sold, x, are related by

$$P(x) = -0.2x^2 + 40x - 100$$

Find the number or radios that should be sold to have the largest weekly profit and the amount of that profit.

• Example 6

An Application Involving a Quadratic Function

A farmer has 3600 ft of fence and wishes to enclose the largest possible rectangular area with that fencing. Find the largest possible area that can be enclosed.

As usual, when dealing with geometric figures, we start by drawing a sketch of the problem.

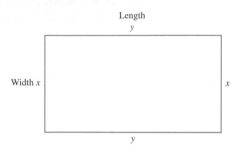

First, we can write the area, A, as

Area = length × width

$$A = xy \qquad (3)$$

Also, since 3600 ft of fence is to be used, we know that

The perimeter of the region is

$2x + 2y$

$$2x + 2y = 3600$$
$$2y = 3600 - 2x$$
$$y = 1800 - x \qquad (4)$$

Substituting for y in equation (3), we have

$$A(x) = x(1800 - x)$$
$$= 1800x - x^2$$
$$= -x^2 + 1800x \qquad (5)$$

Again, the graph for A is a parabola opening downward, and the largest possible area will occur at the vertex. As before,

The width x is 900 ft. Since from (2)

$y = 1800 - 900$

$= 900$ ft

$$x = \frac{-1800}{2(-1)} = \frac{-1800}{-2} = 900$$

and the largest possible area is

The length is also 900 ft. The desired region is a square.

$$A(x) = -(900)^2 + 1800(900) = 810{,}000 \text{ ft}^2$$

● ● ● **CHECK YOURSELF 6**

We want to enclose three sides of the largest possible rectangular area by using 900 ft. of fence. What will be the dimensions of the area?

● ● ● **CHECK YOURSELF ANSWERS**

1. $y = -x^2 - 2x + 3$

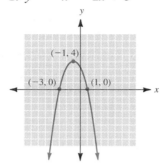

2. $y = x^2 + 4x + 5$

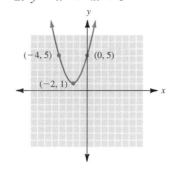

3. $y = \dfrac{1}{2}x^2 - 3x - 1$

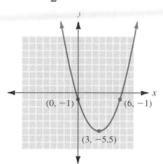

4. $x = -y^2 - 2y + 3$

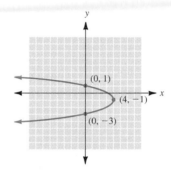

5. 100 radios, $1900.

6. Width 225 ft, length 450 ft.

Electric Power Costs. You and a partner in a small engineering firm have been asked to help calculate the costs involved in running electric power a distance of 3 miles from a public utility company to a nearby community. You have decided to consider three tower heights: 75 feet, 100 feet, and 120 feet. The 75-foot towers cost $850 each to build and install; the 100-foot towers cost $1110 each; and the 120-foot towers cost $1305 each. Spans between towers typically run 300 to 1200 feet. The conducting wires weigh 1900 pounds per 1000 feet, and the tension on the wires is 6000 pounds. To be safe, the conducting wires should never be closer than 45 feet to the ground.

Work with a partner to complete the following.

1. Develop a plan for the community showing all three scenarios and their costs. Remember that hot weather will expand the wires and cause them to sag more than normally. Therefore, allow for about a 10% margin of error when you calculate the sag amount.
2. Write an accompanying letter to the town council explaining your recommendation. Be sure to include any equations you used to help you make your decision. Remember to look at the formula at the beginning of the chapter, which gives the amount of sag for wires given a certain weight per foot, span length, and tension on the conductors.

6.4 Exercises

In Exercises 1 to 8, match each graph with one of the equations at the right.

1.

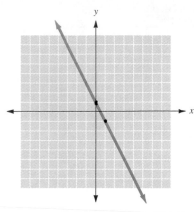

2.

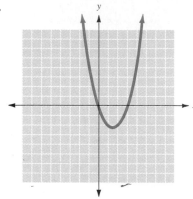

(a) $y = x^2 + 2$
(b) $y = 2x^2 - 1$
(c) $y = 2x + 1$
(d) $y = x^2 - 3x$
(e) $y = -x^2 - 4x$
(f) $y = -2x + 1$
(g) $y = x^2 + 2x - 3$
(h) $y = -x^2 + 6x - 8$

3.

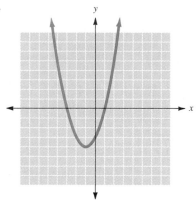

4.

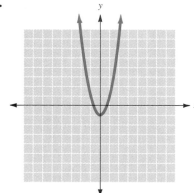

5.

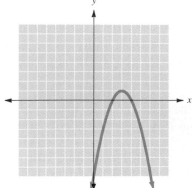

6.

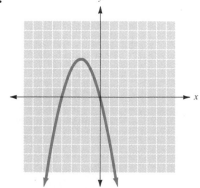

7.

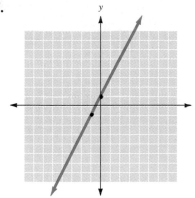

8.

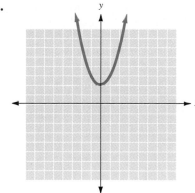

In Exercises 9 to 14, which of the given conditions apply to the graphs of the following equations? Note that more than one condition may apply.

(a) The parabola opens upward.
(b) The parabola opens downward.
(c) The parabola has two x intercepts.
(d) The parabola has one x intercept.
(e) The parabola has no x intercept.

9. $y = x^2 - 3$ **10.** $y = -x^2 + 4x$ **11.** $y = x^2 - 3x - 4$

12. $y = x^2 - 2x + 2$ **13.** $y = -x^2 - 3x + 10$ **14.** $y = x^2 - 8x + 16$

In Exercises 15 to 28, find the equation of the axis of symmetry, the coordinates of the vertex, and the x intercepts. Sketch the graph of each equation.

15. $y = x^2 - 4x$ **16.** $y = x^2 - 1$

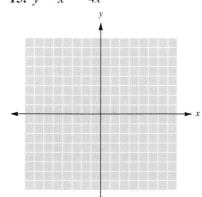   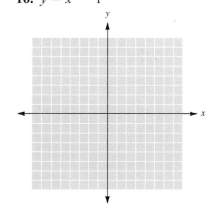

17. $y = -x^2 + 4$

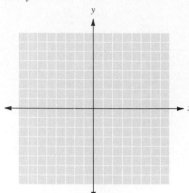

18. $y = x^2 + 2x$

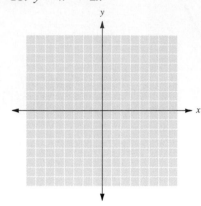

19. $y = -x^2 - 2x$

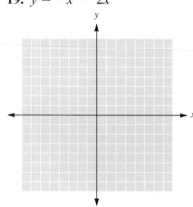

20. $y = -x^2 - 3x$

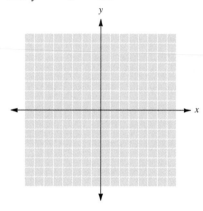

21. $y = x^2 - 6x + 5$

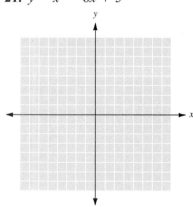

22. $y = x^2 + x - 6$

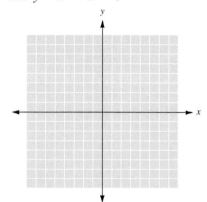

23. $y = x^2 - 5x + 6$

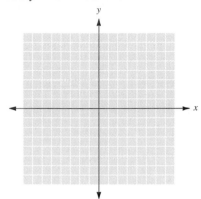

24. $y = x^2 + 6x + 5$

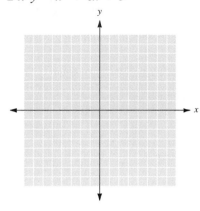

25. $y = x^2 - 6x + 8$

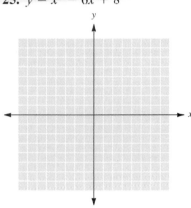

26. $y = -x^2 - 3x + 4$

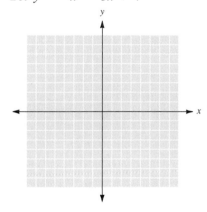

27. $y = -x^2 - 6x - 5$

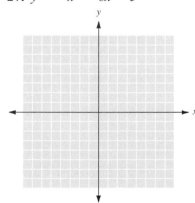

28. $y = -x^2 + 6x - 8$

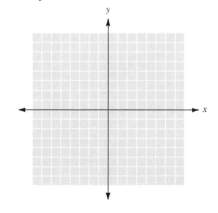

In Exercises 29 to 44, find the equation of the axis of symmetry, the coordinates of the vertex, and at least two symmetric points. Sketch the graph of each equation.

29. $y = x^2 - 2x - 1$

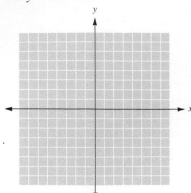

30. $y = x^2 + 4x + 6$

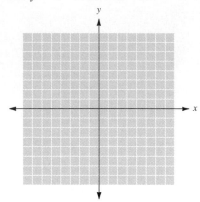

31. $y = x^2 - 4x - 1$

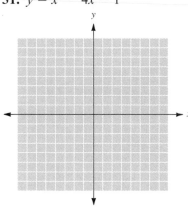

32. $y = -x^2 + 6x - 5$

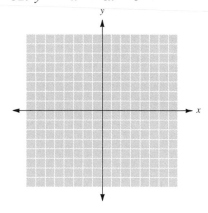

33. $y = -x^2 + 3x - 3$

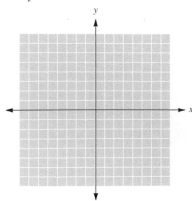

34. $y = x^2 + 5x + 3$

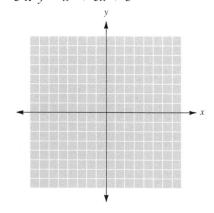

35. $y = 2x^2 + 4x - 1$

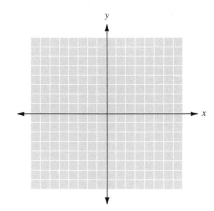

36. $y = \dfrac{1}{2} x^2 - x - 1$

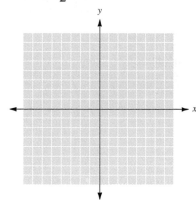

37. $y = -\dfrac{1}{3} x^2 + x - 3$

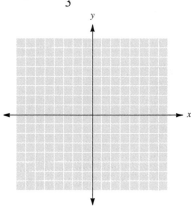

38. $y = -2x^2 - 4x - 1$

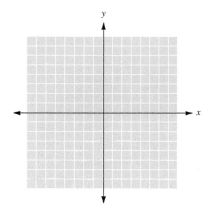

39. $y = 3x^2 + 12x + 5$

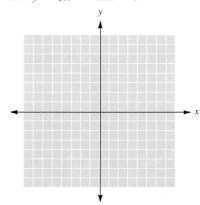

40. $y = -3x^2 + 6x + 1$

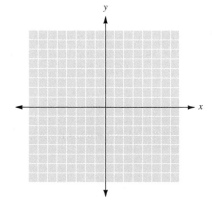

41. $x = y^2 + 4y$

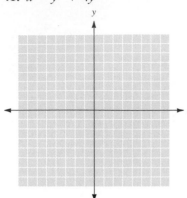

42. $x = y^2 - 3y$

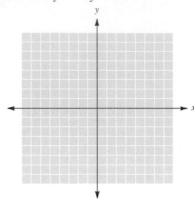

43. $x = y^2 + 8y + 12$

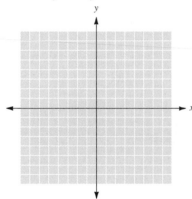

44. $x = -y^2 + 6y - 5$

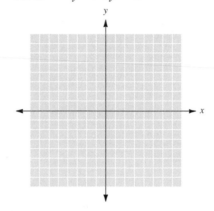

45. Profit A company's weekly profit, P, is related to the number of items sold by $P(x) = -0.3x^2 + 60x - 400$. Find the number of items that should be sold each week in order to maximize the profit. Then find the amount of that weekly profit.

46. Profit A company's monthly profit, P, is related to the number of items sold by $P(x) = -0.2x^2 + 50x - 800$. How many items should be sold each month to obtain the largest possible profit? What is the amount of that profit?

47. Area A builder wants to enclose the largest possible rectangular area with 2000 ft of fencing. What should be the dimensions of the rectangle, and what will the area of the rectangle be?

48. Area A farmer wants to enclose a field along a river on three sides. If 1600 ft of fencing is to be used, what dimensions will give the maximum enclosed area? Find that maximum area.

49. Motion A ball is thrown upward into the air with an initial velocity of 96 ft/s. If h gives the height of the ball at time t, then the equation relating h and t is

$$h(t) = -16t^2 + 96t$$

Find the maximum height the ball will attain.

50. Motion A ball is thrown upward into the air with an initial velocity of 64 ft/s. If h gives the height of ball at time t, then the equation relating h and t is

$$h(t) = -16t^2 + 64t$$

Find the maximum height the ball will attain.

51. Motion In the drawing below, distances d_1 and d_2 are equal by the definition of the parabola. Use the distance formula to express d_1 and d_2 in terms of x, y, and p. Then derive the equation for the parabola.

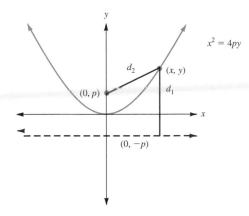

 52. Explain how to determine the domain and range of the function $f(x) = a(x - h)^2 + k$.

In Exercises 53 to 56, describe a viewing window that would include the vertex and all intercepts for the graph of each function.

53. $f(x) = 3x^2 - 25$

54. $f(x) = 9x^2 - 5x - 7$

55. $f(x) = -2x^2 + 5x - 7$

56. $f(x) = -5x^2 + 2x + 7$

Answers

1. (f)   **3.** (g)   **5.** (h)   **7.** (c)   **9.** (a), (c)   **11.** (a), (c)   **13.** (b), (c)

15. $x = 2$; vertex: $(2, -4)$; $(0, 0)$ and $(4, 0)$ **17.** $x = 0$; vertex: $(0, 4)$; $(-2, 0)$ and $(2, 0)$

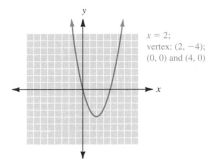

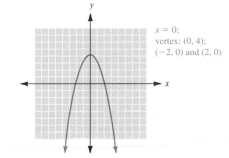

19. $x = -1$; vertex: $(-1, 1)$; $(-2, 0)$ and $(0, 0)$

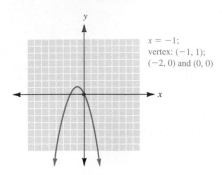

$x = -1$;
vertex: $(-1, 1)$;
$(-2, 0)$ and $(0, 0)$

21. $x = 3$; vertex: $(3, -4)$; $(1, 0)$ and $(3, 0)$

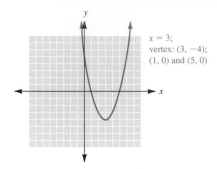

$x = 3$;
vertex: $(3, -4)$;
$(1, 0)$ and $(5, 0)$

23. $x = \dfrac{5}{2}$; vertex: $\left(\dfrac{5}{2}, -\dfrac{1}{4}\right)$; $(3, 0)$ and $(2, 0)$

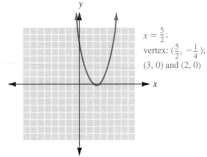

$x = \frac{5}{2}$;
vertex: $(\frac{5}{2}, -\frac{1}{4})$;
$(3, 0)$ and $(2, 0)$

25. $x = 3$; vertex: $(3, -1)$; $(2, 0)$ and $(4, 0)$

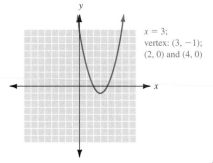

$x = 3$;
vertex: $(3, -1)$;
$(2, 0)$ and $(4, 0)$

27. $x = -3$; vertex: $(-3, 4)$; $(-5, 0)$ and $(-1, 0)$

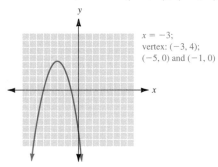

$x = -3$;
vertex: $(-3, 4)$;
$(-5, 0)$ and $(-1, 0)$

29. $x = 1$; vertex: $(1, -2)$

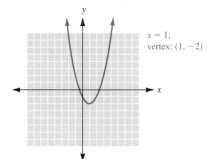

$x = 1$;
vertex: $(1, -2)$

31. $x = 2$; vertex: $(2, -5)$

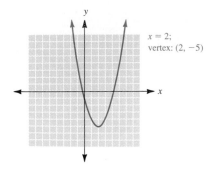

$x = 2$;
vertex: $(2, -5)$

33. $x = \dfrac{3}{2}$; vertex: $\left(\dfrac{3}{2}, -\dfrac{3}{4}\right)$

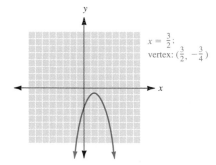

$x = \frac{3}{2}$;
vertex: $(\frac{3}{2}, -\frac{3}{4})$

35. $x = -1$; vertex: $(-1, -3)$

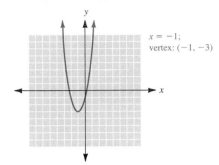

37. $x = \dfrac{3}{2}$; vertex: $\left(\dfrac{3}{2}, -\dfrac{9}{4}\right)$

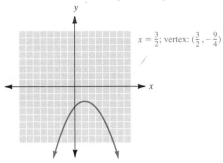

39. $x = -2$; vertex: $(-2, -7)$

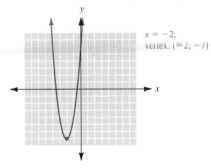

41. $y = -2$; vertex: $(-4, -2)$

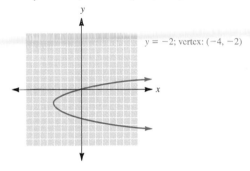

43. $y = -4$; vertex: $(-4, -4)$

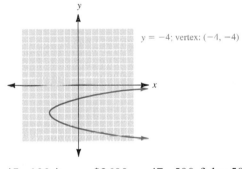

45. 100 items; $2600   **47.** 500 ft by 500 ft; 250,000 ft^2 **49.** 144 ft   **51.** $d_1 = y + p$, $d_2 = \sqrt{(y - p)^2 + x^2}$, $x^2 = 4py$ **53.** $-3 \le x \le 3$; $-25 \le y \le 0$ **55.** $-2 \le x \le 4$; $-10 \le y \le 0$

6.5 Quadratic Inequalities in One Variable

6.5 OBJECTIVES

1. Solve a quadratic inequality graphically
2. Solve a quadratic inequality algebraically

A **quadratic inequality** is an inequality that can be written in the form

$$ax^2 + bx + c < 0 \qquad \text{where } a \neq 0$$

Note that the inequality symbol, $<$ can be replaced by the symbol $>$, $\leq$, or $\geq$ in the above definition.

In Chapter 2, solutions to linear inequalities such as $4x + 2 < 0$ were analyzed graphically. Recall that, given the graph of the function $f(x) = 4x + 2$

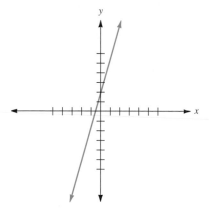

the solution to the inequality was the set of all x values associated with a point on the line that was *below* the x axis.

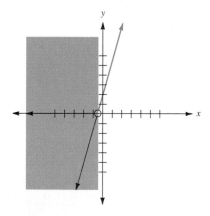

In this case, every x to the left of $-\dfrac{1}{2}$ was associated with a point on the line that was below the x axis. The solution set for the inequality $4x + 2 < 0$ is the set $\left\{ x \mid x < -\dfrac{1}{2} \right\}$.

The same principle can be applied to solving quadratic inequalities. Example 1 illustrates this concept.

472

• Example 1

Solving a Quadratic Inequality Graphically

Solve the inequality

$$x^2 - x - 12 \leq 0$$

First, use the techniques in Section 6.4 to graph the function $f(x) = x^2 - x - 12$.

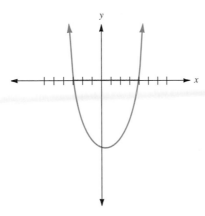

Next, looking at the graph, determine the values of x that make $x^2 - x - 12 < 0$ a true statement. Notice that the graph is below the x axis for values of x between -3 and 4. The graph intercepts the x axis when x is -3 or 4. The solution to the inequality is $\{x \mid -3 \leq x \leq 4\}$.

● ● ● **CHECK YOURSELF 1**

Use a graph to solve the inequality

$$x^2 - 3x - 10 \geq 0$$

Algebraic methods can also be used to find the exact solution to a quadratic inequality. Subsequent examples in this section will discuss algebraic solutions. When solving an equation or inequality algebraically, it is always a good idea to compare the algebraic solution to a graphical one to ensure that the algebraic solution is reasonable.

If we expand the binomial, we get

$$x^2 - 2x - 3 < 0$$

Looking at the graph of

$$y = x^2 - 2x - 3,$$

where is y less than 0 on the graph?

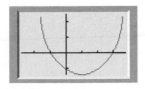

• Example 2

Solving a Quadratic Inequality Algebraically

Solve $(x - 3)(x + 1) < 0$.

We start by finding the solutions of the corresponding quadratic equation. So

$$(x - 3)(x + 1) = 0$$

has solutions 3 and -1, called the *critical points*.

Our solution process depends on determining where each factor is positive or negative. To help visualize that process, we start with a number line and label it as shown below. We begin with our first critical point of -1.

$x + 1$ is negative if x is less than -1. $x + 1$ is positive if x is greater than -1.

Sign of $x + 1$ $----|+++++++++++$
-1

We now continue in the same manner with the second critical point, 3.

$x - 3$ is negative if x is less than 3. $x - 3$ is positive if x is greater than 3.

Sign of $x - 3$ $------------|++++$
3

In practice, we combine the two steps above for the following result.

Sign of $x + 1$ $----|+++++++|+++$
Sign of $x - 3$ $----|-------|+++$
$-1 \qquad 3$
Sign of product $\qquad ++++-------++++$

Examining the signs of the factors, we see that in this case,

Both factors are negative.

For any x less than -1, the product is positive.

The factors have opposite signs.

For any x between -1 and 3, the product is negative.

Both factors are positive.

For any x greater than 3, the product is again positive.

We return to the original inequality:

The product of the two binomials must be negative.

$$(x - 3)(x + 1) < 0$$

We can see that this is true only between -1 and 3. In set notation, the solution can be written as

$$\{x | -1 < x < 3\}$$

On a number line, the graph of the solution is

$-1 \qquad 3$

• • • CHECK YOURSELF 2

Solve and graph the solution set.

$$(x - 2)(x + 4) < 0$$

We now consider an example in which the quadratic member of the inequality must be factored.

• Example 3

Solving a Quadratic Inequality Algebraically

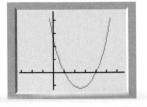

Examine the graph of $y = x^2 - 5x + 4$. For what values of x is y (the graph) greater than zero?

Solve $x^2 - 5x + 4 > 0$.

Factoring the quadratic member, we have

$$(x - 1)(x - 4) > 0$$

The critical points are 1 and 4, and we form the sign graph as before.

Sign of $x - 4$ $--- - -|------ - -|++++$
Sign of $x - 1$ $- - - -|+ + + + + + +|+ + + +$

Sign of
product $\qquad + + + + - - - - - - - -+ + + +$
$\qquad\qquad\quad 1 \qquad\qquad 4$

In this case, we want those values of x for which the product is *positive,* and we can see from the sign graph above that the solution is

$$\{x \mid x < 1 \text{ or } x > 4\}$$

The graph of the solution set is shown below.

$\qquad\qquad 1 \qquad\qquad 4$

CHECK YOURSELF 3

Solve and graph the solution set.

$$2x^2 - x - 3 > 0$$

The method used in the previous examples works *only* when one side of the inequality is factorable and the other is 0. It is sometimes necessary to rewrite the inequality in an equivalent form in order to attain that form, as Example 4 illustrates.

• Example 4

Solving a Quadratic Inequality Algebraically

Use a calculator to graph both $f(x) = x^2 - 3x - 4$ and $g(x) = 6$. Where is $f(x)$ above $g(x)$? Compare this to the algebraic solution.

Solve $(x + 1)(x - 4) \geq 6$.

First, we multiply to clear the parentheses.

$$x^2 - 3x - 4 \geq 6$$

Now we subtract 6 from both sides so that the inequality is *related to 0:*

$$x^2 - 3x - 10 \geq 0$$

Factoring the quadratic member, we have

$$(x - 5)(x + 2) \geq 0$$

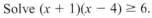

We can now proceed with the sign graph method as before.

Sign of $x - 5$ 　— — — | — — — — — — — — | + + + +
Sign of $x + 2$ 　— — — | + + + + + + + + | + + + +

Sign of 　　　　　　　-2　　　　　5
product 　　+ + + — — — — — — — — — + + + +

Both factors are negative if *x* is less than −2. Both factors are positive if *x* is greater than 5.

From the graph we see that the solution is

$\{x | x \leq -2 \text{ or } x \geq 5\}$

The graph is shown below.

　　　　　　　−2　　　5

● ● ●   **CHECK YOURSELF 4**

Solve and graph the solution set.

$(x - 5)(x + 7) \leq -11$

● ● ●   **CHECK YOURSELF ANSWERS**

1. $\{x | x \leq -2 \text{ or } x \geq 5\}$.   **2.** ◄──○────────○──► $\{x | -4 < x < 2\}$.
　　　　　　　　　　　　　　　　　　　　　−4　　　2

3. ◄────○──○────► $\left\{ x \middle| x < -1 \text{ or } x > \dfrac{3}{2} \right\}$.
　　　　　−1　$\frac{3}{2}$

4. ◄●──────────●──► $\{x | -6 \leq x \leq 4\}$.
　　　−6　　　　　　4

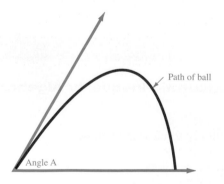

Trajectory and Height. So far in this chapter you have done many exercises involving balls that have been thrown upward with varying velocities. How does trajectory—the angle at which the ball is thrown—affect its height and time in the air?

Path of ball

Angle A

If you throw a ball from ground level with an initial upward velocity of 70 ft per second, the equation

$h = -16t^2 + \partial(70)t$

gives you the height, h, in feet t seconds after the ball has been thrown at a certain angle, A. The value of ∂ in the equation depends on the angle A and is given in the accompanying table.

Measure of Angle A in Degrees	Value of ∂
0	0.000
5	0.087
10	0.174
15	0.259
20	0.342
25	0.423
30	0.500
35	0.574
40	0.643
45	0.707
50	0.766
55	0.819
60	0.866
65	0.906
70	0.940
75	0.966
80	0.985
85	0.996
90 (straight up)	1.000

Investigate the following questions and write your conclusions to each one in complete sentences, showing all charts and graphs. Indicate what initial velocity you are using in each case.

1. Suppose an object is thrown from ground level with an initial upward velocity of 70 ft/s and at an angle of 45 degrees. What will be the height of the ball in 1 s (nearest tenth of a foot)? How long is the ball in the air?
2. Does the ball stay in the air longer if the angle of the throw is greater?
3. If you double the angle of the throw, will the ball stay in the air double the length of time?
4. If you double the angle of the throw, will the ball go twice as high?
5. Is the height of the ball directly related to the angle at which you throw it? That is, does the ball go higher if the angle of the throw is larger?
6. Repeat this exercise using another initial upward velocity.

6.5 Exercises

In Exercises 1 to 8, solve each inequality, and graph the solution set.

1. $(x - 3)(x + 4) < 0$

$\xleftarrow{\hspace{1cm}\underset{0}{+}\hspace{1cm}}\rightarrow$

2. $(x - 2)(x + 5) > 0$

$\xleftarrow{\hspace{1cm}\underset{0}{+}\hspace{1cm}}\rightarrow$

3. $(x - 3)(x + 4) > 0$

$\xleftarrow{\hspace{1cm}\underset{0}{+}\hspace{1cm}}\rightarrow$

4. $(x - 3)(x + 5) < 0$

$\xleftarrow{\hspace{1cm}\underset{0}{+}\hspace{1cm}}\rightarrow$

5. $(x - 3)(x + 4) \leq 0$

$\xleftarrow{\hspace{1cm}\underset{0}{+}\hspace{1cm}}\rightarrow$

6. $(x - 2)(x + 5) \geq 0$

$\xleftarrow{\hspace{1cm}\underset{0}{+}\hspace{1cm}}\rightarrow$

7. $(x - 3)(x + 4) \geq 0$

$\xleftarrow{\hspace{1cm}\underset{0}{+}\hspace{1cm}}\rightarrow$

8. $(x - 2)(x + 5) \leq 0$

$\xleftarrow{\hspace{1cm}\underset{0}{+}\hspace{1cm}}\rightarrow$

In Exercises 9 to 38, solve each inequality, and graph the solution set.

9. $x^2 - 3x - 4 > 0$

$\xleftarrow{\hspace{1cm}\underset{0}{+}\hspace{1cm}}\rightarrow$

10. $x^2 - 2x - 8 < 0$

$\xleftarrow{\hspace{1cm}\underset{0}{+}\hspace{1cm}}\rightarrow$

11. $x^2 + x - 12 \leq 0$

$\xleftarrow{\hspace{1cm}\underset{0}{+}\hspace{1cm}}\rightarrow$

12. $x^2 - 2x - 15 \geq 0$

$\xleftarrow{\hspace{1cm}\underset{0}{+}\hspace{1cm}}\rightarrow$

13. $x^2 - 5x + 6 \geq 0$

$\xleftarrow{\hspace{1cm}\underset{0}{+}\hspace{1cm}}\rightarrow$

14. $x^2 + 7x + 10 \leq 0$

$\xleftarrow{\hspace{1cm}\underset{0}{+}\hspace{1cm}}\rightarrow$

15. $x^2 + 2x \leq 24$

$\xleftarrow{\hspace{1cm}\underset{0}{+}\hspace{1cm}}\rightarrow$

16. $x^2 - 3x > 18$

$\xleftarrow{\hspace{1cm}\underset{0}{+}\hspace{1cm}}\rightarrow$

17. $x^2 > 27 - 6x$

$\xleftarrow{\hspace{1cm}\underset{0}{+}\hspace{1cm}}\rightarrow$

18. $x^2 \leq 7x - 12$

$\xleftarrow{\hspace{1cm}\underset{0}{+}\hspace{1cm}}\rightarrow$

19. $2x^2 + x - 6 \leq 0$

$\xleftarrow{\hspace{1cm}\underset{0}{+}\hspace{1cm}}\rightarrow$

20. $3x^2 - 10x - 8 < 0$

$\xleftarrow{\hspace{1cm}\underset{0}{+}\hspace{1cm}}\rightarrow$

21. $4x^2 + x < 3$

$\xleftarrow{\hspace{1cm}\underset{0}{+}\hspace{1cm}}\rightarrow$

22. $5x^2 - 13x \geq 6$

$\xleftarrow{\hspace{1cm}\underset{0}{+}\hspace{1cm}}\rightarrow$

23. $x^2 - 16 \leq 0$

$\xleftarrow{\hspace{1cm}\underset{0}{+}\hspace{1cm}}\rightarrow$

24. $x^2 - 9 > 0$

$\xleftarrow{\hspace{1cm}\underset{0}{+}\hspace{1cm}}\rightarrow$

25. $x^2 \geq 25$

$\xleftarrow{\hspace{1cm}\underset{0}{+}\hspace{1cm}}\rightarrow$

26. $x^2 < 49$

$\xleftarrow{\hspace{1cm}\underset{0}{+}\hspace{1cm}}\rightarrow$

27. $4 - x^2 < 0$

$\xleftarrow{\hspace{1cm}\underset{0}{+}\hspace{1cm}}\rightarrow$

28. $36 - x^2 \geq 0$

$\xleftarrow{\hspace{1cm}\underset{0}{+}\hspace{1cm}}\rightarrow$

29. $x^2 - 4x \leq 0$

$\xleftarrow{\hspace{1cm}\underset{0}{+}\hspace{1cm}}\rightarrow$

30. $x^2 + 5x > 0$

$\xleftarrow{\hspace{1cm}\underset{0}{+}\hspace{1cm}}\rightarrow$

31. $x^2 \geq 6x$

$\xleftarrow{\hspace{1cm}\underset{0}{+}\hspace{1cm}}\rightarrow$

32. $x^2 < 3x$

$\xleftarrow{\hspace{1cm}\underset{0}{+}\hspace{1cm}}\rightarrow$

33. $4x > x^2$

34. $6x \le x^2$

35. $x^2 - 4x + 4 \le 0$

36. $x^2 + 6x + 9 \ge 0$

37. $(x + 3)(x - 6) \le 10$

38. $(x + 4)(x - 5) > 22$

39. Can a quadratic inequality be solved if the quadratic member of the inequality is not factorable? If so, explain how the solution can be found. If not, explain why not.

40. Is it necessary to relate a quadratic inequality to 0 in order to solve it? Why or why not?

An inequality of the form

$$(x - a)(x - b)(x - c) < 0$$

can be solved by using a sign graph to consider the signs of *all three factors*. In Exercises 41 to 46, use this suggestion to solve each inequality. Then graph the solution set.

41. $x(x - 2)(x + 1) < 0$

42. $x(x + 3)(x - 2) \ge 0$

43. $(x - 3)(x + 2)(x - 1) \ge 0$

44. $(x - 5)(x + 1)(x - 4) < 0$

45. $x^3 - 2x^2 - 15x \le 0$

46. $x^3 + 2x^2 - 24x > 0$

Solve Exercises 47 to 50.

47. A small manufacturer's weekly profit is given by

$$P(x) = -2x^2 + 220x$$

where x is the number of items manufactured and sold. Find the number of items that must be manufactured and sold if the profit is to be greater than or equal to $6000.

48. Suppose that a company's profit is given by

$$P(x) = -2x^2 + 360x$$

How many items must be produced and sold so that the profit will be at least $16,000?

49. If a ball is thrown vertically upward from the ground with an initial velocity of 80 ft/s, its approximate height is given by

$$h(t) = -16t^2 + 80t$$

where t is the time (in seconds) after the ball was released. When will the ball have a height of at least 96 ft?

50. Suppose a ball's height (in meters) is given by

$$h(t) = -5t^2 + 20t$$

When will the ball have a height of at least 15 m?

 In Exercises 51 to 54, use your calculator to find the graph for each equation. Use that graph to estimate the solution of the related inequality.

51. $y = x^2 + 6x$; $x^2 + 6x < 0$

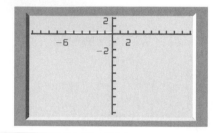

52. $y = x^2 - 49$; $x^2 - 49 \geq 0$

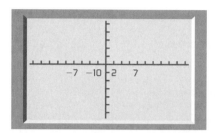

53. $y = x^2 - 5x - 6$; $x^2 - 5x - 6 \leq 0$

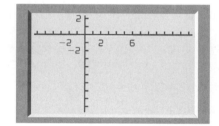

54. $y = 2x^2 + 7x - 15$; $2x^2 + 7x - 15 > 0$

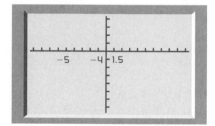

Answers

1.
-4 0 3
$-4 < x < 3$

3.
-4 0 3
$x < -4 \text{ or } x > 3$

5.
-4 0 3
$-4 \leq x \leq 3$

7.
-4 0 3
$x \leq -4 \text{ or } x \geq 3$

9.
-1 0 4
$x < -1 \text{ or } x > 4$

11.
-4 0 3
$-4 \leq x \leq 3$

13.
0 2 3
$x \leq 2 \text{ or } x \geq 3$

15.
-6 0 4
$-6 \leq x \leq 4$

17.
-9 0 3
$x < -9 \text{ or } x > 3$

19.
$-2 \leq x \leq \frac{3}{2}$

21.
-1 0
$-1 < x < \frac{3}{4}$

23.
-4 0 4
$-4 \leq x \leq 4$

25.
-5 0 5
$x \leq -5 \text{ or } x \geq 5$

27.
-2 0 2
$x < -2 \text{ or } x > 2$

29.
0 4
$0 \leq x \leq 4$

31.
0 6
$x \leq 0 \text{ or } x \geq 6$

33.
0 4
$0 < x < 4$

35.
0 2
$x = 2$

37.
-4 0 7
$-4 \leq x \leq 7$

39.

41.
-1 0 2
$x < -1 \text{ or } 0 < x < 2$

43.
-2 0 1 3
$-2 \leq x \leq 1 \text{ or } x \geq 3$

45.
-3 0 5
$x \leq -3 \text{ or } 0 \leq x \leq 5$

47. $50 \leq x \leq 60$ **49.** $2 \leq t \leq 3$ **51.** $-6 < x < 0$ **53.** $-1 \leq x \leq 6$

The Circle

6.6 OBJECTIVES

1. Identify the graph of an equation as a line, a parabola, or a circle
2. Write the equation of a circle in standard form and graph the circle

In Sections 6.1 and 6.2, we examined the parabola. In this section, we turn our attention to another conic section, the circle.

> A **circle** is the set of all points in the plane equidistant from a fixed point, called the **center** of the circle. The distance between the center of the circle and any point on the circle is called the **radius** of the circle.

The distance formula is central to any discussion of conic sections.

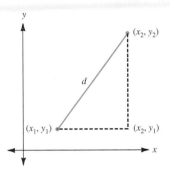

The Distance Formula

The distance, d, between two points (x_1, y_1) and (x_2, y_2) is given by

$$d = \sqrt{(x_2 - x_1)^2 + (y_2 - y_1)^2}$$

We can use the distance formula to derive the algebraic equation of a circle, given its center and its radius.

Suppose a circle has its center at a point with coordinates (h, k) and radius r. If (x, y) represents any point on the circle, then, by its definition, the distance from (h, k) to (x, y) is r. Applying the distance formula, we have

$$r = \sqrt{(x - h)^2 + (y - k)^2}$$

Squaring both sides of the equation gives the equation of the circle

$$r^2 = (x - h)^2 + (y - k)^2$$

In general, we can write the following equation of a circle.

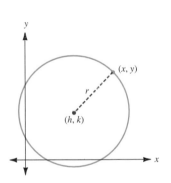

A special case is the circle centered at the origin with radius r. Then $(h, k) = (0, 0)$, and its equation is

$$x^2 + y^2 = r^2$$

Equation of a Circle

The equation of a circle with center (h, k) and radius r is

$$(x - h)^2 + (y - k)^2 = r^2 \qquad (1)$$

Equation (1) can be used in two ways. Given the center and radius of the circle, we can write its equation; or given its equation, we can find the center and radius of a circle.

●Example 1

Finding the Equation of a Circle

Find the equation of a circle with center at $(2, -1)$ and radius 3. Sketch the circle.

Let $(h, k) = (2, -1)$ and $r = 3$. Applying equation (1) yields

481

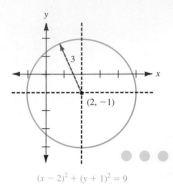

$(x - 2)^2 + (y + 1)^2 = 9$

$$(x - 2)^2 + [y - (-1)]^2 = 3^2$$
$$(x - 2)^2 + (y + 1)^2 = 9$$

To sketch the circle, we locate the center of the circle. Then we determine four points 3 units to the right and left and up and down from the center of the circle. Drawing a smooth curve through those four points completes the graph.

● ● ● **CHECK YOURSELF 1**

Find the equation of the circle with center at $(-2, 1)$ and radius 5. Sketch the circle.

Now, given an equation for a circle, we can also find the radius and center and then sketch the circle. We start with an equation in the special form of equation (1).

● Example 2

Finding the Center and Radius of a Circle

Find the center and radius of the circle with equation

$$(x - 1)^2 + (y + 2)^2 = 9$$

Remember, the general form is

$$(x - h)^2 + (y - k)^2 = r^2$$

Our equation "fits" this form when it is written as

Note: $y + 2 = y - (-2)$

$$(x - 1)^2 + [y - (-2)]^2 = 3^2$$

So the center is at $(1, -2)$, and the radius is 3. The graph is shown.

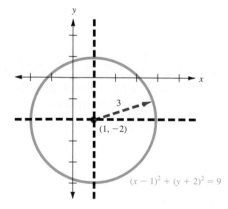

$(x - 1)^2 + (y + 2)^2 = 9$

Just as was true of the parabola in Section 6.4, the circle can be graphed on the calculator by solving for y, then graphing both the upper half and lower half of the circle. In this case,

$$(x - 1)^2 + (y + 2)^2 = 9$$
$$(y + 2)^2 = 9 - (x - 1)^2$$
$$(y + 2) = \pm \sqrt{9 - (x - 1)^2}$$
$$y = -2 \pm \sqrt{9 - (x - 1)^2}$$

Now graph the two functions

$$y = -2 + \sqrt{9 - (x - 1)^2}$$

and

$$y = -2 - \sqrt{9 - (x - 1)^2}$$

on your calculator. (The display screen may need to be squared to obtain the shape of a circle.)

● ● ● **CHECK YOURSELF 2**

Find the center and radius of the circle with equation

$$(x + 3)^2 + (y - 2)^2 = 16$$

Sketch the circle.

To graph the equation of a circle that is not in standard form, we *complete the square*. Let's see how completing the square can be used in graphing the equation of a circle.

● Example 3

To recognize the equation as having the form of a circle, note that the coefficients of x^2 and y^2 are equal.

The linear terms in x and y show a translation of the center away from the origin.

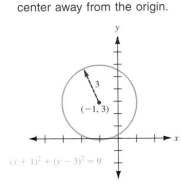

$(x + 1)^2 + (y - 3)^2 = 9$

Finding the Center and Radius of a Circle

Find the center and radius of the circle with equation

$$x^2 + 2x + y^2 - 6y = -1$$

Then sketch the circle.

We could, of course, simply substitute values of x and try to find the corresponding values for y. A much better approach is to rewrite the original equation so that it matches the standard form.

First, add 1 to both sides to complete the square in x.

$$x^2 + 2x + 1 + y^2 - 6y = -1 + 1$$

Then add 9 to both sides to complete the square in y.

$$x^2 + 2x + 1 + y^2 - 6y + 9 = -1 + 1 + 9$$

We can factor the two trinomials on the left (they are both perfect squares) and simplify on the right.

$$(x + 1)^2 + (y - 3)^2 = 9$$

The equation is now in standard form, and we can see that the center is at $(-1, 3)$ and the radius is 3. The sketch of the circle is shown. Note the "translation" of the center to $(-1, 3)$.

● ● ● **CHECK YOURSELF 3**

Find the center and radius of the circle with equation

$$x^2 - 4x + y^2 + 2y = -1$$

Sketch the circle.

● ● ● **CHECK YOURSELF ANSWERS**

1. $(x + 2)^2 + (y - 1)^2 = 25.$

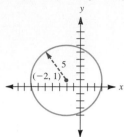

2. $(x + 3)^2 + (y - 2)^2 = 16.$

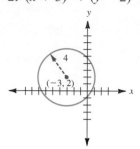

3. $(x - 2)^2 + (y + 1)^2 = 4.$

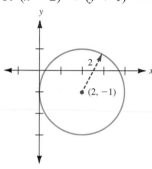

6.6 Exercises

In Exercises 1 to 12, decide whether each equation has as its graph a line, a parabola, a circle, or none of these.

1. $y = x^2 - 2x + 5$

2. $y^2 + x^2 = 64$

3. $y = 3x - 2$

4. $2y - 3x = 12$

5. $(x - 3)^2 + (y + 2)^2 = 10$

6. $y + 2(x - 3)^2 = 5$

7. $x^2 + 4x + y^2 - 6y = 3$

8. $4x = 3$

9. $y^2 - 4x^2 = 36$

10. $x^2 + (y - 3)^2 = 9$

11. $y = -2x^2 + 8x - 3$

12. $2x^2 - 3y^2 + 6y = 13$

In Exercises 13 to 20, find the center and the radius for each circle.

13. $x^2 + y^2 = 25$

14. $x^2 + y^2 = 72$

15. $(x - 3)^2 + (y + 1)^2 = 16$

16. $(x + 3)^2 + y^2 = 81$

17. $x^2 + 2x + y^2 = 15$

18. $x^2 + y^2 - 6y = 72$

19. $x^2 - 6x + y^2 + 8y = 16$ **20.** $x^2 - 5x + y^2 - 3y = 8$

In Exercises 21 to 32, graph each circle by finding the center and the radius.

21. $x^2 + y^2 = 4$

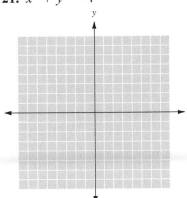

22. $x^2 + y^2 = 25$

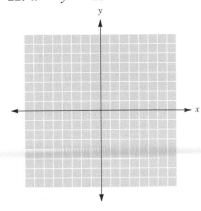

23. $4x^2 + 4y^2 = 36$

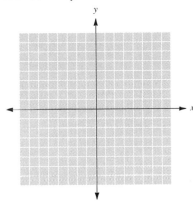

24. $9x^2 + 9y^2 = 144$

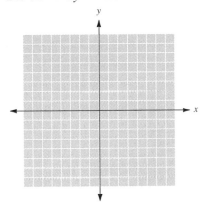

25. $(x - 1)^2 + y^2 = 9$

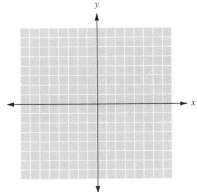

26. $x^2 + (y + 2)^2 = 16$

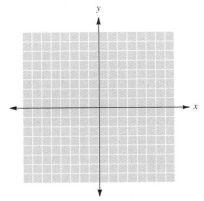

27. $(x - 4)^2 + (y + 1)^2 = 16$

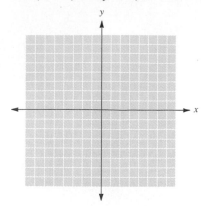

28. $(x + 3)^2 + (y + 2)^2 = 25$

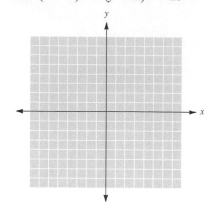

29. $x^2 + y^2 - 4y = 12$

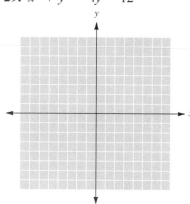

30. $x^2 - 6x + y^2 = 0$

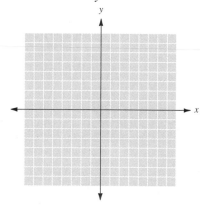

31. $x^2 - 4x + y^2 + 2y = -1$

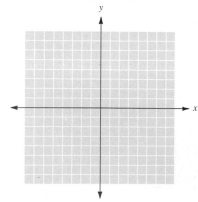

32. $x^2 - 2x + y^2 - 6y = 6$

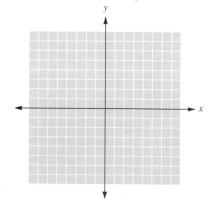

33. Describe the graph of $x^2 + y^2 - 2x - 4y + 5 = 0$.

34. Describe how completing the square is used in graphing circles.

35. A solar oven is constructed in the shape of a hemisphere. If the equation

$$x^2 + y^2 + 500 = 1000$$

describes the circumference of the oven in centimeters, what is its radius?

36. A solar oven in the shape of a hemisphere is to have a diameter of 80 cm. Write the equation that describes the circumference of this oven.

37. A solar water heater is constructed in the shape of a half cylinder, with the water supply pipe at its center. If the water heater has a diameter of $\dfrac{4}{3}$ m, what is the equation that describes its circumference?

38. A solar water heater is constructed in the shape of a half cylinder having a circumference described by the equation

$$9x^2 + 9y^2 - 16 = 0$$

What is its diameter if the units for the equation are meters?

 A circle can be graphed on a calculator by plotting the upper and lower semicircles on the same axes. For example, to graph $x^2 + y^2 = 16$, we solve for y:

$$y = \pm\sqrt{16 - x^2}$$

This is then graphed as two separate functions,

$$y = \sqrt{16 - x^2} \qquad \text{and} \qquad y = -\sqrt{16 - x^2}$$

In Exercises 39 to 42, use that technique to graph each circle.

39. $x^2 + y^2 = 36$

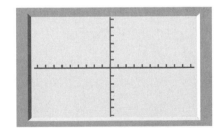

40. $(x - 3)^2 + y^2 = 9$

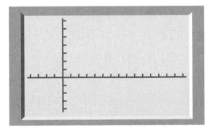

41. $(x + 5)^2 + y^2 = 36$

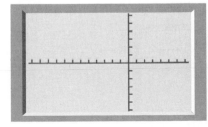

42. $(x - 2)^2 + (y + 1)^2 = 25$

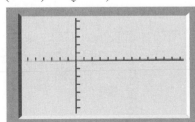

Answers

1. Parabola **3.** Line **5.** Circle **7.** Circle **9.** None of these **11.** Parabola **13.** Center: (0, 0);
radius: 5 **15.** Center: (3, −1); radius: 4 **17.** Center: (−1, 0); radius: 4 **19.** Center: (3, −4); radius: $\sqrt{41}$

21. $x^2 + y^2 = 4$

Center: (0, 0); radius: 2

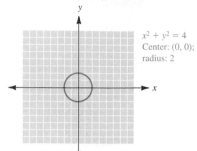

23. $4x^2 + 4y^2 = 36$
$$x^2 + y^2 = 9$$
Center: (0, 0); radius: 3

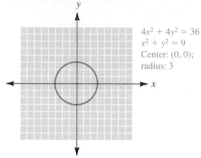

25. $(x − 1)^2 + y^2 = 9$

Center: (1, 0); radius: 3

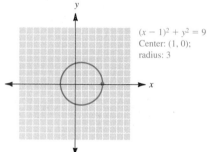

27. $(x − 4)^2 + (y + 1)^2 = 16$

Center: (4, −1); radius: 4

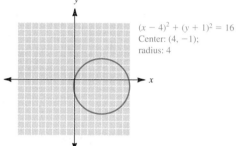

29. $x^2 + y^2 − 4y = 12$

Center: (0, 2); radius: 4

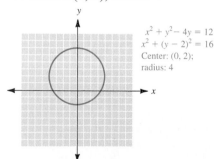

31. $x^2 − 4x + y^2 + 2y = −1$

Center: (2, −1); radius: 2

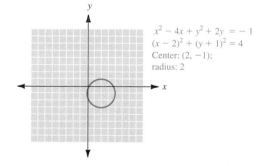

33. 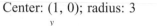 **35.** $\sqrt{500} = 10\sqrt{5}$ cm $\cong$ 22.4 cm **37.** $x^2 + y^2 = \dfrac{4}{9}$

39. $x^2 + y^2 = 36$

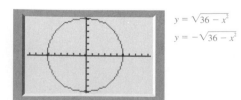

41. $(x + 5)^2 + y^2 = 36$

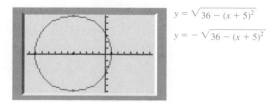

Polynomial Prediction Equations

6.7 OBJECTIVES

1. Use a scatter plot to determine regression type
2. Use a TI-83 to find the polynomial prediction equation

The following illustration shows the location of a ball that was thrown into the air. The pictures were taken every 0.5 s. Use the picture to estimate the height of the ball after 1.25 s.

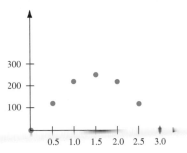

You probably estimated the height of the ball as about 230 ft. You did that by fitting a curve over the pictures. This curve can be interpreted as a prediction equation, which we discussed in Section 1.9. However, the curve presented here is different from the curves presented earlier. It is a **polynomial prediction equation** because the graph is not linear; it appears to be a parabola.

• Example 1

Graphing a Scatter Plot

The following data were collected as part of a physics class experiment. Graph the associated ordered pairs on a scatter plot, then determine whether the points appear to lie on a line.

Time	Height
0	320
2	430
4	405
6	260
8	0

Below is the scatter plot with the five points.

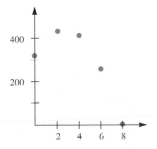

The points are definitely not colinear (on the same line) or even close to the same line. To make a prediction with these data, we will have to find a model that is not linear.

● ● ● CHECK YOURSELF 1

The following data were collected as part of a physics class experiment. Graph the associated ordered pairs on a scatter plot, then determine whether the points appear to lie on a line.

Time	Height
0	520
2	550
4	450
6	214
7	0

Recall, from Section 1.9, that the data could be entered into the TI-82 or TI-83 calculator by using STAT and EDIT. Enter the data from Example 1 in preparation for Example 2.

● Example 2

Finding the Equation of a Quadratic Prediction Equation

A parabola is what we call the graph of a quadratic function.

The scatter plot for the data in Example 1 suggested that the prediction curve was not linear. Its shape is closer to a parabola. Use that information to determine the prediction equation.

With the data in List 1 and List 2, select STAT and use the right arrow to highlight CALC. A parabola is the graph of a quadratic equation. Select 5 (QuadReg) to find the coefficients for the prediction equation. Rounding each to the nearest integer, we have $a = -15$, $b = 82$, and $c = 322$.

The prediction equation is $h(x) = -15\,x^2 + 82\,x + 322$

● ● ● CHECK YOURSELF 2

The scatter plot for the data in Check Yourself 1 suggested that the prediction curve was a parabola. Use that information to determine the prediction equation.

The following algorithm summarizes the method we have used to find the equation of a quadratic prediction equation.

Finding the Equation of a Quadratic Prediction Equation

STEP 1 Enter the data in List 1 and List 2.
STEP 2 Press STAT and select CALC.

STEP 3   Select $\boxed{5}$ (QuadReg).
STEP 4   Use a, b, and c to find the suggested equation,

$$P(x) = ax^2 + bx + c$$

In Example 3, you will use your calculator to graph both the scatter plot and the prediction equation.

• Example 3

Graphing the Prediction Equation

Use the TI-83 calculator to graph the scatter plot and the prediction equation for the data in Example 1.

1. Enter the data in List 1 and List 2.
2. Access $\boxed{\text{STAT}}$ $\boxed{\text{PLOT}}$ with $\boxed{2^{nd}}$ $\boxed{\text{Y}}$ $\boxed{=}$.
3. Turn Plot 1 on and select the first graph type.
4. Use the WINDOW to describe an appropriate viewing window. (x could be 0 to 10 and y could be 0 to 500.)
5. Go to $\boxed{\text{Y}=}$ and enter the prediction equation $Y = -15\,x^2 + 82\,x + 322$.
6. Select $\boxed{\text{GRAPH}}$.

CHECK YOURSELF 3

Use the TI-83 calculator to graph the scatter plot and the prediction equation for the data in Check Yourself 1.

One of the most important steps in determining a prediction equation is identifying the type of curve from the scatter plot. See if you can guess the curve type in Example 4.

• Example 4

Identifying Polynomial Curves

Use the TI calculator to graph the scatter plot for the following data, then identify the type of curve that best fits the points.

x	y
-1	8
0	4
1	2
3	5
4	3
5	-1

Entering the data into List 1 and List 2 by using $\boxed{\text{STAT}}$ $\boxed{\text{EDIT}}$ and then using $\boxed{\text{STAT}}$ $\boxed{\text{PLOT}}$ yields the following scatter plot.

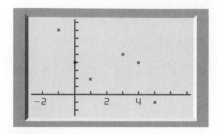

The points go down, up, then down again. Clearly, the curve is neither a line (the graph of a first-degree polynomial) nor a parabola (the graph of a second-degree polynomial). Curves with two turns (down-up-down or up-down-up) are best modeled with a third-degree polynomial, something of the form $f(x) = ax^3 + bx^2 + cx + d$.

● ● ● **CHECK YOURSELF 4**

Use the TI-82 or TI-83 calculator to graph the scatter plot for the following data, then identify the type of curve that best fits the points.

x	y
−1	9
0	5
1	3
3	5
4	7
5	10

In Example 5, we look at a third-degree polynomial, which we call a **cubic prediction equation.**

● Example 5

Finding a Prediction Equation

Find the prediction equation for the data in Example 4. With the data in List 1 and List 2, select $\boxed{\text{STAT}}$ and $\boxed{\text{CALC}}$. Choose option 6 (CubicReg). The coefficients are $a = -0.3$, $b = 1.8$, $c = -2.5$, and $d = 3.5$. The prediction equation is

$$f(x) = -0.3x^3 + 1.8x^2 - 2.5x + 3.5$$

Enter the prediction equation as $\boxed{Y_1}$. Graphing the scatter plot and the prediction equation yields

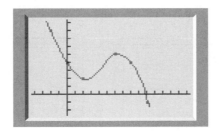

● ● ● **CHECK YOURSELF 5**

Find the prediction equation for the data in Check Yourself 4.

● ● ● **CHECK YOURSELF ANSWERS**

1. No, it is not linear.   **2.** $y = -19x^2 + 59x + 516$.

3.

The curve appears to be a parabola.

4.

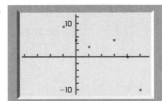

The curve appears to be cubic

5. $y = -0.4x^3 + 2x^2 - 2.3x + 4.4$

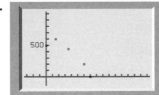

Designing Basketball Boxes. You and your partner have a small business that manufactures packaging materials. You have just received a request to bid on a contract to make boxes for packaging inflated basketballs that are NBA regulation size.

1. What shape box would you design? Draw a picture of the box and write in the dimensions. (You may have to do some research about the size of NBA regulation basketballs.) Make a plan for cutting and folding these boxes out of one piece of cardboard. What size sheet of cardboard will be needed? How will it be cut to allow the box to be folded into the size and shape you have designed? Will you need flaps for gluing, or will your box hold together without glue?

2. What is the surface area of the box? This is important information for both the costs and the design of the logo that is to go on the boxes. If the cost of the cardboard is $1.50 for a 4 × 8 ft sheet, what will be the cost of the cardboard used to produce 100,000 of these boxes? You will want to place the box shapes on the 4 × 8 ft sheet of cardboard in the most efficient way. You may want to reconsider your plan for the box. Would it be less wasteful and more economical to design the box so that it is made out of two separate shapes and then glued together? The costs of gluing each box can be ignored if you were planning to glue your first design.

3. Write a description of your proposal for the box, giving dimensions, a pattern for the box, and showing the way the patterns would be cut from a 4 × 8 ft of cardboard. Summarize the cost of materials in your proposal. Write your proposal in the form of a bid for the manufacturing contract for the boxes.

Designing Pizza Boxes. Godmother's Pizza, a national chain, has asked for a bid from your company for pizza boxes. You order cardboard in 4 × 3 ft sheets and so have decided to make a pizza box from a 36 × 24 in. piece of cardboard according to the following design.

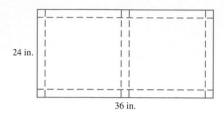

24 in.

36 in.

The solid lines indicate where the flaps will be cut to enable the paper to be folded and cut into a box. Try making a model of this box with a piece of paper so you see how the pizza box is to be made.

The height of the box will vary, depending on the length of the sides of the squares at the corners.

Decide what expressions to use for the width and length of the completed box. These will be in terms of the height.

The width of the pizza box will be _____.

The length of the box will be _____.

The height of the box will be _____.

The volume of the box will be _____.

Now, complete the following.

1. The Godmother's Pizza representative says that they want to be sure that the box will hold their 16-inch deluxe pizza. How tall can the box be and still hold the pizza? Write the representative a memo and explain what the dimensions of the box will be. Draw a picture of the completed box with the dimensions written on it so the rep will be reassured.
2. Make a graph showing what the maximum volume will be, and tell Godmother's what the surface area of this size box will be.
3. The Godmother's rep responds to your memo and says they are concerned because the deluxe pizza is 1.5 inches thick. Will your design work, or is it back to the drawing board?

6.7 Exercises

1. The following data were collected as part of an experiment.

Time (s)	Height (ft)
0	75
2	190
4	260
6	190
8	0

(a) Graph the ordered pairs on a scatter plot, then determine whether the points appear to lie on a line.
(b) Use the information from part **(a)** to determine the prediction equation.

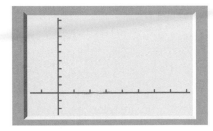

2. The following data were recently collected as part of an experiment with freely falling bodies.

Time (s)	Height (m)
0	17
2	65
4	125
6	110
7	65
8	25
9	0

(a) Graph the points on a scatter plot, and then determine whether the points lie on a line.
(b) Use the information from part **(a)** to determine the prediction equation.

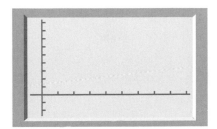

In Exercises 3 and 4, use the TI-82 or TI-83 to graph the scatter plot and determine the prediction equation.

3.

x	y
−3	−37
−2	−11
−1	−3
0	9
2	4
3	−3
4	−9
5	−4

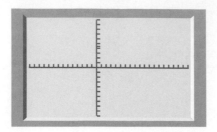

4.

x	y
−2	−33
−1	−10
0	−3
1	−4
2	−2
3	15
4	60

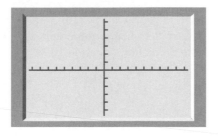

5. Here are some figures gathered for the section of highway discussed in the chapter opener.

Distance in km	Altitude in m
0	75
5	75.75
12	350
12.5	342.6
16	320

Try using your calculator to fit a polynomial to these points. How do your graphs compare with the graph of the road depicted in the introduction to the chapter? Why do the civil engineers use different polynomials for different sections of the road?

Answers

1.

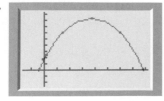

Prediction equation:
$y = -13t^2 + 100t + 66$

3.

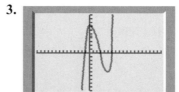

Prediction equation:
$y = .41x^3 - 2.9x^2 + 1.99x + (

5.

Summary

Solving Quadratic Equations by Factoring [6.1]

To solve:

$2x^2 - x = 15$

$2x^2 - x - 15 = 0$

$(2x + 5)(x - 3) = 0$
$2x + 5 = 0 \quad$ or $\quad x - 3 = 0$
$2x = -5 \qquad\qquad x = 3$
$x = -\dfrac{5}{2}$

To Solve a Quadratic Equation by Factoring

You can use the following procedure.

1. Write the equation in standard form:

   $$ax^2 + bx + c = 0 \qquad a \neq 0$$

   where a, b, and c are constants.

2. Factor the polynomial on the left as a product of linear factors.
3. Use the zero-product principle to set each factor equal to zero.
4. Solve the resulting linear equations to obtain solutions for the original quadratic equation.

Solving Quadratic Equations by the Square Root Method and by Completing the Square [6.2]

To solve:

$(x - 3)^2 = 5$
$x - 3 = \pm \sqrt{5}$
$x = 3 \pm \sqrt{5}$

Square Root Property

If $x^2 = k$, where k is a complex number, then $x = \sqrt{k} \qquad$ or $\qquad x = -\sqrt{k}$

To solve:

$2x^2 + 2x - 1 = 0$
$2x^2 + 2x = 1$
$x^2 + x = \dfrac{1}{2}$
$x^2 + x + \left(\dfrac{1}{2}\right)^2 = \dfrac{1}{2} + \left(\dfrac{1}{2}\right)^2$
$x^2 + x + \dfrac{1}{4} = \dfrac{1}{2} + \dfrac{1}{4}$
$\left(x + \dfrac{1}{2}\right)^2 = \dfrac{3}{4}$
$x + \dfrac{1}{2} = \pm\sqrt{\dfrac{3}{4}}$
$x + \dfrac{1}{2} = \pm\dfrac{\sqrt{3}}{2}$
$x = \dfrac{-1 \pm \sqrt{3}}{2}$

Completing the Square

1. Isolate the constant on the right side of the equation.
2. Divide both sides of the equation by the coefficient of the x^2 term if that coefficient is not equal to 1.
3. Add the square of one-half of the coefficient of the linear term to both sides of the equation. This will give a perfect-square trinomial on the left side of the equation.
4. Write the left side of the equation as the square of a binomial, and simplify on the right side.
5. Use the square root property, and then solve the resulting linear equations.

The Quadratic Formula [6.3]

To solve

$$x^2 - 2x = 4$$

Write the equation as

$$x^2 - 2x - 4 = 0$$

$$a = 1 \quad b = -2 \quad c = -4$$

$$x = \frac{-(-2) \pm \sqrt{(-2)^2 - 4(1)(-4)}}{2 \cdot 1}$$

$$= \frac{2 \pm \sqrt{20}}{2}$$

$$= \frac{2 \pm 2\sqrt{5}}{2}$$

$$= 1 \pm \sqrt{5}$$

Any quadratic equation can be solved by using the following algorithm.

1. Write the equation in standard form (set it equal to 0).

$$ax^2 + bx + c = 0$$

2. Determine the values for a, b, and c.
3. Substitute those values into the quadratic formula

$$x = \frac{-b \pm \sqrt{b^2 - 4ac}}{2a}$$

4. Write the solutions in simplest form.

The Discriminant

The expression $b^2 - 4ac$ is called the **discriminant** for a quadratic equation. There are three possibilities:

1. If $b^2 - 4ac < 0$, there are no real solutions (but two complex solutions).
2. If $b^2 - 4ac = 0$, there is one real solution (a double solution).
3. If $b^2 - 4ac > 0$, there are two distinct real solutions.

Graphing Parabolas [6.4]

The graph of

$$f(x) = ax^2 + bx + c \qquad a \neq 0$$

is a parabola.

The parabola opens upward if $a > 0$.   The parabola opens downward if $a < 0$.

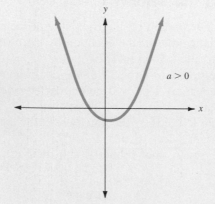

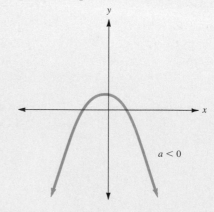

The **vertex** of the parabola (either the highest or the lowest point on the graph) is on the **axis of symmetry** with the equation

$$x = -\frac{b}{2a}$$

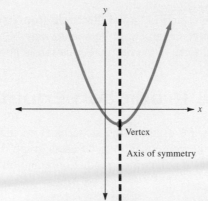

To graph $f(x) = x^2 - 4x + 3$

Axis of symmetry

$$x = -\frac{(-4)}{2(1)}$$

$$x = 2$$

Vertex

$(2, f(2))$

$(2, -1)$

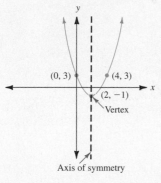

$x = 2$

To graph a parabola:

STEP 1 Find the axis of symmetry.
STEP 2 Find the vertex.
STEP 3 Determine two symmetric points.
 Note: You can use the x intercepts if the quadratic member of the given equation is factorable and the x intercepts are distinct. Otherwise, use the y intercept and its symmetric point.
STEP 4 Draw a smooth curve connecting the points found above to form the parabola. You may choose to find additional pairs of symmetric points at this time.

The graph of

$$x = ay^2 + by + c \quad a \neq 0$$

is a parabola. If $a > 0$, the parabola opens rightward. If $a < 0$, the parabola opens leftward.

Quadratic Inequalities in One Variable [6.5]

To solve

$$x^2 - 3x < 18$$

$$x^2 - 3x - 18 < 0$$

$$(x - 6)(x + 3) < 0$$

Critical points are

-3 and 6

Quadratic inequalities may be solved by first relating the quadratic member to 0. The quadratic member is then factored.

Critical points are determined by noting where the factors are equal to 0, and a sign graph is drawn to indicate where the factors are negative or positive. From this graph and the sense of the original inequality (the quadratic member will be positive or negative), the solution set is determined.

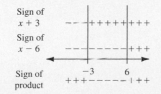

The solution set is

$\{x \mid -3 < x < 6\}$

The Circle [6.6]

The standard form for the circle with center (h, k) and radius r is

Given the equation

$(x - 2)^2 + (y + 3)^2 = 4$

we see that the center is at $(2, -3)$ and the radius is 2.

$$(x - h)^2 + (y - k)^2 = r^2$$

Determining the center and radius of the circle from its equation allows us to easily graph the circle. **Note:** Completing the square may be used to derive an equivalent equation in standard form if the original equation is not in this form.

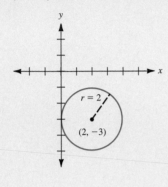

Polynomial Prediction Equations [6.7]

Prediction equations are used to find patterns from incomplete data sets. To determine the type of prediction equation to be used, graph the scatter plot first. If the relation is nonlinear, try a polynomial prediction equation. To use a TI-83 to find such an equation:

1. Enter the data in List 1 and List 2.
2. Press STAT and select CALC.
3. Select 5 (QuadReg).
4. Use a, b, and c to find the suggested equation

$$P(x) = ax^2 + bx + c$$

Summary Exercises

This summary exercise set is provided to give you practice with each of the objectives in the chapter. Each exercise is keyed to the appropriate chapter section. The answers are provided in the *Instructor's Manual.*

[6.1]   Solve each of the following quadratic equations by factoring.

1. $x^2 + 5x - 6 = 0$
2. $x^2 - 2x - 8 = 0$
3. $x^2 + 7x = 30$

4. $x^2 - 6x = 40$
5. $x^2 = 11x - 24$
6. $x^2 = 28 - 3x$

7. $x^2 - 10x = 0$
8. $x^2 - 12x$
9. $x^2 - 25 = 0$

10. $x^2 = 144$
11. $2x^2 - x - 3 = 0$
12. $3x^2 - 4x = 15$

13. $3x^2 + 9x - 30 = 0$
14. $4x^2 + 24x = -32$
15. $x(x - 3) = 18$

16. $(x - 2)(2x + 1) = 33$
17. $x^3 - 2x^2 - 15x = 0$
18. $x^3 + x^2 - 4x - 4 = 0$

19. Suppose that the cost, in dollars, of producing x stereo systems is given by

$$C(x) = 3000 - 60x + 3x^2$$

How many systems can be produced for \$7500?

Solve each of the following applications. Where appropriate, give your answer to the nearest tenth of a unit.

20. The demand equation for a certain type of computer paper is predicted to be

$$D = -3p + 60$$

The supply equation is predicted to be

$$S = -p^2 + 24p - 3$$

Find the equilibrium price.

[6.2]   Solve each of the following equations, using the square root method.

21. $x^2 - 8 = 0$
22. $3y^2 - 15 = 0$
23. $(x - 2)^2 = 20$

24. $(2x + 1)^2 - 10 = 0$

Find the constant that must be added to each of the following binomials to form a perfect-square trinomial.

25. $x^2 - 12x$ **26.** $y^2 + 3y$

Solve the following equations by completing the square.

27. $x^2 - 4x - 5 = 0$ **28.** $x^2 + 8x + 12 = 0$ **29.** $w^2 - 10w - 3 = 0$

30. $y^2 + 3y - 1 = 0$ **31.** $2x^2 - 8x - 5 = 0$ **32.** $3x^2 + 3x - 1 = 0$

[6.3]   Solve each of the following equations by using the quadratic formula.

33. $x^2 - 5x - 24 = 0$ **34.** $w^2 + 10w + 25 = 0$ **35.** $x^2 = 3x + 3$

36. $2y^2 - 5y + 2 = 0$ **37.** $3y^2 + 4y = 1$ **38.** $2y^2 + 5y + 4 = 0$

39. $(x - 5)(x + 3) = 13$ **40.** $\dfrac{1}{x^2} - \dfrac{4}{x} + 1 = 0$ **41.** $3x^2 + 2x + 5 = 0$

42. $(x - 1)(2x + 3) = -5$

For each of the following quadratic equations, use the discriminant to determine the number of real solutions.

43. $x^2 - 3x + 3 = 0$ **44.** $x^2 + 4x = 2$ **45.** $4x^2 - 12x + 9 = 0$

46. $2x^2 + 3 = 3x$

47. Number Problem The sum of two integers is 12, and their product is 32. Find the two integers.

48. Number Problem The product of two consecutive, positive, even integers is 80. What are the two integers?

49. Number Problem Twice the square of a positive integer is 10 more than 8 times that integer. Find the integer.

50. Rectangles The length of a rectangle is 2 ft more than its width. If the area of the rectangle is 80 ft^2, what are the dimensions of the rectangle?

51. Rectangles The length of a rectangle is 3 cm less than twice its width. The area of the rectangle is 35 cm^2. Find the length and width of the rectangle.

52. **Rectangles** An open box is formed by cutting 3-in. squares from each corner of a rectangular piece of cardboard which is 3-in. longer than it is wide. If the box is to have a volume of 120 in.^{3}, what must be the size of the original piece of cardboard?

53. **Profit** Suppose that a manufacturer's weekly profit P is given by

$$P(x) = -3x^2 + 240x$$

where x is the number of items manufactured and sold. Find the number of items that must be manufactured and sold if the profit is to be at least $4500.

54. **Thrown Ball** If a ball is thrown vertically upward from the ground with an initial velocity of 64 ft/s, its approximate height is given by

$$h(t) = -16t^2 + 64t$$

When will the ball reach a height of at least 48 ft?

55. **Rectangle** The length of a rectangle is 1 cm more than twice its width. If the length is doubled, the area of the new rectangle is 36 cm^2 more than that of the old. Find the dimensions of the original rectangle.

56. **Triangle** One leg of a right triangle is 4 in. longer than the other. The hypotenuse of the triangle is 8 in. longer than the shorter leg. What are the lengths of the three sides of the triangle?

57. **Rectangle** The diagonal of a rectangle is 9 ft longer than the width of the rectangle, and the length is 7 ft more than its width. Find the dimensions of the rectangle.

58. **Thrown Ball** If a ball is thrown vertically upward from the ground, the height, h, after t seconds is given by

$$h(t) = 128t - 16t^2$$

(a) How long does it take the ball to return to the ground?

(b) How long does it take the ball to reach a height of 240 ft on the way up?

59. **Triangle** One leg of a right triangle is 2 m longer than the other. If the length of the hypotenuse is 8 m, find the length of the other two legs.

60. **Thrown Ball** Suppose that the height (in meters) of a golf ball, hit off a raised tee, is approximated by

$$h(t) = -5t^2 + 10t + 10$$

t seconds after the ball is hit. When will the ball hit the ground?

[6.1–6.3]   Find the zeros of the following functions.

61. $f(x) = x^2 - x - 2$ **62.** $f(x) = 6x^2 + 7x + 2$

63. $f(x) = -2x^2 - 7x - 6$ **64.** $f(x) = -x^2 - 1$

[6.4]   Find the equation of the axis of symmetry and the coordinates for the vertex of each of the following.

65. $f(x) = x^2$ **66.** $f(x) = x^2 + 2$ **67.** $f(x) = x^2 - 5$

68. $f(x) = (x - 3)^2$ **69.** $f(x) = (x + 2)^2$ **70.** $f(x) = -(x - 3)^2$

71. $f(x) = (x + 3)^2 + 1$ **72.** $f(x) = -(x + 2)^2 - 3$ **73.** $f(x) = -(x - 5)^2 - 2$

74. $f(x) = 2(x - 2)^2 - 5$ **75.** $f(x) = -x^2 + 2x$ **76.** $f(x) = x^2 - 4x + 3$

77. $f(x) = -x^2 - x + 6$ **78.** $f(x) = x^2 + 4x + 5$ **79.** $f(x) = -x^2 - 6x + 4$

80. $f(x) = -x^2 + 2x - 2$ **81.** $f(x) = 2x^2 - 4x + 1$ **82.** $x = y^2 - 4y$

83. $x = -y^2 + 4y$ **84.** $x = y^2 - 5y + 6$

Graph each of the following.

85. $f(x) = x^2$

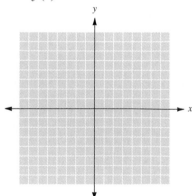

86. $f(x) = x^2 + 2$

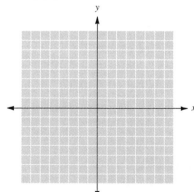

87. $f(x) = x^2 - 5$

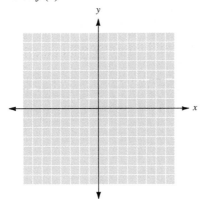

88. $f(x) = (x - 3)^2$

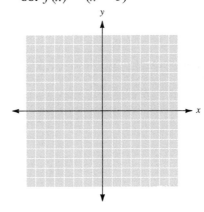

89. $f(x) = (x + 2)^2$

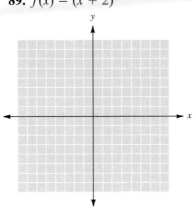

90. $f(x) = -(x - 3)^2$

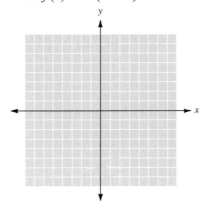

91. $f(x) = (x + 3)^2 + 1$

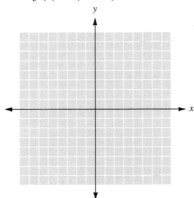

92. $f(x) = -(x + 2)^2 - 3$

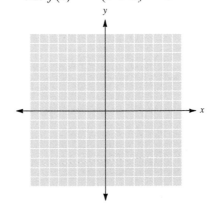

93. $f(x) = x^2 - 4x$

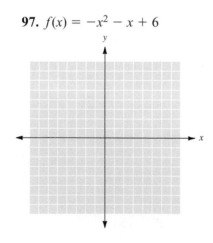

94. $f(x) = -x^2 + 2x$

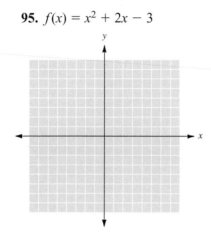

95. $f(x) = x^2 + 2x - 3$

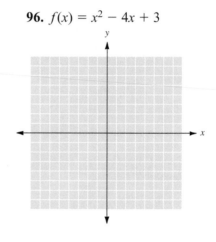

96. $f(x) = x^2 - 4x + 3$

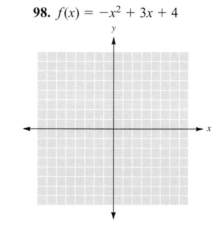

97. $f(x) = -x^2 - x + 6$

98. $f(x) = -x^2 + 3x + 4$

99. $f(x) = x^2 + 4x + 5$

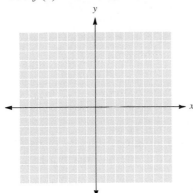

100. $f(x) = x^2 - 6x + 4$

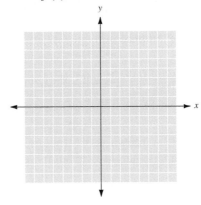

101. $f(x) = x^2 - 2x + 4$

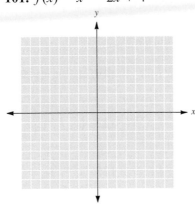

102. $f(x) = -x^2 + 2x - 2$

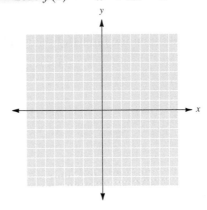

103. $f(x) = 2x^2 - 4x + 1$

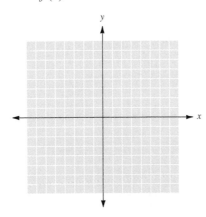

104. $f(x) = \dfrac{1}{2}x^2 - 4x$

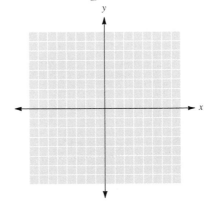

[6.5]   Solve the following inequalities and graph the solution set.

105. $(x - 2)(x + 5) > 0$ ⟵ ┼┼┼┼┼┼┼ ⟶
                              0

106. $(x - 1)(x - 6) < 0$ ⟵ ──── ┼┼┼┼┼┼ ⟶
                              0

107. $(x + 1)(x + 3) \leq 0$ ⟵ ┼┼┼┼ ⟶
                              0

108. $(x + 4)(x - 5) \geq 0$ ⟵ ┼┼┼┼┼┼┼┼┼┼ ⟶
                              0

109. $x^2 - 5x - 24 \leq 0$

110. $x^2 + 4x \geq 21$

111. $x^2 \geq 64$

112. $x^2 + 5x \geq 0$

113. $(x + 2)(x - 6) < 9$

114. $(x - 1)(x + 2) \geq 4$

115. If a ball is thrown vertically upward from the ground with an initial velocity of 64 ft/s its approximate height is given by $h(t) = -16t^2 + 64t$. When will the ball reach a height of at least 48 ft?

116. Suppose that the cost, in dollars, of producing x stereo systems is given by the equation $C(x) = 3000 - 60x + 3x^2$. How many systems can be produced if the cost cannot exceed \$7500?

[6.6]   Find the center and the radius of the graph of each equation.

117. $x^2 + y^2 = 16$

118. $x^2 + y^2 = 50$

119. $4x^2 + 4y^2 = 36$

120. $3x^2 + 3y^2 = 36$

121. $(x - 3)^2 + y^2 = 36$

122. $(x - 2)^2 + y^2 = 9$

123. $(x - 1)^2 + (y - 2)^2 = 16$

124. $x^2 + 6x + y^2 + 4y = 12$

125. $x^2 + 8x + y^2 + 10y = 23$

126. $x^2 - 6x + y^2 + 6y = 18$

Graph each of the following.

127. $x^2 + y^2 = 16$

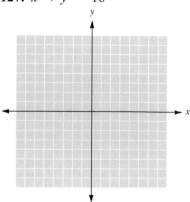

128. $4x^2 + 4y^2 = 36$

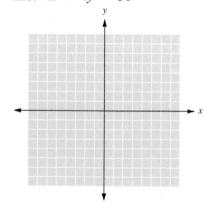

129. $x^2 + (y + 3)^2 = 25$

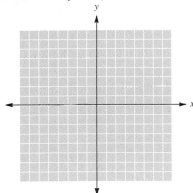

130. $(x - 2)^2 + y^2 = 9$

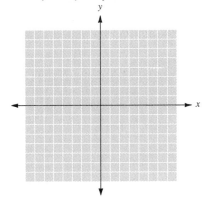

131. $(x - 1)^2 + (y - 2)^2 = 16$

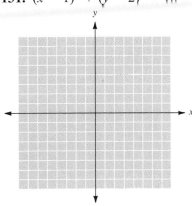

132. $(x + 3)^2 + (y + 3)^2 = 25$

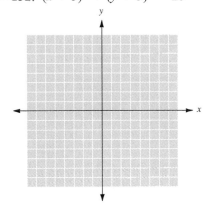

133. $x^2 + y^2 - 4y - 5 = 0$

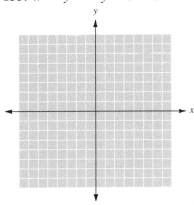

134. $x^2 - 2x + y^2 - 6y = 6$

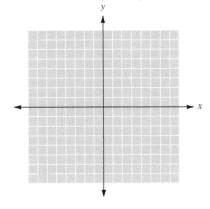

For each of the following equations decide whether its graph is a line, parabola, circle, or none of these.

135. $x + y = 16$

136. $x + y^2 = 5$

137. $4x^2 + 4y^2 = 36$

138. $3x + 3y = 36$

139. $y = (x - 3)^2$

140. $(x - 2)^2 + y^2 = 9$

141. $y = (x - 1)^2 + 1$

142. $x = y^2 + 4y + 4$

143. $\dfrac{x^2}{4} + \dfrac{y^2}{25} = 1$

144. $x^2 - 6x + y^2 + 6y = 18$

145. $\dfrac{x^2}{9} - \dfrac{y^2}{25} = 1$

146. $9x^2 - 4y^2 = 36$

147. $16x^2 + 4y^2 = 64$

148. $x^2 = -y^2 + 18$

149. $\dfrac{x^2}{4} - \dfrac{y^2}{9} = 1$

150. $4x^2 + 4y^2 = 36$

151. $4x - 6y = 12$

152. $3x^2 - y + 4 = 0$

[6.7] **153.** The following data were collected as part of an experiment.

Time	Height (ft)
1	2.5
2	3.1
3	4.8
4	9.2
5	17
6	23

Graph the ordered pairs on a scatter diagram, and determine the prediction equation.

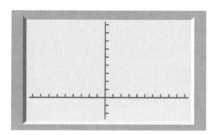

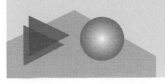

The purpose of this self-test is to help you check your progress and to review for a chapter test in class. Allow yourself about 1 hour to take the test. When you are done, check your answers in the back of the book. If you missed any answers, be sure to go back and review the appropriate sections in the chapter and the exercises that are provided.

Solve each of the following equations by factoring.

1. $2x^2 + 7x + 3 = 0$

2. $6x^2 = 10 - 11x$

3. $4x^3 - 9x = 0$

Solve each of the following equations, using the square root method.

4. $4w^2 - 20 = 0$

5. $(x - 1)^2 = 10$

6. $4(x - 1)^2 = 23$

Solve each of the following equations by completing the square.

7. $m^2 + 3m - 1 = 0$

8. $2x^2 - 10x + 3 = 0$

Solve each of the following equations, using the quadratic formula.

9. $x^2 - 5x - 3 = 0$

10. $x^2 + 4x = 7$

11. Find the zeros of the function $f(x) = 3x^2 - 10x - 8$.

Solve.

12. The product of two consecutive, positive, odd integers is 63. Find the two integers.

13. Suppose that the height (in feet) of a ball thrown upward from a raised platform is approximated by

$$h(t) = -16t^2 + 32t + 32$$

t seconds after the ball has been released. How long will it take the ball to hit the ground?

Find the equation of the axis of symmetry and the coordinates of the vertex of each of the following.

14. $y = -3(x + 2)^2 + 1$

15. $y = x^2 - 4x - 5$

16. $y = -2x^2 + 6x - 3$

17. $x = (y - 3)^2 - 2$

18. $x = y^2 - 6y + 2$

Graph each of the following functions.

19. $f(x) = (x - 5)^2$

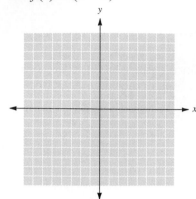

20. $f(x) = (x + 2)^2 - 3$

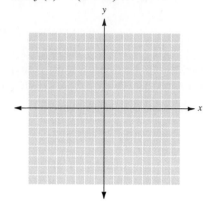

21. $f(x) = -2(x - 3)^2 - 1$

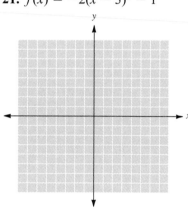

22. $f(x) = 3x^2 + 9x + 2$

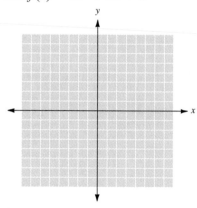

Graph each of the following equations.

23. $x = \dfrac{1}{2}(y - 4)^2 + 2$

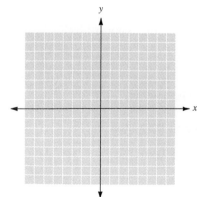

24. $x = 2y^2 - 10y + 3$

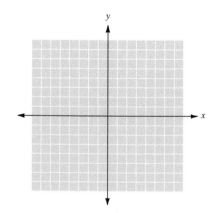

Solve the following equations and graph the solution set.

25. $(x + 3)(x - 1) > 0$

26. $x^2 + 5x - 14 < 0$

27. $x^2 - 3x \geq 18$

28. $3x^2 + x - 10 \leq 0$

Find the coordinates for the center and the radius and then graph each equation.

29. $(x - 3)^2 + (y + 2)^2 = 36$

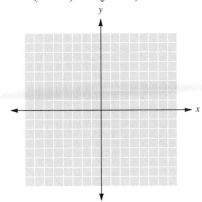

30. $x^2 + 2x + y^2 - 4y - 21 = 0$

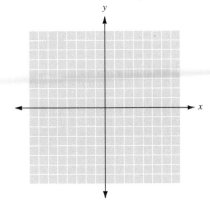

31. $(x - 2)^2 + (y + 3)^2 = 9$

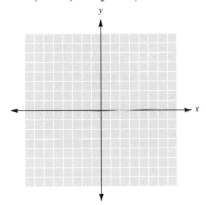

32. The following data were collected as part of an experiment.

Time, h	Distance, m
2	9
3	27
5	82
7	140
9	253
11	375

Determine the prediction equation.

This test is provided to help you in the process of reviewing the previous chapters. Answers are provided in the back of the book. If you missed any answers, be sure to go back and review the appropriate chapter section.

Graph each of the following equations.

1. $2x - 3y = 6$

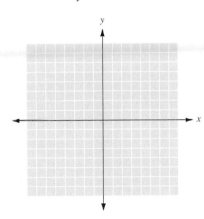

2. $y = -\dfrac{1}{3}x - 2$

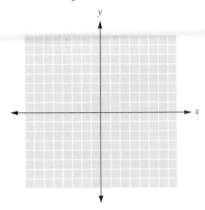

3. $y = 4$

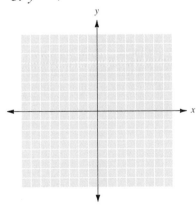

Find the slope of the line determined by each set of points.

4. $(-4, 7)$ and $(-3, 4)$

5. $(-2, 3)$ and $(-5, -1)$

6. Let $f(x) = 6x^2 - 5x + 1$. Evaluate $f(-2)$.

7. Simplify the function $f(x) = (x^2 - 1)(x + 3)$.

8. Completely factor the expression $x^3 + x^2 - 6x$.

9. Simplify the expression $\dfrac{2}{x+2} - \dfrac{3x-2}{x^2-x-6}$.

10. Simplify the expression $(\sqrt{7} - \sqrt{2})(\sqrt{3} + \sqrt{6})$.

Solve each equation.

11. $2x - 7 = 0$ **12.** $3x - 5 = 5x + 3$ **13.** $0 = (x-3)(x+5)$

14. $x^2 - 3x + 2 = 0$ **15.** $x^2 + 7x - 30 = 0$ **16.** $x^2 - 3x - 3 = 0$

17. $(x-3)^2 = 5$

Solve the following word problems. Show the equation used for the solution.

18. Five times a number decreased by 7 is -72. Find the number.

19. One leg of a right triangle 4 ft longer than the shorter leg. If the hypotenuse is 28 ft, how long is each leg?

20. Suppose that a manufacturer's weekly profit P is given by

$P(x) = -4x^2 + 320x$

where x is the number of units manufactured and sold. Find the number of items that must be manufactured and sold to guarantee a profit of at least \$4956.

SYSTEMS OF EQUATIONS AND INEQUALITIES

INTRODUCTION

Successful businesses juggle many factors, including workers' schedules, available machine time, and costs of raw material and storage. Getting the most efficient mix of these factors is crucial if the business is to make money. Systems of linear equations can be used to find the most efficient ways to combine these costs.

For example, the owner of a small bakery must decide how much of several kinds of bread to make based on the time it takes to produce each kind and the profit to be made on each one. The owner knows the bakery requires 0.75 hours of oven time and 1.25 hours of preparation time for every 8 loaves of bread, and that a coffee cake takes 1 hour of oven time and 1.25 hours of preparation time for every 6 coffee cakes. In 1 day the bakery has 12 hours of oven time and 16 hours of preparation time available. The owner knows that she clears a profit of $0.50 for every loaf of bread sold and $1.75 for every coffee cake sold. But, she has found that on a regular day she sells no more than 12 coffee cakes.

Given these constraints, how many of each type of product can be made in a day? Which combination of products would give the highest profit if all the products are sold?

These questions can be answered by graphing a system of equations that model the constraints, where b is the number of loaves of bread, and c is the number of coffee cakes. We begin by solving equations for baking and preparation times, and coffee cake sales.

© Frank Siteman/The Picture Cube

The graph of these three inequalities shows where all of these constraints overlap and indicate what is possible.

Baking time:

$$b\left(\frac{0.75}{8}\right) + c\left(\frac{1}{6}\right) \le 12\ h$$

Preparation time:

$$b\left(\frac{1.25}{8}\right) + c\left(\frac{1.25}{6}\right) \le 16\ h$$

Coffee cake sales:

$$c < 20$$

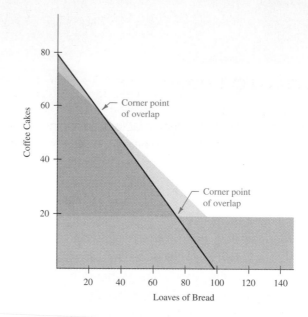

There are two points of interest; the point (26, 58) indicates that 26 loaves and 58 cakes would make the highest daily profit, about $114.50. The other point, at (92, 20), indicates that if no more than 20 coffee cakes can be sold, then 92 loaves of break and 20 coffee cakes is the combination that fits the constraints and makes the most profit. In this case, the profit will be 92($.50) + 20($1.75) or $81.00 for 1 day.

This graph not only tells the baker what combination gives the most profit but also indicates that she has room to market more of the coffee cakes. She might consider a sales promotion. To solve this problem, she would use a system of equations, which is the topic of this chapter. _____ ■

 **7.1**

Solving Systems of Linear Equations in Two Variables

7.1 OBJECTIVES

1. Find ordered pairs associated with each equation
2. Solve a system by graphing
3. Solve a system by the addition method
4. Solve a system by the substitution method
5. Use a system of equations to solve an application

It is helpful at this point to review Section 1.5 on graphing linear equations.

Of course, there are an infinite number of solutions for an equation of this type. You might want to verify that (2, −2) and (6, 0) are also solutions.

Our work in this chapter focuses on systems of equations and the various solution techniques available for your work with such systems. First, let's consider what we mean by a system of equations.

In many applications, you will find it helpful to use two variables when labeling the quantities involved. Often this leads to a **linear equation in two variables.** A typical equation might be

$$x - 2y = 6$$

A solution for such an equation is any ordered pair of real numbers (x, y) that satisfies the equation. For example, the ordered pair $(4, -1)$ is a solution for the equation since substituting 4 for x and -1 for y results in a true statement.

$$4 - 2(-1) \stackrel{?}{=} 6$$
$$4 + 2 \stackrel{?}{=} 6$$
$$6 \stackrel{?}{=} 6 \qquad \text{True}$$

Whenever two or more equations are considered together, they form a **system of equations.** If the equations of the system are linear, the system is called a **linear system.** Our work here involves finding solutions for such systems. We present three methods for solving such systems: the graphing method, the addition method, and the substitution method.

THE GRAPHING METHOD

We begin our discussion with a definition.

> A **solution** for a linear system of equations in two variables is an ordered pair of real numbers (x, y) that satisfies *both* equations in the system.

For instance, given the linear system

$$x - 2y = -1$$
$$2x + y = 8$$

the pair $(3, 2)$ is a solution because after substituting 3 for x and 2 for y in the two equations of the system, we have the *two* true statements

Both equations are satisfied by (3, 2).

$$3 - 2(2) \stackrel{?}{=} -1 \qquad \text{and} \qquad 2(3) + 2 \stackrel{?}{=} 8$$
$$-1 = -1 \qquad \text{and} \qquad 8 = 8$$

Since a solution to a system of equations represents a point on both lines, one approach to finding the solution for a system is to **graph** each equation on the same set of coordinate axes and then identify the point of intersection. This is shown in Example 1.

519

• Example 1

Solving a System by Graphing

Solve the system by graphing.

$$2x + y = 4$$
$$x - y = 5$$

Solve each equation for *y* and then graph.

$$y = -2x + 4$$

and

$$y = x - 5$$

We can *approximate* the solution by tracing the curves near their intersection.

We graph the lines corresponding to the two equations of the system.

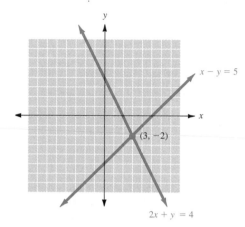

Each equation has an infinite number of solutions (ordered pairs) corresponding to points on a line. The point of intersection, here $(3, -2)$, is the *only* point lying on both lines, and so $(3, -2)$ is the only ordered pair satisfying both equations, and $(3, -2)$ is the solution for the system.

● ● ● **CHECK YOURSELF 1**

Solve the system by graphing.

$$3x - y = 2$$
$$x + y = 6$$

In Example 1, the two lines are nonparallel and intersect at only one point. The system has a unique solution corresponding to that point. Such a system is called a **consistent system.** In Example 2, we examine a system representing two lines that have no point of intersection.

• Example 2

Solving a System by Graphing

Solve the system by graphing.

$$2x - y = 4$$
$$6x - 3y = 18$$

The lines corresponding to the two equations are graphed below.

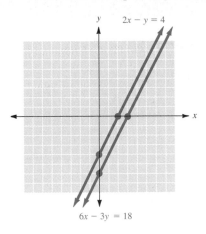

The lines are distinct and parallel. There is no point at which they intersect, so the system has no solution. We call such a system an **inconsistent system.**

● ● ●   **CHECK YOURSELF 2**

Solve the system, if possible.

$3x - y = 1$

$6x - 2y = 3$

Sometimes the equations in a system have the same graph.

•Example 3

Solving a System by Graphing

Solve the system by graphing.

$2x - y = 2$

$4x - 2y = 4$

The equations are graphed, as follows.

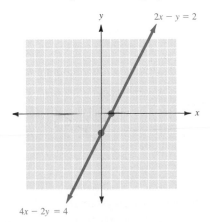

The lines have the same graph, so they have an infinite number of solutions in common.

● ● ● **CHECK YOURSELF 3**

Solve the system by graphing.

$6x - 3y = 12$

$\qquad y = 2x - 4$

THE ADDITION METHOD

A second method for solving systems of linear equations in two variables is by the **addition method.** This method of solving systems is based on the following result regarding equivalent systems.

> An equivalent system is formed whenever
>
> **1.** One of the equations is multiplied by a nonzero number.
> **2.** One of the equations is replaced by the sum of a constant multiple of another equation and that equation.

Example 4 illustrates the addition method of solution.

● Example 4

Solving a System by the Addition Method

Solve the system by the addition method.

$$5x - 2y = 12 \qquad\qquad (1)$$
$$3x + 2y = 12 \qquad\qquad (2)$$

The addition method is sometimes called **solution by elimination** for this reason.

In this case, adding the equations will eliminate variable y, and we have

$$8x = 24$$
$$x = 3 \qquad\qquad (3)$$

Now equation (3) can be paired with either of the original equations to form an equivalent system. We let $x = 3$ in equation (1):

The solution should be checked by substituting these values into equation (2). Here

$$5(3) - 2y = 12$$
$$15 - 2y = 12$$
$$-2y = -3$$
$$y = \frac{3}{2}$$

$$3(3) + 2\left(\frac{3}{2}\right) \stackrel{?}{=} 12$$

$$9 + 3 \stackrel{?}{=} 12$$

$$12 = 12$$

is a true statement.

and $\left(3, \dfrac{3}{2}\right)$ is the solution for our system.

● ● ●   **CHECK YOURSELF 4**

Solve the system by the addition method.

$$4x - 3y = 19$$
$$-4x + 5y = -25$$

Remember that multiplying one or both of the equations by a nonzero constant produces an equivalent system.

Example 4 and Check Yourself 4 were straightforward in that adding the equations of the system immediately eliminated one of the variables. Example 5 illustrates a common situation in which we must multiply one or both of the equations by a nonzero constant before the addition method is applied.

All these solutions can be approximated by graphing the lines and tracing near the intersection. This is particularly useful when the solutions are not integers (the technical term for such solutions is "ugly").

● Example 5

Solving a System by the Addition Method

Solve the system by the addition method.

$$3x - 5y = 19 \tag{4}$$
$$5x + 2y = 11 \tag{5}$$

It is clear that adding the equations of the given system will *not* eliminate one of the variables. Therefore, we must use multiplication to form an equivalent system. The choice of multipliers depends on which variable we decide to eliminate. Here we have decided to eliminate y. We multiply equation (4) by 2 and equation (5) by 5. We then have

Note that the coefficients of y are now *opposites* of each other.

$$6x - 10y = 38$$
$$25x + 10y = 55$$

Adding now eliminates y and yields

$$31x = 93$$
$$x = 3 \tag{6}$$

Pairing equation (6) with equation (4) gives an equivalent system, and we can substitute 3 for x in equation (4):

$$3 \cdot 3 - 5y = 19$$
$$9 - 5y = 19$$
$$-5y = 10$$
$$y = -2$$

Again, the solution should be checked by substitution in equation (5).

The solution for the system is $(3, -2)$.

● ● ● ● **CHECK YOURSELF 5**

Solve the system by the addition method.

$2x + 3y = -18$

$3x - 5y = 11$

The following algorithm summarizes the addition method of solving linear systems of two equations in two variables.

Solving by the Addition Method

STEP 1   If necessary, multiply one or both of the equations by a constant so that one of the variables can be eliminated by addition.
STEP 2   Add the equations of the equivalent system formed in step 1.
STEP 3   Solve the equation found in step 2.
STEP 4   Substitute the value found in step 3 into either of the equations of the original system to find the corresponding value of the remaining variable. The ordered pair formed is the solution to the system.
STEP 5   Check the solution by substituting the pair of values found in step 4 into the other equation of the original system.

Example 6 illustrates two special situations you may encounter while applying the addition method.

● Example 6

Solving a System by the Addition Method

Solve each system by the addition method.

(a) $4x + 5y = 20$ (7)
     $8x + 10y = 19$ (8)

Multiply equation (7) by -2. Then

$-8x - 10y = -40$ We add the two left sides to get 0 and the
$\underline{8x + 10y = 19}$ two right sides to get -21.
        $0 = -21$

The result $0 = -21$ is a *false* statement, which means that there is no point of intersection. Therefore, it is inconsistent, and there is no solution.

(b) $5x - 7y = 9$ (9)
     $15x - 21y = 27$ (10)

Multiply equation (9) by -3. We then have

$$-15x + 21y = -27 \qquad \text{We add the two equations.}$$
$$\underline{15x - 21y = 27}$$
$$0 = 0$$

The solution set could be written $\{(x, y)|5x - 7y = 9\}$. This means the set of all ordered pairs (x, y) that make $5x - 7y = 9$ a true statement.

Both variables have been eliminated, and the result is a *true* statement. The two lines coincide, and there are an infinite number of solutions, one for each point on that line. We call this a **dependent system.**

CHECK YOURSELF 6

Solve each system by the addition method, if possible.

(a) $3x + 2y = 8$
$9x + 6y = 11$

(b) $x - 2y = 8$
$3x - 6y = 24$

The results of Example 6 can be summarized as follows.

When a system of two linear equations is solved:

1. If a false statement such as $3 = 4$ is obtained, then the system is inconsistent and has no solution.
2. If a true statement such as $8 = 8$ is obtained, then the system is dependent and has an infinite number of solutions.

THE SUBSTITUTION METHOD

A third method for finding the solutions of linear systems in two variables is called the **substitution method.** You may very well find the substitution method more difficult to apply in solving certain systems than the addition method, particularly when the equations involved in the substitution lead to fractions. However, the substitution method does have important extensions to systems involving higher-degree equations, as you will see in later mathematics classes.

To outline the technique, we solve one of the equations from the original system for one of the variables. That expression is then substituted into the *other* equation of the system to provide an equation in a single variable. That equation is solved, and the corresponding value for the other variable is found as before, as Example 7 illustrates.

• Example 7

Solving a System by the Substitution Method

(a) Solve the system by the substitution method.

$$2x - 3y = -3 \tag{11}$$
$$y = 2x - 1 \tag{12}$$

Since equation (12) is already solved for y, we substitute $2x - 1$ for y in equation (11).

We now have an equation in the single variable x.

$$2x - 3(2x - 1) = -3$$

Solving for x gives

$$2x - 6x + 3 = -3$$
$$-4x = -6$$
$$x = \frac{3}{2}$$

To check this result, we substitute these values in equation (11) and have

$$2\left(\frac{3}{2}\right) - 3 \cdot 2 \stackrel{?}{=} -3$$
$$3 - 6 \stackrel{?}{=} -3$$
$$-3 = -3$$

A true statement!

We now substitute $\frac{3}{2}$ for x in equation (12).

$$y = 2\left(\frac{3}{2}\right) - 1$$
$$= 3 - 1 = 2$$

The solution for our system is $\left(\frac{3}{2}, 2\right)$.

(b) Solve the system by the substitution method.

$$2x + 3y = 16 \tag{13}$$
$$3x - y = 2 \tag{14}$$

Why did we choose to solve for y in equation (14)? We could have solved for x, so that

$$x = \frac{y + 2}{3}$$

We simply chose the easier case to avoid fractions.

We start by solving equation (14) for y.

$$3x - y = 2$$
$$-y = -3x + 2 \tag{15}$$
$$y = 3x - 2$$

Substituting in equation (13) yields

$$2x + 3(3x - 2) = 16$$
$$2x + 9x - 6 = 16$$
$$11x = 22$$
$$x = 2$$

We now substitute 2 for x in equation (15).

$$y = 3 \cdot 2 - 2$$
$$= 6 - 2 = 4$$

The solution should be checked in *both* equations of the original system.

The solution for the system is $(2, 4)$. We leave the check of this result to you.

● ● ● **CHECK YOURSELF 7**

Solve each system by the substitution method.

(a) $2x + 3y = 6$
$x = 3y + 6$

(b) $3x + 4y = -3$
$x + 4y = 1$

The following algorithm summarizes the substitution method for solving linear systems of two equations in two variables.

Solving by the Substitution Method

STEP 1 If necessary, solve one of the equations of the original system for one of the variables.

STEP 2 Substitute the expression obtained in step 1 into the *other* equation of the system to write an equation in a single variable.

STEP 3 Solve the equation found in step 2.

STEP 4 Substitute the value found in step 3 into the equation derived in step 1 to find the corresponding value of the remaining variable. The ordered pair formed is the solution for the system.

STEP 5 Check the solution by substituting the pair of values found in step 4 into *both* equations of the original system.

A natural question at this point is, How do you decide which solution method to use? First, the graphical method can generally provide only approximate solutions. When exact solutions are necessary, one of the algebraic methods must be applied. Which method to use depends totally on the given system.

If you can easily solve for a variable in one of the equations, the substitution method should work well. However, if solving for a variable in either equation of the system leads to fractions, you may find the addition approach more efficient.

SOLVING APPLICATIONS

We are now ready to apply our equation-solving skills to solving various applications or word problems. Being able to extend these skills to problem solving is an important goal, and the procedures developed here are used throughout the rest of the book.

Although we consider applications from a variety of areas in this section, all are approached with the same five-step strategy presented here to begin the discussion.

Solving Applications

STEP 1 Read the problem carefully to determine the unknown quantities.

STEP 2 Choose a variable to represent the unknown. Express all other unknowns in terms of this variable.

STEP 3 Translate the problem to the language of algebra to form a system of equations.

STEP 4 Solve the system of equations, and answer the question of the original problem.

STEP 5 Verify your solution by returning to the original problem.

•Example 8

Solving a Mixture Problem

A coffee merchant has two types of coffee beans, one selling for $3 per pound and the other for $5 per pound. The beans are to be mixed to provide 100 lb of a mixture selling for $4.50 per pound. How much of each type of coffee bean should be used to form 100 lb of the mixture?

Step 1 The unknowns are the amounts of the two types of beans.

Step 2 We use two variables to represent the two unknowns. Let x be the amount of $3 beans and y the amount of $5 beans.

Step 3 We now want to establish a system of two equations. One equation will be based on the *total amount* of the mixture, the other on the mixture's *value*.

Since we use *two* variables, we must form *two* equations.

$$x + y = 100 \qquad \text{The mixture must weigh 100 lb.} \qquad (16)$$

$$3x + 5y = 450 \qquad\qquad (17)$$

Value of   Value of   Total value
$3 beans $5 beans

Step 4 An easy approach to the solution of the system is to multiply equation (16) by -3 and add to eliminate x.

$$-3x - 3y = -300$$
$$3x + 5y = 450$$
$$2y = 150$$
$$y = 75 \text{ lb}$$

By substitution in equation (16), we have

$$x = 25 \text{ lb}$$

Step 5 To check the result, show that the value of the $3 beans, added to the value of the $5 beans, equals the desired value of the mixture.

● ● ●   CHECK YOURSELF 8

Peanuts, which sell for $2.40 per pound, and cashews, which sell for $6 per pound, are to be mixed to form a 60-lb mixture selling for $3 per pound. How much of each type of nut should be used?

A related problem is illustrated in Example 9.

• Example 9

Solving a Mixture Problem

A chemist has a 25% and a 50% acid solution. How much of each solution should be used to form 200 mL of a 35% acid solution?

Step 1 The unknowns in this case are the amounts of the 25% and 50% solutions to be used in forming the mixture.

Step 2 Again we use two variables to represent the two unknowns. Let x be the amount of the 25% solution and y the amount of the 50% solution. Let's draw a picture before proceeding to form a system of equations.

Step 3 Now, to form our two equations, we want to consider two relationships: the *total amounts* combined and the *amounts of acid* combined.
From our sketch of the problem, we have

Drawing a sketch of a problem is often a valuable part of the problem-solving strategy.

Total amounts combined.

$$x + \quad y = 200 \tag{18}$$

Amounts of acid combined.

$$0.25x + 0.50y = 0.35(200) \tag{19}$$

Step 4 Now, clear equation (19) of decimals by multiplying equation (19) by 100. The solution then proceeds as before, with the result

$$x = 120 \text{ mL} \qquad (25\% \text{ solution})$$

$$y = 80 \text{ mL} \qquad (50\% \text{ solution})$$

Step 5 To check, show that the amount of acid in the 25% solution, $(0.25)(120)$, added to the amount in the 50% solution, $(0.50)(80)$, equals the correct amount in the mixture, $(0.35)(200)$. We leave that to you.

● ● ● CHECK YOURSELF 9

A pharmacist wants to prepare 300 mL of a 20% alcohol solution. How much of a 30% solution and a 15% solution should be used to form the desired mixture?

Applications that involve a constant rate of travel, or speed, require the use of the distance formula seen earlier:

$$d = rt$$

where d = distance traveled
r = rate, or speed
t = time

Example 10 illustrates this approach.

• Example 10

Solving a Distance-Rate-Time Problem

A boat can travel 36 mi downstream in 2 h. Coming back upstream, the boat takes 3 h. What is the rate of the boat in still water? What is the rate of the current?

Step 1 We want to find the two rates.

Step 2 Let x be the rate of the boat in still water and y the rate of the current.

Downstream the rate is then

$x + y$

Upstream, the rate is

$x - y$

Step 3 To form a system, think about the following. Downstream, the rate of the boat is *increased* by the effect of the current. Upstream, the rate is *decreased*.

In many applications, it helps to lay out the information in tabular form. Let's try that strategy here.

	d	r	t
Downstream	36	$x + y$	2
Upstream	36	$x - y$	3

Since $d = rt$, from the table we can easily form two equations:

$$36 = (x + y)(2) \tag{20}$$
$$36 = (x - y)(3) \tag{21}$$

Step 4 We clear equations (20) and (21) of parentheses and simplify, to write the equivalent system

$$x + y = 18$$
$$x - y = 12$$

Solving, we have

$$x = 15 \text{ mi/h}$$
$$y = 3 \text{ mi/h}$$

Step 5 To check, verify the $d = rt$ equation in *both* the upstream and the downstream cases. We leave that to you.

● ● ● **CHECK YOURSELF 10**

A plane flies 480 mi in an easterly direction, with the wind, in 4 h. Returning westerly along the same route, against the wind, the plane takes 6 h. What is the rate of the plane in still air? What is the rate of the wind?

The use of systems of equations in problem solving have many applications in a business setting. Example 11 illustrates one such application.

● Example 11

Solving a Business-Based Application

A manufacturer produces a standard model and a deluxe model of a 13-inch (in.) television set. The standard model requires 12 h of labor to produce, while the deluxe model requires 18 h. The company has 360 h of labor available per week. The plant's capacity is a total of 25 sets per week. If all the available time and capacity are to be used, how many of each type of set should be produced?

Step 1 The unknowns in this case are the number of standard and deluxe models that can be produced.

The choices for *x* and *y* could have been reversed.

Step 2 Let x be the number of standard models and y the number of deluxe models.

Step 3 Our system will come from the two given conditions that fix the total number of sets that can be produced and the total labor hours available.

$$x + y = 25 \leftarrow \text{Total number of sets}$$
$$12x + 18y = 360 \leftarrow \text{Total labor hours available}$$

Labor hours— standard sets
Labor hours— deluxe sets

Step 4 Solving the system in step 3, we have

$$x = 15 \quad \text{and} \quad y = 10$$

which tells us that to use all the available capacity, the plant should produce 15 standard sets and 10 deluxe sets per week.

Step 5 We leave the check of this result to the reader.

● ● ● **CHECK YOURSELF 11**

A manufacturer produces standard cassette players and compact disc players. The cassette players require 2 h of electronic assembly and the CDs 3 h. The cassette players require 4 h of case assembly and the CDs 2 h. The company has 120 h of electronic assembly time available per week and 160 h of case assembly time. How many of each type of unit can be produced each week if all available assembly time is to be used?

Let's look at one final application that leads to a system of two equations.

● Example 12

Solving a Business-Based Application

Two car rental agencies have the following rate structures for a subcompact car. Company A charges $20 per day plus 15¢ per mile. Company B charges $18 per day plus

16¢ per mile. If you rent a car for 1 day, for what number of miles will the two companies have the same total charge?

Letting c represent the total a company will charge and m the number of miles driven, we calculate the following.

For company A:

You first saw this type of linear model in exercises in Section 1.8.

$$c(m) = 20 + 0.15m \qquad\qquad (22)$$

For company B:

$$c(m) = 18 + 0.16m \qquad\qquad (23)$$

The system can be solved most easily by substitution. Substituting $18 + 0.16m$ for $c(m)$ in equation (22) gives

$$18 + 0.16m = 20 + 0.15m$$
$$0.01m = 2$$
$$m = 200 \text{ mi}$$

The graph of the system is shown below.

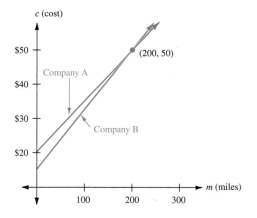

From the graph, how would you make a decision about which agency to use?

● ● ● CHECK YOURSELF 12

For a compact car, the same two companies charge $27 per day plus 20¢ per mile and $24 per day plus 22¢ per mile. For a 2-day rental, when will the charges be the same?

● ● ● CHECK YOURSELF ANSWERS

1.

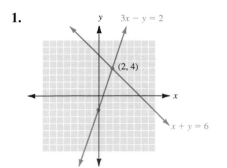

2.

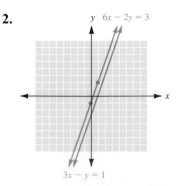

3.

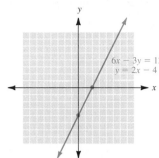

$6x - 3y = 12$
$y = 2x - 4$

4. $\left(\dfrac{5}{2}, -3\right)$.   **5.** $(-3, -4)$.

6. (a) Inconsistent system: no solution; **(b)** dependent system: an infinite number of solutions.   **7. (a)** $\left(4, \dfrac{-2}{3}\right)$; **(b)** $\left(-2, \dfrac{3}{4}\right)$.   **8.** 50 lb of peanuts and 10 lb of cashews.   **9.** 100 mL of the 30% and 200 mL of the 15%.   **10.** 100 mi/h plane and 20 mi/h wind.   **11.** 30 cassette players and 20 CDs.
12. At 300 mi, \$114 charge.

7.1 Exercises

In Exercises 1 to 8, solve each system by graphing. If a unique solution does not exist, state whether the system is dependent or inconsistent.

1. $x + y = 6$
    $x - y = 4$

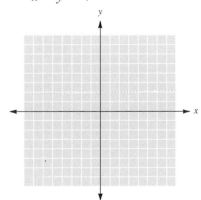

2. $x - y = 8$
    $x + y = 2$

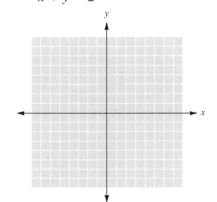

3. $x + 2y = 4$
   $x - y = 1$

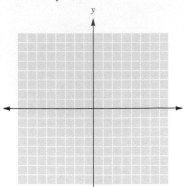

4. $x - 2y = 2$
   $x + 2y = 6$

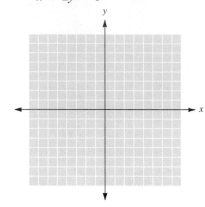

5. $3x - y = 3$
   $3x - y = 6$

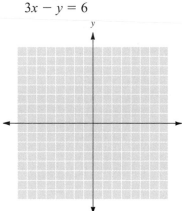

6. $3x + 2y = 12$
   $y = 3$

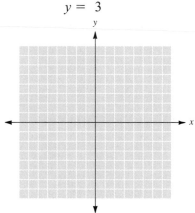

7. $x + 3y = 12$
   $2x - 3y = 6$

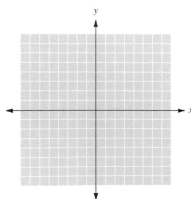

8. $3x - 6y = 9$
   $x - 2y = 3$

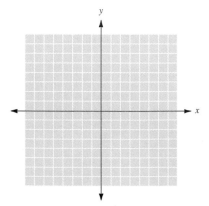

In Exercises 9 to 22, solve each system by the addition method. If a unique solution does not exist, state whether the system is inconsistent or dependent.

9. $2x - y = 1$
   $-2x + 3y = 5$

10. $x + 3y = 12$
   $2x - 3y = 6$

11. $x + 2y = -2$
   $3x + 2y = -12$

12. $2x + 3y = 1$
   $5x + 3y = 16$

13. $x + y = 3$
$3x - 2y = 4$

14. $x - y = -2$
$2x + 3y = 21$

15. $2x + y = 8$
$-4x - 2y = -16$

16. $3x - 4y = 2$
$4x - y = 20$

17. $5x - 2y = 31$
$4x + 3y = 11$

18. $2x - y = 4$
$6x - 3y = 10$

19. $3x - 2y = 7$
$-6x + 4y = -15$

20. $3x + 4y = 0$
$5x - 3y = -29$

21. $-2x + 7y = 2$
$3x - 5y = -14$

22. $5x - 2y = 3$
$10x - 4y = 6$

In Exercises 23 to 34, solve each system by the substitution method. If a unique solution does not exist, state whether the system is inconsistent or dependent.

23. $x - y = 7$
$y = 2x - 12$

24. $x - y = 4$
$x = 2y - 2$

25. $3x + 2y = -18$
$x = 3y + 5$

26. $3x - 18y = 4$
$x = 6y + 2$

27. $10x - 2y = 4$
$y = 5x - 2$

28. $4x + 5y = 6$
$y = 2x - 10$

29. $3x + 4y = 9$
$y = 3x + 1$

30. $6x - 5y = 27$
$x = 5y + 2$

31. $x - 7y = 3$
$2x - 5y = 15$

32. $4x + 3y = -11$
$5x + y = -11$

33. $4x - 12y = 5$
$-x + 3y = -1$

34. $5x - 6y = 21$
$x - 2y = 5$

In Exercises 35 to 40, solve each system by any method discussed in this section.

35. $2x - 3y = 4$
$x = 3y + 6$

36. $5x + y = 2$
$5x - 3y = 6$

37. $4x - 3y = 0$
$5x + 2y = 23$

38. $7x - 2y = -17$
$x + 4y = 4$

39. $3x - y = 17$
$5x + 3y = 5$

40. $7x + 3y = -51$
$y = 2x + 9$

In Exercises 41 to 44, solve each system by any method discussed in this section. *Hint:* You should multiply to clear fractions as your first step.

41. $\frac{1}{2}x - \frac{1}{3}y = 8$
$\frac{1}{3}x + y = -2$

42. $\frac{1}{5}x - \frac{1}{2}y = 0$
$x - \frac{3}{2}y = 4$

43. $\frac{2}{3}x + \frac{3}{5}y = -3$
$\frac{1}{3}x + \frac{2}{5}y = -3$

44. $\frac{3}{8}x - \frac{1}{2}y = -5$
$\frac{1}{4}x + \frac{3}{2}y = 4$

Each application in Exercises 45 to 52 can be solved by the use of a system of linear equations. Match the application with the appropriate system below.

(a) $12x + 5y = 116$
  $8x + 12y = 112$

(b) $x + y = 8000$
  $0.06x + 0.09y = 600$

(c) $x + y = 200$
  $0.20x + 0.60y = 90$

(d) $x + y = 36$
  $y = 3x - 4$

(e) $2(x + y) = 36$
  $3(x - y) = 36$

(f) $x + y = 200$
  $5.50x + 4y = 980$

(g) $L = 2W + 3$
  $2L + 2W = 36$

(h) $x + y = 120$
  $2.20x + 5.40y = 360$

45. One number is 4 less than 3 times another. If the sum of the numbers is 36, what are the two numbers?

46. Suppose a movie theater sold 200 adult and student tickets for a showing with a revenue of $980. If the adult tickets were $5.50 and the student tickets were $4, how many of each type of ticket were sold?

47. The length of a rectangle is 3 cm more than twice its width. If the perimeter of the rectangle is 36 cm, find the dimensions of the rectangle.

48. An order of 12 dozen roller-ball pens and 5 dozen ballpoint pens cost $116. A later order for 8 dozen roller-ball pens and 12 dozen ballpoint pens cost $112. What was the cost of 1 dozen of each type of pen?

49. A candy merchant wants to mix peanuts selling at $2.20 per pound with cashews selling at $5.40 per pound to form 120 lb of a mixed-nut blend that will sell for $3 per pound. What amount of each type of nut should be used?

50. Donald has investments totaling $8000 in two accounts—one a savings account paying 6% interest and the other a bond paying 9%. If the annual interest from the two investments was $600, how much did he have invested at each rate?

51. A chemist wants to combine a 20% alcohol solution with a 60% solution to form 200 mL of a 45% solution. How much of each solution should be used to form the mixture?

52. Xian was able to make a downstream trip of 36 mi in 2 h. Returning upstream, he took 3 h to make the trip. How fast can his boat travel in still water? What was the rate of the river's current?

In Exercises 53 to 74, solve by choosing a variable to represent each unknown quantity and writing a system of equations.

53. Mixture Problem. Suppose 750 tickets were sold for a concert with a total revenue of $5300. If adult tickets were $8 and student tickets were $4.50, how many of each type of ticket were sold?

54. **Mixture Problem.** Theater tickets sold for $7.50 on the main floor and $5 in the balcony. The total revenue was $3250, and there were 100 more main-floor tickets sold than balcony tickets. Find the number of each type of ticket sold.

55. **Geometry.** The length of a rectangle is 3 in. less than twice its width. If the perimeter of the rectangle is 84 in., find the dimensions of the rectangle.

56. **Geometry.** The length of a rectangle is 5 cm more than 3 times its width. If the perimeter of the rectangle is 74 cm, find the dimensions of the rectangle.

57. **Mixture Problem.** A garden store sold 8 bags of mulch and 3 bags of fertilizer for $24. The next purchase was for 5 bags of mulch and 5 bags of fertilizer. The cost of that purchase was $25. Find the cost of a single bag of mulch and a single bag of fertilizer.

58. **Mixture Problem.** The cost of an order for 10 computer disks and 3 packages of paper was $22.50. The next order was for 30 disks and 5 packages of paper, and its cost was $53.50. Find the price of a single disk and a single package of paper.

59. **Mixture Problem.** A coffee retailer has two grades of decaffeinated beans—one selling for $4 per pound and the other for $6.50 per pound. She wishes to blend the beans to form a 150-lb mixture that will sell for $4.75 per pound. How many pounds of each grade of bean should be used in the mixture?

60. **Mixture Problem.** A candy merchant sells jelly beans at $3.50 per pound and gumdrops at $4.70 per pound. To form a 200-lb mixture that will sell for $4.40 per pound, how many pounds of each type of candy should be used?

61. **Investment.** Cheryl decided to divide $12,000 into two investments—one a time deposit that pays 8% annual interest and the other a bond that pays 9%. If her annual interest was $1010, how much did she invest at each rate?

62. **Investment.** Miguel has $2000 more invested in a mutual fund paying 10% interest than in a savings account paying 7%. If he received $880 in interest for 1 year, how much did he have invested in the two accounts?

63. **Science.** A chemist mixes a 10% acid solution with a 50% acid solution to form 400 mL of a 40% solution. How much of each solution should be used in the mixture?

64. **Science.** A laboratory technician wishes to mix a 70% saline solution and a 20% solution to prepare 500 mL of a 40% solution. What amount of each solution should be used?

65. **Motion.** A boat traveled 36 mi up a river in 3 h. Returning downstream, the boat took 2 h. What is the boat's rate in still water, and what is the rate of the river's current?

66. **Motion.** A jet flew east a distance of 1800 mi with the jetstream in 3 h. Returning west, against the jetstream, the jet took 4 h. Find the jet's speed in still air and the rate of the jetstream.

67. **Number Problem.** The sum of the digits of a two-digit number is 8. If the digits are reversed, the new number is 36 more than the original number. Find the original number. *Hint:* If u represents the units digit of the number and t the tens digit, the original number can be represented by $10t + u$.

68. **Number Problem.** The sum of the digits of a two-digit number is 10. If the digits are reversed, the new number is 54 less than the original number. What was the original number?

69. **Business.** A manufacturer produces a battery-powered calculator and a solar model. The battery-powered model requires 10 min of electronic assembly and the solar model 15 min. There are 450 min of assembly time available per day. Both models require 8 min for packaging, and 280 min of packaging time are available per day. If the manufacturer wants to use all the available time, how many of each unit should be produced per day?

70. **Business.** A small tool manufacturer produces a standard- and a cordless-model power drill. The standard model takes 2 h of labor to assembly and the cordless model 3 h. There are 72 h of labor available per week for the drills. Material costs for the standard drill are $10, and for the cordless drill they are $20. The company wishes to limit material costs to $420 per week. How many of each model drill should be produced in order to use all the available resources?

71. **Economics.** In economics, a demand equation gives the quantity D that will be demanded by consumers at a given price p, in dollars. Suppose that $D = 210 - 4p$ for a particular product.

    A supply equation gives the supply S that will be available from producers at price p. Suppose also that for the same product $S = 10p$.

    The equilibrium point is that point where the supply equals the demand (here, where $S = D$). Use the given equations to find the equilibrium point.

72. **Economics.** Suppose the demand equation for a product is $D = 150 - 3p$ and the supply equation is $S = 12p$. Find the equilibrium point for the product.

73. **Consumer Affairs.** Two car rental agencies have the following rate structure for compact cars.

    Company A: $30/day and 22¢/mi.

    Company B: $28/day and 26¢/mi.

    For a 2-day rental, at what number of miles will the charges be the same?

74. **Construction.** Two construction companies submit the following bid.

    Company A: $5000 plus $15/square foot of building.

    Company B: $7000 plus $12.50/square foot of building.

    For what number of square feet of building will the bids of the two companies be the same?

Certain systems that are not linear can be solved with the methods of this section if we first substitute to change variables. For instance, the system

$$\frac{1}{x} + \frac{1}{y} = 4$$

$$\frac{1}{x} - \frac{3}{y} = -6$$

can be solved by the substitutions $u = \dfrac{1}{x}$ and $v = \dfrac{1}{y}$. That gives the system $u + v = 4$ and $u - 3v = -6$. The system is then solved for u and v, and the corresponding values for x and y are found. Use this method to solve the systems in Exercises 75 to 78.

75. $\dfrac{1}{x} + \dfrac{1}{y} = 4$

$\dfrac{1}{x} - \dfrac{3}{y} = -6$

76. $\dfrac{1}{x} + \dfrac{3}{y} = 1$

$\dfrac{4}{x} + \dfrac{3}{y} = 3$

77. $\dfrac{2}{x} + \dfrac{3}{y} = 4$

$\dfrac{2}{x} - \dfrac{6}{y} = 10$

70. $\dfrac{4}{x} - \dfrac{3}{y} = -1$

$\dfrac{12}{x} - \dfrac{1}{y} = 1$

Writing the equation of a line through two points can be done by the following method. Given the coordinates of two points, substitute each pair of values into the equation $y = mx + b$. This gives a system of two equations in variables m and b, which can be solved as before.

In Exercises 79 and 80, write the equation of the line through each of the following pairs of points, using the method outlined above.

79. $(2, 1)$ and $(4, 4)$

80. $(-3, 7)$ and $(6, 1)$

In Exercises 81 and 82, use your calculator to approximate the solution to each system. Express your answer to the nearest tenth.

81. $y = 2x - 3$
$2x + 3y = 1$

82. $3x - 4y = -7$
$2x + 3y = -1$

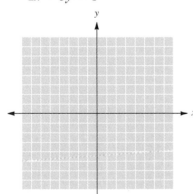

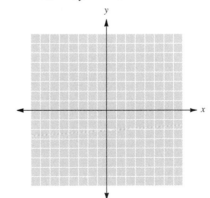

For Exercises 83 and 84, adjust the viewing window on your calculator so that you can see the point of intersection for the two lines representing the equations in the system. Then approximate the solution.

83. $5x - 12y = 8$
    $7x + 2y = 44$

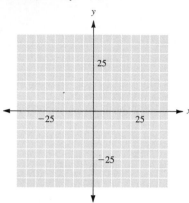

84. $9x - 3y = 10$
    $x + 5y = 58$

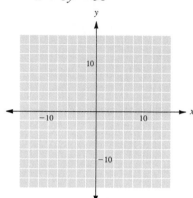

Answers

1. Solution: (5, 1)

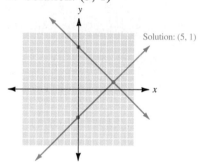

3. Solution: (2, 1)

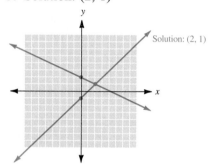

5. Inconsistent

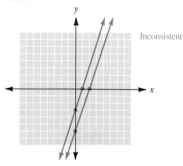

7. Solution: (6, 2)

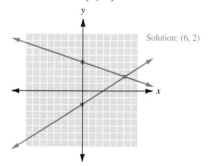

9. (2, 3)　　**11.** $\left(-5, \dfrac{3}{2}\right)$　　**13.** (2, 1)　　**15.** Dependent　　**17.** (5, −3)　　**19.** Inconsistent

21. (−8, −2)　　**23.** (5, −2)　　**25.** (−4, −3)　　**27.** Dependent　　**29.** $\left(\dfrac{1}{3}, 2\right)$　　**31.** (10, 1)

33. Inconsistent　　**35.** $\left(-2, -\dfrac{8}{3}\right)$　　**37.** (3, 4)　　**39.** (4, −5)　　**41.** (12, −6)　　**43.** (9, −15)

45. (d)　　**47.** (g)　　**49.** (h)　　**51.** (c)　　**53.** 550 adult, 200 student tickets　　**55.** 27 in. × 15 in.

57. Mulch: \$1.80; fertilizer: \$3.20　　**59.** 105 lb of \$4 beans, 45 lb of \$6.50 beans　　**61.** \$7000 time deposit, \$5000 bond　　**63.** 100 mL of 10%, 300 mL of 50%　　**65.** 15 mi/h boat, 3 mi/h current

67. 26   **69.** 15 battery powered, 20 solar models   **71.** 15   **73.** 100 mi   **75.** $\left(\dfrac{2}{3}, \dfrac{3}{5}\right)$ **77.** $\left(\dfrac{1}{3}, -\dfrac{3}{2}\right)$

79. $y = \dfrac{3}{2}x - 2$ **81.** $(1.3, -0.5)$ **83.** $(6, 2)$

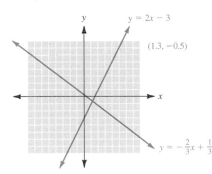

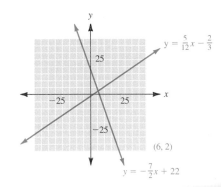

 7.2

Solving Systems of Linear Equations in Three Variables

7.2 OBJECTIVES

1. Find ordered pairs associated with each equation
2. Solve by the addition method
3. Look at a graphic interpretation of a solution
4. Solve an application of a system with three variables

Suppose an application involves three quantities that we want to label x, y, and z. A typical equation used for the solution might be

$$2x + 4y - z = 8$$

This is called a **linear equation in three variables.** The solution for such an equation is an **ordered triple** (x, y, z) of real numbers that satisfies the equation. For example, the ordered triple $(2, 1, 0)$ is a solution for the equation above since substituting 2 for x, 1 for y, and 0 for z results in the following true statement.

$$2 \cdot 2 + 4 \cdot 1 \stackrel{?}{=} 8$$
$$4 + 4 \stackrel{?}{=} 8$$
$$8 = 8 \qquad \text{True}$$

Of course, other solutions, in fact infinitely many, exist. You might want to verify that $(1, 1, -2)$ and $(3, 1, 2)$ are also solutions. To extend the concepts of the last section, we want to consider systems of three linear equations in three variables such as

$$x + y + z = 5$$
$$2x - y + z - 9$$
$$x - 2y + 3z = 16$$

For a unique solution to exist, when *three variables* are involved, we must have *three equations*.

The choice of which variable to eliminate is yours. Generally, you should pick the variable that allows the easiest computation.

The solution for such a system is the set of all ordered triples that satisfy each equation of the system. In the case on page 541, you should verify that $(2, -1, 4)$ is a solution for the system since that ordered triple makes each equation a true statement.

Let's turn now to the solution process itself. In this section, we will consider the addition method. We will then apply what we have learned to solving applications.

THE ADDITION METHOD

The central idea is to choose *two pairs* of equations from the system and, by the addition method, to eliminate the *same variable* from each of those pairs. The method is best illustrated by example. So let's proceed to see how the solution for the previous system was determined.

• **Example 1**

Solving a Linear System in Three Variables

Solve the system.

$$x + y + z = 5 \tag{1}$$
$$2x - y + z = 9 \tag{2}$$
$$x - 2y + 3z = 16 \tag{3}$$

First we choose two of the equations and the variable to eliminate. Variable y seems convenient in this case. Pairing equations (1) and (2) and then adding, we have

Any pair of equations could have been selected.

$$
\begin{aligned}
x + y + z &= 5 \\
\underline{2x - y + z} &= \underline{9} \\
3x \quad\;\; + 2z &= 14
\end{aligned} \tag{4}
$$

We now want to choose a different pair of equations to eliminate y. Using equations (1) and (3) this time, we multiply equation (1) by 2 and then add the result to equation (3):

$$
\begin{aligned}
2x + 2y + 2z &= 10 \\
\underline{x - 2y + 3z} &= \underline{16} \\
3x \quad\;\; + 5z &= 26
\end{aligned} \tag{5}
$$

We now have equations (4) and (5) in variables x and z.

$$3x + 2z = 14$$
$$3x + 5z = 26$$

Since we are now dealing with a system of two equations in two variables, any of the methods of the previous section apply. We have chosen to multiply equation (4) by -1 and then add that result to equation (5). This yields

$$3z = 12$$
$$z = 4$$

Substituting $z = 4$ in equation (4) gives

$$3x + 2 \cdot 4 = 14$$
$$3x + 8 = 14$$
$$3x = 6$$
$$x = 2$$

Any of the original equations could have been used.

Finally, letting $x = 2$ and $z = 4$ in equation (1) gives

$$2 + y + 4 = 5$$
$$y = -1$$

To check, substitute these values into the other equations of the original system.

and $(2, -1, 4)$ is shown to be the solution for the system.

● ● ● **CHECK YOURSELF 1**

Solve the system.

$$x - 2y + z = 0$$
$$2x + 3y - z = 16$$
$$3x - y - 3z = 23$$

One or more of the equations of a system may already have a missing variable. The elimination process is simplified in that case, as Example 2 illustrates.

● Example 2

Solving a Linear System in Three Variables

Solve the system.

$$2x + y - z = -3 \tag{6}$$
$$y + z = 2 \tag{7}$$
$$4x - y + z = 12 \tag{8}$$

Noting that equation (7) involves only y and z, we must simply find another equation in those same two variables. Multiply equation (6) by -2 and add the result to equation (8) to eliminate x.

$$-4x - 2y + 2z = 6$$
$$\underline{4x - y + z = 12}$$

We now have a *second* equation in y and z.

$$-3y + 3z = 18$$
$$y - z = -6 \tag{9}$$

We now form a system consisting of equations (7) and (9) and solve as before.

$$y + z = 2$$
$$\underline{y - z = -6} \quad \text{Adding eliminates } z.$$
$$2y = -4$$
$$y = -2$$

From equation (7), if $y = -2$,

$$-2 + z = 2$$
$$z = 4$$

and from equation (6), if $y = -2$ and $z = 4$,

$$2x - 2 - 4 = -3$$
$$2x = 3$$
$$x = \frac{3}{2}$$

The solution for the system is

$$\left(\frac{3}{2}, -2, 4\right)$$

● ● ● **CHECK YOURSELF 2**

Solve the system.

$$x + 2y - z = -3$$
$$x - y + z = 2$$
$$x - z =, 3$$

The following algorithm summarizes the procedure for finding the solutions for a linear system of three equations in three variables.

Solving a System of Three Equations in Three Unknowns

STEP 1   Choose a pair of equations from the system, and use the addition method to eliminate one of the variables.

STEP 2   Choose a *different* pair of equations, and eliminate the *same* variable.

STEP 3   Solve the system of two equations in two variables determined in steps 1 and 2.

STEP 4   Substitute the values found above into one of the original equations, and solve for the remaining variable.

STEP 5   The solution is the ordered triple of values found in steps 3 and 4. It can be checked by substituting into the other equations of the original system.

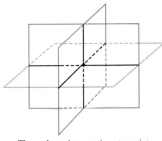

Three planes intersecting at a point

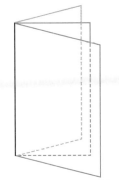

Three planes intersecting in a line

Systems of three equations in three variables may have (1) exactly one solution, (2) infinitely many solutions, or (3) no solution. Before we look at an algebraic approach in the second and third cases, let's discuss the geometry involved.

The graph of a linear equation in three variables is a plane (a flat surface) in three dimensions. Two distinct planes either will be parallel or will intersect in a line.

If three distinct planes intersect, that intersection will be either a single point (as in our first example) or a line (think of three pages in an open book—they intersect along the binding of the book).

Let's look at an example of how the solution proceeds in these cases.

• Example 3

Solving a Dependent Linear System in Three Variables

Solve the system.

$$x + 2y - z = 5 \tag{10}$$
$$x - y + z = -2 \tag{11}$$
$$-5x - 4y + z = -11 \tag{12}$$

We begin as before by choosing two pairs of equations from the system and eliminating the same variable from each of the pairs. Adding equations (10) and (11) gives

$$2x + y = 3 \tag{13}$$

Adding equations (10) and (12) gives

$$-4x - 2y = -6 \tag{14}$$

Now consider the system formed by equations (13) and (14). We multiply equation (13) by 2 and add again:

$$
\begin{array}{r}
4x + 2y = 6 \\
-4x - 2y = -6 \\
\hline
0 = 0
\end{array}
$$

This true statement tells us that the system has an infinite number of solutions (lying along a straight line). Again, such a system is dependent.

There are ways of representing the solutions, as you will see in later courses.

● ● ● **CHECK YOURSELF 3**

Solve the system.

$$2x - y + 3z = 3$$
$$-x + y - 2z = 1$$
$$y - z = 5$$

There is a third possibility for the solutions of systems in three variables, as Example 4 illustrates.

• Example 4

Solving an Inconsistent Linear System in Three Variables

Solve the system.

$$3x + y - 3z = 1 \tag{15}$$
$$-2x - y + 2z = 1 \tag{16}$$
$$-x - y + z = 2 \tag{17}$$

This time we eliminate variable y. Adding equations (15) and (16), we have

$$x - z = 2 \tag{18}$$

Adding equations (15) and (17) gives

$$2x - 2z = 3 \tag{19}$$

Now, multiply equation (18) by -2 and add the result to equation (19).

$$\begin{array}{r} -2x + 2z = -4 \\ \underline{2x - 2z = 3} \\ 0 = -1 \end{array}$$

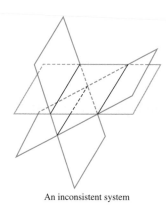

An inconsistent system

All the variables have been eliminated, and we have arrived at a contradiction, $0 = -1$. This means that the system is *inconsistent* and has no solutions. There is *no* point common to all three planes.

● ● ● CHECK YOURSELF 4

Solve the system.

$$\begin{array}{r} x - y - z = 0 \\ -3x + 2y + z = 1 \\ 3x - y + z = -1 \end{array}$$

As a closing note, we have by no means illustrated all possible types of inconsistent and dependent systems. Other possibilities involve either distinct parallel planes or planes that coincide. The solution techniques in these additional cases are, however, similar to those illustrated above.

SOLVING APPLICATIONS

In many instances, if an application involves three unknown quantities, you will find it useful to assign three variables to those quantities and then build a system of three equations from the given relationships in the problem. The extension of our problem-solving strategy is natural, as Example 5 illustrates.

• Example 5

Solving a Number Problem

The sum of the digits of a three-digit number is 12. The tens digit is 2 less than the hundreds digit, and the units digit is 4 less than the sum of the other two digits. What is the number?

Step 1 The three unknowns are, of course, the three digits of the number.

Sometimes it helps to choose variable letters that relate to the words as is done here.

Step 2 We now want to assign variables to each of the three digits. Let u be the units digit, t be the tens digit, and h be the hundreds digit.

Step 3 There are three conditions given in the problem that allow us to write the necessary three equations. From those conditions

$$h + t + u = 12$$
$$t = h - 2$$
$$u = h + t - 4$$

Take a moment now to go back to the original problem and pick out those conditions. That skill is a crucial part of the problem-solving strategy.

Step 4 There are various ways to approach the solution. To use addition, write the system in the equivalent form

$$h + t + u = \ \ 12$$
$$-h + t \qquad = -2$$
$$-h - t + u = -4$$

and solve by our earlier methods. The solution, which you can verify, is $h = 5$, $t = 3$, and $u = 4$. The desired number is 534.

Step 5 To check, you should show that the digits of 534 meet each of the conditions of the original problem.

● ● ●   **CHECK YOURSELF 5**

The sum of the measures of the angles of a triangle is $180°$. In a given triangle, the measure of the second angle is twice the measure of the first. The measure of the third angle is $30°$ less than the sum of the measures of the first two. Find the measure of each angle.

Let's continue with a slightly different application that will lead to a system of three equations.

•Example 6

Solving an Investment Application

Monica decided to divide a total of $42,000 into three investments: a savings account paying 5% interest, a time deposit paying 7%, and a bond paying 9%. Her total annual interest from the three investments was $2600, and the interest from the bank account was $200 less than the total interest from the other two investments. How much did she invest at each rate?

Step 1 The three amounts are the unknowns.

Again, we choose letters that suggest the unknown quantities—*s* for savings, *t* for time deposit, and *b* for bond.

Step 2 We let s be the amount invested at 5%, t the amount at 7%, and b the amount at 9%. Note that the interest from the savings account is then 0.05s, and so on.

A table will help with the next step.

For 1 year, the interest formula is

$I = Pr$

(interest equals principal times rate).

	5%	7%	9%
Principal	s	t	b
Interest	$0.05s$	$0.07t$	$0.09b$

Step 3 Again there are three conditions in the given problem. By using the table above, they lead to the following equations.

Total invested.

$$s + t + b = 42,000$$

Total interest.

$$0.05s + 0.07t + 0.09b = 2,600$$

The savings interest was $200 *less than* that from the other two investments.

$$0.05s = 0.07t + 0.09b - 200$$

Step 4 We clear of decimals and solve as before, with the result

Find the interest earned from each investment, and verify that the conditions of the problem are satisfied.

$$s = \$24,000 \qquad t = \$11,000 \qquad b = \$7000$$

Step 5 We leave the check of these solutions to you.

● ● ● **CHECK YOURSELF 6**

Glenn has a total of $11,600 invested in three accounts: a savings account paying 6% interest, a stock paying 8%, and a mutual fund paying 10%. The annual interest from the stock and mutual fund is twice that from the savings account, and the mutual fund returned $120 more than the stock. How much did Glenn invest in each account?

● ● ● **CHECK YOURSELF ANSWERS**

1. $(5, 1, -3)$.   2. $(1, -3, -2)$.
3. The system is dependent (there are an infinite number of solutions).
4. The system is inconsistent (there are no solutions).
5. The three angles are 35°, 70°, and 75°.
6. $5000 in savings, $3000 in stocks, and $3600 in mutual funds.

7.2 Exercises

In Exercises 1 to 20, solve each system of equations. If a unique solution does not exist, state whether the system is inconsistent or has an infinite number of solutions.

1.
$$x - y + z = 3$$
$$2x + y + z = 8$$
$$3x + y - z = 1$$

2.
$$x - y - z = 2$$
$$2x + y + z = 8$$
$$x + y + z = 6$$

3.
$$x + y + z = 1$$
$$2x - y + 2z = -1$$
$$-x - 3y + z = 1$$

4.
$$x - y - z = 6$$
$$-x + 3y + 2z = -11$$
$$3x + 2y + z = 1$$

5.
$$x + y + z = 1$$
$$-2x + 2y + 3z = 20$$
$$2x - 2y - z = -16$$

6.
$$x + y + z = -3$$
$$3x + y - z = 13$$
$$3x + y - 2z = 18$$

7.
$$2x + y - z = 2$$
$$-x - 3y + z = -1$$
$$-4x + 3y + z = -4$$

8.
$$x + 4y - 6z = 8$$
$$2x - y + 3z = -10$$
$$3x - 2y + 3z = 18$$

9.
$$3x - y + z = 5$$
$$x + 3y + 3z = -6$$
$$x + 4y - 2z = 12$$

10.
$$2x - y + 3z = 2$$
$$x - 2y + 3z = 1$$
$$4x - y + 5z = 5$$

11.
$$x + 2y + z = 2$$
$$2x + 3y + 3z = -3$$
$$2x + 3y + 2z = 2$$

12.
$$x - 4y - z = -3$$
$$x + 2y + z = 5$$
$$3x - 7y - 2z = -6$$

13.
$$x + 3y - 2z = 8$$
$$3x + 2y - 3z = 15$$
$$4x + 2y + 3z = -1$$

14.
$$x + y - z = 2$$
$$3x + 5y - 2z = -5$$
$$5x + 4y - 7z = -7$$

15.
$$x + y - z = 2$$
$$x - 2z = 1$$
$$2x - 3y - z = 8$$

16.
$$x + y + z = 6$$
$$x - 2y = -7$$
$$4x + 3y + z = 7$$

17.
$$x - 3y + 2z = 1$$
$$16y - 9z = 5$$
$$4x + 4y - z = 8$$

18.
$$x - 4y + 4z = -1$$
$$y - 3z = 5$$
$$3x - 4y + 6z = 1$$

19.
$$x + 2y - 4z = 13$$
$$3x + 4y - 2z = 19$$
$$3x + 2z = 3$$

20.
$$x + 2y - z = 6$$
$$-3x - 2y + 5z = -12$$
$$x - 2z = 3$$

Solve Exercises 21 to 36 by choosing a variable to represent each unknown quantity and writing a system of equations.

21. Number Problem. The sum of three numbers is 16. The largest number is equal to the sum of the other two, and 3 times the smallest number is 1 more than the largest. Find the three numbers.

22. Number Problem. The sum of three numbers is 24. Twice the smallest number is 2 less than the largest number, and the largest number is equal to the sum of the other two. What are the three numbers?

23. Coin Problem. A cashier has 25 coins consisting of nickels, dimes, and quarters with a value of $4.90. If the number of dimes is 1 less than twice the number of nickels, how many of each type of coin does she have?

24. Recreation. A theater has tickets at $6 for adults, $3.50 for students, and $2.50 for children under 12 years old. A total of 278 tickets were sold for one showing with a total revenue of $1300. If the number of adult tickets sold was 10 less than twice the number of student tickets, how many of each type of ticket were sold for the showing?

25. Geometry. The perimeter of a triangle is 19 cm. If the length of the longest side is twice that of the shortest side and 3 cm less than the sum of the lengths of the other two sides, find the lengths of the three sides.

26. **Geometry.** The measure of the largest angle of a triangle is 10° more than the sum of the measures of the other two angles and 10° less than 3 times the measure of the smallest angle. Find the measures of the three angles of the triangle.

27. **Investments.** Jovita divides $17,000 into three investments: a savings account paying 6% annual interest, a bond paying 9%, and a money market fund paying 11%. The annual interest from the three accounts is $1540, and she has 3 times as much invested in the bond as in the savings account. What amount does she have invested in each account?

28. **Investments.** Adrienne has $10,000 invested in a savings account paying 5%, a time deposit paying 7%, and a bond paying 10%. She has $1000 less invested in the bond than in her savings account, and she earned $700 in annual interest. What has she invested in each account?

29. **Number Problem.** The sum of the digits of a three-digit number is 9, and the tens digit of the number is twice the hundreds digit. If the digits are reversed in order, the new number is 99 more than the original number. What is the original number?

30. **Number Problem.** The sum of the digits of a three-digit number is 9. The tens digit is 3 times the hundreds digit. If the digits are reversed in order, the new number is 99 less than the original number. Find the original three-digit number.

31. **Business.** A manufacturer can produce and sell x items per week at a cost of $C = 20x + 3600$. The revenue from selling those items is given by $R = 140x$. Find the break-even point for this product.

32. **Business.** If the cost for a second product is given by $C = 30x + 3600$ and the revenue by $R = 130x$, find the break-even point for that product.

33. **Consumer Affairs.** To encourage carpooling, a city charges $10 per single driver or $4 per person for carpools of two or more people in its city parking lots. If one parking lot took in $2020 and 340 cars used that lot for 1 day, how many of each type commuter—single driver or carpool rider—used that lot that day?

34. **Consumer Affairs.** To encourage carpooling, a city charges $12 per single driver or $5 per person for carpools of two or more people in its city parking lots. If one parking lot took in $3220 and 420 cars used that lot for 1 day, how many of each type commuter—single driver or carpool rider—used that lot that day?

35. **Motion.** Roy, Sally, and Jeff drive a total of 50 mi to work each day. Sally drives twice as far as Roy, and Jeff drives 10 mi farther than Sally. Use a system of three equations in three unknowns to find how far each person drives each day.

36. **Consumer Affairs.** A parking lot has spaces reserved for motorcycles, cars, and vans. There are 5 more spaces reserved for vans than for motorcycles. There are 3 times as many car spaces as van and motorcycle spaces combined. If the parking lot has 180 total reserved spaces, how many of each type are there?

The solution process illustrated in this section can be extended to solving systems of more than three variables in a natural fashion. For instance, if four variables are involved, eliminate one variable in the

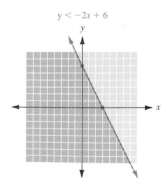

system and then solve the resulting system in three variables as before. Substituting those three values into one of the original equations will provide the value for the remaining variable and the solution for the system.

In Exercises 37 and 38, use this procedure to solve the system.

37.
$$\begin{aligned} x + 2y + 3z + w &= 0 \\ -x - y - 3z + w &= -2 \\ x - 3y + 2z + 2w &= -11 \\ -x + y - 2z + w &= 1 \end{aligned}$$

38.
$$\begin{aligned} x + y - 2z - w &= 4 \\ x - y + z + 2w &= 3 \\ 2x + y - z - w &= 7 \\ x - y + 2z + w &= 2 \end{aligned}$$

In some systems of equations there are more equations than variables. We can illustrate this situation with a system of three equations in two variables. To solve this type of system, pick any two of the equations and solve this system. Then substitute the solution obtained into the third equation. If a true statement results, the solution used is the solution to the entire system. If a false statement occurs, the system has no solution.

In Exercises 39 and 40, use this procedure to solve each system.

39.
$$\begin{aligned} x - y &= 5 \\ 2x + 3y &= 20 \\ 4x + 5y &= 38 \end{aligned}$$

40.
$$\begin{aligned} 3x + 2y &= 6 \\ 5x + 7y &= 35 \\ 7x + 9y &= 8 \end{aligned}$$

41. Experiments have shown that cars (C), trucks (T), and buses (B) emit different amounts of air pollutants. In one such experiment, a truck emitted 1.5 pounds (lb) of carbon dioxide (CO_2) per passenger-mile and 2 grams (g) of nitrogen oxide (NO) per passenger-mile. A car emitted 1.1 lb of CO_2 per passenger-mile and 1.5 g of NO per passenger-mile. A bus emitted 0.4 lb of CO_2 per passenger-mile and 1.8 g of NO per passenger-mile. A total of 85 mi was driven by the three vehicles, and 73.5 lb of CO_2 and 149.5 g of NO were collected. Use the following system of equations to determine the miles driven by each vehicle.

$$\begin{aligned} T + C + B &= 85.0 \\ 1.5T + 1.1C + 0.4B &= 73.5 \\ 2T + 1.5C + 1.8B &= 149.5 \end{aligned}$$

42. Experiments have shown that cars (C), trucks (T), and trains (R) emit different amounts of air pollutants. In one such experiment, a truck emitted 0.8 lb of carbon dioxide per passenger-mile and 1 g of nitrogen oxide per passenger-mile. A car emitted 0.7 lb of CO_2 per passenger-mile and 0.9 g of NO per passenger-mile. A train emitted 0.5 lb of CO_2 per passenger-mile and 4 g of NO per passenger-mile. A total of 141 mi was driven by the three vehicles, and 82.7 lb of CO_2 and 424.4 g of NO were collected. Use the following system of equations to determine the miles driven by each vehicle.

$$\begin{aligned} T + C + R &= 141.0 \\ 0.8T + 0.7C + 0.5R &= 82.7 \\ T + 0.9C + 4R &= 424.4 \end{aligned}$$

Answers

1. (1, 2, 4)   **3.** $(-2, 1, 2)$ **5.** $(-4, 3, 2)$ **7.** Infinite number of solutions   **9.** $\left(3, \dfrac{1}{2}, -\dfrac{7}{2}\right)$

11. $(3, 2, -5)$ **13.** $(2, 0, -3)$ **15.** $\left(4, -\dfrac{1}{2}, \dfrac{3}{2}\right)$ **17.** Inconsistent   **19.** $\left(2, \dfrac{5}{2}, -\dfrac{3}{2}\right)$ **21.** 3, 5, 8

23. 3 nickels, 5 dimes, 17 quarters   **25.** 4 cm, 7 cm, 8 cm   **27.** $3000 savings, $9000 bond, $5000 money market   **29.** 243   **31.** 30   **33.** 110 single, 230 carpool   **35.** Roy 8 mi, Sally 16 mi, Jeff 26 mi   **37.** $(1, 2, -1, -2)$ **39.** $(7, 2)$ **41.** $T = 20, C = 25, B = 40$

7.3 Graphing Linear Inequalities in Two Variables

7.3 OBJECTIVES

1. Graph linear inequalities in two variables
2. Graph a region defined by linear inequalities

What does the solution set look like when we are faced with an inequality in two variables? Again, we will see that it is a set of ordered pairs best represented by a shaded region. Recall that the general form for a linear inequality in two variables is

$$ax + by < c$$

where a and b cannot both be 0. The symbol $<$ can be replaced with $>$, $\leq$, or $\geq$. Some examples are

$$y \leq -2x + 6 \qquad x - 2y = 4 \qquad \text{or} \qquad 2x - 3y = x + 5y$$

As was the case with an equation, the solution set of a linear inequality is a set of ordered pairs of real numbers. However, in the case of the linear inequalities, we will find that the solution sets will be all the points in an entire region of the plane, called a **half plane**.

To determine such a solution set, let's start with the first inequality listed above. To graph the solution set of

$$y < -2x + 6$$

we begin by writing the corresponding linear equation

$$y = -2x + 6$$

First, note that the graph of $y = -2x + 6$ is simply a straight line.

Now, to graph the solution set of $y < -2x + 6$, we must include all ordered pairs that satisfy that inequality. For instance, if $x = 1$, we have

$$y < -2 \cdot 1 + 6$$

$$y < 4$$

So we want to include all points of the form $(1, y)$, where $y < 4$. Of course, since $(1, 4)$ is *on* the corresponding line, this means that we want all points *below* the line along the vertical line $x = 1$. The result will be similar for any choice of x, and our solution set will then contain all points below the line $y = -2x + 6$. We can then graph the solution set as the shaded region shown. We have the following definition.

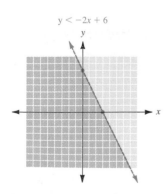

The line is dashed to indicate that the equality is *not* included.

We call the graph of the equation

$ax + by = c$

the **boundary line** at the half planes.

> In general, the solution set of an inequality of the form
>
> $ax + by < c$ or $ax + by > c$
>
> will be a half plane either above or below the corresponding line determined by
>
> $ax + by = c$

How do we decide which half plane represents the desired solution set? The use of a **test point** provides an easy answer. Choose any point *not* on the line. Then substitute the coordinates of that point into the given inequality. If the coordinates satisfy the inequality (result in a true statement), then shade the region or half plane that includes the test point, if not, shade the opposite half plane. Example 1 illustrates the process.

•Example 1

Graphing a Linear Inequality

Graph the linear inequality

$x - 2y < 4$

First, we graph the corresponding equation

$x - 2y = 4$

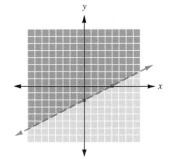

to find the boundary line. Now to decide on the appropriate half plane, we need a test point *not* on the line. As long as the line *does not pass through the origin,* we can always use $(0, 0)$ as a test point. It provides the easiest computation.

Here letting $x = 0$ and $y = 0$, we have

$0 - 2 \cdot 0 < 4$

$0 < 4$

Since this is a true statement, we proceed to shade the half plane including the origin (the test point), as shown.

● ● ● **CHECK YOURSELF 1**

Graph the solution set of $3x + 4y > 12$.

The graphs of some linear inequalities will include the boundary line. That will be the case whenever equality is included with the inequality statement, as illustrated in Example 2.

● Example 2

Graphing a Linear Inequality

Graph the inequality

$2x + 3y \geq 6$

$2x + 3y \geq 6$

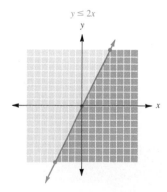

First, we graph the boundary line, here corresponding to $2x + 3y = 6$. Note that we use a solid line in this case since equality is included in the original statement.

Again, we choose a convenient test point not on the line. As before, the origin will provide the simplest computation.

Substituting $x = 0$ and $y = 0$, we have

$2 \cdot 0 + 3 \cdot 0 \geq 6$

$0 \geq 6$

A solid boundary line means that equality *is included.*

This is a *false* statement. Hence the graph will consist of all points on the *opposite* side from the origin. The graph will then be the upper half plane, as shown.

● ● ● **CHECK YOURSELF 2**

Graph the solution set of $x - 3y \leq 6$.

● Example 3

Graphing a Linear Inequality

Graph the solution set of

$y \leq 2x$

$y \leq 2x$

We proceed as before by graphing the boundary line (it is a solid since equality is included). The only difference between this and previous examples is that we *cannot use the origin* as a test point. Do you see why?

Choosing $(1, 1)$ as our test point gives the statement

$1 \leq 2 \cdot 1$

$1 \leq 2$

The choice of $(1, 1)$ is arbitrary. We simply want *any* point *not* on the line.

Since the statement is *true,* we shade the half plane *including* the test point $(1, 1)$.

● ● ● **CHECK YOURSELF 3**

Graph the solution set of $3x + y > 0$.

Let's consider a special case of graphing linear inequalities in the rectangular coordinate system.

• Example 4

Graphing a Linear Inequality

Here we specify the rectangular coordinate system to indicate we want a two-dimensional graph.

Graph the solution set of $x > 3$.

First, we draw the boundary line (a dashed line since equality is not included) corresponding to

$$x = 3$$

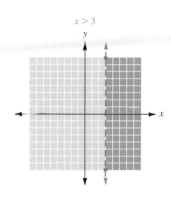

We can choose the origin as a test point in this case, and that results in the false statement

$$0 > 3$$

We then shade the half plane *not* including the origin. In this case, the solution set is represented by the half plane to the right of the vertical boundary line.

As you may have observed, in this special case choosing a test point is not really necessary. Since we want values of x that are *greater than* 3, we want those ordered pairs that are to the *right* of the boundary line.

● ● ● **CHECK YOURSELF 4**

Graph the solution set of

$$y \leq 2$$

in the rectangular coordinate system.

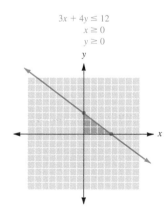

Applications of linear inequalities will often involve more than one inequality condition. Consider Example 5.

• Example 5

Graphing a Region Defined by Linear Inequalities

Graph the region satisfying the following conditions.

$$3x + 4y \leq 12$$
$$x \geq 0$$
$$y \geq 0$$

The solution set in this case must satisfy *all three conditions*. As before, the solution set of the first inequality is graphed as the half plane *below* the boundary line. The second and third inequalities mean that x and y must also be nonnegative. Therefore, our solution set is restricted to the first quadrant (and the appropriate segments of the x and y axes), as shown.

● ● ● **CHECK YOURSELF 5**

Graph the region satisfying the following conditions.

$$3x + 4y < 12$$
$$x \geq 0$$
$$y \geq 0$$

The following algorithm summarizes our work in graphing linear inequalities in two variables.

To Graph a Linear Inequality

1. Replace the inequality symbol with an equality symbol to form the equation of the boundary line of the solution set.
2. Graph the boundary line. Use a dashed line if equality is not included ($<$ or $>$). Use a solid line if equality is included ($\leq$ or $\geq$).
3. Choose any convenient test point *not* on the boundary line.
4. If the inequality is *true* for the test point, shade the half plane *including* the test point. If the inequality is *false* for the test point, shade the half plane *not including* the test point.

● ● ● **CHECK YOURSELF ANSWERS**

1.

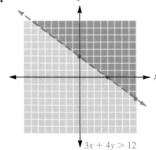

$3x + 4y > 12$

2.

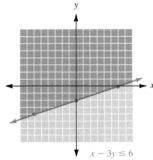

$x - 3y \leq 6$

3.

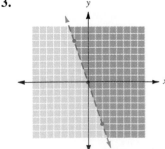

$3x + y > 0$

4.

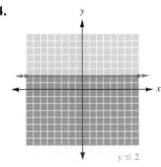

$y \leq 2$

5.
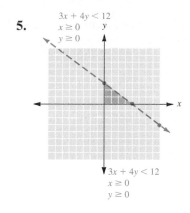
$3x + 4y < 12$
$x \geq 0$
$y \geq 0$

$3x + 4y < 12$
$x \geq 0$
$y \geq 0$

7.3 Exercises

In Exercises 1 to 24, graph the solution sets of the linear inequalities.

1. $x + y < 4$

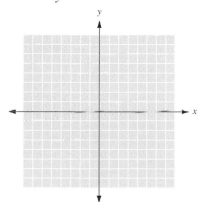

2. $x + y \geq 6$

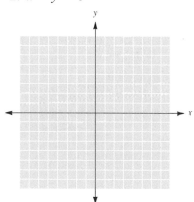

3. $x - y \geq 3$

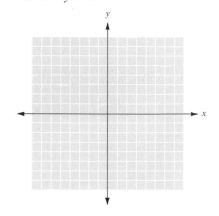

4. $x - y < 5$

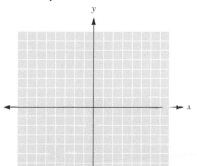

5. $y \geq 2x + 1$

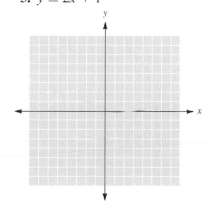

6. $y < 3x - 4$

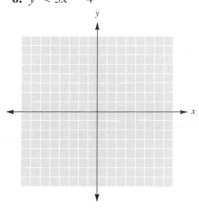

7. $2x + 3y < 6$

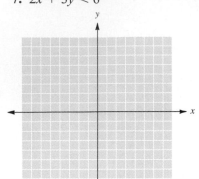

8. $3x - 4y \geq 12$

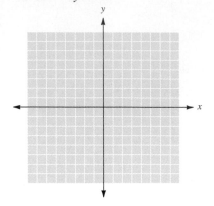

9. $x - 4y > 8$

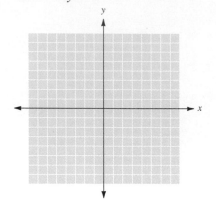

10. $2x + 5y \leq 10$

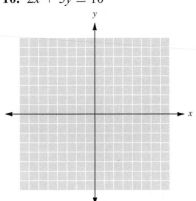

11. $y \geq 3x$

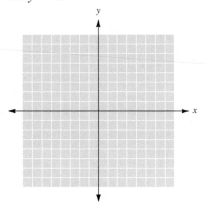

12. $y \leq -2x$

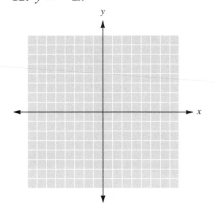

13. $x - 2y > 0$

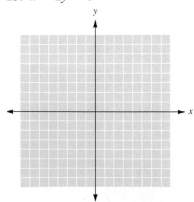

14. $x + 4y \leq 0$

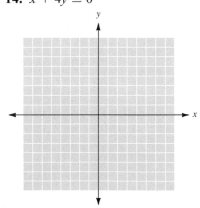

15. $x < 3$

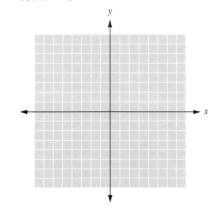

16. $y < -2$

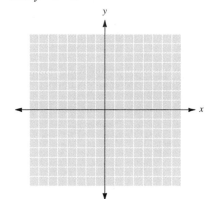

17. $y > 3$

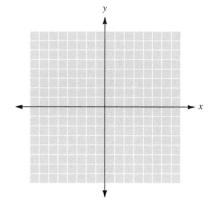

18. $x \leq -4$

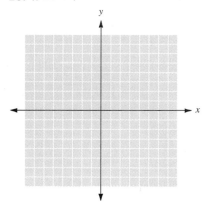

19. $3x - 6 \leq 0$

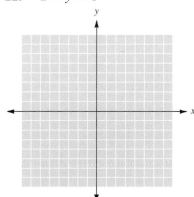

20. $-2y > 6$

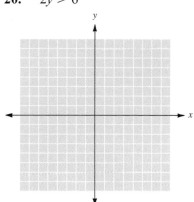

21. $0 < x < 1$

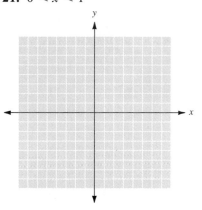

22. $-2 \leq y \leq 1$

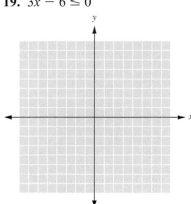

23. $1 \leq x \leq 3$

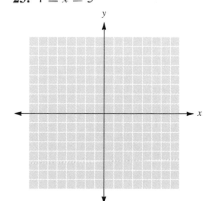

24. $1 < y < 5$

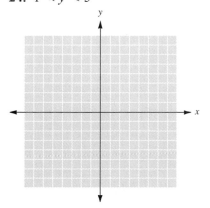

In Exercises 25 to 28, graph the region satisfying each set of conditions.

25. $0 \leq x \leq 3$
$2 \leq y \leq 4$

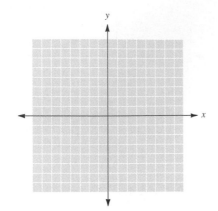

26. $1 \leq x \leq 5$
$0 \leq y \leq 3$

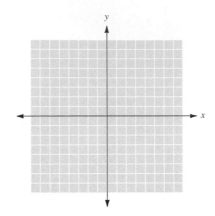

27. $x + 2y \leq 4$
$x \geq 0$
$y \geq 0$

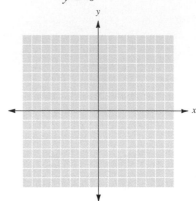

28. $2x + 3y \leq 6$
$x \geq 0$
$y \geq 0$

29. Assume that you are working only with the variable x. Describe the solution to the statement $x > -1$.

30. Now, assume that you are working in two variables, x and y. Describe the solution to the statement $x > -1$.

31. Manufacturing. A manufacturer produces a standard model and a deluxe model of a 13-in. television set. The standard model requires 12 h to produce, while the deluxe model requires 18 h. The labor available is limited to 360 h per week.

If x represents the number of standard-model sets produced per week and y represents the number of deluxe models, draw a graph of the region representing the feasible values for x and y. Keep in mind that the values for x and y must be nonnegative since they represent a quantity of items. (This will be the solution set for the system of inequalities.)

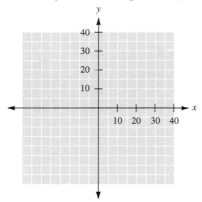

32. **Manufacturing.** A manufacturer produces standard record turntables and CD players. The turntables require 10 h of labor to produce while CD players require 20 h. Let x represent the number of turntables produced and y the number of CD players.

If the labor hours available are limited to 300 h per week, graph the region representing the feasible values for x and y.

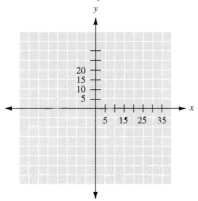

33. **Serving Capacity.** A hospital food service can serve at most 1000 meals per day. Patients on a normal diet receive 3 meals per day and patients on a special diet receive 4 meals per day. Write a linear inequality that describes the number of patients that can be served per day and draw its graph.

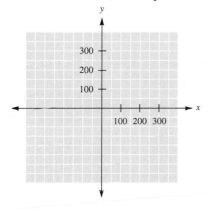

34. Time on Job. The movie and TV critic for the local radio station spends 3 to 7 hours daily reviewing movies and fewer than 4 hours reviewing TV shows. Let x represent the hours watching movies and y represent the time spent watching TV. Write two inequalities that model the situation, and graph their intersection.

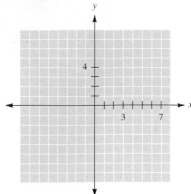

Answers

1. $x + y < 4$

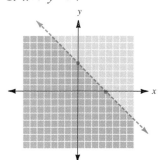

3. $x - y \geq 3$

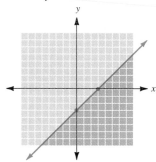

5. $y \geq 2x + 1$

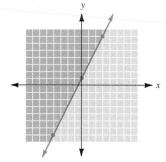

7. $2x + 3y \leq 6$

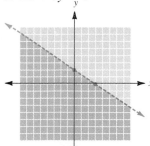

9. $x - 4y > 8$

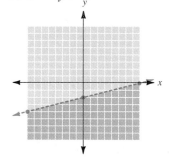

11. $y \geq 3x$

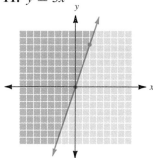

13. $x - 2y > 0$

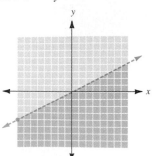

15. $x < 3$

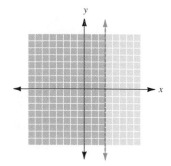

17. $y > 3$

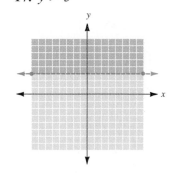

19. $3x - 6 \leq 0$

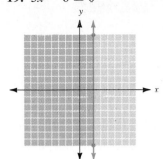

21. $0 < x < 1$

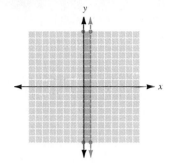

23. $1 \leq x \leq 3$

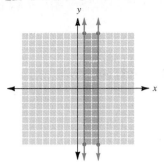

25. $0 \leq x \leq 3$
$\quad\ 2 \leq y \leq 4$

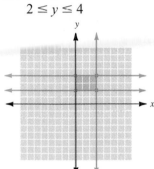

27. $x + 2y \leq 4$
$\qquad x \geq 0$
$\qquad y \geq 0$

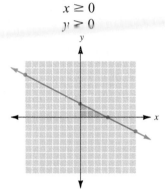

29.

31.

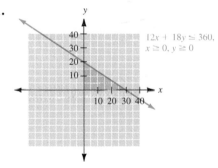

$12x + 18y \geq 360,$
$x \geq 0, y \geq 0$

33.

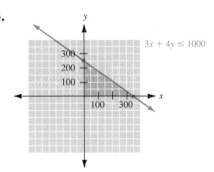

$3x + 4y \leq 1000$

7.4 Graphing Systems of Linear Inequalities in Two Variables

7.4 OBJECTIVES

1. Graph systems of linear inequalities
2. Find the solution to an application of a system of linear inequalities

Our previous work in this chapter dealt with finding the solution set of a system of linear equations. That solution set represented the points of intersection of the graphs of the equations in the system. In this section, we extend that idea to include systems of linear inequalities.

In this case, the solution set is all ordered pairs that satisfy each inequality. **The graph of the solution set of a system of linear inequalities** is then the intersection of the graphs of the individual inequalities. Let's look at an example.

● Example 1

Solving a System by Graphing

Solve the following system of linear inequalities by graphing

$x + y > 4$

$x - y < 2$

We start by graphing each inequality separately. The boundary line is drawn, and using $(0, 0)$ as a test point, we see that we should shade the half plane above the line in both graphs.

Note that the boundary line is dashed to indicate it is *not* included in the graph.

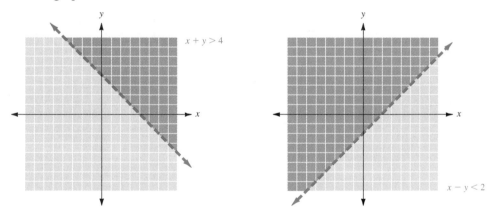

In practice, the graphs of the two inequalities are combined on the same set of axes, as is shown below. The graph of the solution set of the original system is the intersection of the graphs drawn above.

Points on the lines are not included in the solution.

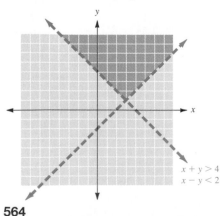

● ● ● **CHECK YOURSELF 1**

Solve the following system of linear inequalities by graphing.

$2x - y < 4$

$x + y < 3$

Most applications of systems of linear inequalities lead to **bounded regions.** This requires a system of three or more inequalities, as shown in Example 2.

● Example 2

Solving a System by Graphing

Solve the following system of linear inequalities by graphing.

$x + 2y \leq 6$

$x + y \leq 5$

$x \geq 2$

$y \geq 0$

On the same set of axes, we graph the boundary line of each of the inequalities. We then choose the appropriate half planes (indicated by the arrow that is perpendicular to the line) in each case, and we locate the intersection of those regions for our graph.

The vertices of the shaded region are given because they have particular significance in later applications of this concept. Can you see how the coordinates of the vertices were determined?

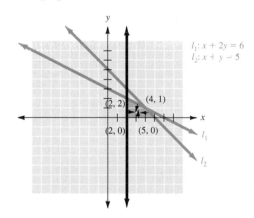

$l_1: x + 2y = 6$
$l_2: x + y = 5$

● ● ● **CHECK YOURSELF 2**

Solve the following system of linear inequalities by graphing.

$2x - y \leq 8 \qquad x \geq 0$

$x + y \leq 7 \qquad y \geq 0$

Let's expand on Example 11, Section 7.1, to see an application of our work with systems of linear inequalities. Consider Example 3.

• Example 3

Solving a Business-Based Application

A manufacturer produces a standard model and a deluxe model of a 13-in. television set. The standard model requires 12 h of labor to produce, while the deluxe model requires 18 h. The labor available is limited to 360 h per week. Also, the plant capacity is limited to producing a total of 25 sets per week. Draw a graph of the region representing the number of sets that can be produced, given these conditions.

As suggested earlier, we let x represent the number of standard-model sets produced and y the number of deluxe-model sets. Since the labor is limited to 360 h, we have

The total labor is limited to (or less than or equal to) 360 h.

$$12x \quad + \quad 18y \quad \leq \quad 360 \tag{1}$$

12 h per standard set   18 h per deluxe set

The total production, here $x + y$ sets, is limited to 25, so we can write

$$x + y \leq 25 \tag{2}$$

We have $x \geq 0$ and $y \geq 0$ since the number of sets produced cannot be negative.

For convenience in graphing, we divide both members of inequality (1) by 6, to write the equivalent system

$$2x + 3y \leq 60$$
$$x + y \leq 25$$
$$x \geq 0$$
$$y \geq 0$$

We now graph the system of inequalities as before. The shaded area represents all possibilities in terms of the number of sets that can be produced.

The shaded area is called the **feasible region.** All points in the region meet the given conditions of the problem and represent possible production options.

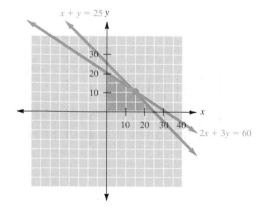

● ● ●   CHECK YOURSELF 3

A manufacturer produces standard record turntables and CD players. The turntables require 10 h of labor to produce while the CD players require 20 h. The labor hours available are limited to 300 h per week. Existing orders require that at least 10 turntables and at least 5 CD players be produced per week. Draw a graph of the region representing the possible production options.

● ● ●   CHECK YOURSELF ANSWERS

1. $2x - y < 4$
$x + y < 3$

2. $2x - y \leq 8$
$x + y \leq 7$
$x \geq 0$
$y \geq 0$

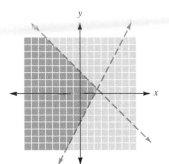
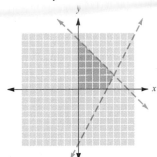

3. Let x be the number of turntables and y be the number of players. The system is

$10x + 20y \leq 300$
$x \geq 10$
$y \geq 5$

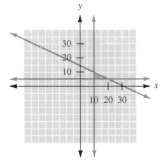

7.4 Exercises

In Exercises 1 to 18, solve each system of linear inequalities graphically.

1. $x + 2y \leq 4$
    $x - y \geq 1$

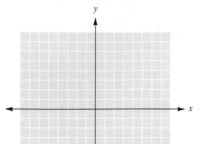

2. $3x - y > 6$
    $x + y < 6$

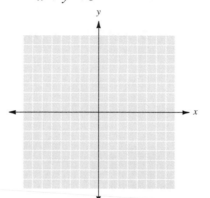

3. $3x + y < 6$
    $x + y > 4$

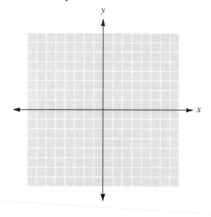

4. $2x + y \geq 8$
    $x + y \geq 4$

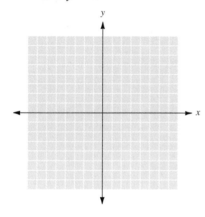

5. $x + 3y \leq 12$
    $2x - 3y \leq 6$

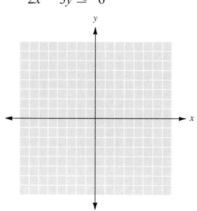

6. $x - 2y > 8$
    $3x - 2y > 12$

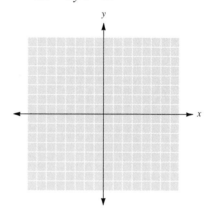

7. $3x + 2y \leq 12$
    $x \geq 2$

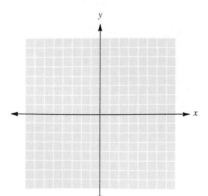

8. $2x + y \leq 6$
    $y \geq 1$

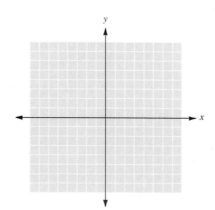

9. $2x + y < 8$
    $x > 1$
    $y > 2$

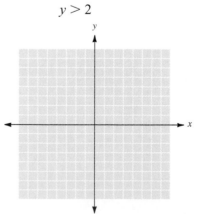

10. $3x - y \leq 6$
$ x \geq 1$
$ y \leq 3$

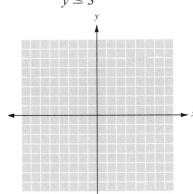

11. $x + 2y \leq 8$
$ 2 \leq x \leq 6$
$ y \geq 0$

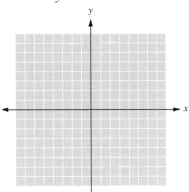

12. $x + y < 6$
$ 0 \leq y \leq 3$
$ x \geq 1$

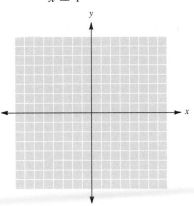

13. $3x + y \leq 6$
$ x + y \leq 4$
$ x \geq 0$
$ y \geq 0$

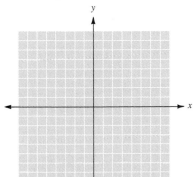

14. $x - 2y \geq -2$
$ x + 2y \leq 6$
$ y \geq 0$
$ y \geq 0$

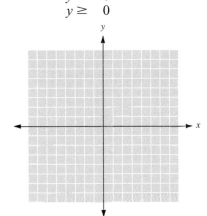

15. $4x + 3y \leq 12$
$ x + 4y \leq 8$
$ x \geq 0$
$ y \geq 0$

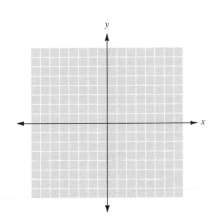

16. $2x + y \leq 8$
$ x + y \geq 3$
$ x \geq 0$
$ y \geq 0$

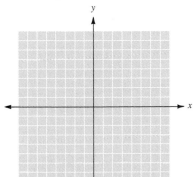

17. $x - 4y \leq -4$
$x + 2y \leq 8$
$ x \geq 2$

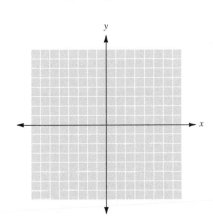

18. $x - 3y \geq -6$
$x + 2y \geq 4$
$ x \leq 4$

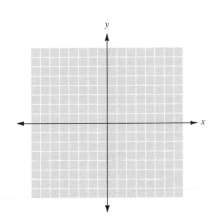

In Exercises 19 and 20, draw the appropriate graph.

19. **Manufacturing.** A manufacturer produces both two-slice and four-slice toasters. The two-slice toaster takes 6 h of labor to produce and the four-slice toaster 10 h. The labor available is limited to 300 h per week, and the total production capacity is 40 toasters per week. Draw a graph of the feasible region, given these conditions, where x is the number of two-slice toasters and y is the number of four-slice toasters.

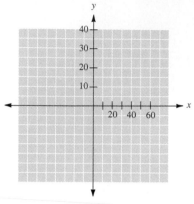

20. **Production.** A small firm produces both AM and AM/FM car radios. The AM radios take 15 h to produce, and the AM/FM radios take 20 h. The number of production hours is limited to 300 h per week. The plant's capacity is limited to a total of 18 radios per week, and existing orders require that at least 4 AM radios and at least 3 AM/FM radios be produced per week. Draw a graph of the feasible region, given these conditions, where x is the number of AM radios and y the number of AM/FM radios.

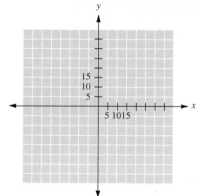

21. When one solves a system of linear inequalities, it is often easier to shade the region that is not part of the solution, rather than the region that is. Try this method, then describe its benefits.

22. Describe a system of linear inequalities for which there is no solution.

23. Write the system of inequalities whose graph is the shaded region.

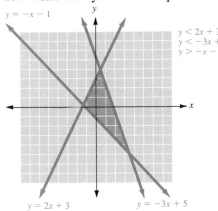

$y = -x - 1$

$y < 2x + 3$
$y < -3x + 5$
$y > -x - 1$

$y = 2x + 3$ $y = -3x + 5$

24. Write the system of inequalities whose graph is the shaded region.

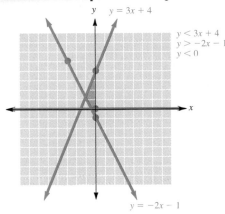

y $y = 3x + 4$

$y < 3x + 4$
$y > -2x - 1$
$y < 0$

$y = -2x - 1$

Answers

1.

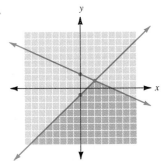

3.

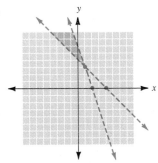

5.

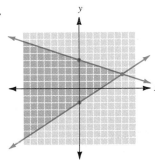

7.

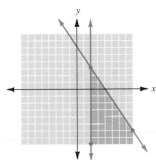

9.

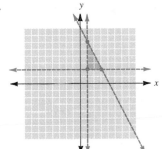

11.

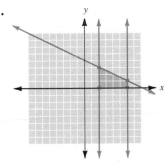

13.

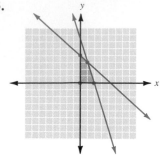

15.

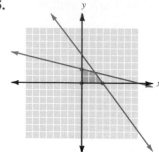

17.

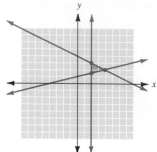

19.

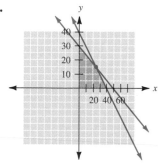

21.

23.

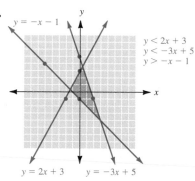

$y = -x - 1$

$y < 2x + 3$
$y < -3x + 5$
$y > -x - 1$

$y = 2x + 3$ $\quad$ $y = -3x + 5$

Graphing Systems of Nonlinear Equations and Inequalities

7.5

7.5 OBJECTIVES

1. Graph systems of non-linear equations
2. Find ordered pairs associated with each system
3. Graph systems of non-linear inequalities
4. Identify the solution of a system of nonlinear inequalities

When we first looked at the solution to a system of linear equations, we graphed the lines and looked at the point of intersection. That point was the solution to the system. When we have a system that involves one or more nonlinear equations, the graph of the system can be extremely valuable. A system with two conic curves can have zero, one, two, three, or four solutions. The following graphs demonstrate each of these possibilities.

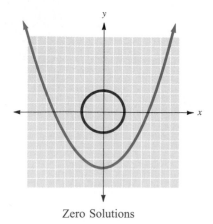

Zero Solutions

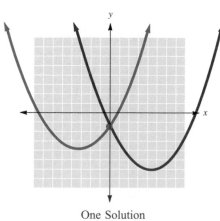

One Solution

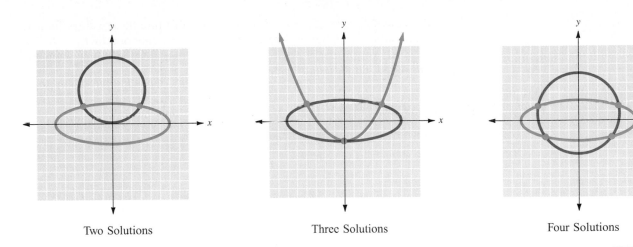

Two Solutions   Three Solutions   Four Solutions

For the remainder of this section, we will restrict our discussion to a system that has as its graph a line and a parabola. Such a system has either zero, one, or two solutions.

• Example 1

Solving a System of Nonlinear Equations

Solve the following system of equations.

$$y = x^2 - 3x + 2$$

$$y = 6$$

First we will graph the system. From this graph we will be able to see the number of solutions. The graph will also give us a way to check the reasonableness of our algebraic solution(s).

Use your calculator to approximate the solutions for the system.

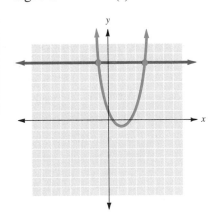

Let's use the method of substitution to solve the system. Substituting 6, from the second equation, for y in the first equation, we get

$$6 = x^2 - 3x + 2$$

$$0 = x^2 - 3x - 4$$

$$0 = (x - 4)(x + 1)$$

The x values for the solutions are -1 and 4. We can substitute these values for x in either equation to solve for y, but we know from the second equation that $y = 6$. The solution set is $\{(-1, 6), (4, 6)\}$. Looking at the graph, we see that this is a reasonable solution to the system.

● ● ● **CHECK YOURSELF 1**

Solve the following system of equations.

$$y = x^2 - 5x + 4$$
$$y = 10$$

Of course, not every quadratic expression is factorable. In Example 2, we must use the quadratic formula.

● **Example 2**

Solving a Nonlinear System

Solve the following system of equations.

$$y = x^2 + x + 3$$
$$y = 7$$

Let's look at the graph of the system.

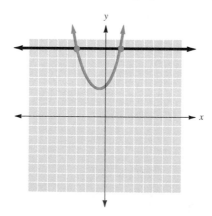

We see two points of intersection, but neither seems to be an integer value for x. Let's solve the system algebraically. Using the method of substitution, we find

$$7 = x^2 + 2x + 3$$
$$0 = x^2 + 2x - 4$$

The result is not factorable, so we use the quadratic formula to find the solutions.

$$x = \frac{-1 \pm \sqrt{1 + 16}}{2} = \frac{-1 \pm \sqrt{17}}{2}$$

The two points of intersection are $\left(\dfrac{-1-\sqrt{17}}{2}, 7\right)$ and $\left(\dfrac{-1+\sqrt{17}}{2}, 7\right)$. It is difficult to check these points against the graph, so we will approximate them. The approximate solutions (to the nearest tenth) are $(-2.6, 7)$ and $(1.6, 7)$. The graph indicates that these are reasonable answers.

CHECK YOURSELF 2

Solve the following system of equations.

$$y = x^2 + x + 5$$
$$y = 8$$

As was stated earlier, not every system has two solutions. In Eample 3, we will see a system with no real solution.

● Example 3

Solving a System of Nonlinear Equations

Solve the following system of equations.

$$y = x^2 - 2x + 1$$
$$y = -2$$

As we did with the previous systems, we will first look at the graph of the system.

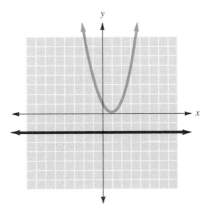

Using the method of substitution, we get

$$-2 = x^2 - 2x + 1$$
$$0 = x^2 - 2x + 3$$

Using the quadratic formula, we can confirm that there are no real solutions to this system.

● ● ●   **CHECK YOURSELF 3**

Solve the following system of equations.

$y = x^2 + 3x + 5$

$y = 2$

NONLINEAR SYSTEMS OF INEQUALITIES

Recall that a system of inequalities has as its solution the set of all ordered pairs that make every inequality in the system a true statement. We almost always express the solution to a system of inequalities graphically. We will do the same thing with non-linear systems.

● Example 4

Solving a System of Nonlinear Inequalities

Solve the following system.

$y \geq x^2 - 3x + 2$

$y \leq 6$

From Example 1, we have the graph of the related system of equations.

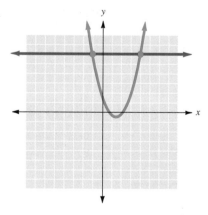

The first inequality has as its solution every ordered pair with a y value that is greater than (above) the graph of the parabola. The second statement has as its solution every ordered pair with a y value that is less than (below) the graph of the line. The solution set to the system is the set of ordered pairs that meet both of those criteria. Here is the graph of the solution set.

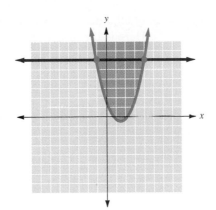

The solution set is the shaded area above the parabola and below the line.

● ● ● **CHECK YOURSELF 4**

Solve the following system.

$$y \geq x^2 - 5x + 4$$
$$y \leq 10$$

Example 5 demonstrates that, even if the related system of equations has no solution, the system of inequalities could have a solution.

● Example 5

Solving a System of Nonlinear Inequalities

Solve the following system.

$$y \leq x^2 - 2x + 1$$
$$y \geq -2$$

As we did with the previous systems, we will first look at the graph of the related system of equations (from Example 3.)

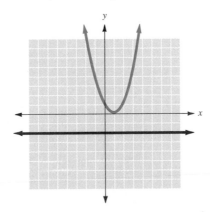

The solution set is now the set of all ordered pairs *below* the parabola ($y < x^2 - 2x + 1$) and *above* the line ($y > -2$). Here is the graph of the solution set.

The solution continued beyond the borders of the grid.

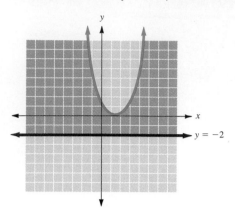

● ● ● **CHECK YOURSELF 5**

Solve the following system.

$$y \le x^2 + 3x + 5$$

$$y \ge 2$$

● ● ● **CHECK YOURSELF ANSWERS**

1. $\{(-1, 10), (6, 10)\}$.   **2.** $\left(\dfrac{-1 \pm \sqrt{13}}{2}, 8\right) \approx \{(1.3, 8), (-2.3, 8)\}$.

3. No real solution.

4.

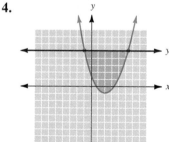

5.

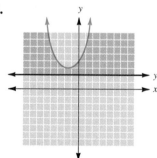

In Exercises 1 to 8, the graph of a system of equations is given. Determine how many real solutions each system has.

1.

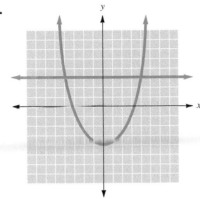

2.

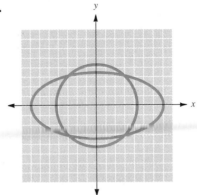

3.

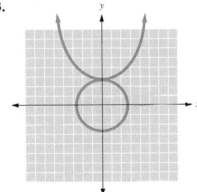

4.

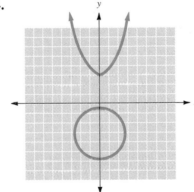

5.

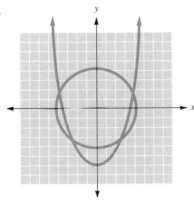

6.

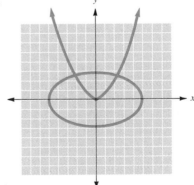

7.

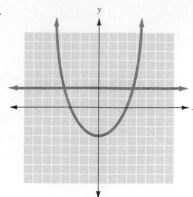

8.

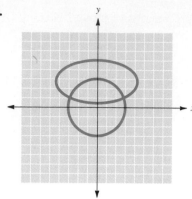

In Exercises 9 to 12, draw the graph of a system that has the indicated number of solutions. Use the conic sections indicated.

9. 0 solutions: **(a)** use a circle and an ellipse, and **(b)** use a parabola and a line.

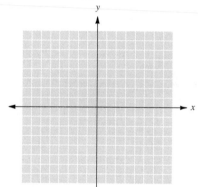

10. 1 solution: **(a)** use a parabola and a circle, and **(b)** use a line and an ellipse.

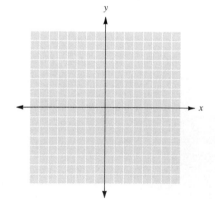

11. 2 solutions: **(a)** use a parabola and a circle, and **(b)** use an ellipse and a parabola.

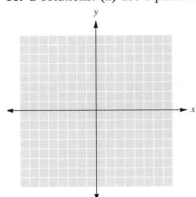

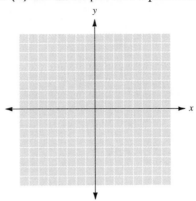

12. 4 solutions: **(a)** use a circle and an ellipse, and **(b)** use a parabola and a circle.

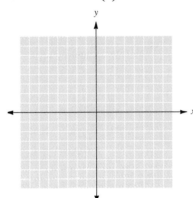

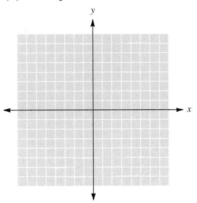

In Exercises 13 to 24, graph each system and estimate the solution.

13. $y = x^2 - x - 2$
   $y = 4$

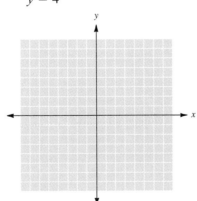

14. $y = x^2 - 2x - 8$
   $y = 7$

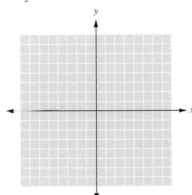

15. $y = x^2 - 5x + 7$
$y = 3$

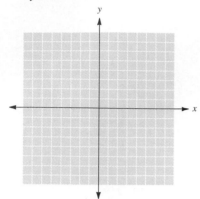

16. $y = x^2 - 8x + 18$
$y = 6$

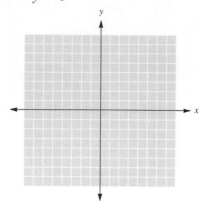

17. $y = x^2 + 4x + 7$
$y = 4$

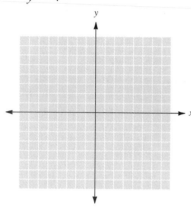

18. $y = x^2 - 7x + 14$
$y = 2$

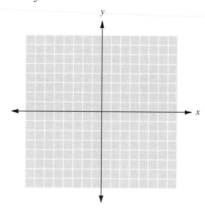

19. $y = x^2 + x + 5$
$y = 6$

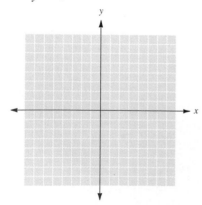

20. $y = x^2 - 5x + 9$
$y = 7$

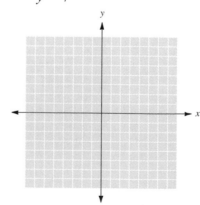

21. $y = x^2 - 7x + 11$
    $y = 6$

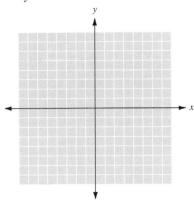

22. $y = x^2 - 2x + 2$
    $y = 6$

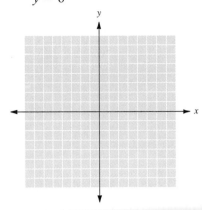

23. $y = x^2 + 5$
    $y = 4$

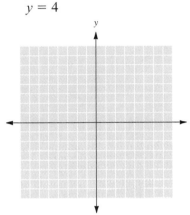

24. $y = x^2 - 4x + 9$
    $y = 2$

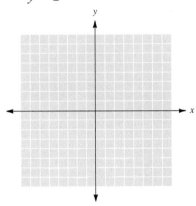

In Exercises 25 to 32, solve using algebraic methods. (Note: These exercises have been solved graphically in Exercises 13 to 24.)

25. $y = x^2 - x - 2$
    $y = 4$
    (See Exercise 13.)

26. $y = x^2 - 2x - 8$
    $y = 7$
    (See Exercise 14.)

27. $y = x^2 - 5x + 7$
    $y = 3$
    (See Exercise 15.)

28. $y = x^2 - 8x + 18$
    $y = 6$
    (See Exercise 16.)

29. $y = x^2 + x + 5$
    $y = 6$
    (See Exercise 19.)

30. $y = x^2 - 5x + 9$
    $y = 7$
    (See Exercise 20.)

31. $y = x^2 + x + 5$
    $y = 4$
    (See Exercise 23.)

32. $y = x^2 - 4x + 9$
    $y = 2$
    (See Exercise 24.)

In Exercises 33 to 40, solve the systems of inequalities graphically. (*Note:* These have already been graphed as systems of equations in Exercises 13 to 24.)

33. $y \geq x^2 - x - 2$
$y \leq 4$
(See Exercise 13.)

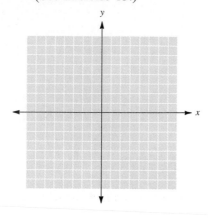

34. $y \geq x^2 - 2x - 8$
$y \leq 7$
(See Exercise 14.)

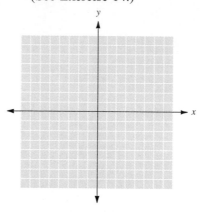

35. $y \geq x^2 - 5x + 7$
$y \leq 3$
(See Exercise 15.)

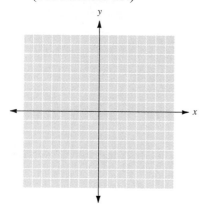

36. $y \geq x^2 - 8x + 18$
$y \leq 6$
(See Exercise 16.)

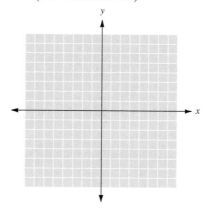

37. $y \geq x^2 + 4x + 7$
$y \geq 4$
(See Exercise 17.)

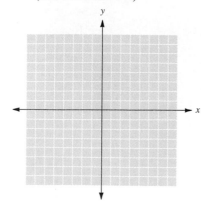

38. $y \geq x^2 - 7x + 14$
$y \geq 2$
(See Exercise 18.)

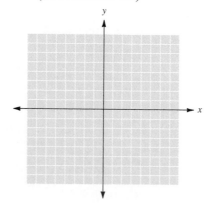

39. $y \leq x^2 + x + 5$
$y \geq 4$
(See Exercise 23.)

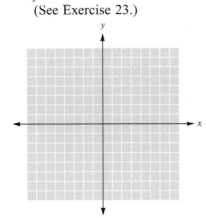

40. $y \leq x^2 - 4x + 9$
$y \geq 2$
(See Exercise 24.)

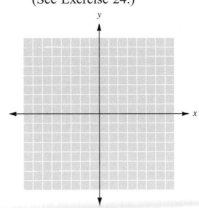

In Exercises 41 to 44, **(a)** graph each system and estimate the solution, and **(b)** use algebraic methods to solve each system.

41. $y = x^2$
$x + y = 2$

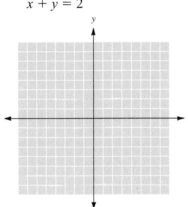

42. $y = 6x + x^2$
$3x - 2y = 10$

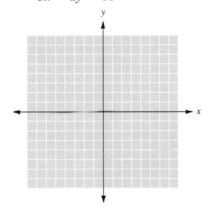

43. $x^2 + y^2 = 5$
$-3x + 4y = 2$

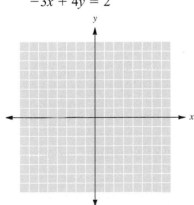

44. $x^2 + y^2 = 9$
$x + y = -3$

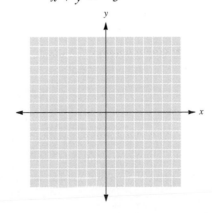

Solve the following applications.

45. The manager of a large apartment complex has found that the profit, in dollars, is given by the equation

$$P = 120x - x^2$$

where x is the number of apartments rented. How many apartments must be rented in order to produce a profit of $3600?

46. The manager of a bicycle shop has found that the revenue (in dollars) from the sale of x bicycles is given by the following equation.

$$R = x^2 - 200x$$

How many bicycles must be sold in order to produce a revenue of $12,500?

47. Find the equation of the line passing through the points of intersection of the graphs $y = x^2$ and $x^2 + y^2 = 90$.

48. Write a system of inequalities to describe the following set of points: The points are in the interior of a circle whose center is the origin with a radius of 4, and above the line $y = 2$.

49. We are asked to solve the following system of equations.

$$x^2 - y = 5$$
$$x + y = -3.$$

Explain how we can determine, before doing any work, that this system cannot have more than two solutions.

50. Without graphing, how can you tell that the following system of inequalities has no solution?

$$x^2 + y^2 < 9$$
$$y > 4$$

Answers

1. 2   **3.** 1   **5.** 4   **7.** 2   **9. (a)**

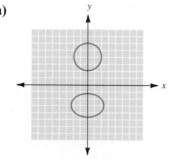

(b)

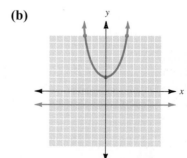

11. (a)

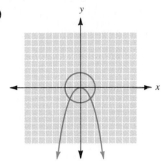

(b)

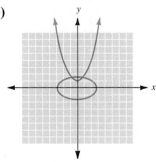

13. (3, 4) and (−2, 4)

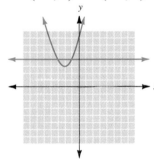

15. (4, 3) and (1, 3)

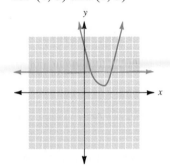

17. (−3, 4) and (−1, 4)

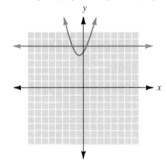

19. (0.6, 6) and (−1.6, 6)

21. (6.2, 6) and (0.8, 6)

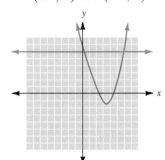

23. No solution

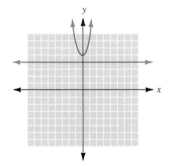

25. (3, 4) and (−2, 4)   **27.** (4, 3) and (1, 3)

29. $\left(\dfrac{-1 + \sqrt{5}}{2}, 6\right)$ and $\left(\dfrac{-1 - \sqrt{5}}{2}, 6\right)$ or (0.618, 6) and (−1.62, 6)   **31.** No solution

33.

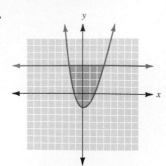

35.

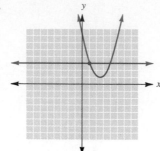

37.

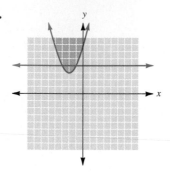

39.

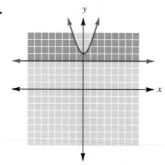

41. $(1, 1)$ and $(-2, 4)$

43. $(-2, -1)$ and $\left(\dfrac{38}{25}, \dfrac{41}{25}\right)$

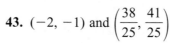

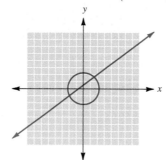

45. 60 **47.** $y = 9$ **49.**

Solving Systems of Linear Equations in Two Variables [7.1]

The solution for the system

$2x - y = 7$

$x + y = 2$

is $(3, -1)$. It is the only ordered pair that will satisfy each equation.

A **system of linear equations** is two or more linear equations considered together. The solution for a linear system in two variables is an ordered pair of real numbers (x, y) that satisfies both equations in the system.

There are three solution techniques: the graphing method, the addition method, and the substitution method.

To solve

$2x - y = 7$

$x + y = 2$

graphically.

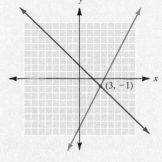

One solution—a consistent system.
No solutions—an inconsistent system.
An infinite number of solutions—a dependent system.

Solving by the Graphing Method

Graph each equation of the system on the same set of coordinate axes. If a solution exists, it will correspond to the point of intersection of the two lines. Such a system is called a **consistent system.** If a solution does not exist, there is no point at which the two lines intersect. Such lines are parallel, and the system is called an **inconsistent system.** If there are an infinite number of solutions, the lines coincide. Such a system is called a **dependent system.** You may or may not be able to determine exact solutions for the system of equations with this method.

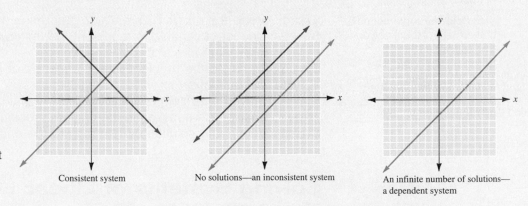

Consistent system

No solutions—an inconsistent system

An infinite number of solutions—a dependent system

To solve

$5x - 2y = 11$ (1)

$2x + 3y = 12$ (2)

Multiply equation (1) by 3 and equation (2) by 2. Then add to eliminate y.

$19x = 57$

$x = 3$

Substituting 3 for x in equation (1), we have

$15 - 2y = 11$

$y = 2$

$(3, 2)$ is the solution.

Solving by the Additional Method

STEP 1   If necessary, multiply one or both of the equations by a constant so that one of the variables can be eliminated by addition.

STEP 2   Add the equations of the equivalent system formed in step 1.

STEP 3   Solve the equation found in step 2.

STEP 4   Substitute the value found in step 3 into either of the equations of the original system to find the corresponding value of the remaining variable. The ordered pair formed is the solution to the system.

STEP 5   Check the solution by substituting the pair of values found in step 4 into the other equation of the original system.

589

To solve

$$3x - 2y = 6 \qquad (3)$$

$$6x + y = 2 \qquad (4)$$

by substitution, solve (4) for y.

$$y = -6x + 2 \qquad (5)$$

Substituting in (3) gives

$$3x - 2(-6x + 2) = 6$$

and

$$x = \frac{2}{3}$$

Substituting $\frac{2}{3}$ for x in (5) gives

$$y = (-6)\left(\frac{2}{3}\right) + 2$$

$$= -4 + 2 = -2$$

The solution is

$$\left(\frac{2}{3}, -2\right)$$

Solving by the Substitution Method

STEP 1 If necessary, solve one of the equations of the original system for one of the variables.

STEP 2 Substitute the expression obtained in step 1 into the other equation of the system to write an equation in a single variable.

STEP 3 Solve the equation found in step 2.

STEP 4 Substitute the value found in step 3 into the equation derived in step 1 to find the corresponding value of the remaining variable. The ordered pair formed is the solution for the system.

STEP 5 Check the solution by substituting the pair of values found in step 4 in *both* equations of the original system.

Solving Applications

Also determine the condition that relates the unknown quantities.

Use a different letter for each variable.

A table or a sketch often helps in writing the equations of the system.

STEP 1 Read the problem carefully to determine the unknown quantities.

STEP 2 Choose a variable to represent the unknowns. Express all other unknowns in terms of this variable.

STEP 3 Translate the problem to the language of algebra to form a system of equations.

STEP 4 Solve the system of equations by any of the methods discussed, and answer the question in the original problem.

STEP 5 Verify your solution by returning to the original problem.

Solving Systems of Linear Equations in Three Variables [7.2]

The solution for a linear system of three equations in three variables is an ordered triple of numbers (x, y, z) that satisfies each equation in the system.

To solve

$x + y - z = 6$ (6)

$2x - 3y + z = -9$ (7)

$3x + y + 2z = 2$ (8)

Adding (6) and (7) gives

$3x - 2y = -3$ (9)

Multiplying (6) by 2 and adding the result to (8) gives

$5x + 3y = 14$ (10)

The system consisting of (9) and (10) is solved as before and

$x = 1$ $y = 3$

Substituting these values into (6) gives

$z = -2$

The solution is $(1, 3, -2)$.

Solving a System of Three Equations in Three Unknowns

STEP 1 Choose a pair of equations from the system, and use the addition method to eliminate one of the variables.

STEP 2 Choose a different pair of equations, and eliminate the same variable.

STEP 3 Solve the system of two equations in two variables determined in steps 1 and 2.

STEP 4 Substitute the values found above into one of the original equations, and solve for the remaining variable.

STEP 5 The solution is the ordered triple of values found in steps 3 and 4. It can be checked by substituting into the other equations of the original system.

Graphing Linear Inequalities in Two Variables [7.3]

In general, the solution set of an inequality of the form

$$ax + by < c \qquad \text{or} \qquad ax + by > c$$

will be a **half plane** either above or below the **boundary line** determined by

$$ax + by = c$$

The boundary line is included in the graph if equality is included in the statement of the original inequality. Such a line is dashed. The boundary line is solid if it is not included in the graph.

To graph a linear inequality:

1. Replace the inequality symbol with an equality symbol to form the equation of the boundary line of the solution set.
2. Graph the boundary line. Use a dashed line if equality is not included ($<$ or $>$). Use a solid line if equality is included ($\leq$ or $\geq$).
3. Choose any convenient test point *not* on the boundary line.
4. If the inequality is *true* for the test point, shade the half plane *including* the test point. If the inequality is *false* for the test point, shade the half plane *not including* the test point.

To graph

$x - 2y < 4$

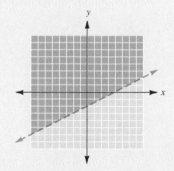

Graphing Systems of Linear Inequalities in Two Variables [7.4]

To solve

$x + 2y \leq 8$

$x + y \leq 6$

$x \geq 0$

$y \geq 0$

graphically

A **system of linear inequalities** is two or more linear inequalities considered together. The **graph of the solution set** of a system of linear inequalities is the intersection of the graphs of the individual inequalities.

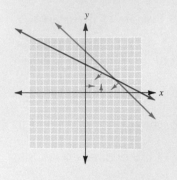

Solving Systems of Linear Inequalities Graphically

1. Graph each inequality, shading the appropriate half plane, on the same set of coordinate axes.
2. The graph of the system is the intersection of the regions shaded in step 1.

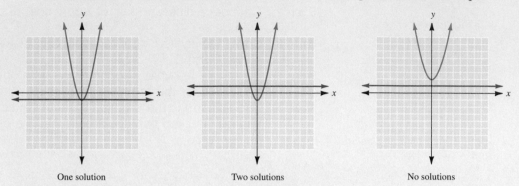

| One solution | Two solutions | No solutions |

To solve:

$y = x^2 + x - 5$

$y = 7$

let

$7 = x^2 + x - 5$

$0 = x^2 + x - 12$

$0 = (x + 4)(x - 3)$

$x = 3, -4$

$(x, y) = (3, 7)(-4, 7)$

Graphing Systems of Nonlinear Equations and Inequalities [7.5]

A system with two conic curves can have zero, one, two, three, or four solutions.

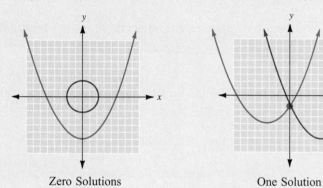

| Zero Solutions | One Solution |

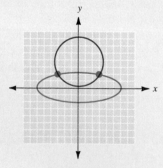

Two Solutions

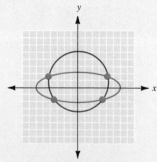

| Three Solutions | Four Solutions |

A system that has as its graph a line and a parabola has, at most, two solutions. To solve such a system, use the following steps:

1. Solve both equations for y.
2. Create a new equation by setting the two right-hand expressions equal to each other.
3. Solve the equation for x.
4. Find the associated ordered pair.

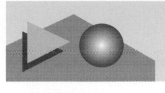

Summary Exercises

This summary exercise set will give you practice with each of the objectives in the chapter. The answers are provided in the *Instructor's Manual*.

[7.1]   Solve each of the following systems by graphing.

1. $x + y = 8$
$x - y = 4$

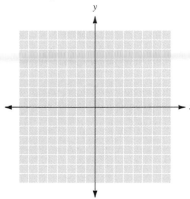

2. $x + 2y = 8$
$x - y = 5$

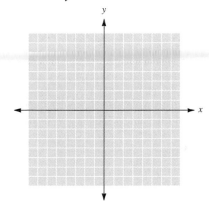

3. $2x + 3y = 12$
$2x + y = 8$

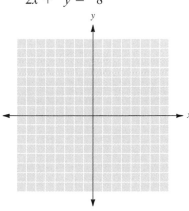

4. $x + 4y = 8$
$y = 1$

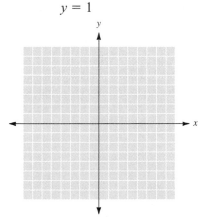

Solve each of the following systems by the addition method. If a unique solution does not exist, state whether the given system is inconsistent or dependent.

5. $x + 2y = 7$
$x - y - 1$

6. $x + 3y = 14$
$4x + 3y = 29$

7. $3x - 5y = 5$
$-x + y = -1$

8. $x - 4y = 12$
$2x - 8y = 24$

9. $6x + 5y - -9$
$-5x + 4y = 32$

10. $3x + y = -17$
$5x - 3y = -19$

11. $3x + y = 8$
$-6x - 2y = -10$

12. $5x - y = -17$
$4x + 3y = -6$

13. $7x - 4y = 27$
$5x + 6y = 6$

14. $4x - 3y = 1$
$6x + 5y = 30$

15. $x - \dfrac{1}{2}y = 8$
$\dfrac{2}{3}x + \dfrac{3}{2}y = -2$

16. $\dfrac{1}{5}x - 2y = 4$
$\dfrac{3}{5}x + \dfrac{2}{3}y = -8$

Solve each of the following systems by the substitution method. If a unique solution does not exist, state whether the given system is inconsistent or dependent.

17. $2x + y = 23$
$x = y + 4$

18. $x - 5y = 26$
$y = x - 10$

19. $3x + y = 7$
$y = -3x + 5$

20. $2x - 3y = 13$
$x = 3y + 9$

21. $5x - 3y = 13$
$x - y = 3$

22. $4x - 3y = 6$
$x + y = 12$

23. $3x - 2y = -12$
$6x + y = 1$

24. $x - 4y = 8$
$-2x + 8y = -16$

Solve each of the following problems by choosing a variable to represent each unknown quantity. Then, write a system of equations that will allow you to solve for each variable.

25. Number Problem. One number is 2 more than 3 times another. If the sum of the two numbers is 30, find the two numbers.

26. Money Value. Suppose that a cashier has 78 $5 and $10 dollar bills with a value of $640. How many of each type of bill does she have?

27. Ticket Sales. Tickets for a basketball game sold at $7 for an adult ticket and $4.50 for a student ticket. If the revenue from 1200 tickets was $7400, how many of each type of ticket were sold?

28. Purchase Price. A purchase of 8 blank cassette tapes and 4 blank videotapes costs $36. A second purchase of 4 cassette tapes and 5 videotapes costs $30. What is the price of a single cassette tape and of a single videotape?

29. Rectangles. The length of a rectangle is 4 cm less than twice its width. If the perimeter of the rectangle is 64 cm, find the dimensions of the rectangle.

30. Mixture. A grocer in charge of bulk foods wishes to combine peanuts selling for $2.25 per pound and cashews selling for $6 per pound. What amount of each nut should be used to form a 120-lb mixture selling for $3 per pound?

31. Investments. Reggie has two investments totaling $17,000—one a savings account paying 6%, the other a time deposit paying 8%. If his annual interest is $1200, what does he have invested in each account?

32. Mixtures. A pharmacist mixes a 20% alcohol solution and a 50% alcohol solution to form 600 mL of a 40% solution. How much of each solution should she use in forming the mixture?

33. **Motion.** A jet flying east, with the wind, makes a trip of 2200 mi in 4 h. Returning, against the wind, the jet can travel only 1800 mi in 4 h. What is the plane's rate in still air? What is the rate of the wind?

34. **Number Problem.** The sum of the digits of a two-digit number is 9. If the digits are reversed, the new number is 45 more than the original number. What was the original number?

35. **Work.** A manufacturer produces $5\frac{1}{4}$-in. computer disk drives and $3\frac{1}{2}$-in. drives. The $5\frac{1}{4}$-in. drives require 20 min of component assembly time; the $3\frac{1}{2}$-in. drives, 25 min. The manufacturer has 500 min of component assembly time available per day. Each drive requires 30 min for packaging and testing, and 690 min of that time is available per day. How many of each of the drives should be produced daily to use all the available time?

36. **Equilibrium Price.** If the demand equation for a product is $D = 270 - 5p$ and the supply equation is $S = 13p$, find the equilibrium point.

37. **Rental Charges.** Two car rental agencies have the following rates for the rental of a compact automobile:

    Company A: $18 per day plus 12¢ per mile.

    Company B: $20 per day plus 10¢ per mile.

    For a 3-day rental, at what number of miles will the charges from the two companies be the same?

Solve each of the following systems by the addition method. If a unique solution does not exist, state whether the given system is inconsistent or dependent.

[7.2]
38. $\begin{aligned} x - y + z &= 0 \\ x + 4y - z &= 14 \\ x + y - z &= 6 \end{aligned}$
39. $\begin{aligned} x - y + z &= 3 \\ 3x + y + 2z &= 15 \\ 2x - y + 2z &= 7 \end{aligned}$
40. $\begin{aligned} x - y - z &= 2 \\ -2x + 2y + z &= -5 \\ -3x + 3y + z &= -10 \end{aligned}$
41. $\begin{aligned} x - y &= 3 \\ 2y + z &= 5 \\ x \quad + 2z &= 7 \end{aligned}$

42. $\begin{aligned} x + y + z &= 2 \\ x + 3y - 2z &= 13 \\ y - 2z &= 7 \end{aligned}$
43. $\begin{aligned} x + y - z &= -1 \\ x - y + 2z &= 2 \\ -5x - y - z &= -1 \end{aligned}$
44. $\begin{aligned} 2x + 3y + z &= 7 \\ -2x - 9y + 2z &= 1 \\ 4x - 6y + 3z &= 10 \end{aligned}$

Solve each of the following problems by choosing a variable to represent each unknown quantity. Then, write a system of equations that will allow you to solve for each variable.

45. **Number Problem.** The sum of three numbers is 15. The largest number is 4 times the smallest number, and it is also 1 more than the sum of the other two numbers. Find the three numbers.

46. **Number Problem.** The sum of the digits of a three-digit number is 16. The tens digit is 3 times the hundreds digit, and the units digit is 1 more than the hundreds digit. What is the number?

47. **Tickets Sold.** A theater has orchestra tickets at $10, box-seat tickets at $7, and balcony tickets at $5. For one performance, a total of 360 tickets were sold, and the total revenue was $3040. If the number of orchestra tickets sold was 40 more than that of the other two types combined, how many of each type of ticket were sold for the performance?

48. **Triangles.** The measure of the largest angle of a triangle is 15° less than 4 times the measure of the smallest angle and 30° more than the sum of the measures of the other two angles. Find the measures of the three angles of the triangle.

49. **Investments.** Rachel divided $12,000 into three investments: a savings account paying 5%, a stock paying 7%, and a mutual fund paying 9%. Her annual interest from the investments was $800, and the amount that she had invested at 5% was equal to the sum of the amounts invested in the other accounts. How much did she have invested in each type of account?

50. **Number Problem.** The difference of two positive numbers is 3, and the sum of those numbers is 41. Find the two numbers.

51. **Number Problem.** The sum of two integers is 144, and the difference 42. What are the two integers?

52. **Rectangle.** A rectangular building lot is $1\frac{1}{2}$ times as wide as it is long. The perimeter of the lot is 400 ft. Find the length and width of the lot.

53. **Break-Even Analysis.** A manufacturer's cost for producing x units of a product is given by

$$C = 10x + 3600$$

The revenue from selling x units of that product is given by

$$R = 100x$$

Find the break-even point for this product.

[7.3] Graph the solution set for each of the following linear inequalities.

54. $y < 2x + 1$

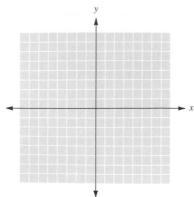

55. $y \geq -2x + 3$

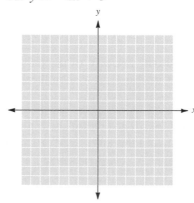

56. $3x + 2y \geq 6$

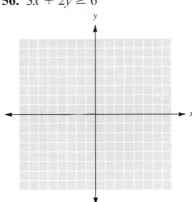

57. $3x - 5y < 15$

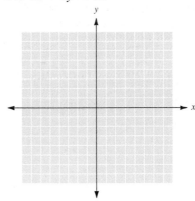

58. $y < -2x$

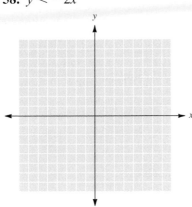

59. $4x - y \geq 0$

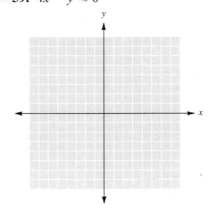

60. $y \geq -3$

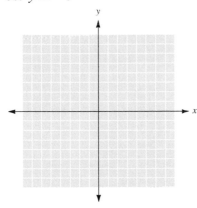

61. $x < 4$

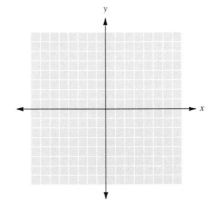

[7.4] Solve each of the following linear inequalities graphically.

62. $x - y < 7$
$x + y > 3$

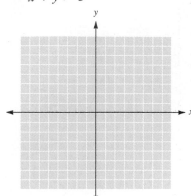

63. $x - 2y \leq -2$
$x + 2y \leq 6$

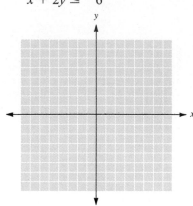

64. $x - 6y < 6$
$-x + y < 4$

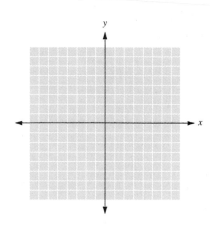

65. $2x + y \leq 8$
$x \geq 1$
$y \geq 0$

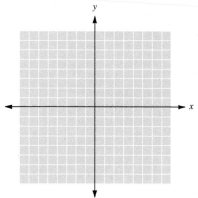

66. $2x + y \leq 6$
$x \geq 1$
$y \geq 0$

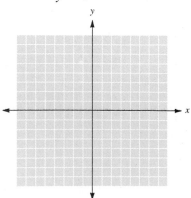

67. $4x + y \leq 8$
$x \geq 0$
$y \geq 2$

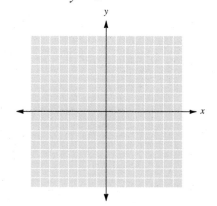

68. $4x + 2y \leq 8$
$x + y \leq 3$
$x \geq 0$
$y \geq 0$

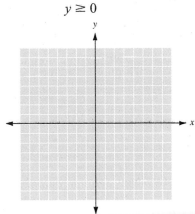

69. $3x + y \leq 6$
$x + y \leq 4$
$x \geq 0$
$y \geq 0$

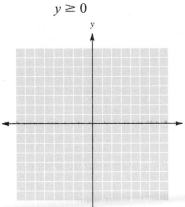

[7.5]   Solve each of the following systems graphically.

70. $y = x^2 - x + 4$
$y = 4$

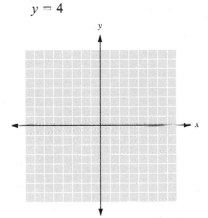

71. $y = 2x^2 - 5x + 9$
$y = 6$

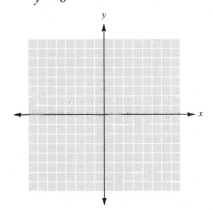

72. $y = 6x^2 + x + 5$
$y = 7$

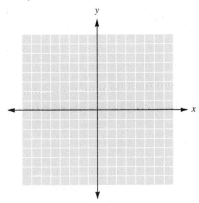

73. $y = x^2 + 12x + 40$
$y = 5$

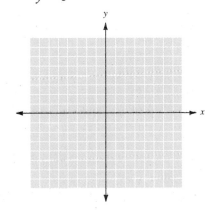

74. $y = 8x^2 + 6x + 7$
   $y = 12$

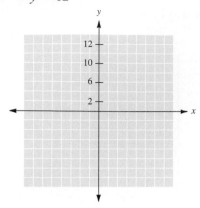

Solve the following systems algebraically.

75. $y = x^2 - x + 4$
   $y = 4$
   (See Exercise 70.)

76. $y = 2x^2 - 5x + 9$
   $y = 6$
   (See Exercise 71.)

77. $y = 6x^2 + x + 5$
   $y = 7$
   (See Exercise 72.)

78. $y = x^2 + 12x + 40$
   $y = 5$
   (See Exercise 73.)

79. $y = 8x^2 + 6x + 7$
   $y = 12$
   (See Exercise 74.)

Solve each of the following systems of inequalities.

80. $y \geq x^2 - x + 4$
   $y \geq 4$
   (See Exercise 70.)

81. $y \geq 2x^2 - 5x + 9$
   $y \leq 6$
   (See Exercise 71.)

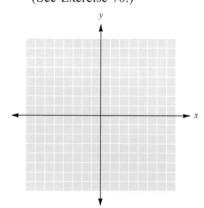

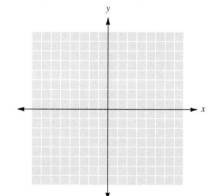

82. $y \geq 6x^2 + x + 5$
    $y \leq 7$
    (See Exercise 72.)

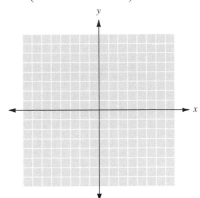

83. $y \geq x^2 + 12x + 40$
    $y \leq 5$
    (See Exercise 73.)

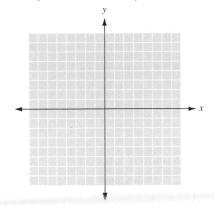

84. $y \leq 8x^2 + 6x + 7$
    $y \leq 12$
    (See Exercise 74.)

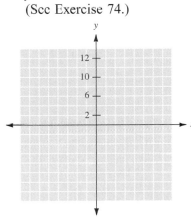

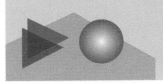

The purpose of this self-test is to help you check your progress and to review for a chapter test in class. Allow yourself about 1 hour to take the test. When you are done, check your answers in the back of the book. If you missed any answers, be sure to go back and review the appropriate sections in the chapter and the exercises that are provided.

Solve each of the following systems. If a unique solution does not exist, state whether the given system is inconsistent or dependent.

1. $3x + y = -5$
$5x - 2y = -23$

2. $4x - 2y = -10$
$y = 2x + 5$

3. $9x - 3y = 4$
$-3x + y = -1$

4. $5x - 3y = 5$
$3x + 2y = -16$

5. $x - 2y = 5$
$2x + 5y = 10$

6. $5x - 3y = 20$
$4x + 9y = -3$

Solve each of the following systems.

7. $x - y + z = 1$
$-2x + y + z = 8$
$x \qquad + 5z = 19$

8. $x + 3y - 2z = -6$
$3x - y + 2z = 8$
$-2x + 3y - 4z = -11$

Solve each of the following problems by choosing a variable to represent each unknown quantity. Then write a system of equations that will allow you to solve for each variable.

9. An order for 30 computer disks and 12 printer ribbons totaled $147. A second order for 12 more disks and 6 additional ribbons cost $66. What was the cost per individual disk and ribbon?

10. A candy dealer wants to combine jawbreakers selling for $2.40 per pound and licorice selling for $3.90 per pound to form a 100-lb mixture that will sell for $3 per pound. What amount of each type of candy should be used?

11. A small electronics firm assembles 5-in. portable television sets and 12-in. models. The 5-in. set requires 9 h of assembly time; the 12-in. set, 6 h. Each unit requires 5 h for packaging and testing. If 72 h of assembly time and 50 h of packaging and testing time are available per week, how many of each type of set should be finished if the firm wishes to use all its available capacity?

12. Hans decided to divide $14,000 into three investments: a savings account paying 6% annual interest, a bond paying 9%, and a mutual fund paying 13%. His annual interest from the three investments was $1100, and he had twice as much invested in the bond as in the mutual fund. What amount did he invest in each type?

13. The fence around a rectangular yard requires 260 ft of fencing. The length is 20 ft less than twice the width. Find the dimensions of the yard.

Graph the solutions set in each of the following.

14. $5x + 6y \leq 30$

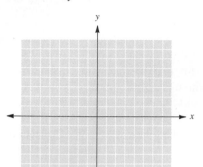

15. $x + 3y > 6$

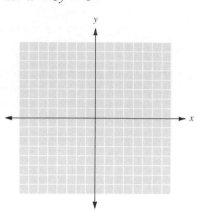

16. $4x - 8 \leq 0$

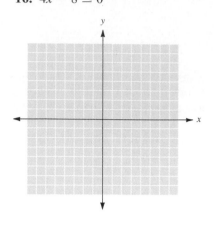

Solve each of the following systems of linear inequalities graphically.

17. $x - 2y < 6$
   $x + \ \ y < 3$

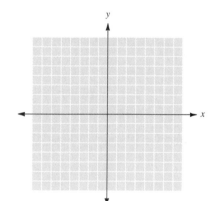

18. $3x + 4y \geq 12$
   $x \geq \ \ 1$

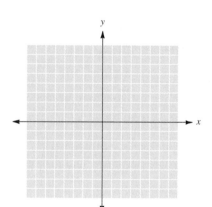

19. $x + 2y \leq 8$
   $x + \ \ y \leq 6$
   $\qquad x \geq 0$
   $\qquad y \geq 0$

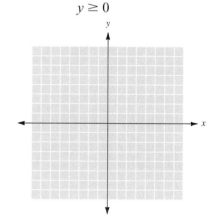

Solve each of the following systems graphically.

20. $y = 2x^2 - x + 6$
   $y = 16$

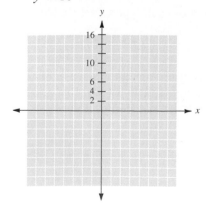

21. $y = 12x^2 + 17x + 3$
   $y = 8$

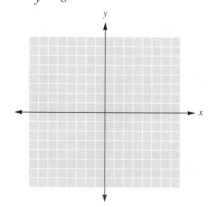

Solve each of the following systems algebraically.

22. $y = 2x^2 - x + 6$
$y = 16$
(See Exercise 20.)

23. $y = 12x^2 + 17x + 3$
$y = 8$
(See Exercise 21.)

Solve each of the following systems of inequalities.

24. $y > 2x^2 - x + 6$
$y < 16$
(See Exercise 20.)

25. $y > 12x^2 + 17x + 3$
$y > 8$
(See Exercise 21.)

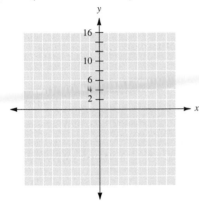

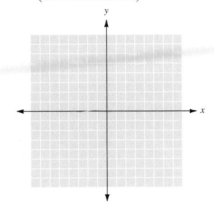

This test is provided to help you in the process of reviewing the previous chapters. Answers are provided in the back of the book. If you missed any answers, be sure to go back and review the appropriate sections.

Solve each of the following.

1. $3x - 2(x + 5) = 12 - 3x$

2. $2x - 7 < 3x - 5$

3. $|2x - 3| = 5$

4. $|3x + 5| \leq 7$

5. $|5x - 4| > 21$

6. $x^2 - 5x - 24 = 0$

Graph each of the following.

7. $5x + 7y = 35$

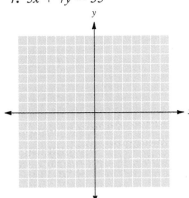

8. $2x + 3y < 6$

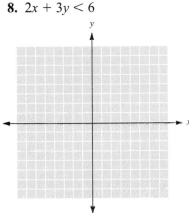

9. Find the distance between the points $(-1, 2)$ and $(4, -22)$.

10. Find the slope of the line connecting $(4, 6)$ and $(3, -1)$.

11. Write the function form of the equation of line that passes through the points $(-1, 4)$ and $(5, -2)$.

Simplify the following polynomial functions.

12. $f(x) = (2x + 1)(x - 3)$

13. $f(x) = (3x - 2)^2$

14. Completely factor the function $f(x) = x^3 - 3x^2 - 5x + 15$.

607

Graph the following.

15. $y = x^2 - 6x + 5$

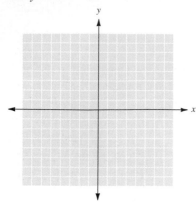

16. $(x + 1)^2 + (y - 2)^2 = 25$

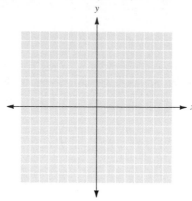

Solve each of the following systems of equations.

17. $2x + 3y = 6$
$5x + 3y = -24$

18. $x + y + z = 3$
$2x - y + 2z = 0$
$-x - 3y + z = -9$

Solve each of the following applications.

19. The length of a rectangle is 3 cm more than twice its width. If the perimeter of the rectangle is 54 cm, find the dimensions of the rectangle.

20. The sum of the digits of a two-digit number is 10. If the digits are reversed, the new number is 36 less than the original number. What was the original number?

EXPONENTIAL AND LOGARITHMIC FUNCTIONS

INTRODUCTION

Pharmacologists researching the effects of drugs use exponential and logarithmic functions to model drug absorption and elimination. After a drug is taken orally, it is distributed throughout the body via the circulatory system. Once in the bloodstream, the drug is carried to the body's organs, where it is first absorbed and then eliminated again into the bloodstream. For a medicine or drug to be effective, there must be enough of the substance in the body to achieve the desired effect but not enough to cause harm. This therapeutic level is maintained by taking the proper dosage at timed intervals determined by the rate the body absorbs or eliminates the medicine.

The rate at which the body eliminates the drug is proportional to the amount of the drug present. That is, the more drug there is, the faster the drug is eliminated. The amount of a drug dosage, P, still left after a number of hours, t, is affected by the **half-life** of the drug. In this case, the half-life is how many hours it takes for the body to use up or eliminate half the drug dosage.

If P is the amount of an initial dose, and H is the time it takes the body to eliminate half a dose of a drug, then the amount of the drug still remaining in the system after t units of time is

$$A(t) = Pe^{t\left(\frac{-\ln 2}{H}\right)}$$

If the amount of an initial dose of a drug is 30 mg and if the half-life of the drug in the body is 4 h, the amount in mg of the drug still in the body t h after one dose is given by the following formula:

$$A(t) = 30e^{-0.173t}$$

© 1990 Chris Jones Photo/The Stock Market

The graph of this function shows how much of the drug will be in the body at each hour after the dose is administered.

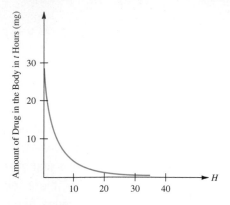

If the therapeutic level of this drug is 20 mg for a certain size person, and it is necessary to give 8 doses over regular 4-h periods, the amount of each dose is given by the equation

$$D = \frac{20(1 - e^{-0.173(4)})}{e^{-0.173(4)}\left(1 - e^{8(-0.173)(4)}\right)} = 20 \text{ mg}$$

Although these formulas model how drugs behave in a living organism, the administration of drugs is not an exact science. The half-life of a drug can vary among different people and can include such factors as the age of the patient, the patient's general health, and the health of vital organs such as the liver and kidneys. Whether or not a person smokes can have an effect on how quickly a drug is cleared from the body. A medical practitioner must use knowledge of the person as well as the data given by this equation and adjust the dosage and time intervals accordingly.

Each formula in this introduction is an example of an *exponential function.* Such functions form the foundation of this chapter. ■

8.1 Inverse Relations and Functions

8.1 OBJECTIVES

1. Find the inverse of a relation
2. Graph a relation and its inverse
3. Find the inverse of a function
4. Graph a function and its inverse
5. Identify a one-to-one function

Let's consider an extension of the concepts of relations and functions discussed in Chapter 2.

Suppose we are given the relation

$$\{(1, 2), (2, 4), (3, 6)\} \tag{1}$$

If we *interchange* the first and second components (the x and y values) of each of the ordered pairs in relation (1), we have

$$\{(2, 1), (4, 2), (6, 3)\} \tag{2}$$

which is another relation. Relations (1) and (2) are called **inverse relations,** and in general we have the following definition.

Inverse of a Relation

The *inverse* of a relation is formed by interchanging the components of each of the ordered pairs in the given relation.

Since we know that relations are often specified by equations, it is natural for us to want to work with the concept of the inverse relation in that setting. We form the inverse relation by interchanging the roles of x and y in the defining equation. Example 1 illustrates this concept.

● Example 1

Finding the Inverse of a Relation

Find the inverse of the relation.

$$f = \{(x, y)| y = 2x - 4\} \tag{3}$$

First interchange variables x and y to obtain

Note that x and y have been interchanged from the original equation.

$$x = 2y - 4$$

We now solve the defining equation for y.

$$2y = x + 4$$

$$y = \frac{1}{2}x + 2$$

Then, we rewrite the relation in the equivalent form.

$$f^{-1} = \left\{(x, y)\Big| y = \frac{1}{2}x + 2\right\} \tag{4}$$

Note: The notation f^{-1} has a *different meaning* from the negative exponent, as in x^{-1} or $\dfrac{1}{x}$.

The inverse of the original relation (3) is now shown in (4) with the defining equation "solved for y." That inverse is denoted f^{-1} (this is read as "f inverse").

We use the notation f^{-1} to indicate the inverse of f when that inverse is *also a function.*

● ● ● **CHECK YOURSELF 1**

Write the inverse relation for $g = \{(x, y)|y = 3x + 6\}$.

The graphs of relations and their inverses are related in an interesting way. First, note that the graphs of the ordered pairs (a, b) and (b, a) always have symmetry about the line $y = x$.

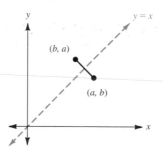

Now, with the above symmetry in mind, let's consider Example 2.

● Example 2

Graphing a Relation and Its Inverse

Graph the relation f from Example 1 along with its inverse.
    Recall that

$$f = \{(x, y)|y = 2x - 4\}$$

and

$$f^{-1} = \left\{(x, y)\middle|y = \frac{1}{2}x + 2\right\}$$

The graphs of f and f^{-1} are shown below.

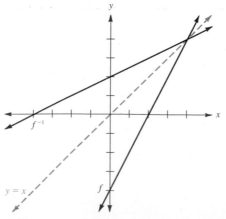

Note that the graphs of f and f^{-1} are symmetric about the line $y = x$. That symmetry follows from our earlier observation about the pairs (a, b) and (b, a) since we simply reversed the roles of x and y in forming the inverse relation.

CHECK YOURSELF 2

Graph the relation g from the Check Yourself 1 exercise along with its inverse.

From our work thus far, it should be apparent that every relation has an inverse. However, that inverse may or may not be a function.

• Example 3

Finding the Inverse of a Function

Find the inverses of the following functions.

(a) $f = \{(1, 3), (2, 4), (3, 9)\}$

Its inverse is

The elements of the ordered pairs have been interchanged.

$\{(3, 1), (4, 2), (9, 3)\}$

which is also a function.

(b) $g = \{(1, 3), (2, 6), (3, 6)\}$

Its inverse is

It is not a function because 6 is mapped to both 2 and 3.

$\{(3, 1), (6, 2), (6, 3)\}$

which is *not* a function.

CHECK YOURSELF 3

Write the inverses for each of the following relations. Which of the inverses are also functions?

(a) $\{(-1, 2), (0, 3), (1, 4)\}$ **(b)** $\{(2, 5), (3, 7), (4, 5)\}$

Can we predict in advance whether the inverse of a function will also be a function? The answer is yes.

We already know that for a relation to be a function, no element in its domain can be associated with more than one element in its range.

In addition, if the inverse of a function is to be a function, no element in the range can be associated with more than one element in the domain—that is, no two distinct ordered pairs in the function can have the same second component. A function that satisfies this additional restriction is called a **one-to-one function.**

The function in Example 3(*a*)

$$f = \{(1, 3), (2, 4), (3, 9)\}$$

is a one-to-one function and its inverse is also a function. However, the function in Example 3(*b*)

$$g = \{(1, 3), (2, 6), (3, 6)\}$$

is *not* a one-to-one function, and its inverse is *not* a function.

From those observations we can state the following general result.

Inverse of a Function

A function *f* has an inverse f^{-1}, which is also a function, if and only if *f* is a one-to-one function.

Because the statement is an "if and only if" statement, it can be turned around without changing the meaning. Here we use the same statement as a definition for a one-to-one function.

One-To-One Function

A function *f* is a *one-to-one function* if and only if it has an inverse f^{-1}, which is also a function.

Our result regarding a one-to-one function and its inverse also has a convenient graphical interpretation, as Example 4 illustrates.

● Example 4

Graphing a Function and Its Inverse

Graph each function and its inverse. State which inverses are functions.

(*a*) $f = \{(x, y)|y = 4x - 8\}$

Since *f* is a one-to-one function (no value for *y* can be associated with more than one value for *x*), its inverse is also a function. Here,

$$f^{-1} = \left\{(x, y)\Big|y = \frac{1}{4}x + 2\right\}$$

This is a **linear function** of the form $f = \{(x, y)|y = mx + b\}$. Its graph is a straight line. A linear function, where $m \neq 0$, is always one-to-one.

The graphs of f and f^{-1} are shown below.

The vertical-line test tells us that *both f* and f^{-1} are functions.

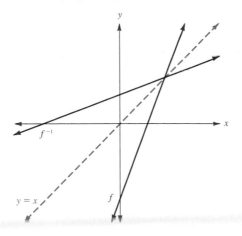

(b) $a = \{(x, y) | y = x^2\}$

This is a **quadratic function** of the form

$g = \{(x, y) | y = ax^2 + bx + c\}$ where $a \neq 0$

Its graph is always a parabola, and a quadratic function is *not* a one-to-one function.

For instance, 4 in the range is associated with both 2 and -2 from the domain. It follows that the inverse of g

$\{(x, y) | x = y^2\}$

or

$\{(x, y) | y = \pm\sqrt{x}\}$

By the vertical-line test, we see that the inverse of *g* is *not* a function because *g* was *not* one-to-one.

is *not* a function. The graphs of g and its inverse are shown below.

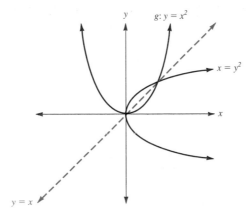

Note: When a function is not one-to-one, as in Example 4(*b*), we can restrict the domain of the function so that it will be one-to-one. In this case, if we redefine function g as

The domain is now restricted to nonnegative values for *x*.

$g = \{(x, y) | y = x^2, x \geq 0\}$

it will be one-to-one and its inverse

$$g^{-1} = \{(x, y)|y = \sqrt{x}\}$$

will be a function, as shown in the following graph.

The function g is now one-to-one, its inverse g^{-1} is also a function.

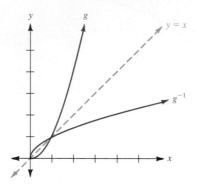

●●● **CHECK YOURSELF 4**

Graph each function and its inverse. Which inverses are functions?

(a) $f = \{(x, y)|y = 2x - 2\}$ **(b)** $g = \{(x, y)|y = 2x^2\}$

It is easy to tell from the graph of a function whether that function is one-to-one. If any horizontal line can meet the graph of a function in at most one point, the function is one-to-one. Example 5 illustrates this approach.

●**Example 5**

Identifying a One-to-One Function

Which of the following graphs represent one-to-one functions?

(a)

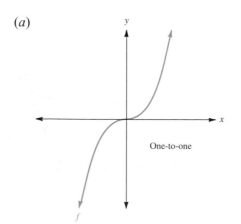

Since no horizontal line passes through any two points of the graph, f is one-to-one.

(b)

Since a horizontal line can meet the graph of function g at two points, g is *not* a one-to-one function.

Not one-to-one

● ● ● **CHECK YOURSELF 5**

Consider the graphs of the functions of Check Yourself 4. Which functions are one-to-one?

The following algorithm summarizes our work in this section.

Finding Inverse Relations and Functions

1. Interchange the x and y components of the ordered pairs of the given relation or the roles of x and y in the defining equation.
2. If the relation was described in equation form, solve the defining equation of the inverse for y.
3. If desired, graph the relation and its inverse on the same set of axes. The two graphs will be symmetric about the line $y = x$.

● ● ● **CHECK YOURSELF ANSWERS**

1. $g^{-1} = \left\{ (x, y) \middle| y = \dfrac{1}{3}x - 2 \right\}$ **2.**

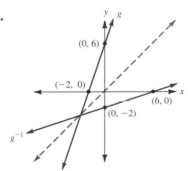

3. (a) $\{(2, -1), (3, 0), (4, 1)\}$—a function; **(b)** $\{(5, 2), (7, 3), (5, 4)\}$—*not* a function.

4. (a) $f = \{(x, y)| y = 2x - 2\}, f^{-1} = \left\{(x, y)| y = \dfrac{1}{2}x + 1\right\}$, the inverse is a function.

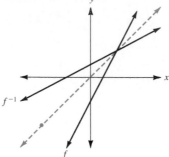

(b) $g = \{(x, y)| y = 2x^2\}, g^{-1} = \left\{(x, y)| y = \pm\sqrt{\dfrac{x}{2}}\right\}$, the inverse is not a function.

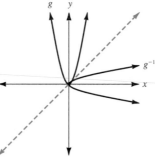

5. (a) Is one-to-one, **(b)** is not one-to-one.

8.1 Exercises

In Exercises 1 to 6, write the inverse relation for each function. In each case, decide whether the inverse relation is also a function.

1. $\{(2, 3), (3, 4), (4, 5)\}$

2. $\{(2, 3), (3, 4), (4, 3)\}$

3. $\{(1, 2), (2, 2), (3, 2)\}$

4. $\{(5, 9), (3, 7), (7, 5)\}$

5. $\{(2, 4), (3, 9), (4, 16)\}$

6. $\{(-1, 2), (0, 3), (1, 2)\}$

In Exercises 7 to 16, write an equation for the inverse of the relation defined by each equation.

7. $y = 2x + 8$

8. $y = -2x - 4$

9. $y = \dfrac{x - 1}{2}$

10. $y = \dfrac{x + 1}{3}$

11. $y = x^2 - 1$ **12.** $y = -x^2 + 2$ **13.** $x^2 + 4y^2 = 36$

14. $4x^2 + y^2 = 36$ **15.** $x^2 - y^2 = 9$ **16.** $4y^2 - x^2 = 4$

In Exercises 17 to 22, write an equation for the inverse of the relation defined by each of the following, and graph the relation and its inverse on the same set of axes. Determine which inverse relations are also functions.

17. $y = 3x - 6$

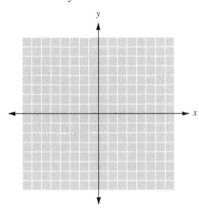

18. $y = 4x + 8$

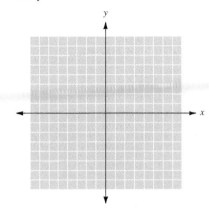

19. $2x - 3y = 6$

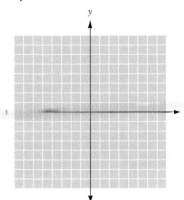

20. $y = 3$

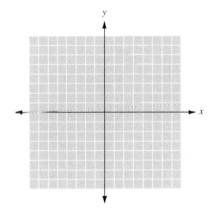

21. $y = x^2 + 1$

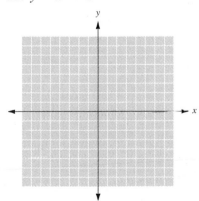

22. $y = -x^2 + 1$

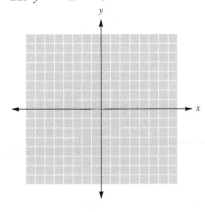

23. An inverse process is an operation that undoes a procedure. If the procedure is wrapping a present, describe in detail the inverse process.

24. If the procedure is the series of steps that take you from home to your classroom, describe the inverse process.

If $f(x) = 3x - 6$, then $f^{-1}(x) = \dfrac{1}{3}x + 2$. Given these two functions, in Exercises 25 to 30, find each of the following.

25. $f(6)$ **26.** $f^{-1}(6)$ **27.** $f(f^{-1}(6))$

28. $f^{-1}(f(6))$ **29.** $f(f^{-1}(x))$ **30.** $f^{-1}(f(x))$

If $g(x) = \dfrac{x + 1}{2}$, then $g^{-1}(x) = 2x - 1$. Given these two functions, in Exercises 31 to 36, find each of the following.

31. $g(3)$ **32.** $g^{-1}(3)$ **33.** $g(g^{-1}(3))$

34. $g^{-1}(g(3))$ **35.** $g(g^{-1}(x))$ **36.** $g^{-1}(g(x))$

Given $h(x) = 2x + 8$, in Exercises 37 to 42, find each of the following.

37. $h(4)$ **38.** $h^{-1}(4)$ **39.** $h(h^{-1}(4))$

40. $h^{-1}(h(4))$ **41.** $h(h^{-1}(x))$ **42.** $h^{-1}(h(x))$

Answers

1. $\{(3, 2), (4, 3), (5, 4)\}$; function   **3.** $\{(2, 1), (2, 2), (2, 3)\}$; not a function   **5.** $\{(4, 2), (9, 3), (16, 4)\}$; function   **7.** $y = \dfrac{1}{2}x - 4$ **9.** $y = 2x + 1$ **11.** $y = \pm\sqrt{x + 1}$ or $x = y^2 - 1$ **13.** $4x^2 + y^2 = 36$ **15.** $y^2 - x^2 = 9$ **17.** Inverse is a function.   **19.** Inverse is a function.

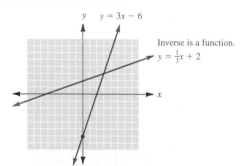

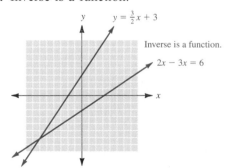

21. Inverse is not a function.

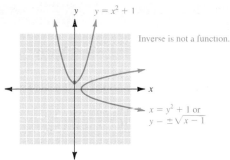

$y \quad y = x^2 + 1$

Inverse is not a function.

x

$x = y^2 + 1$ or
$y = \pm\sqrt{x-1}$

23.   **25.** 12   **27.** 6

29. x **31.** 2   **33.** 3   **35.** x **37.** 16   **39.** 4   **41.** x

Exponential Functions

8.2 OBJECTIVES

1. Graph an exponential function
2. Solve an application of exponential functions
3. Solve an elementary exponential equation

Up to this point in the book, we have worked with polynomial functions and other functions in which the variable was used as a base. We now want to turn to a new classification of functions, the **exponential function.**

Exponential functions are functions whose defining equations involve the variable as an *exponent.* The introduction of these functions will allow us to consider many further applications, including population growth and radioactive decay.

Exponential Functions

An *exponential function* is a function that can be expressed in the form

$f(x) = b^x$

where $b > 0$ and $b \neq 1$. We call b the *base* of the exponential function.

The following are examples of exponential functions.

$$f(x) = 2^x \qquad g(x) = 3^x \qquad h(x) = \left(\frac{1}{2}\right)^x$$

As we have done with other new functions, we begin by finding some function values. We then use that information to graph the function.

● **Example 1**

Graphing an Exponential Function

Graph the exponential function

$$f(x) = 2^x$$

First, choose convenient values for x.

Note:

$$2^{-2} = \frac{1}{2^2} = \frac{1}{4}$$

$$f(0) = 2^0 = 1 \qquad f(-1) = 2^{-1} = \frac{1}{2} \qquad f(1) = 2^1 = 2$$

$$f(-2) = 2^{-2} = \frac{1}{4} \qquad f(2) = 2^2 = 4 \qquad f(-3) = 2^{-3} = \frac{1}{8}$$

Next, form a table from these values. Then, plot the corresponding points, and connect them with a smooth curve for the desired graph.

x	$f(x)$
-3	0.125
-2	0.25
-1	0.5
0	1
1	2
2	4
3	8

A vertical line will cross the graph at one point at most. The same is true for a horizontal line.

There is no value for x such that

$$2^x = 0$$

so the graph never touches the x axis.

We call $y = 0$ (or the x axis) the **horizontal asymptote.**

Let's examine some characteristics of the graph of the exponential function. First, the vertical-line test shows that this is indeed the graph of a function. Also note that the horizontal-line test shows that the function is one-to-one.

The graph *approaches* the x axis on the left, but it does *not intersect* the x axis. The y intercept is 1 (because $2^0 = 1$ by definition). To the right the functional values get larger. We say that the values *grow without bound.*

● ● ● **CHECK YOURSELF 1**

Sketch the graph of the exponential function

$$g(x) = 3^x$$

Let's look at an example in which the base of the function is less than 1.

● Example 2

Graphing an Exponential Function

Graph the exponential function

Recall that

$$\left(\frac{1}{2}\right)^x = 2^{-x}$$

$$f(x) = \left(\frac{1}{2}\right)^x$$

First, choose convenient values for x.

$$f(0) = \left(\frac{1}{2}\right)^0 = 1 \qquad f(-1) = \left(\frac{1}{2}\right)^{-1} = 2 \qquad f(1) = \left(\frac{1}{2}\right)^1 = \frac{1}{2}$$

$$f(-2) = \left(\frac{1}{2}\right)^{-2} = 4 \qquad f(2) = \left(\frac{1}{2}\right)^2 = \frac{1}{4} \qquad f(-3) = \left(\frac{1}{2}\right)^{-3} = 8$$

$$f(3) = \left(\frac{1}{2}\right)^3 = \frac{1}{8}$$

Again, form a table of values and graph the desired function.

Again, by the vertical- and horizontal-line tests, this is the graph of a one-to-one function.

x	$f(x)$
-3	8
-2	4
-1	2
0	1
1	0.5
2	0.25
3	0.125

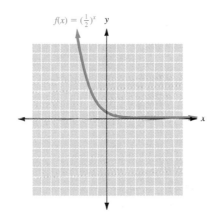

Let's compare this graph and that in Example 1. Clearly, the graph also represents a one-to-one function. As was true in the first example, the graph does not intersect the x axis but approaches that axis, here on the right. The values for the function again grow without bound, but this time on the left. The y intercept for both graphs occurs at 1.

The base of a *growth* function is *greater than* 1.

Note that the graph of Example 1 was *increasing* (going up) as we moved from left to right. That function is an example of a **growth function.**

The base of a *decay function* is *less than* 1 but greater than 0.

The graph of Example 2 was *decreasing* (going down) as we moved from left to right. It is an example of a **decay function.**

● ● ● ● **CHECK YOURSELF 2**

Sketch the graph of the exponential function

$$g(x) = \left(\frac{1}{3}\right)^x$$

The following algorithm summarizes our work thus far in this section.

Graphing an Exponential Function

STEP 1 Establish a table of values by considering the function in the form $y = b^x$.

STEP 2 Plot points from that table of values and connect them with a smooth curve to form the graph.

STEP 3 If $b > 1$, the graph increases from left to right. If $0 < b < 1$, the graph decreases from left to right.

STEP 4 All graphs will have the following in common:
  (a) The y intercept will be 1.
  (b) The graphs will approach, but not touch, the x axis.
  (c) The graphs will represent one-to-one functions.

The use of the letter e as a base originated with Leonhard Euler (1707–1783), and e is sometimes called *Euler's number* for that reason.

We used bases of 2 and $\dfrac{1}{2}$ for the exponential functions of our examples because they provided convenient computations. A far more important base for an exponential function is an irrational number named e. In fact, when e is used as a base, the function defined by

$$f(x) = e^x$$

is called *the* exponential function.

The significance of this number will be made clear in later courses, particularly calculus. For our purposes, e can be approximated as

$$e \approx 2.71828$$

Graph $y = e^x$ on your calculator. You may find the $\boxed{e^x}$ key to be the 2nd (or inverse) function to the ln x key. Note that e^1 is approximately 2.71828.

The graph of $f(x) = e^x$ is shown below. Of course, it is very similar to the graphs seen earlier in this section.

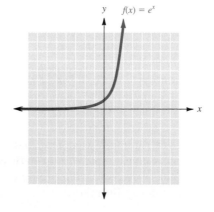

Exponential expressions involving base e occur frequently in real-world applications. Example 3 illustrates this approach.

• Example 3

A Population Application

Be certain that you enclose the multiplication (0.05 × 5) in parentheses or the calculator will misinterpret your intended order of operation.

(*a*) Suppose that the population of a city is presently 20,000 and that the population is expected to grow at a rate of 5% per year. The equation

$$P(t) = 20,000e^{(0.05)t}$$

gives the town's population after *t* years. Find the population in 5 years.

Let $t = 5$ in the original equation to obtain

$$P(5) = 20,000e^{(0.05)(5)} \approx 25,681$$

which is the population expected 5 years from now.

Continuous compounding will give the highest accumulation of interest at any rate. However, daily compounding will result in an amount of interest that is only slightly less.

(*b*) Suppose $1000 is invested at an annual rate of 8%, compounded continuously. The equation

$$A(t) = 1000e^{0.08t}$$

gives the amount in the account after *t* years. Find the amount in the account after 9 years.

Let $t = 9$ in the original equation to obtain

$$A(9) = 1000e^{(0.08)(9)} \approx 2054$$

Note that in 9 years the amount in the account is a little more than *double* the original principal.

which is the amount in the account after 9 years.

● ● ● CHECK YOURSELF 3

If $1000 is invested at an annual rate of 6%, compounded continuously, then the equation for the amount in the account after *t* years is

$$A(t) = 1000e^{0.06t}$$

Use your calculator to find the amount in the account after 12 years.

As we observed in this section, the exponential function is always one-to-one. This yields an important property that can be used to solve certain types of equations involving exponents.

> If $b > 0$ and $b \neq 1$, then
>
> $b^m = b^n$ if and only if $m = n$ (1)
>
> where *m* and *n* are any real numbers.

The usefulness of this property is illustrated in Example 4.

• Example 4

Solving an Exponential Equation

(*a*) Solve $2^x = 8$ for x.

We recognize that 8 is a power of 2, and we can write the equation as

$2^x = 2^3$ Write with equal bases.

Applying property (1) on page 625, we have

$x = 3$ Set exponents equal.

and 3 is the solution.

(*b*) Solve $3^{2x} = 81$ for x.

Since $81 = 3^4$, we can write

$3^{2x} = 3^4$

$2x = 4$

$x = 2$

We see that 2 is the solution for the equation.

The answer can easily be checked by substitution. Letting $x = 2$ gives

$3^{2(2)} = 3^4 = 81$

(*c*) Solve $2^{x+1} = \dfrac{1}{16}$ for x.

Again, we write $\dfrac{1}{16}$ as a power of 2, so that

$2^{x+1} = 2^{-4}$

Note:

$\dfrac{1}{16} = \dfrac{1}{2^4} = 2^{-4}$

To verify the solution.

$2^{-5-1} \stackrel{?}{=} 2^{-4}$

$2^{-4} \stackrel{?}{=} 2^{-4}$

$\dfrac{1}{16} = \dfrac{1}{16}$

Then

$x + 1 = -4$

$x = -5$

The solution is -5.

● ● ● **CHECK YOURSELF 4**

Solve each of the following equations for x.

(a) $2^x = 16$ **(b)** $4^{x+1} = 64$ **(c)** $3^{2x} = \dfrac{1}{81}$

● ● ● **CHECK YOURSELF ANSWERS**

1. $y = g(x) = 3^x$.

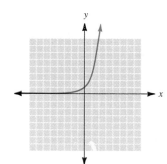

2. $y = \left(\dfrac{1}{3}\right)^x$.

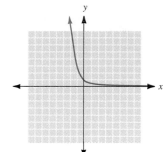

3. $2054.43. **4. (a)** $x = 4$, **(b)** $x = 2$, **(c)** $x = -2$

8.2 Exercises

Match the graphs in Exercises 1 to 8 with the appropriate equation.

(a) $y = \left(\dfrac{1}{2}\right)^x$ **(b)** $y = 2x - 1$ **(c)** $y = 2^x$ **(d)** $y = x^2$

(e) $y - 1^x$ **(f)** $y = 5^x$ **(g)** $x = 2^y$ **(h)** $x = y^2$

1.

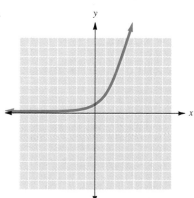

2.

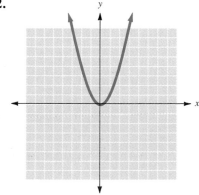

3.

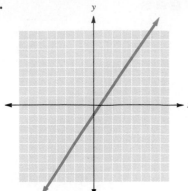

4.

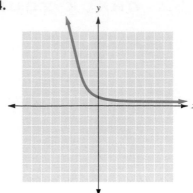

5.

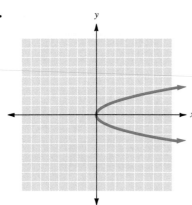

6.

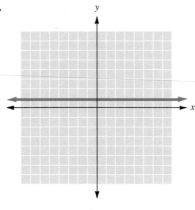

7.

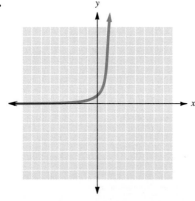

8.

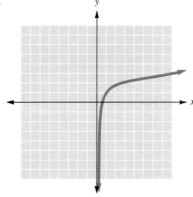

In Exercises 9 to 12, let $f(x) = 4^x$ and find each of the following.

9. $f(0)$ **10.** $f(1)$ **11.** $f(2)$ **12.** $f(-2)$

In Exercises 13 to 16, let $g(x) = 4^{x+1}$ and find each of the following.

13. $g(0)$ **14.** $g(1)$ **15.** $g(2)$ **16.** $g(-2)$

In Exercises 17 to 20, let $h(x) = 4^x + 1$ and find each of the following.

17. $h(0)$ **18.** $h(1)$ **19.** $h(2)$ **20.** $h(-2)$

In Exercises 21 to 24, let $f(x) = \left(\dfrac{1}{4}\right)^x$ and find each of the following.

21. $f(1)$ **22.** $f(-1)$ **23.** $f(-2)$ **24.** $f(2)$

In Exercises 25 to 36, graph each exponential function.

25. $y = 4^x$

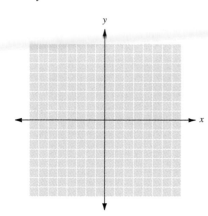

26. $y = \left(\dfrac{1}{4}\right)^x$

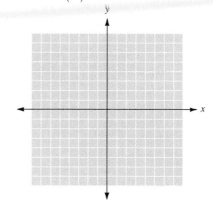

27. $y = \left(\dfrac{2}{3}\right)^x$

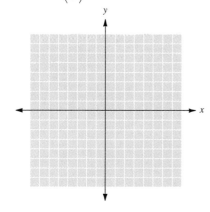

28. $y = \left(\dfrac{3}{2}\right)^x$

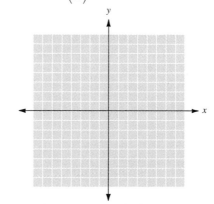

29. $y = 3 \cdot 2^x$

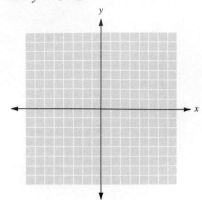

30. $y = 2 \cdot 3^x$

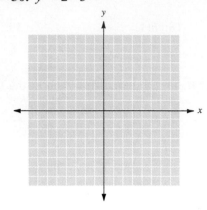

31. $y = 3^x$

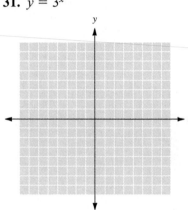

32. $y = 2^{x-1}$

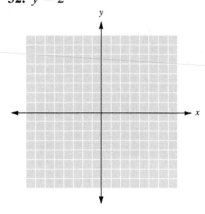

33. $y = 2^{2x}$

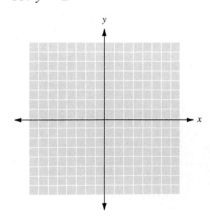

34. $y = \left(\dfrac{1}{2}\right)^{2x}$

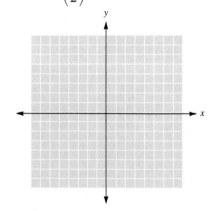

35. $y = e^{-x}$

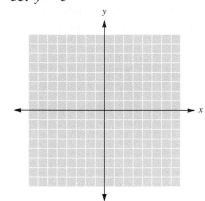

36. $y = e^{2x}$

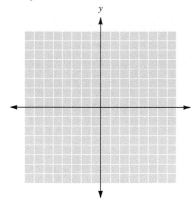

In Exercises 37 to 48, solve each exponential equation for x.

37. $2^x = 32$

38. $4^x = 64$

39. $10^x = 10,000$

40. $5^x = 125$

41. $3^x = \dfrac{1}{9}$

42. $2^x = \dfrac{1}{16}$

43. $2^{2x} = 64$

44. $3^{2x} = 81$

45. $2^{x+1} = 64$

46. $4^{x-1} = 16$

47. $3^{x-1} = \dfrac{1}{27}$

48. $2^{x+2} = \dfrac{1}{8}$

Suppose it takes 1 h for a certain bacterial culture to double by dividing in half. If there are 100 bacteria in the culture to start, then the number of bacteria in the culture after x hours is given by $N(x) = 100 \cdot 2^x$. In Exercises 49 to 51, use this function to find each of the following.

49. The number of bacteria in the culture after 2 h

50. The number of bacteria in the culture after 3 h

51. The number of bacteria in the culture after 5 h

52. Graph the relationship between the number of bacteria in the culture and the number of hours. Be sure to choose an appropriate scale for the N axis.

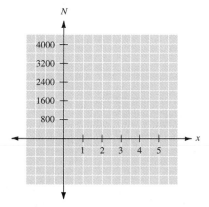

The half-life of radium is 1690 years. That is after a 1690-year period, one-half of the original amount of radium will have decayed into another substance. If the original amount of radium was 64 grams (g), the formula relating the amount of radium left after time t is given by $R(t) = 64 \cdot 2^{-t/1690}$. In Exercises 53 to 55, use that formula to find each of the following.

53. The amount of radium left after 1690 years

54. The amount of radium left after 3380 years

55. The amount of radium left after 5070 years

56. Graph the relationship between the amount of radium remaining and time. Be sure to use appropriate scales for the R and t axes.

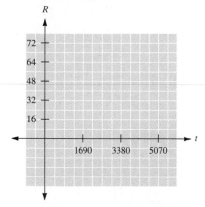

 If $1000 is invested in a savings account with an interest rate of 8%, compounded annually, the amount in the account after t years is given by $A(t) = 1000(1 + 0.08)^t$. In Exercises 57 to 59, use a calculator to find each of the following.

57. The amount in the account after 2 years

58. The amount in the account after 5 years

59. The amount in the account after 9 years

60. Graph the relationship between the amount in the account and time. Be sure to choose appropriate scales for the A and t axes.

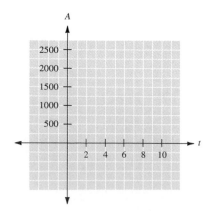

The so-called learning curve in psychology applies to learning a skill, such as typing, in which the performance level progresses rapidly at first and then levels off with time. One can approximate N, the number of words per minute that a person can type after t weeks of training, with the equation $N = 80(1 - e^{-0.06t})$. Use a calculator to find the following.

61. **(a)** N after 10 weeks, **(b)** N after 20 weeks, **(c)** N after 30 weeks.

62. Graph the relationship between the number of words per minute N and the number of weeks of training t.

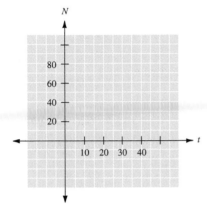

63. Find two different calculators that have $\boxed{e^x}$ keys. Describe how to use the function on each of the calculators.

64. Are there any values of x for which e^x produces an exact answer on the calculator? Why are other answers not exact?

A possible calculator sequence for evaluating the expression

$$\left(1 + \frac{1}{n}\right)^n$$

where $n = 10$ is

$$\boxed{(}\ 1\ \boxed{+}\ 1\ \boxed{\div}\ 10\ \boxed{)}\ \boxed{\wedge}\ 10\ \boxed{=}$$

In Exercises 65 to 69, use that sequence to find $\left(1 + \dfrac{1}{n}\right)^n$ for the following values of n.

65. $n = 100$ **66.** $n = 1000$ **67.** $n = 10,000$

68. $n = 100,000$ **69.** $n = 1,000,000$

70. What did you observe from the experiment above?

71. Graph the exponential function defined by $y = 2^x$.

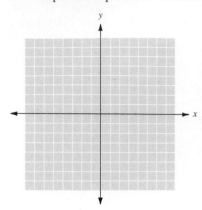

72. Graph the function defined by $x = 2^y$ on the same set of axes as the previous graph. What do you observe? *Hint:* To graph $x = 2^y$, choose convenient values for y and then the corresponding values for x.

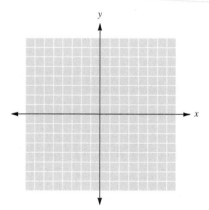

Answers

1. (c)   **3.** (b)   **5.** (h)   **7.** (f)   **9.** 1   **11.** 16   **13.** 4   **15.** 64   **17.** 2   **19.** 17

21. $\dfrac{1}{4}$ **23.** 16   **25.** $y = 4^x$ **27.** $y = \left(\dfrac{2}{3}\right)^x$

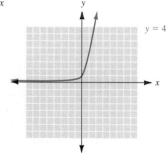

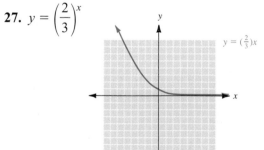

29. $y = 3 \cdot 2^x$

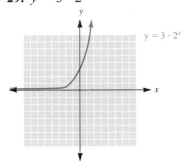

31. $y = 3^x$

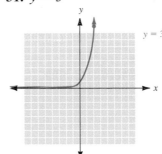

33. $y = 2^{2x}$

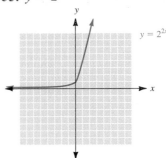

35. $y = e^{-x}$

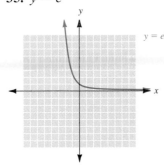

37. 5   **39.** 4   **41.** -2 **43.** 3   **45.** 5   **47.** -2 **49.** 400

51. 3200   **53.** 32 g   **55.** 8 g   **57.** $1166.40   **59.** $1999

61. **(a)** 36, **(b)** 56, **(c)** 67   **63.**   **65.** 2.7048   **67.** 2.71815

69. 2.71828   **71.** See text for graph of $y = 2^x$.

8.3 Logarithmic Functions

8.3 OBJECTIVES

1. Graph a logarithmic function
2. Convert between logarithmic and exponential equations
3. Evaluate a logarithmic expression
4. Solve an elementary logarithmic equation

Napier also coined the word "logarithm" from the Greek words "logos"—a ratio—and "arithmos"—a number.

Given our experience with the exponential function in Section 8.2 and our earlier work with the inverse of a function, we now can introduce the logarithmic function.

John Napier (1550–1617), a Scotsman, is credited with the invention of logarithms. The development of the logarithm grew out of a desire to ease the work involved in numerical computations, particularly in the field of astronomy. Today the availability of inexpensive scientific calculators has made the use of logarithms as a computational tool unnecessary.

However, the concept of the logarithm and the properties of the logarithmic function that we describe in a later section still are very important in the solutions of particular equations, in calculus, and in the applied sciences.

Again, the applications for this new function are numerous. The Richter scale for measuring the intensity of an earthquake and the decibel scale for measuring the intensity of sound both make use of logarithms.

To develop the idea of a logarithmic function, we must return to the exponential function

$$f = \{(x, y) \mid y = b^x, b > 0, b \neq 1\} \tag{1}$$

Interchanging the roles of x and y, we have the inverse function

Recall that f is a one-to-one function, so its inverse is also a function.

$$f^{-1} = \{(x, y) | x = b^y\} \tag{2}$$

Presently, we have no way to solve the equation $x = b^y$ for y. So, to write the inverse (2) in a more useful form, we offer the following definition.

> The *logarithm of x to base b* is denoted
>
> $log_b\ x$
>
> and
>
> $y = log_b\ x$ if and only if $x = b^y$

We can now write our inverse function, using this new notation, as

Note that the restrictions on the base are the same as those used for the exponential function.

$$f^{-1} = \{(x, y) | y = log_b\ x,\ b > 0,\ b \neq 1\} \tag{3}$$

In general, any function defined in this form is called a **logarithmic function.**

At this point we should stress the meaning of this new relationship. Consider the equivalent forms illustrated here.

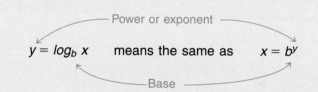

The logarithm y is the power to which we must raise b to get x. In other words, *a logarithm is simply a power or an exponent.* We return to this thought later when using the exponential and logarithmic forms of equivalent equations.

We begin our work by graphing a typical logarithmic function.

• Example 1

Graphing a Logarithmic Function

Graph the logarithmic function

$$y = log_2 x$$

Since $y = log_2 x$ is equivalent to the exponential form

The base is 2, and the logarithm or power is y.

$$x = 2^y$$

we can find ordered pairs satisfying this equation by choosing convenient values for y and calculating the corresponding values for x.

Letting y take on values from -3 to 3 yields the table of values shown below. As before, we plot points from the ordered pairs and connect them with a smooth curve to form the graph of the function.

What do the vertical- and horizontal-line tests tell you about this graph?

Use your calculator to compare the graphs of $y = 2^x$ and $y = \log_2 x$. Are they inverse functions? How can you tell?

x	y
$\dfrac{1}{8}$	-3
$\dfrac{1}{4}$	-2
$\dfrac{1}{2}$	-1
1	0
2	1
4	2
8	3

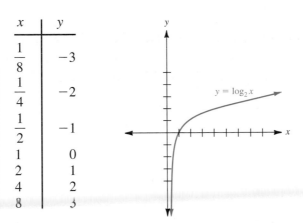

We observe that the graph represents a one-to-one function whose domain is $\{x \mid x > 0\}$ and whose range is the set of all real numbers.

For base 2 (or for any base greater than 1) the function will always be increasing over its domain.

Recall from Section 8.1 that the graphs of a function and its inverse are always reflections of each other about the line $y = x$. Since we have defined the logarithmic function as the inverse of an exponential function, we can anticipate the same relationship.

The graphs of

$$f(x) = 2^x \qquad \text{and} \qquad f^{-1}(x) = \log_2 x$$

are shown below.

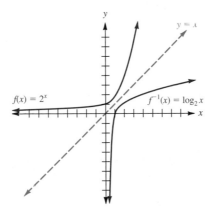

We see that the graphs of f and f^{-1} are indeed reflections of each other about the line $y = x$. In fact, this relationship provides an alternate method of sketching $y = \log_b x$. We can sketch the graph of $y = b^x$ and then reflect that graph about the line $y = x$ to form the graph of the logarithmic function.

● ● ● **CHECK YOURSELF 1**

Graph the logarithmic function defined by

$$y = \log_3 x$$

Hint: Consider the equivalent form $x = 3^y$.

For our later work in this chapter, it will be necessary for us to be able to convert back and forth between exponential and logarithmic forms. The conversion is straightforward. You need only keep in mind the basic relationship

Again, this tells us that a logarithm is an exponent or a power.

| $y = \log_b x$ | means the same as | $x = b^y$ |

Look at the following example.

● Example 2

Writing Equations in Logarithmic Form

Convert to logarithmic form.

The base is 3, the exponent or power is 4.

(*a*) $3^4 = 81$ is equivalent to $\log_3 81 = 4$.

(*b*) $10^3 = 1000$ is equivalent to $\log_{10} 1000 = 3$.

(*c*) $2^{-3} = \dfrac{1}{8}$ is equivalent to $\log_2 \dfrac{1}{8} = -3$.

(*d*) $9^{1/2} = 3$ is equivalent to $\log_9 3 = \dfrac{1}{2}$.

● ● ● **CHECK YOURSELF 2**

Convert each statement to logarithmic form.

(a) $4^3 = 64$ **(b)** $10^{-2} = 0.01$ **(c)** $3^{-3} = \dfrac{1}{27}$ **(d)** $27^{1/3} = 3$

Example 3 shows how to write a logarithmic expression in exponential form.

•Example 3

Writing Equations in Exponential Form

Convert to exponential form.

Here, the base is 2; the logarithm, which is the power, is 3.

(a) $\log_2 8 = 3$ is equivalent to $2^3 = 8$.

(b) $\log_{10} 100 = 2$ is equivalent to $10^2 = 100$.

(c) $\log_3 \dfrac{1}{9} = -2$ is equivalent to $3^{-2} = \dfrac{1}{9}$.

(d) $\log_{25} 5 = \dfrac{1}{2}$ is equivalent to $25^{1/2} = 5$.

● ● ● CHECK YOURSELF 3

Convert to exponential form.

(a) $\log_2 32 = 5$ **(b)** $\log_{10} 1000 = 3$ **(c)** $\log_4 \dfrac{1}{16} = -2$ **(d)** $\log_{27} 3 = \dfrac{1}{3}$

Certain logarithms can be directly calculated by changing an expression to the equivalent exponential form, as Example 4 illustrates.

•Example 4

Evaluating Logarithmic Expressions

(a) Evaluate $\log_3 27$.

If $x = \log_3 27$, in exponential form we have

Recall that $b^m = b^n$ if and only if $m = n$.

$3^x = 27$

$3^x = 3^3$

$x = 3$

We then have $\log_3 27 = 3$.

(b) Evaluate $\log_{10} \dfrac{1}{10}$.

If $x = \log_{10} \dfrac{1}{10}$, we can write

Rewrite each side as a power of the same base.

$10^x = \dfrac{1}{10}$

$= 10^{-1}$

We then have $x = -1$ and

$$\log_{10} \frac{1}{10} = -1$$

● ● ● **CHECK YOURSELF 4**

Evaluate each logarithm.

(a) $\log_2 64$ **(b)** $\log_3 \dfrac{1}{27}$

The relationship between exponents and logarithms also allows us to solve certain equations involving logarithms where two of the quantities in the equation $y = \log_b x$ are known, as Example 5 illustrates.

● Example 5

Solving Logarithmic Equations

(a) Solve $\log_5 x = 3$ for x.

Since $\log_5 x = 3$, in exponential form we have

$x = 5^3$

$ = 125$

(b) Solve $y = \log_4 \dfrac{1}{16}$ for y.

The original equation is equivalent to

$4^y = \dfrac{1}{16}$

$ = 4^{-2}$

We then have $y = -2$ as the solution.

(c) Solve $\log_b 81 = 4$ for b.

In exponential form the equation becomes

Keep in mind that the base must be *positive*, so we do not consider the possible solution $b = -3$.

$b^4 = 81$

$b = 3$

● ● ● **CHECK YOURSELF 5**

Solve each of the following equations for the variable cited.

(a) $\log_4 x = 4$ for x **(b)** $\log_b \dfrac{1}{8} = -3$ for b **(c)** $y = \log_9 3$ for y

---

Loudness can be measured in **bels (B),** a unit named for Alexander Graham Bell. This unit is rather large, so a more practical unit is the **decibel (dB),** a unit one-tenth as large.

To conclude this section, we turn to two common applications of the logarithmic function. The **decibel scale** is used in measuring the loudness of various sounds.

If I represents the intensity of a given sound and I_0 represents the intensity of a "threshold sound," then the decibel (dB) rating of the given sound is given by

$$L = 10 \log_{10} \frac{I}{I_0}$$

Variable I_0 is the intensity of the minimum sound level detectable by the human ear.

where $I_0 = 10^{-16}$ watt per square centimeter (W/cm^2). Consider Example 6.

• Example 6

A Decibel Application

(*a*) A whisper has intensity $I = 10^{-14}$. Its decibel rating is

$$L = 10 \log_{10} \frac{10^{-14}}{10^{-16}}$$

$$= 10 \log_{10} 10^2$$

$$= 10 \cdot 2$$

$$= 20$$

(*b*) A rock concert has intensity $I = 10^{-4}$. Its decibel rating is

$$L = 10 \log_{10} \frac{10^{-4}}{10^{-16}}$$

$$= 10 \log_{10} 10^{12}$$

$$= 10 \cdot 12$$

$$= 120$$

CHECK YOURSELF 6

Ordinary conversation has intensity $I = 10^{-12}$. Find its rating on the decibel scale.

The scale was named after Charles Richter, a U.S. geologist.

Another commonly used logarithmic scale is the **Richter scale.** Geologists use that scale to convert seismographic readings, which give the intensity of the shock waves of an earthquake, to a measure of the magnitude of that earthquake.

The magnitude M of an earthquake is given by

$$M = \log_{10} \frac{a}{a_0}$$

A "zero-level" earthquake is the quake of least intensity that is measurable by a seismograph.

where a is the intensity of its shock waves and a_0 is the intensity of the shock wave of a zero-level earthquake.

• Example 7

A Richter Scale Application

How many times stronger is an earthquake measuring 5 on the Richter scale than one measuring 4 on the Richter scale?

Suppose a_1 is the intensity of the earthquake with magnitude 5 and a_2 is the intensity of the earthquake with magnitude 4. Then

On your calculator, the $\boxed{\log}$ key is actually $\log_{10} x$.

$$5 = \log_{10} \frac{a_1}{a_0} \quad \text{and} \quad 4 = \log_{10} \frac{a_2}{a_0}$$

We convert these logarithmic expressions to exponential form.

$$10^5 = \frac{a_1}{a_0} \quad \text{and} \quad 10^4 = \frac{a_2}{a_0}$$

or

$$a_1 = a_0 \cdot 10^5 \quad \text{and} \quad a_2 = a_0 \cdot 10^4$$

The ratio of a_1 to a_2 is

$$\frac{a_1}{a_2}$$

We want the ratio of the intensities of the two earthquakes, so

$$\frac{a_1}{a_2} = \frac{a_0 \cdot 10^5}{a_0 \cdot 10^4} = 10^1 = 10$$

The earthquake of magnitude 5 is *10 times stronger* than the earthquake of magnitude 4.

● ● ● CHECK YOURSELF 7

How many times stronger is an earthquake of magnitude 6 than one of magnitude 4?

● ● ● CHECK YOURSELF ANSWERS

1. $y = \log_3 x$.

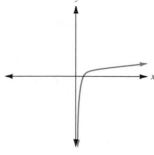

2. (a) $\log_4 64 = 3$, **(b)** $\log_{10} 0.01 = -2$, **(c)** $\log_3 \dfrac{1}{27} = -3$, **(d)** $\log_{27} 3 = \dfrac{1}{3}$.

3. (a) $2^5 = 32$, **(b)** $10^3 = 1000$, **(c)** $4^{-2} = \dfrac{1}{16}$, **(d)** $27^{1/3} = 3$.

4. (a) $\log_2 64 = 6$, **(b)** $\log_3 \dfrac{1}{27} = -3$. **5. (a)** $x = 256$, **(b)** $b = 2$, **(c)** $y = \dfrac{1}{2}$.

6. 40 dB. **7.** 100 times.

8.3 Exercises

In Exercises 1 to 6, sketch the graph of the function defined by each equation.

1. $y = \log_4 x$

2. $y = \log_{10} x$

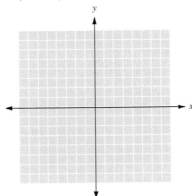

3. $y = \log_2 (x - 1)$

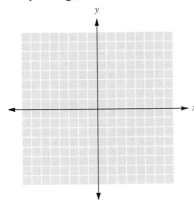

4. $y = \log_3 (x + 1)$

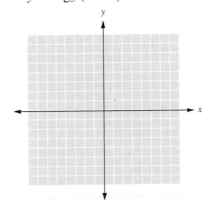

5. $y = \log_8 x$ (Use your calculator.)

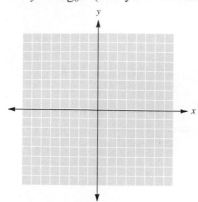

6. $y = \log_3 x + 1$

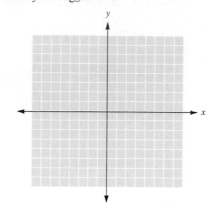

In Exercises 7 to 24, convert each statement to logarithmic form.

7. $2^4 = 16$ **8.** $3^5 = 243$ **9.** $10^2 = 100$ **10.** $4^3 = 64$

11. $3^0 = 1$ **12.** $10^0 = 1$ **13.** $4^{-2} = \dfrac{1}{16}$ **14.** $3^{-4} = \dfrac{1}{81}$

15. $10^{-3} = \dfrac{1}{1000}$ **16.** $2^{-5} = \dfrac{1}{32}$ **17.** $16^{1/2} = 4$ **18.** $125^{1/3} = 5$

19. $64^{-1/3} = \dfrac{1}{4}$ **20.** $36^{-1/2} = \dfrac{1}{6}$ **21.** $8^{2/3} = 4$ **22.** $9^{3/2} = 27$

23. $27^{-2/3} = \dfrac{1}{9}$ **24.** $16^{-3/2} = \dfrac{1}{64}$

In Exercises 25 to 42, convert each statement to exponential form.

25. $\log_2 16 = 4$ **26.** $\log_3 3 = 1$ **27.** $\log_5 1 = 0$ **28.** $\log_3 27 = 3$

29. $\log_{10} 10 = 1$ **30.** $\log_2 32 = 5$ **31.** $\log_5 125 = 3$ **32.** $\log_{10} 1 = 0$

33. $\log_3 \dfrac{1}{27} = -3$ **34.** $\log_5 \dfrac{1}{25} = -2$ **35.** $\log_{10} 0.01 = -2$ **36.** $\log_{10} \dfrac{1}{1000} = -3$

37. $\log_{16} 4 = \dfrac{1}{2}$ **38.** $\log_{125} 5 = \dfrac{1}{3}$ **39.** $\log_8 4 = \dfrac{2}{3}$ **40.** $\log_9 27 = \dfrac{3}{2}$

41. $\log_{25} \dfrac{1}{5} = -\dfrac{1}{2}$ **42.** $\log_{64} \dfrac{1}{16} = -\dfrac{2}{3}$

In Exercises 43 to 52, evaluate each logarithm.

43. $\log_2 32$ **44.** $\log_3 81$ **45.** $\log_4 64$ **46.** $\log_{10} 1000$

47. $\log_3 \dfrac{1}{81}$ **48.** $\log_4 \dfrac{1}{64}$ **49.** $\log_{10} \dfrac{1}{100}$ **50.** $\log_5 \dfrac{1}{25}$

51. $\log_{25} 5$ **52.** $\log_{27} 3$

In Exercises 53 to 74, solve each equation for the unknown variable.

53. $y = \log_5 25$ **54.** $\log_2 x = 4$ **55.** $\log_b 64 = 3$ **56.** $y = \log_3 1$

57. $\log_{10} x = 2$ **58.** $\log_b 125 = 3$ **59.** $y = \log_5 5$ **60.** $y = \log_3 81$

61. $\log_{3/2} x = 3$ **62.** $\log_b \dfrac{4}{9} = 2$ **63.** $\log_b \dfrac{1}{25} = -2$ **64.** $\log_3 x = -3$

65. $\log_{10} x = -3$ **66.** $y = \log_2 \dfrac{1}{16}$ **67.** $y = \log_8 \dfrac{1}{64}$ **68.** $\log_b \dfrac{1}{100} = -2$

69. $\log_{27} x = \dfrac{1}{3}$ **70.** $y = \log_{100} 10$ **71.** $\log_b 5 = \dfrac{1}{2}$ **72.** $\log_{64} x = \dfrac{2}{3}$

73. $y = \log_{27} \dfrac{1}{9}$ **74.** $\log_b \dfrac{1}{8} = -\dfrac{3}{4}$

Use the decibel formula

$$L = 10 \log_{10} \dfrac{I}{I_0}$$

to solve Exercises 75 to 78.

75. Sound A television commercial has a volume with intensity $I = 10^{-11}$ W/cm^2. Find its rating in decibels.

76. Sound The sound of a jet plane on takeoff has an intensity $I = 10^{-2}$ W/cm^2. Find its rating in decibels.

77. Sound The sound of a computer printer has an intensity of $I = 10^{-9}$ W/cm^2. Find its rating in decibels.

78. Sound The sound of a busy street has an intensity if $I = 10^{-8}$ W/cm^2. Find its rating in decibels.

The formula for the decibel rating L can be solved for the intensity of the sound as $I = l_0 \cdot 10^{L/10}$. Use this formula in Exercises 79 to 83.

79. Sound Find the intensity of the sound in an airport waiting area if the decibel rating is 80.

80. Sound Find the intensity of the sound of conversation in a crowded room if the decibel rating is 70.

81. Sound What is the ratio of intensity of the sound of 80 dB to that of 70 dB?

82. Sound What is the ratio of intensity of a sound of 60 dB to one measuring 40 dB?

83. Sound What is the ratio of intensity of a sound of 70 dB to one measuring 40 dB?

84. Derive the formula for intensity provided above. *Hint:* First divide both sides of the decibel formula by 10. Then write the equation in exponential form.

Use the earthquake formula

$$M = \log_{10} \frac{a}{a_0}$$

to solve Exercises 85 to 88.

85. Earthquakes An earthquake has an intensity a of $10^6 \cdot a_0$, where a_0 is the intensity of the zero-level earthquake. What was its magnitude?

86. Earthquakes The great San Francisco earthquake of 1906 had an intensity of $10^{8.3} \cdot a_0$. What was its magnitude?

87. Earthquakes An earthquake can begin causing damage to buildings with a magnitude of 5 on the Richter scale. Find its intensity in terms of a_0.

88. Earthquakes An earthquake may cause moderate building damage with a magnitude of 6 on the Richter scale. Find its intensity in terms of a_0.

89. The **learning curve** describes the relationship between learning and time. Its graph is a logarithmic curve in the first quadrant. Describe that curve as it relates to learning.

90. In which scientific fields would you expect to again encounter a discussion of logarithms?

The *half-life* of a radioactive substance is the time it takes for half the original amount of the substance to decay to a nonradioactive element. The half-life of radioactive waste is very important in figuring how long the waste must be kept isolated from the environment in some sort of storage facility. Half-lives of various radioactive waste products vary from a few seconds to millions of years. It usually takes at least 10 half-lives for a radioactive waste product to be considered safe.

The half-life of a radioactive substance can be determined by the following formula.

$$\ln \frac{1}{2} = -\lambda x$$

where λ = radioactive decay constant

   x = half-life

In Exercises 91 to 95, find the half-lives of the following important radioactive waste products given the radioactive decay constant (RDC).

91. Plutonium 239. RDC – 0.000029

92. Strontium 90. RDC = 0.024755

93. Thorium 230. RDC = 0.000009

94. Cesium 135. RDC = 0.00000035

95. How many years will it be before each waste product will be considered safe?

Answers

1.

3.

5.

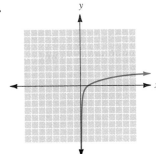 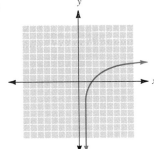 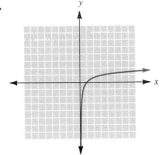

7. $\log_2 16 = 4$ **9.** $\log_{10} 100 = 2$ **11.** $\log_3 1 = 0$ **13.** $\log_4 \frac{1}{16} = -2$ **15.** $\log_{10} \frac{1}{1000} = -3$

17. $\log_{16} 4 = \frac{1}{2}$ **19.** $\log_{64} \frac{1}{4} = -\frac{1}{3}$ **21.** $\log_8 4 = \frac{2}{3}$ **23.** $\log_{27} \frac{1}{9} = -\frac{2}{3}$ **25.** $2^4 = 16$

27. $5^0 = 1$ **29.** $10^1 = 10$ **31.** $5^3 = 125$ **33.** $3^{-3} = \frac{1}{27}$ **35.** $10^{-2} = 0.01$ **37.** $16^{1/2} = 4$

39. $8^{2/3} = 4$ **41.** $25^{-1/2} = \frac{1}{5}$ **43.** 5   **45.** 3   **47.** -4 **49.** -2 **51.** $\frac{1}{2}$ **53.** 2   **55.** 4

57. 100   **59.** 1   **61.** $\frac{27}{8}$ **63.** 5   **65.** $\frac{1}{100}$ **67.** -2 **69.** 3   **71.** 25   **73.** $-\frac{2}{3}$

75. 50 dB   **77.** 70 dB   **79.** 10^{-8} **81.** 10   **83.** 1000   **85.** 6   **87.** $10^5 \cdot a_0$ **89.**

91. 24,000 yr   **93.** 77,000 yr   **95.** Pu239: 240,000 yr, Sr90: 280 yr, Th230: 770,000 yr, Cs135: 20,000,000 yr

As we mentioned earlier, logarithms were developed as aids to numerical computations. The early utility of the logarithm was due to the properties that we will discuss in this section. Even with the advent of the scientific calculator, that utility remains important today. We can apply these same properties to applications in a variety of areas that lead to exponential or logarithmic equations.

Since a logarithm is, by definition, an exponent, it seems reasonable that our knowledge of the properties of exponents should lead to useful properties for logarithms. That is, in fact, the case.

LOGARITHMIC PROPERTIES

We start with two basic facts that follow immediately from the definition of the logarithm.

The properties follow from the facts that

$$b^1 = b \quad \text{and} \quad b^0 = 1$$

> For $b > 0$ and $b \neq 1$,
>
> **1.** $\log_b b = 1$
> **2.** $\log_b 1 = 0$

We know that the logarithmic function $y = \log_b x$ and the exponential function $y = b^x$ are inverses of each other. So, for $f(x) = b^x$, we have $f^{-1}(x) = \log_b x$.

Also recall from our work in Section 8.1 that for any one-to-one function f,

The inverse has "undone" whatever f did to x.

$$f^{-1}(f(x)) = x \qquad \text{for any } x \text{ in domain of } f$$

and

$$f(f^{-1}(x)) = x \qquad \text{for any } x \text{ in domain of } f^{-1}$$

Since $f(x) = b^x$ is a one-to-one function, we can apply the above to the case where

$$f(x) = b^x \quad \text{and} \quad f^{-1}(x) = \log_b x$$

to derive the following.

For Property 3,

$$f^{-1}(f(x)) = f^{-1}(b^x) = \log_b b^x$$

But in general, for any one-to-one function f,

$$f^{-1}(f(x)) = x$$

> **3.** $\log_b b^x = x$
> **4.** $b^{\log_b x} = x \qquad$ for $x > 0$

Since logarithms are exponents, we can again turn to the familiar exponent rules to derive some further properties of logarithms. Consider the following.

We know that

$$\log_b M = x \qquad \text{if and only if} \qquad M = b^x$$

and

$$\log_b N = y \qquad \text{if and only if} \qquad N = b^y$$

Then

$$M \cdot N = b^x \cdot b^y = b^{x+y} \qquad (1)$$

From equation (1) we see that $x + y$ is the power to which we must raise b to get the product MN. In logarithmic form, that becomes

$$\log_b MN = x + y \qquad (2)$$

Now, since $x = \log_b M$ and $y = \log_b N$, we can substitute in (2) to write

$$\log_b MN = \log_b M + \log_b N \qquad (3)$$

This is the first of the basic logarithmic properties presented here. The remaining properties may all be proved by arguments similar to those presented in equations (1) to (3).

In all cases, M, $N > 0$, $b > 0$, $b \neq 1$, and p is any real number.

Properties of Logarithms

PRODUCT PROPERTY

$$\log_b MN = \log_b M + \log_b N$$

QUOTIENT PROPERTY

$$\log_b \frac{M}{N} = \log_b M - \log_b N$$

POWER PROPERTY

$$\log_b M^p = p \log_b M$$

Many applications of logarithms require using these properties to write a single logarithmic expression as the sum or difference of simpler expressions, as Example 1 illustrates.

● Example 1

Using the Properties of Logarithms

Expand, using the properties of logarithms.

(a) $\log_b xy = \log_b x + \log_b y$ Product property.

(b) $\log_b \dfrac{xy}{z} = \log_b xy - \log_b z$ Quotient property.

   $\phantom{\log_b \dfrac{xy}{z}} - \log_b x + \log_b y - \log_b z$ Product property.

(c) $\log_{10} x^2 y^3 = \log_{10} x^2 + \log_{10} y^3$ Product property.

   $\phantom{\log_{10} x^2 y^3} = 2 \log_{10} x + 3 \log_{10} y$ Power property.

Recall $\sqrt{a} = a^{1/2}$.

$(d)\ \log_b \sqrt{\dfrac{x}{y}} = \log_b \left(\dfrac{x}{y}\right)^{1/2}$ Definition of exponent.

$= \dfrac{1}{2} \log_b \dfrac{x}{y}$ Power property.

$= \dfrac{1}{2} (\log_b x - \log_b y)$ Quotient property.

● ● ● **CHECK YOURSELF 1**

Expand each expression, using the properties of logarithms.

(a) $\log_b x^2 y^3 z$ **(b)** $\log_{10} \sqrt{\dfrac{xy}{z}}$

In some cases, we will reverse the process and use the properties to write a single logarithm, given a sum or difference of logarithmic expressions.

● Example 2

Rewriting Logarithmic Expressions

Write each expression as a single logarithm with coefficient 1.

$(a)\ 2 \log_b x + 3 \log_b y$

$= \log_b x^2 + \log_b y^3$ Power property.

$= \log_b x^2 y^3$ Product property.

$(b)\ 5 \log_{10} x + 2 \log_{10} y - \log_{10} z$

$= \log_{10} x^5 y^2 - \log_{10} z$

$= \log_{10} \dfrac{x^5 y^2}{z}$ Quotient property.

$(c)\ \dfrac{1}{2}(\log_2 x - \log_2 y)$

$= \dfrac{1}{2}\left(\log_2 \dfrac{x}{y}\right)$

$= \log_2 \left(\dfrac{x}{y}\right)^{1/2}$ Power property.

$= \log_2 \sqrt{\dfrac{x}{y}}$

● ● ● ● **CHECK YOURSELF 2**

Write each expression as a single logarithm with coefficient 1.

(a) $3 \log_b x + 2 \log_b y - 2 \log_b z$ **(b)** $\dfrac{1}{3}(2 \log_2 x - \log_2 y)$

Example 3 illustrates the basic concept of the use of logarithms as a computational aid.

● Example 3

Evaluating Logarithmic Expressions

We have written the logarithms correct to three decimal places and will follow this practice throughout the remainder of this chapter.

Suppose $\log_{10} 2 = 0.301$ and $\log_{10} 3 = 0.477$. Given these values, find the following.

(*a*) $\log_{10} 6$

Since $6 = 2 \cdot 3$,

$$\log_{10} 6 = \log_{10} (2 \cdot 3)$$
$$= \log_{10} 2 + \log_{10} 3$$
$$= 0.301 + 0.477$$
$$= 0.778$$

Keep in mind, however, that this is an approximation and that $10^{0.301}$ will only approximate 2. Verify this with your calculator.

(*b*) $\log_{10} 18$

Since $18 = 2 \cdot 3 \cdot 3$,

$$\log_{10} 18 = \log_{10} (2 \cdot 3 \cdot 3)$$
$$= \log_{10} 2 + \log_{10} 3 + \log_{10} 3$$
$$= 1.255$$

We have extended the product rule for logarithms.

(*c*) $\log_{10} \dfrac{1}{9}$

Since $\dfrac{1}{9} = \dfrac{1}{3^2}$,

$$\log_{10} \dfrac{1}{9} = \log_{10} \dfrac{1}{3^2}$$
$$= \log_{10} 1 - \log_{10} 3^2$$
$$= 0 - 2 \log_{10} 3$$
$$= -0.954$$

Note that $\log_b 1 = 0$ for any base b.

(*d*) $\log_{10} 16$

Since $16 = 2^4$,

$$\log_{10} 16 = \log_{10} 2^4 = 4 \log_{10} 2$$
$$= 1.204$$

Verify each answer with your calculator.

(*e*) $\log_{10} \sqrt{3}$

Since $\sqrt{3} = 3^{1/2}$,

$$\log_{10} \sqrt{3} = \log_{10} 3^{1/2} = \frac{1}{2} \log_{10} 3$$

$$= 0.239$$

● ● ● **CHECK YOURSELF 3**

Given the values above for $\log_{10} 2$ and $\log_{10} 3$, find each of the following.

(a) $\log_{10} 12$ **(b)** $\log_{10} 27$ **(c)** $\log_{10} \sqrt[3]{2}$

LOGARITHMS TO PARTICULAR BASES

You can easily check the results in Example 3 by using the $\boxed{\log}$ key on your calculator. For instance, in Example 3(*d*), to find $\log_{10} 16$, enter

16 $\boxed{\log}$

and the result (to three decimal places) will be 1.204. As you can see, the $\boxed{\log}$ key on your calculator provides logarithms to base 10, which is one of two types of logarithms used most frequently in mathematics:

Logarithms to base 10

Logarithms to base e

Of course, the use of logarithms to base 10 is convenient because our number system has base 10. We call logarithms to base 10 **common logarithms,** and it is customary to omit the base in writing a common (or base-10) logarithm. So

Note: When no base for "log" is written, it is assumed to be 10.

$\log N$	means	$\log_{10} N$

The following table shows the common logarithms for various powers of 10.

Exponential Form		Logarithmic Form		
10^3	$= 1000$	$\log 1000$	$=$	3
10^2	$= 100$	$\log 100$	$=$	2
10^1	$= 10$	$\log 10$	$=$	1
10	$= 1$	$\log 1$	$=$	0
10^{-1}	$= 0.1$	$\log 0.1$	$= -1$	
10^{-2}	$= 0.01$	$\log 0.01$	$= -2$	
10^{-3}	$= 0.001$	$\log 0.001$	$= -3$	

• Example 4

Approximating Logarithms with a Calculator

Verify each of the following with a calculator.

The number 4.8 lies between 1 and 10, so log 4.8 lies between 0 and 1.

Note that

$$480 = 4.8 \times 10^2$$

and

$$\log (4.8 \times 10^2)$$
$$= \log 4.8 + \log 10^2$$
$$= \log 4.8 + 2$$
$$= 2 + \log 4.8$$

The value of log 0.48 is really $-1 + 0.681$. Your calculator will combine the signed numbers.

(a) $\log 4.8 = 0.681$
(b) $\log 48 = 1.681$
(c) $\log 480 = 2.681$
(d) $\log 4800 = 3.681$
(e) $\log 0.48 = -0.319$

● ● ●

CHECK YOURSELF 4

Use your calculator to find each of the following logarithms, correct to three decimal places.

(a) $\log 2.3$ **(b)** $\log 23$ **(c)** $\log 230$

(d) 2300 **(e)** $\log 0.23$ **(f)** $\log 0.023$

Let's look at an application of common logarithms from chemistry. Common logarithms are used to define the pH of a solution. This is a scale that measures whether the solution is acidic or basic.

The pH of a solution is defined as

Note: A solution is **neutral** with pH = 7, **acidic** if the pH is less than 7, and **basic** if the pH is greater than 7.

$$pH = -\log [H^+]$$

where $[H^+]$ is the hydrogen ion concentration, in moles per liter (mol/L), in the solution.

• Example 5

A pH Application

Find the pH of each of the following. Determine whether each is a base or an acid.

(a) Rainwater: $[H^+] = 1.6 \times 10^{-7}$

From the definition,

Note the use of the product rule here.

Also, in general, $\log_b b^x = x$, so $\log 10^{-7} = -7$.

$$pH = -\log [H^+]$$
$$= -\log (1.6 \times 10^{-7})$$
$$= -(\log 1.6 + \log 10^{-7})$$
$$= -[0.204 + (-7)]$$
$$= -(-6.796) = 6.796$$

The rain is just slightly acidic.

(b) Household ammonia: $[H^+] = 2.3 \times 10^{-8}$

$$
\begin{aligned}
pH &= -\log (2.3 \times 10^{-8}) \\
&= -(\log 2.3 + \log 10^{-8}) \\
&= -[0.362 + (-8)] \\
&= 7.638
\end{aligned}
$$

The ammonia is slightly basic.

(c) Vinegar: $[H^+] = 2.9 \times 10^{-3}$

$$
\begin{aligned}
pH &= -\log (2.9 \times 10^{-3}) \\
&= -(\log 2.9 + \log 10^{-3}) \\
&= 2.538
\end{aligned}
$$

The vinegar is very acidic.

● ● ● CHECK YOURSELF 5

Find the pH for the following solutions. Are they acidic or basic?

(a) Orange juice: $[H^+] = 6.8 \times 10^{-5}$
(b) Drain cleaner: $[H^+] = 5.2 \times 10^{-13}$

Many applications require reversing the process. That is, given the logarithm of a number, we must be able to find that number. The process is straightforward.

● Example 6

Using a Calculator to Estimate Antilogarithms

Suppose that $\log x = 2.1567$. We want to find a number x whose logarithm is 2.1567. Using a calculator requires one of the following sequences:

Because it is a one-to-one function, the logarithmic function has an inverse.

$$2.1567 \;\boxed{10^x} \qquad \text{or} \qquad 2.1567 \;\boxed{\text{INV}}\;\boxed{\log}$$

Both give the result 143.45, often called the **antilogarithm** of 2.1567.

● ● ● CHECK YOURSELF 6

Find the value of the antilogarithm of x.

(a) $\log x = 0.828$ **(b)** $\log x = 1.828$
(c) $\log x = 2.828$ **(d)** $\log x = -0.172$

Let's return to the application from chemistry for an example requiring the use of the antilogarithm.

• Example 7

A pH Application

Suppose that the pH for tomato juice is 6.2. Find the hydrogen ion concentration $[H^+]$. Recall from our earlier formula that

$$pH = -\log [H^+]$$

In this case, we have

$$6.2 = -\log [H^+]$$

or

$$\log [H^+] = -6.2$$

The desired value for $[H^+]$ is then the antilogarithm of -6.2, and we use the following calculator sequence:

6.2 $\boxed{+/-}$ $\boxed{\text{INV}}$ $\boxed{\log}$

The result is 0.00000063, and we can write

$$[H^+] = 6.3 \times 10^{-7}$$

● ● ● **CHECK YOURSELF 7**

The pH for eggs is -7.8. Find $[H^+]$ for the same solution.

Natural logarithms are also called **napierian logarithms** after Napier. The importance of this system of logarithms was not fully understood until later developments in the calculus.

The restrictions on the domain of the natural logarithmic function are the same as before. The function is defined only if $x > 0$.

As we mentioned, there are two systems of logarithms in common use. The second type of logarithm uses the number e as a base, and we call logarithms to base e the **natural logarithms.** As with common logarithms, a convenient notation has developed, as the following definition shows.

> The *natural logarithm* is a logarithm to base e, and it is denoted in x, where
>
> $\ln x = \log_e x$

By the general definition of a logarithm,

$$y = \ln x \qquad \text{means the same as} \qquad x = e^y$$

and this leads us directly to the following rules.

$$\ln 1 = 0 \qquad \text{since } e^0 = 1$$

In general,

$\log_b b^x = x \qquad b \neq 1$

$$\ln e = 1 \qquad \text{since } e^1 = e$$

$$\ln e^2 = 2 \qquad \text{and} \qquad \ln e^{-3} = -3$$

• Example 8

Estimating Natural Logarithms

To find other natural logarithms, we can again turn to a calculator. To find the value of ln 2, use the sequence

2 $\boxed{\text{ln}}$

The result is 0.693 (to three decimal places).

● ● ●   **CHECK YOURSELF 8**

Use a calculator to find each of the following.

(a) ln 3   **(b)** ln 6   **(c)** ln 4   **(d)** ln $\sqrt{3}$

Of course, the properties of logarithms are applied in an identical fashion, no matter what the base.

• Example 9

Evaluating Logarithms

If ln 2 = 0.693 and ln 3 = 1.099, find the following.

Recall that

$\log_b MN = \log_b M + \log_b N$

$\log_b M^p = p \log_b M$

(a) ln 6 = ln (2 · 3) = ln 2 + ln 3 = 1.792

(b) ln 4 = ln 2^2 = 2 ln 2 = 1.386

(c) ln $\sqrt{3}$ = ln $3^{1/2}$ = $\dfrac{1}{2}$ ln 3 = 0.549

Again, verify these results with your calculator.

● ● ●   **CHECK YOURSELF 9**

Use ln 2 = 0.693 and ln 3 = 1.099 to find the following.

(a) ln 12   **(b)** ln 27

The natural logarithm function plays an important role in both theoretical and applied mathematics. Example 10 illustrates just one of the many applications of this function.

• Example 10

A Learning Curve Application

A class of students took a final mathematics examination and received an average score of 76. In a psychological experiment, the students are retested at weekly intervals over the same material. If t is measured in weeks, then the new average score after t weeks is given by

$$S(t) = 76 - 5 \ln (t + 1)$$

Complete the following.

(*a*) Find the score after 10 weeks.

$$S(t) = 76 - 5 \ln (10 + 1)$$
$$= 76 - 5 \ln 11 \approx 64$$

(*b*) Find the score after 20 weeks.

$$S(t) = 76 - 5 \ln (20 + 1) \approx 61$$

(*c*) Find the score after 30 weeks.

$$S(t) = 76 - 5 \ln (30 + 1) \approx 59$$

Recall that we read $S(t)$ as "S of t" which means that S is a function of t.

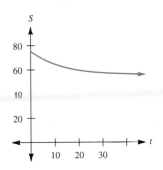

This is an example of a **forgetting curve.** Note how it drops more rapidly at first. Compare this curve to the learning curve drawn in Section 8.2, Exercise 62.

● ● ● **CHECK YOURSELF 10**

The average score for a group of biology students, retested after time t (in months), is given by

$$S(t) = 83 - 9 \ln (t + 1)$$

Find the average score after

(a) 3 months.   **(b)** 6 months.

We conclude this section with one final property of logarithms. This property will allow us to quickly find the logarithm of a number to any base. Although work with logarithms with bases other than 10 or e is relatively infrequent, the relationship between logarithms of different bases is interesting in itself. Consider the following argument.

Suppose that

$$x = \log_2 5$$

or

$$2^x = 5 \tag{4}$$

Taking the logarithm to base 10 of both sides of equation (4) yields

$$\log 2^x = \log 5$$

or

$$x \log 2 = \log 5 \qquad \text{Use the power property of logarithms.} \tag{5}$$

(Note that we omit the 10 for the base and write log 2, for example.) Now, dividing both sides of equation (5) by log 2, we have

$$x = \frac{\log 5}{\log 2}$$

We can now find a value for x with the calculator. Dividing with the calculator log 5 by log 2, we get an approximate answer of 2.3219.

Since $x = \log_2 5$ and $x = \dfrac{\log 5}{\log 2}$, then

$$\log_2 5 = \frac{\log 5}{\log 2}$$

Generalizing our result, we find the following.

Change-of-Base Formula

For the positive real numbers a and x,

$$\log_a x = \frac{\log x}{\log a}$$

Note that the logarithm on the left side has base a while the logarithms on the right side have base 10. This allows us to calculate the logarithm to base a of any positive number, given the corresponding logarithms to base 10 (or any other base), as Example 11 illustrates.

• Example 11

Evaluating Logarithms

Find $\log_5 15$.
    From the change-of-base formula with $a = 5$ and $b = 10$,

We have written $\log_{10} 15$ rather than log 15 to emphasize the change-of-base formula.

$$\log_5 15 = \frac{\log_{10} 15}{\log_{10} 5}$$

$$= 1.683$$

Note: $\log_5 5 = 1$ and $\log_5 25 = 2$, so the result for $\log_5 15$ must be between 1 and 2.

The calculator sequence for the above computation is

$$\boxed{\log}\ 15\ \boxed{\div}\ \boxed{\log}\ 5\ \boxed{\text{ENTER}}$$

CAUTION

A *common error* is to write

$$\frac{\log 15}{\log 5} = \log 15 - \log 5$$

This is *not* a logarithmic property. A true statement would be

$$\log \frac{15}{5} = \log 15 - \log 5$$

but

$$\log \frac{15}{5} \quad \text{and} \quad \frac{\log 15}{\log 5}$$

are *not* the same.

CHECK YOURSELF 11

Use the change-of-base formula to find $\log_8 32$.

Note: The $\log_e x$ is called the **natural log** of x. We use "ln x" to designate the natural log of x. A special case of the change-of-base formula allows us to find natural logarithms in terms of common logarithms:

$$\ln x = \frac{\log x}{\log e}$$

so

$$\ln x \approx \frac{\log x}{0.434} \quad \text{or, since} \quad \frac{1}{0.434} \approx 2.304, \text{ then } \ln x \approx 2.304 \log x$$

Of course, since all modern calculators have both the log function key and the ln function key, this conversion formula is now rarely used.

CHECK YOURSELF ANSWERS

1. (a) $2 \log_b x + 3 \log_b y + \log_b z$, (b) $\dfrac{1}{2} (\log_{10} x + \log_{10} y - \log_{10} z)$.

2. (a) $\log_b \dfrac{x^3 y^2}{z^2}$, (b) $\log_2 \sqrt[3]{\dfrac{x^2}{y}}$. 3. (a) 1.079, (b) 1.431, (c) 0.100.

4. (a) 0.362, (b) 1.362, (c) 2.362, (d) 3.362, (e) -0.638, (f) -1.638.

5. (a) 4.17, acidic; (b) 12.28, basic.   6. (a) 6.73, (b) 67.3, (c) 673, (d) 0.673.

7. $[\text{H}^+] = 1.6 \times 10^{-8}$.   8. (a) 1.099, (b) 1.792, (c) 1.386, (d) 0.549.

9. (a) 2.485, (b) 3.297.   10. (a) 70.5, (b) 65.5.

11. $\log_8 32 = \dfrac{\log 32}{\log 8} \approx 1.667$.

8.4 Exercises

In Exercises 1 to 18, use the properties of logarithms to expand each expression.

1. $\log_b 5x$

2. $\log_3 7x$

3. $\log_4 \dfrac{x}{3}$

4. $\log_b \dfrac{2}{y}$

5. $\log_3 a^2$

6. $\log_5 y^4$

7. $\log_5 \sqrt{x}$

8. $\log \sqrt[3]{z}$

9. $\log_b x^3 y^2$

10. $\log_5 x^2 z^4$

11. $\log_4 y^2 \sqrt{x}$

12. $\log_b x^3 \sqrt[3]{z}$

13. $\log_b \dfrac{x^2 y}{z}$

14. $\log_5 \dfrac{3}{xy}$

15. $\log \dfrac{xy^2}{\sqrt{z}}$

16. $\log_4 \dfrac{x^3 \sqrt{y}}{z^2}$

17. $\log_5 \sqrt[3]{\dfrac{xy}{z^2}}$

18. $\log_b \sqrt[4]{\dfrac{x^2 y}{z^3}}$

In Exercises 19 to 30, write each expression as a single logarithm.

19. $\log_b x + \log_b y$

20. $\log_5 x - \log_5 y$

21. $2 \log_2 x - \log_2 y$

22. $3 \log_b x + \log_b z$

23. $\log_b x + \dfrac{1}{2} \log_b y$

24. $\dfrac{1}{3} \log_b x - 2 \log_b z$

25. $\log_b x + 2 \log_b y - \log_b z$

26. $2 \log_5 x - (3 \log_5 y + \log_5 z)$

27. $\dfrac{1}{2} \log_6 y - 3 \log_6 z$

28. $\log_b x - \dfrac{1}{3} \log_b y - 4 \log_b z$

29. $\dfrac{1}{3} (2 \log_b x + \log_b y - \log_b z)$

30. $\dfrac{1}{5}(2 \log_4 x - \log_4 y + 3 \log_4 z)$

In Exercises 31 to 38, given that $\log 2 = 0.301$ and $\log 3 = 0.477$, find each logarithm.

31. $\log 24$

32. $\log 36$

33. $\log 8$

34. $\log 81$

35. $\log \sqrt{2}$

36. $\log \sqrt[3]{3}$

37. $\log \dfrac{1}{4}$

38. $\log \dfrac{1}{27}$

In Exercises 39 to 44, use your calculator to find each logarithm.

39. $\log 6.8$

40. $\log 68$

41. $\log 680$

42. $\log 6800$

43. $\log 0.68$

44. $\log 0.068$

In Exercises 45 and 46, find the pH, given the hydrogen ion concentration $[H^+]$ for each solution. Use the formula

$$pH = -\log [H^+]$$

Are the solutions acidic or basic?

45. Blood: $[H^+] = 3.8 \times 10^{-8}$

46. Lemon juice: $[H^+] = 6.4 \times 10^{-3}$

In Exercises 47 to 50, use your calculator to find the antilogarithm for each logarithm.

47. 0.749

48. 1.749

49. 3.749

50. -0.251

In Exercises 51 and 52, given the pH of the solutions, find the hydrogen ion concentration $[H^+]$.

51. Wine: $pH = 4.7$

52. Household ammonia: $pH = 7.8$

In Exercises 53 to 56, use your calculator to find each logarithm.

53. $\ln 2$

54. $\ln 3$

55. $\ln 10$

56. $\ln 30$

The average score on a final examination for a group of psychology students, retested after time t (in weeks), is given by

$$S = 85 - 8 \ln (t + 1)$$

In Exercises 57 and 58, find the average score on the retests:

57. After 3 weeks

58. After 12 weeks

In Exercises 59 and 60, use the change-of-base formula to find each logarithm.

59. $\log_3 25$

60. $\log_5 30$

The amount of a radioactive substance remaining after a given amount of time t is given by the following formula:

$$A = e^{\lambda t + \ln A_0}$$

where A is the amount remaining after time t, variable A_0 is the original amount of the substance, and λ is the radioactive decay constant.

61. How much plutonium 239 will remain after 50,000 years if 24 kg was originally stored? Plutonium 239 has a radioactive decay constant of -0.000029.

62. How much plutonium 241 will remain after 100 years if 52 kg was originally stored? Plutonium 241 has a radioactive decay constant of -0.053319.

63. How much strontium 90 was originally stored if after 56 years it is discovered that 15 kg still remains? Strontium 90 has a radioactive decay constant of -0.024755.

64. How much cesium 137 was originally stored if after 90 years it is discovered that 20 kg still remains? Cesium 137 has a radioactive decay constant of -0.023105.

65. Which keys on your calculator are function keys and which are operation keys? What is the difference?

66. How is the pH factor relevant to your selection of a hair care product?

Answers

1. $\log_b 5 + \log_b x$ **3.** $\log_4 x - \log_4 3$ **5.** $2 \log_3 a$ **7.** $\frac{1}{2} \log_5 x$ **9.** $3 \log_b x + 2 \log_b y$

11. $2 \log_4 y + \frac{1}{2} \log_4 x$ **13.** $2 \log_b x + \log_b y - \log_b z$ **15.** $\log x + 2 \log y - \frac{1}{2} \log z$

17. $\frac{1}{3} (\log_5 x + \log_5 y - 2 \log_5 z)$ **19.** $\log_b xy$ **21.** $\log_2 \frac{x^2}{y}$ **23.** $\log_b x \sqrt{y}$ **25.** $\log_b \frac{xy^2}{z}$

27. $\log_6 \frac{\sqrt{y}}{z^3}$ **29.** $\log_b \sqrt[3]{\frac{x^2y}{z}}$ **31.** 1.380   **33.** 0.903   **35.** 0.151   **37.** -0.602 **39.** 0.833

41. 2.833   **43.** -0.167 **45.** 7.42, basic   **47.** 5.61   **49.** 5610   **51.** 2×10^{-5} **53.** 0.693

55. 2.303   **57.** 74   **59.** 2.930   **61.** 5.6 kg   **63.** 60 kg   **65.**

8.5 Logarithmic and Exponential Equations

8.5 OBJECTIVES

1. Solve a logarithmic equation
2. Solve an exponential equation
3. Solve an application involving an exponential equation

Much of the importance of the properties of logarithms developed in the previous section lies in the application of those properties to the solution of equations involving logarithms and exponentials. Our work in this section will consider solution techniques for both types of equations. Let's start with a definition.

> A **logarithmic equation** is an equation that contains a logarithmic expression.

We solved some simple examples in Section 8.3. Let's review for a moment. To solve $\log_3 x = 4$ for x, recall that we simply convert the logarithmic equation to exponential form. Here,

$$x = 3^4$$

so

$$x = 81$$

and 81 is the solution to the given equation.

Now, what if the logarithmic equation involves more than one logarithmic term? Example 1 illustrates how the properties of logarithms must then be applied.

• Example 1

Solving a Logarithmic Equation

Solve each logarithmic equation.

(a) $\log_5 x + \log_5 3 = 2$

The original equation can be written as

We apply the product rule for logarithms:

$\log_b M + \log_b N = \log_b MN$

$$\log_5 3x = 2$$

Now, since only a single logarithm is involved, we can write the equation in the equivalent exponential form:

$$3x = 5^2$$
$$3x = 25$$
$$x = \frac{25}{3}$$

(b) $\log x + \log (x - 3) = 1$

Write the equation as

Since no base is written, it is assumed to be 10.

$$\log x(x - 3) = 1$$

or

Given the base of 10, this is the equivalent exponential form.

$$x(x - 3) = 10^1$$

We now have

$$x^2 - 3x = 10$$

$$x^2 - 3x - 10 = 0$$

$$(x - 5)(x + 2) = 0$$

Possible solutions are $x = 5$ or $x = -2$.

Note that substitution of -2 into the original equation gives

Checking possible solutions is particularly important here.

$$\log (-2) + \log (-5) = 1$$

Since logarithms of negative numbers are *not* defined, -2 is an extraneous solution and we must reject it. The only solution for the original equation is 5.

● ● ● **CHECK YOURSELF 1**

Solve $\log_2 x + \log_2 (x + 2) = 3$ for x.

The quotient property is used in a similar fashion for solving logarithmic equations. Consider Example 2.

●Example 2

Solving a Logarithmic Equation

Solve each equation for x.

(a) $\log_5 x - \log_5 2 = 2$

Rewrite the original equation as

We apply the quotient rule for logarithms:

$$\log_5 \frac{x}{2} = 2$$

$$\log_b M - \log_b N = \log_b \frac{M}{N}$$

Now,

$$\frac{x}{2} = 5^2$$

$$\frac{x}{2} = 25$$

$$x = 50$$

(b) $\log_3 (x + 1) - \log_3 x = 3$

$$\log_3 \frac{x + 1}{x} = 3$$

$$\frac{x + 1}{x} = 27$$

$$x + 1 = 27x$$

$$1 = 26x$$

Again, you should verify that substituting $\dfrac{1}{26}$ for x leads to a positive value in each of the original logarithms.

$$x = \frac{1}{26}$$

● ● ● ● **CHECK YOURSELF 2**

Solve $\log_5 (x + 3) - \log_5 x = 2$ for x.

The solution of certain types of logarithmic equations calls for the one-to-one property of the logarithmic function.

> If $\qquad \log_b M = \log_b N$
>
> then $\qquad M = N$

● Example 3

Solving a Logarithmic Equation

Solve the following equation for x.

$\log (x + 2) - \log 2 = \log x$

Again, we rewrite the left-hand side of the equation. So

$$\log \frac{x + 2}{2} = \log x$$

Since the logarithmic function is one-to-one, this is equivalent to

$$\frac{x + 2}{2} = x$$

or

$$x = 2$$

● ● ● **CHECK YOURSELF 3**

Solve for x.

$$\log (x + 3) - \log 3 = \log x$$

The following algorithm summarizes our work in solving logarithmic equations.

> **Solving Logarithmic Equations**
>
> **Step 1** Use the properties of logarithms to combine terms containing logarithmic expressions into a single term.
> **Step 2** Write the equation formed in step 1 in exponential form.
> **Step 3** Solve for the indicated variable.
> **Step 4** Check your solutions to make sure that possible solutions do not result in the logarithms of negative numbers.

Let's look now at **exponential equations,** which are equations in which the variable appears as an exponent.

We solved some particular exponential equations in Section 8.2. In solving an equation such as

$$3^x = 81$$

we wrote the right-hand member as a power of 3, so that

Again, we want to write both sides as a power of the same base, here 3.

$$3^x = 3^4$$

or

$$x = 4$$

The technique here will work only when both sides of the equation can be conveniently expressed as powers of the same base. If that is not the case, we must use logarithms for the solution of the equation, as illustrated in Example 4.

● **Example 4**

Solving an Exponential Equation

Solve $3^x = 5$ for x.

We begin by taking the common logarithm of both sides of the original equation.

Again:

if $M = N$, then

$\log_b M = \log_b N$

$$\log 3^x = \log 5$$

Now we apply the power property so that the variable becomes a coefficient on the left.

$$x \log 3 = \log 5$$

CAUTION

This is *not* log 5 − log 3, a common error.

Dividing both sides of the equation by log 3 will isolate x, and we have

$$x = \frac{\log 5}{\log 3}$$

$$= 1.465 \qquad \text{(to three decimal places)}$$

Note: You can verify the approximate solution by using the $\boxed{y^x}$ key on your calculator. Raise 3 to power 1.465.

● ● ● **CHECK YOURSELF 4**

Solve $2^x = 10$ for x.

Example 5 shows how to solve an equation with a more complicated exponent.

● Example 5

On the left, we apply

$\log_b M^p = p \log_b M$

On a graphing calculator, the sequence would be

$\boxed{(}$ $\boxed{\log}$ 8 ÷ $\boxed{\log}$ 5 −

1 $\boxed{)}$ $\boxed{-}$ 2 $\boxed{=}$

Another calculator sequence to find x would be

8 $\boxed{\log}$ $\boxed{÷}$ 5 $\boxed{\log}$ $\boxed{-}$ 1

$\boxed{=}$ $\boxed{÷}$ 2 $\boxed{=}$

Solving an Exponential Equation

Solve $5^{2x+1} = 8$ for x.
    The solution begins as in Example 4.

$$\log 5^{2x+1} = \log 8$$

$$(2x + 1) \log 5 = \log 8$$

$$2x + 1 = \frac{\log 8}{\log 5}$$

$$2x = \frac{\log 8}{\log 5} - 1$$

$$x = \frac{1}{2}\left(\frac{\log 8}{\log 5} - 1\right)$$

$$x \approx 0.146$$

● ● ● **CHECK YOURSELF 5**

Solve $3^{2x-1} = 7$ for x.

The procedure is similar if the variable appears as an exponent in more than one term of the equation.

• Example 6

Solving an Exponential Equation

Solve $3^x = 2^{x+1}$ for x.

$$\log 3^x = \log 2^{x+1}$$

Use the power property to write the variables as coefficients.

$$x \log 3 = (x + 1) \log 2$$

$$x \log 3 = x \log 2 + \log 2$$

We now isolate x on the left.

$$x \log 3 - x \log 2 = \log 2$$

$$x(\log 3 - \log 2) = \log 2$$

To check the reasonableness of this result, use your calculator to verify that

$$3^{1.710} = 2^{2.710}$$

$$x = \frac{\log 2}{\log 3 - \log 2}$$

$$\approx 1.710$$

● ● ● ● CHECK YOURSELF 6

Solve $5^{x+1} = 3^{x+2}$ for x.

The following algorithm summarizes our work with solving exponential equations.

Solving Exponential Equations

STEP 1 Try to write each side of the equation as a power of the same base. Then equate the exponents to form an equation.
STEP 2 If the above procedure is not applicable, take the common logarithm of both sides of the original equation.
STEP 3 Use the power rule for logarithms to write an equivalent equation with the variables as coefficients.
STEP 4 Solve the resulting equation.

There are many applications of our work with exponential equations. Consider the following.

• Example 7

An Interest Application

If an investment of P dollars earns interest at an annual interest rate r and the interest is compounded n times per year, then the amount in the account after t years is given by

$$A = P\left(1 + \frac{r}{n}\right)^{nt} \tag{1}$$

If $1000 is placed in an account with an annual interest rate of 6%, find out how long it will take the money to double when interest is compounded annually and quarterly.

(*a*) Compounding interest annually.

Since the interest is compounded *once* a year, $n = 1$

Using equation (1) with $A = 2000$ (we want the original 1000 to double). $P = 1000$, $r = 0.06$, and $n = 1$, we have

$$2000 = 1000(1 + 0.06)^t$$

Dividing both sides by 1000 yields

$$2 = (1.06)^t$$

We now have an exponential equation that can be solved by our earlier techniques.

$$\log 2 = \log (1.06)^t$$
$$= t \log 1.06$$

From accounting, we have the **rule of 72,** which states that the doubling time is approximately 72 divided by the interest rate as a percentage. Here $\dfrac{72}{6} = 12$ years.

or

$$t = \frac{\log 2}{\log 1.06}$$
$$\approx 11.9 \text{ years}$$

It takes just a little less than 12 years for the money to double.

(*b*) Compounding interest quarterly.

Since the interest is compounded 4 times per year, $n = 4$.

Now $n = 4$ in equation (1), so

$$2000 = 1000\left(1 + \frac{0.06}{4}\right)^{4t}$$
$$2 = (1.015)^{4t}$$
$$\log 2 = \log (1.015)^{4t}$$
$$\log 2 = 4t \log 1.015$$
$$\frac{\log 2}{4 \log 1.015} = t$$
$$t \approx 11.6 \text{ years}$$

Note that the doubling time is reduced by approximately 3 months by the more frequent compounding.

● ● ● **CHECK YOURSELF 7**

Find the doubling time in Example 7 if the interest is compounded monthly.

Problems involving rates of growth or decay can also be solved by using exponential equations.

● Example 8

A Population Application

A town's population is presently 10,000. Given a projected growth rate of 7% per year, t years from now the population P will be given by

$$P = 10,000e^{0.07t}$$

In how many years will the town's population double?
    We want the time t when P will be 20,000 (doubled in size). So

$$20,000 = 10,000e^{0.07t}$$

or

Divide both sides by 10,000. $2 = e^{0.07t}$

In this case, we take the *natural logarithm* of both sides of the equation. This is because e is involved in the equation.

$$\ln 2 = \ln e^{0.07t}$$

Apply the power property. $\ln 2 = 0.07t \ln e$

Note: $\ln e = 1$ $\ln 2 = 0.07t$

$$\frac{\ln 2}{0.07} = t$$

$$t \approx 9.9 \text{ years}$$

The population will double in approximately 9.9 years.

● ● ●   **CHECK YOURSELF 8**

If $1000 is invested in an account with an annual interest rate of 6%, compounded continuously, the amount A in the account after t years is given by

$$A = 1000e^{0.06t}$$

Find the time t that it will take for the amount to double ($A = 2000$). Compare this time with the result of the Check Yourself 7 exercise. Which is shorter? Why?

● ● ●   **CHECK YOURSELF ANSWERS**

1. 2.   **2.** $\frac{1}{8}$.   **3.** $\frac{3}{2}$.   **4.** 3.322.   **5.** 1.386.   **6.** 1.151.   **7.** 11.58

years.   **8.** 11.55 years. The doubling time is shorter, because interest is compounded more frequently.

8.5 Exercises

In Exercises 1 to 20, solve each logarithmic equation for x.

1. $\log_4 x = 3$

2. $\log_3 x = -2$

3. $\log (x + 1) = 2$

4. $\log_5 (2x - 1) = 2$

5. $\log_2 x + \log_2 8 = 6$

6. $\log 5 + \log x = 2$

7. $\log_3 x - \log_3 6 = 3$

8. $\log_4 x - \log_4 8 = 3$

9. $\log_2 x + \log_2 (x + 2) = 3$

10. $\log_3 x + \log_3 (2x + 3) = 2$

11. $\log_7 (x + 1) + \log_7 (x - 5) = 1$

12. $\log_2 (x + 2) + \log_2 (x - 5) = 3$

13. $\log x - \log (x - 2) = 1$

14. $\log_5 (x + 5) - \log_5 x = 2$

15. $\log_3 (x + 1) - \log_3 (x - 2) = 2$

16. $\log (x + 2) - \log (2x - 1) = 1$

17. $\log (x + 5) - \log (x - 2) = \log 5$

18. $\log_3 (x + 12) - \log_3 (x - 3) = \log_3 6$

19. $\log_2 (x^2 - 1) - \log_2 (x - 2) = 3$

20. $\log (x^2 + 1) - \log (x - 2) = 1$

In Exercises 21 to 38, solve each exponential equation for x. Give your solutions in decimal form, correct to three decimal places.

21. $5^x = 625$

22. $4^x = 64$

23. $2^{x+1} = \dfrac{1}{8}$

24. $9^x = 3$

25. $8^x = 2$

26. $3^{2x-1} = 27$

27. $3^x = 7$

28. $5^x = 30$

29. $4^{x+1} = 12$

30. $3^{2x} = 5$

31. $7^{3x} = 50$

32. $6^{x-3} = 21$

33. $5^{3x-1} = 15$

34. $8^{2x+1} = 20$

35. $4^x = 3^{x+1}$

36. $5^x = 2^{x+2}$

37. $2^{x+1} = 3^{x-1}$

38. $3^{2x+1} = 5^{x+1}$

Use the formula

$$A = P\left(1 + \frac{r}{n}\right)^{nt}$$

to solve Exercises 39 to 42.

39. Interest If $5000 is placed in an account with an annual interest rate of 9%, how long will it take the amount to double if the interest is compounded annually?

40. Repeat Exercise 39 if the interest is compounded semiannually.

41. Repeat Exercise 39 if the interest is compounded quarterly.

42. Repeat Exercise 39 if the interest is compounded monthly.

Suppose the number of bacteria present in a culture after t hours is given by $N(t) = N_0 \cdot 2^{t/2}$, where N_0 is the initial number of bacteria. Use the formula to solve Exercises 43 to 46.

43. How long will it take the bacteria to increase from 12,000 to 20,000?

44. How long will it take the bacteria to increase from 12,000 to 50,000?

45. How long will it take the bacteria to triple? *Hint:* Let $N = 3N_0$.

46. How long will it take the culture to increase to 5 times its original size? *Hint:* Let $N = 5N_0$.

The radioactive element strontium 90 has a half-life of approximately 28 years. That is, in a 28-year period, one-half of the initial amount will have decayed into another substance. If A_0 is the initial amount of the element, then the amount A remaining after t years is given by

$$A(t) = A_0\left(\frac{1}{2}\right)^{t/28}$$

Use the formula to solve Exercises 47 to 50.

47. If the initial amount of the element is 100 g, in how many years will 60 g remain?

48. If the initial amount of the element is 100 g, in how many years will 20 g remain?

49. In how many years will 75% of the original amount remain? *Hint:* Let $A = 0.75A_0$.

50. In how many years will 10% of the original amount remain? *Hint:* Let $A = 0.1A_0$.

Given projected growth, t years from now a city's population P can be approximated by $P(t) = 25,000e^{0.045t}$. Use the formula to solve Exercises 51 and 52.

51. How long will it take the city's population to reach 35,000?

52. How long will it take the population to double?

The number of bacteria in a culture after t hours can be given by $N(t) = N_0e^{0.03t}$, where N_0 is the initial number of bacteria in the culture. Use the formula to solve Exercises 53 and 54.

53. In how many hours will the size of the culture double?

54. In how many hours will the culture grow to 4 times its original population?

The atmospheric pressure P, in inches of mercury (inHg), at an altitude h feet above sea level is approximated by $P(t) = 30e^{-0.00004h}$. Use the formula to solve Exercises 55 and 56.

55. Find the altitude if the pressure at that altitude is 25 inHg.

56. Find the altitude if the pressure at that altitude is 20 inHg.

Carbon 14 dating is used to measure the age of specimens and is based on the radioactive decay of the element carbon 14. This decay begins once a plant or animal dies. If A_0 is the initial amount of carbon 14, then the amount remaining after t years is $A(t) = A_0e^{-0.000124t}$. Use the formula to solve Exercises 57 and 58.

57. Estimate the age of a specimen if 70% of the original amount of carbon 14 remains.

58. Estimate the age of a specimen if 20% of the original amount of carbon 14 remains.

59. In some of the earlier exercises, we talked about bacteria cultures that double in size every few minutes. Can this go on forever? Explain.

60. The population of the United States has been doubling every 45 years. Is it reasonable to assume that this rate will continue? What factors will start to limit that growth?

In Exercises 61 to 64, use your calculator to find the graph for each equation, then explain the result.

61. $y = \log 10^x$

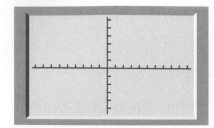

62. $y = 10^{\log x}$

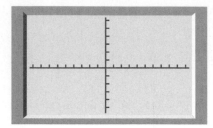

63. $y = \ln e^x$

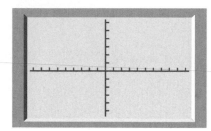

64. $y = e^{\ln x}$

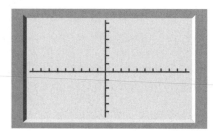

Answers

1. 64   **3.** 99   **5.** 8   **7.** 162   **9.** 2   **11.** 6   **13.** $\dfrac{20}{9}$ **15.** $\dfrac{19}{8}$ **17.** $\dfrac{15}{4}$ **19.** 5, 3

21. 4   **23.** -4 **25.** $\dfrac{1}{3}$ **27.** 1.771   **29.** 0.792   **31.** 0.670   **33.** 0.894   **35.** 3.819

37. 4.419   **39.** 8.04 yr   **41.** 7.79 yr   **43.** 1.47 h   **45.** 3.17 h   **47.** 20.6 yr   **49.** 11.6 yr

51. 7.5 yr   **53.** 23.1 h   **55.** 4558 ft   **57.** 2876 yr   **59.**

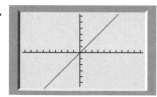

61.

63.

Given projected growth, t years from now a city's population P can be approximated by $P(t) = 25,000e^{0.045t}$. Use the formula to solve Exercises 51 and 52.

51. How long will it take the city's population to reach 35,000?

52. How long will it take the population to double?

The number of bacteria in a culture after t hours can be given by $N(t) = N_0e^{0.03t}$, where N_0 is the initial number of bacteria in the culture. Use the formula to solve Exercises 53 and 54.

53. In how many hours will the size of the culture double?

54. In how many hours will the culture grow to 4 times its original population?

The atmospheric pressure P, in inches of mercury (inHg), at an altitude h feet above sea level is approximated by $P(t) = 30e^{-0.00004h}$. Use the formula to solve Exercises 55 and 56.

55. Find the altitude if the pressure at that altitude is 25 inHg.

56. Find the altitude if the pressure at that altitude is 20 inHg.

Carbon 14 dating is used to measure the age of specimens and is based on the radioactive decay of the element carbon 14. This decay begins once a plant or animal dies. If A_0 is the initial amount of carbon 14, then the amount remaining after t years is $A(t) = A_0e^{-0.000124t}$. Use the formula to solve Exercises 57 and 58.

57. Estimate the age of a specimen if 70% of the original amount of carbon 14 remains.

58. Estimate the age of a specimen if 20% of the original amount of carbon 14 remains.

59. In some of the earlier exercises, we talked about bacteria cultures that double in size every few minutes. Can this go on forever? Explain.

60. The population of the United States has been doubling every 45 years. Is it reasonable to assume that this rate will continue? What factors will start to limit that growth?

In Exercises 61 to 64, use your calculator to find the graph for each equation, then explain the result.

61. $y = \log 10^x$

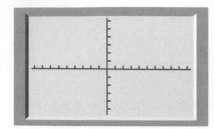

62. $y = 10^{\log x}$

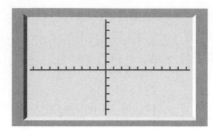

63. $y = \ln e^x$

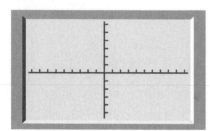

64. $y = e^{\ln x}$

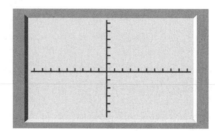

Answers

1. 64   **3.** 99   **5.** 8   **7.** 162   **9.** 2   **11.** 6   **13.** $\dfrac{20}{9}$ **15.** $\dfrac{19}{8}$ **17.** $\dfrac{15}{4}$ **19.** 5, 3

21. 4   **23.** -4 **25.** $\dfrac{1}{3}$ **27.** 1.771   **29.** 0.792   **31.** 0.670   **33.** 0.894   **35.** 3.819

37. 4.419   **39.** 8.04 yr   **41.** 7.79 yr   **43.** 1.47 h   **45.** 3.17 h   **47.** 20.6 yr   **49.** 11.6 yr

51. 7.5 yr   **53.** 23.1 h   **55.** 4558 ft   **57.** 2876 yr   **59.**

61.

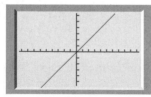

63.

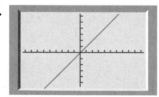

Summary

Inverse Relations and Functions [8.1]

The inverse of the relation

$\{(1, 2), (2, 3), (4, 3)\}$ is
$\{(2, 1), (3, 2), (3, 4)\}$

The **inverse** of a relation is formed by interchanging the components of each ordered pair in the given relation.

If a relation (or function) is specified by an equation, interchange the roles of x and y in the defining equation to form the inverse.

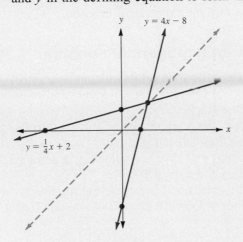

If $f(x) = 4x - 8$, then

$f^{-1}(x) = \dfrac{1}{4}x + 2$

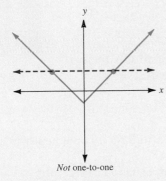

Not one-to-one

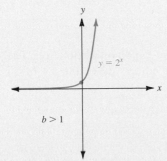

$b > 1$

The inverse of a function f may or may not be a function. If the inverse *is* also a function, we denote that inverse as f^{-1}, read "the inverse of f."

A function f has an inverse f^{-1}, which is also a function, if and only if f is a **one-to-one** function. That is, no two ordered pairs in the function have the same second component.

A function f is a one-to-one function if and only if it has an inverse f^{-1}, which is also a function.

The **horizontal-line test** can be used to determine whether a function is one-to-one.

Finding Inverse Relations and Functions

1. Interchange the x and y components of the ordered pairs of the given relation or the roles of x and y in the defining equation.
2. If the relation was described in equation form, solve the defining equation of the inverse for y.
3. If desired, graph the relation and its inverse on the same set of axes. The two graphs will be symmetric about the line $y = x$.

Exponential Functions [8.2]

An **exponential function** is any function defined by an equation of the form

$$y = f(x) = b^x \qquad b > 0, b \neq 1$$

If b is greater than 1, the function is always increasing (a **growth function**). If b is less than 1, the function is always decreasing (a **decay function**).

In both cases, the exponential function is one-to-one. The domain is the set of all real numbers, and the range is the set of positive real numbers.

The function defined by $f(x) = e^x$, where e is an irrational number (approximately 2.71828), is called *the* exponential function.

Graphing an Exponential Function

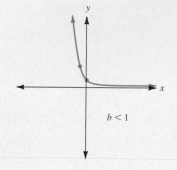

1. Establish a table of values by considering the function in the form $y = b^x$.
2. Plot points from that table of values and connect them with a smooth curve to form the graph.
3. If $b > 1$, the graph increases from left to right. If $0 < b < 1$, the graph decreases from left to right.
4. All graphs will have the following in common:
   (a) The y intercept will be 1.
   (b) The graphs will approach, but not touch, the x axis.
   (c) The graphs will represent one-to-one functions.

log₃ 9 = 2 is in logarithmic form.

$\log_3 9 = 2$ is in logarithmic form.

$3^2 = 9$ is the exponential form.

$\log_3 9 = 2$ is equivalent to $3^2 = 9$

2 is the power to which we must raise 3 in order to get 9.

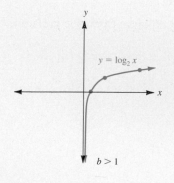

Logarithmic Functions [8.3]

In the expression

$$y = \log_b x$$

y is called the *logarithm of x to base b*, where $b > 0$ and $b \neq 1$.

An expression such as $y = \log_b x$ is said to be in **logarithmic form.**

An expression such as $x = b^y$ is said to be in **exponential form.**

$$y = \log_b x \qquad \text{means the same as} \qquad x = b^y$$

A logarithm is an exponent or a power. The logarithm of x to base b is the power to which we must raise b in order to get x.

A **logarithmic function** is any function defined by an equation of the form

$$y = f(x) = \log_b x \qquad b > 0, b \neq 1$$

The logarithm function is the inverse of the corresponding exponential function. The function is one-to-one with domain $\{x \mid x > 0\}$ and range composed of the set of all real numbers.

Properties of Logarithms [8.4]

If M, N, and b are positive real numbers with $b \neq 1$ and if p is any real number, then we can state the following properties of logarithms:

$\log 10 = 1$

$\log_2 1 = 0$

$3^{\log_3 2} = 2$

$\log_5 5^x = x$

1. $\log_b b = 1$

2. $\log_b 1 = 0$

3. $b^{\log_b x} = x$

4. $\log_b b^x = x$

Product Property

$\log_3 x + \log_3 y = \log_3 xy$

$\log_b MN = \log_b M + \log_b N$

Quotient Property

$\log_5 8 - \log_5 3 = \log_5 \dfrac{8}{3}$

$\log_b \dfrac{M}{N} = \log_b M - \log_b N$

Power Property

$\log 3^2 = 2 \log 3$

$\log_b M^p = p \log_b M$

Common logarithms are logarithms to base 10. For convenience, we omit the base in writing common logarithms:

$\log_{10} 1000 = \log 1000$

$\qquad = \log 10^3 = 3$

$\log M = \log_{10} M$

Natural logarithms are logarithms to base e. By custom we also omit the base in writing natural logarithms:

$\ln 3 = \log_e 3$

$\ln M = \log_e M$

Logarithmic and Exponential Equations [8.5]

To solve $\log_2 x = 5$:
Write the equation in the equivalent exponential form to solve

$x = 2^5 \quad$ or $\quad x = 32$

A **logarithmic equation** is an equation that contains a logarithmic expression.

$\log_2 x = 5$

is a logarithmic equation.

Solving Logarithmic Equations

To solve

$\log_4 x + \log_4 (x - 6) = 2$

$\log_4 x(x - 6) = 2$

$x(x - 6) = 4^2$

$x^2 - 6x - 16 = 0$

$(x - 8)(x + 2) = 0$

$x = 8 \quad \text{or} \quad x = -2$

Since substituting -2 for x in the original equation results in the logarithm of a negative number, we reject that answer. The only solution is 8.

STEP 1 Use the properties of logarithms to combine terms containing logarithmic expressions into a single term.

STEP 2 Write the equation formed in step 1 in exponential form.

STEP 3 Solve for the indicated variable.

STEP 4 Check your solutions to make sure that possible solutions do not result in the logarithms of negative numbers.

To solve $4^x = 64$:

Since $64 = 4^3$, write

$4^x = 4^3 \quad \text{or} \quad x = 3$

An **exponential equation** is an equation in which the variable appears as an exponent.

The following algorithm summarizes the steps in solving any exponential equation.

Solving Exponential Equations

$2^{x+3} = 5^x$

$\log 2^{x+3} = \log 5^x$

$(x + 3) \log 2 = x \log 5$

$x \log 2 + 3 \log 2 = x \log 5$

$x \log 2 - x \log 5 = -3 \log 2$

$x(\log 2 - \log 5) = -3 \log 2$

$x = \dfrac{-3 \log 2}{\log 2 - \log 5} \approx 2.269$

STEP 1 Try to write each side of the equation as a power of the same base. Then, equate the exponents to form an equation.

STEP 2 If the above procedure is not applicable, take the common logarithm of both sides of the original equation.

STEP 3 Use the power rule for logarithms to write an equivalent equation with the variables as coefficients.

STEP 4 Solve the resulting equation.

This summary exercise set is provided to give you practice with each of the objectives in the chapter. Each chapter is keyed to the appropriate chapter section. The answers are provided in the *Instructor's Manual.*

[8.1]   Write the inverse relation for each of the following functions. Which inverses are also functions?

1. $\{(1, 5), (2, 7), (3, 9)\}$ **2.** $\{(3, 1), (5, 1), (7, 1)\}$ **3.** $\{(2, 4), (4, 3), (6, 4)\}$

Write an equation for the inverse of the relation defined by each of the following equations.

4. $y = 3x - 6$ **5.** $y = \dfrac{x + 1}{2}$ **6.** $y = x^2 - 2$

Write an equation for the inverse of the relation defined by each of the following equations. Which inverses are also functions?

7. $y = 3x + 6$ **8.** $y = -x^2 + 3$ **9.** $4x^2 + 9y^2 = 36$

[8.2]   Graph the exponential functions defined by each of the following equations.

10. $y = 3^x$ **11.** $y = \left(\dfrac{3}{4}\right)^x$

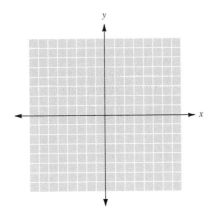   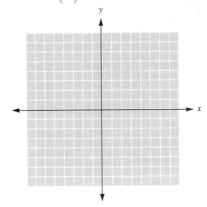

[8.3]   Solve each of the following exponential equations for x.

12. $5^x = 125$ **13.** $2^{2x+1} = 32$ **14.** $3^{x-1} = \dfrac{1}{9}$

If it takes 2 h for the population of a certain bacteria culture to double (by dividing in half), then the number N of bacteria in the culture after t hours is given by $N = 1000 \cdot 2^{t/2}$, where the initial population of the culture was 1000. Using this formula, find the number in the culture:

15. After 4 h.   **16.** After 12 h.   **17.** After 15 h.

Graph the logarithmic functions defined by each of the following equations.

18. $y = \log_3 x$

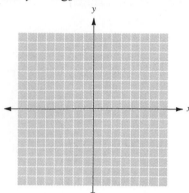

19. $y = \log_2 (x - 1)$

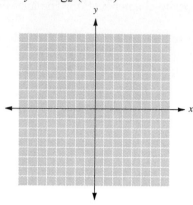

Convert each of the following statements to logarithmic form.

20. $3^4 = 81$ **21.** $10^3 = 1000$ **22.** $5^0 = 1$

23. $5^{-2} = \dfrac{1}{25}$ **24.** $25^{1/2} = 5$ **25.** $16^{3/4} = 8$

Convert each of the following statements to exponential form.

26. $\log_3 81 = 4$ **27.** $\log 1 = 0$ **28.** $\log_{81} 9 = \dfrac{1}{2}$

29. $\log_5 25 = 2$ **30.** $\log 0.001 = -3$ **31.** $\log_{32} \dfrac{1}{2} = -\dfrac{1}{5}$

Solve each of the following equations for the unknown variable.

32. $y = \log_5 125$ **33.** $\log_b \dfrac{1}{9} = -2$ **34.** $\log_8 x = 2$

35. $y = \log_5 1$ **36.** $\log_b 3 = \dfrac{1}{2}$ **37.** $y = \log_{16} 2$

38. $y = \log_8 2$

The decibel (dB) rating for the loudness of a sound is given by

$$L = 10 \log \frac{I}{I_0}$$

where I is the intensity of that sound in watts per square centimeter and I_0 is the intensity of the "threshold" sound $I_0 = 10^{-16}$ W/cm². Find the decibel rating of each of the given sounds.

39. A table saw in operation with intensity $I = 10^{-6}$ W/cm²

40. The sound of a passing car horn with intensity $I = 10^{-8}$ W/cm²

The formula for the decibel rating of a sound can be solved for the intensity of the sound as

$$I = I_0 \cdot 10^{L/10}$$

where L is the decibel rating of the given sound.

41. What is the ratio of intensity of a 60-dB sound to one of 50 dB?

42. What is the ratio of intensity of a 60-dB sound to one of 40 dB?

The magnitude of an earthquake on the Richter scale is given by

$$M = \log \frac{a}{a_0}$$

where a is the intensity of the shock wave of the given earthquake and a_0 is the intensity of the shock wave of a zero-level earthquake. Use that formula to solve the following.

43. The Alaskan earthquake of 1964 had an intensity of $10^{8.4}\, a_0$. What was its magnitude on the Richter scale?

44. Find the ratio of intensity of an earthquake of magnitude 7 to an earthquake of magnitude 6.

[8.4]   Use the properties of logarithms to expand each of the following expressions.

45. $\log_b x^2 y$

46. $\log_4 \dfrac{y^3}{5}$

47. $\log_3 \dfrac{xy^2}{z}$

48. $\log_5 x^3 y z^2$

49. $\log \dfrac{xy}{\sqrt{z}}$

50. $\log_b \sqrt[3]{\dfrac{x^2 y}{z}}$

Use the properties of logarithms to write each of the following expressions as a single logarithm.

51. $\log x + 2 \log y$

52. $3 \log_b x - 2 \log_b z$

53. $\log_b x + \log_b y - \log_b z$

54. $2 \log_5 x - 3 \log_5 y - \log_5 z$

55. $\log x - \dfrac{1}{2} \log y$

56. $\dfrac{1}{3} (\log_b x - 2 \log_b y)$

Given that $\log 2 = 0.301$ and $\log 3 = 0.477$, find each of the following logarithms. Verify your results with a calculator.

57. $\log 18$

58. $\log 16$

59. $\log \dfrac{1}{8}$

60. $\log \sqrt{3}$

Use your calculator to find the pH of each of the following solutions, given the hydrogen ion concentration $[H^+]$ for each solution, where

$$pH = -\log [H^+]$$

Are the solutions acidic or basic?

61. Coffee: $[H^+] = 5 \times 10^{-6}$

62. Household detergent: $[H^+] = 3.2 \times 10^{-10}$

Given the pH of the following solutions, find the hydrogen ion concentration $[H^+]$.

63. Lemonade: pH = 3.5

64. Ammonia: pH = 10.2

The average score on a final examination for a group of chemistry students, retested after time t (in weeks), is given by

$$S(t) = 81 - 6 \ln (t + 1)$$

Find the average score on the retests after the given times.

65. After 5 weeks

66. After 10 weeks

67. After 15 weeks

68. Graph these results.

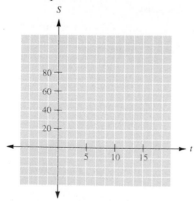

The formula for converting from a logarithm with base b to a logarithm with base a is

$$\log_a x = \frac{\log_b x}{\log_b a}$$

Use that formula to find each of the following logarithms.

69. $\log_4 20$

70. $\log_8 60$

[8.5]   Solve each of the following logarithmic equations for x.

71. $\log_3 x + \log_3 5 = 3$

72. $\log_5 x - \log_5 10 = 2$

73. $\log_3 x + \log_3 (x + 6) = 3$

74. $\log_5 (x + 3) + \log_5 (x - 1) = 1$

75. $\log x - \log (x - 1) = 1$

76. $\log_2 (x + 3) - \log_2 (x - 1) = \log_2 3$

Solve each of the following exponential equations for x. Give your results correct to three decimal places.

77. $3^x = 243$

78. $5^x = \dfrac{1}{25}$

79. $5^x = 10$

80. $4^{x-1} = 8$

81. $6^x = 2^{2x+1}$

82. $2^{x+1} = 3^{x-1}$

If an investment of P dollars earns interest at an annual rate of 12% and the interest is compounded n times per year, then the amount A in the account after t years is

$$A(t) = P\left(1 + \frac{0.12}{n}\right)^{nt}$$

Use that formula to solve each of the following.

83. If $1000 is invested and the interest is compounded quarterly, how long will it take the amount in the account to double?

84. If $3000 is invested and the interest is compounded monthly, how long will it take the amount in the account to reach $8000?

A certain radioactive element has a half-life of 50 years. The amount A of the substance remaining after t years is given by

$$A(t) = A_0 \cdot 2^{-t/50}$$

where A_0 is the initial amount of the substance. Use this formula to solve each of the following.

85. If the initial amount of the substance is 100 milligrams (mg), after how long will 40 mg remain?

86. After how long will only 10% of the original amount of the substance remain?

A city's population is presently 50,000. Given the projected growth, t years from now the population P will be given by $P(t) = 50{,}000e^{0.08t}$. Use this formula to solve each of the following.

87. How long will it take the population to reach 70,000?

88. How long will it take the population to double?

The atmospheric pressure, in inches of mercury, at an altitude h miles above the surface of the earth, is approximated by $P(h) = 30e^{-0.021h}$. Use this formula to solve the following exercises.

89. Find the altitude at the top of Mt. McKinley in Alaska if the pressure is 27.7 inHg.

90. Find the altitude outside an airliner in flight if the pressure is 26.1 inHg.

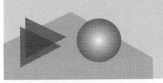
The purpose of this self-test is to help you check your progress and to review for a chapter test in class. Allow yourself about 1 hour to take the test. When you are done, check your answers in the back of the book. If you missed any answers, be sure to go back and review the appropriate sections in the chapter and the exercises that are provided.

1. Use $f(x) = 4x - 2$ and $g(x) = x^2 + 1$ in each of the following.

   (a) Find the inverse of f. Is the inverse also a function?

   (b) Find the inverse of g. Is the inverse also a function?

   (c) Graph f and its inverse on the same set of axes.

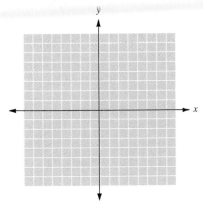

Graph the exponential functions defined by each of the following equations.

2. $y = 4^x$

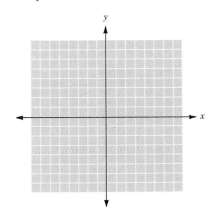

3. $y = \left(\dfrac{2}{3}\right)^x$

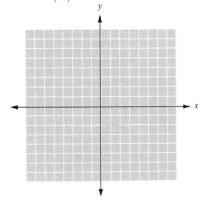

4. Solve each of the following exponential equations for x.

   (a) $5^x = \dfrac{1}{25}$

   (b) $3^{2x-1} = 81$

5. Graph the logarithmic function defined by the following equation.

$$y = \log_4 x$$

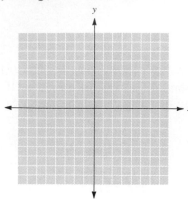

Convert each of the following statements to logarithmic form.

6. $10^4 = 10{,}000$ **7.** $27^{2/3} = 9$

Convert each of the following statements to exponential form.

8. $\log_5 125 = 3$ **9.** $\log 0.01 = -2$

Solve each of the following equations for the unknown variable.

10. $y = \log_2 64$ **11.** $\log_b \dfrac{1}{16} = -2$ **12.** $\log_{25} x = \dfrac{1}{2}$

Use the properties of logarithms to expand each of the following expressions.

13. $\log_b x^2 y z^3$ **14.** $\log_5 \sqrt{\dfrac{xy^2}{z}}$

Use the properties of logarithms to write each of the following expressions as a single logarithm.

15. $\log x + 3 \log y$ **16.** $\dfrac{1}{3} (\log_b x - 2 \log_b z)$

Solve each of the following logarithmic equations for x.

17. $\log_6 (x + 1) + \log_6 (x - 4) = 2$ **18.** $\log (2x + 1) - \log (x - 1) = 1$

Solve each of the following exponential equations for x. Give your results correct to three decimal places.

19. $3^{x+1} = 4$ **20.** $5^x = 3^{x+1}$

This test is provided to help you in the process of reviewing the previous chapters. Answers are provided in the back of the book. If you missed any answers, be sure to go back and review the appropriate chapter section.

Solve each of the following.

1. $2x - 3(x + 2) = 4(5 - x) + 7$

2. $|3x - 7| > 5$

3. $\log x - \log (x - 1) = 1$

Graph each of the following.

4. $5x - 3y = 15$

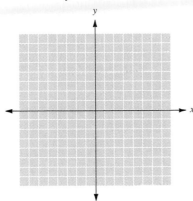

5. $8(2 - x) \geq y$

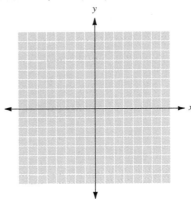

6. Find the equation of the line that passes through the points $(2, -1)$ and $(-3, 5)$.

7. Solve the linear inequality

$3x - 2(x - 5) \geq 20$

Simplify each of the following expressions.

8. $4x^2 - 3x + 8 - 2(x^2 + 5) - 3(x - 1)$

9. $(3x + 1)(2x - 5)$

Factor each of the following completely.

10. $2x^2 - x - 10$

11. $25x^3 - 16xy^2$

Perform the indicated operations.

12. $\dfrac{2}{x - 4} - \dfrac{3}{x - 5}$

13. $\dfrac{x^2 - x - 6}{x^2 + 2x - 15} \div \dfrac{x - 2}{x + 5}$

687

Simplify each of the following radical expressions.

14. $\sqrt{18} + \sqrt{50} - 3\sqrt{32}$

15. $(3\sqrt{2} + 2)(3\sqrt{2} + 2)$

16. $\dfrac{5}{\sqrt{5} - \sqrt{2}}$

17. Find three consecutive odd integers whose sum is 237.

Solve each of the following equations.

18. $x^2 + x - 2 = 0$

19. $2x^2 - 6x - 5 = 0$

20. Solve the following inequality:

$2x^2 + x - 3 \leq 0$

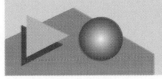

Appendix A Adding and Subtracting Unlike Fractions

Adding or subtracting **unlike fractions** (fractions that do not have the same denominator) requires a bit more work than adding or subtracting like fractions. When the denominators are not the same, we must use the idea of the **lowest common denominator (LCD)**. Each fraction is "built up" to an equivalent fraction having the LCD as a denominator. You can then add or subtract as before.

Let's review with an example from arithmetic.

• Example 1

Adding Unlike Arithmetic Fractions

Add $\dfrac{5}{9} + \dfrac{1}{6}$.

Step 1 Find the LCD. Factor each denominator.

$9 = 3 \cdot 3 \leftarrow$ 3 appears twice.

$6 = 2 \cdot 3$

To form the LCD, include each factor the greatest number of times it appears in any single denominator. Use one 2, since 2 appears only once in the factorization of 6. Use two 3s, since 3 appears twice in the factorization of 9. Thus the LCD for the fractions is $2 \cdot 3 \cdot 3 = 18$.

Step 2 "Build up" each fraction to an equivalent fraction with the LCD as the denominator. Do this by multiplying the numerator and denominator of the given fractions by the same number.

Do you see that this uses the fundamental principle in the following form?

$$\frac{P}{Q} = \frac{PR}{QR}$$

$$\frac{5}{9} = \frac{5 \cdot 2}{9 \cdot 2} = \frac{10}{18}$$

$$\frac{1}{6} = \frac{1 \cdot 3}{6 \cdot 3} = \frac{3}{18}$$

Step 3 Add the fractions.

$$\frac{5}{9} + \frac{1}{6} = \frac{10}{18} + \frac{3}{18} = \frac{13}{18}$$

Since $\dfrac{13}{18}$ is in its simplest form, we are done!

● ● ● **CHECK YOURSELF 1**

Add.

(a) $\dfrac{1}{6} + \dfrac{3}{8}$ **(b)** $\dfrac{3}{10} + \dfrac{4}{15}$

The process is exactly the same in algebra. We can summarize the steps with the following rule:

Adding Unlike Arithmetic Fractions

STEP 1 Find the lowest common denominator of all the fractions.
STEP 2 Convert each fraction to an equivalent fraction with the LCD as a denominator.
STEP 3 Add or subtract the like fractions formed in step 2.
STEP 4 Write the sum or difference in simplified form.

● Example 2

Adding and Subtracting Unlike Algebraic Fractions

(*a*) Add $\dfrac{3}{2x} + \dfrac{4}{x^2}$.

Step 1 Factor the denominators.

$2x = 2 \cdot x$

$x^2 = x \cdot x$

The original denominators are not multiplied (together) to form the LCD.

The LCD must contain the factors 2 and x. The factor x must appear *twice* because it appears twice as a factor in the second denominator.

Step 2

$$\dfrac{2}{2x} = \dfrac{3 \cdot x}{2x \cdot x} = \dfrac{3x}{2x^2}$$

$$\dfrac{4}{x^2} = \dfrac{4 \cdot 2}{x^2 \cdot 2} = \dfrac{8}{2x^2}$$

Step 3

$$\dfrac{3}{2x} + \dfrac{4}{x^2} = \dfrac{3x}{2x^2} + \dfrac{8}{2x^2} = \dfrac{3x + 8}{2x^2}$$

The sum is in the simplest form.

(*b*) Subtract $\dfrac{4}{3x^2} - \dfrac{3}{2x^3}$.

Step 1 Factor the denominators.

$3x^2 = 3 \cdot x \cdot x$

$2x^3 = 2 \cdot x \cdot x \cdot x$

The LCD must contain the factors 2, 3, and x. The factor x must appear 3 times. Do you see why? The LCD is

$2 \cdot 3 \cdot x \cdot x \cdot x$ or $6x^3$

Step 2

Both the numerator and the denominator must be multiplied by the same quantity.

$\dfrac{4}{3x^2} = \dfrac{4 \cdot 2x}{3x^2 \cdot 2x} = \dfrac{8x}{6x^3}$

$\dfrac{3}{2x^3} = \dfrac{3 \cdot 3}{2x^3 \cdot 3} = \dfrac{9}{6x^3}$

Step 3

$\dfrac{4}{3x^2} - \dfrac{3}{2x^3} = \dfrac{8x}{6x^3} - \dfrac{9}{6x^3} = \dfrac{8x - 9}{6x^3}$

The difference is in the simplest form.

● ● ● **CHECK YOURSELF 2**

Add or subtract as indicated.

(a) $\dfrac{5}{x^2} + \dfrac{3}{x^3}$ (b) $\dfrac{3}{5x} - \dfrac{1}{4x^2}$

We can also add fractions with more than one variable in the denominator. Example 3 shows this.

● **Example 3**

Adding Unlike Fractions with Several Variables

Add $\dfrac{2}{3x^2y} + \dfrac{3}{4x^3}$.

Step 1 Factor the denominators.

$$3x^2y = 3 \cdot x \cdot x \cdot y$$

$$4x^3 = 2 \cdot 2 \cdot x \cdot x \cdot x$$

The LCD is $12x^3y$. Do you see why?

Step 2

$$\frac{2}{3x^2y} = \frac{2 \cdot 4x}{3x^2y \cdot 4x} = \frac{8x}{12x^3y}$$

$$\frac{3}{4x^3} = \frac{3 \cdot 3y}{4x^3 \cdot 3y} = \frac{9y}{12x^3y}$$

Step 3

The y in the numerator and the y in the denominator cannot be divided out since they are not factors.

$$\frac{2}{3x^2y} + \frac{3}{4x^3} = \frac{8x}{12x^3y} + \frac{9y}{12x^3y}$$

$$= \frac{8x + 9y}{12x^3y}$$

● ● ● **CHECK YOURSELF 3**

Add.

$$\frac{2}{3x^2y} + \frac{1}{6xy^2}$$

Fractions with binomials in the denominator can also be added by taking the same approach. Example 4 illustrates this approach.

● **Example 4**

Adding and Subtracting Unlike Fractions with Binomial Denominators

(*a*) Add $\dfrac{5}{x} + \dfrac{2}{x-1}$.

Step 1 The LCD must have factors of x and $x - 1$. The LCD is $x(x - 1)$.

Step 2

$$\frac{5}{x} = \frac{5(x-1)}{x(x-1)}$$

$$\frac{2}{x-1} = \frac{2x}{x(x-1)}$$

Step 3

$$\frac{5}{x} + \frac{2}{x-1} = \frac{5(x-1)}{x(x-1)} + \frac{2x}{x(x-1)}$$

$$= \frac{5x - 5 + 2x}{x(x-1)}$$

$$= \frac{7x - 5}{x(x-1)}$$

(b) Subtract $\dfrac{3}{x-2} - \dfrac{4}{x+2}$.

Step 1 The LCD must have factors of $x - 2$ and $x + 2$. The LCD is $(x-2)(x+2)$.

Step 2

Multiply numerator and denominator by $x + 2$.

$$\frac{3}{x-2} = \frac{3(x+2)}{(x-2)(x+2)}$$

Multiply numerator and denominator by $x - 2$.

$$\frac{4}{x+2} = \frac{4(x-2)}{(x+2)(x-2)}$$

Step 3

$$\frac{3}{x-2} - \frac{4}{x+2} = \frac{3(x+2) - 4(x-2)}{(x+2)(x-2)}$$

Note the sign change.

$$= \frac{3x + 6 - 4x + 8}{(x+2)(x-2)}$$

$$= \frac{-x + 14}{(x+2)(x-2)}$$

● ● ● **CHECK YOURSELF 4**

Add or subtract as indicated.

(a) $\dfrac{3}{x+2} + \dfrac{5}{x}$

(b) $\dfrac{4}{x+3} - \dfrac{2}{x-3}$

Example 5 will show how factoring must sometimes be used in forming the LCD.

• Example 5

Adding and Subtracting Unlike Fractions by Factoring

(a) Add $\dfrac{3}{2x - 2} + \dfrac{5}{3x - 3}$.

Step 1 Factor the denominators.

$$2x - 2 = 2(x - 1)$$
$$3x - 3 = 3(x - 1)$$

CAUTION

$x - 1$ is not used twice in forming the LCD.

The LCD must have factors of 2, 3, and $x - 1$. The LCD is $2 \cdot 3(x - 1)$, or $6(x - 1)$.

Step 2

$$\frac{3}{2x - 2} = \frac{3}{2(x - 1)} = \frac{3 \cdot 3}{2(x - 1) \cdot 3} = \frac{9}{6(x - 1)}$$

$$\frac{5}{3x - 3} = \frac{5}{3(x - 1)} = \frac{5 \cdot 2}{3(x - 1) \cdot 2} = \frac{10}{6(x - 1)}$$

Step 3

$$\frac{3}{2x - 2} + \frac{5}{3x - 3} = \frac{9}{6(x - 1)} + \frac{10}{6(x - 1)}$$

$$= \frac{9 + 10}{6(x - 1)}$$

$$= \frac{19}{6(x - 1)}$$

(b) Subtract $\dfrac{3}{2x - 4} - \dfrac{6}{x^2 - 4}$.

Step 1 Factor the denominators.

$$2x - 4 = 2(x - 2)$$
$$x^2 - 4 = (x + 2)(x - 2)$$

The LCD must have factors of 2, $x - 2$, and $x + 2$. The LCD is $2(x - 2)(x + 2)$.

Step 2

Multiply the numerator and denominator by $x \cdot 2$.

$$\frac{3}{2x - 4} = \frac{3}{2(x - 2)} = \frac{3(x + 2)}{2(x - 2)(x + 2)}$$

Multiply the numerator and denominator by 2.

$$\frac{6}{x^2 - 4} = \frac{6}{(x + 2)(x - 2)} = \frac{6 \cdot 2}{2(x + 2)(x - 2)} = \frac{12}{(x + 2)(x - 2)}$$

Step 3

$$\frac{3}{2x - 4} - \frac{6}{x^2 - 4} = \frac{3(x + 2) - 12}{2(x - 2)(x + 2)}$$

Remove the parentheses and combine like terms in the numerator.

$$= \frac{3x + 6 - 12}{2(x - 2)(x + 2)}$$

$$= \frac{3x - 6}{2(x - 2)(x + 2)}$$

Step 4 Simplify the difference.

Factor the numerator and denominator; the common factor is 2.

$$\frac{3x - 6}{2(x - 2)(x + 2)} = \frac{3(x - 2)}{2(x - 2)(x + 2)} = \frac{3}{2(x + 2)}$$

(*c*) Subtract $\dfrac{5}{x^2 - 1} - \dfrac{2}{x^2 + 2x + 1}$.

Step 1 Factor the denominators.

$$x^2 - 1 = (x - 1)(x + 1)$$

$$x^2 + 2x + 1 = (x + 1)(x + 1)$$

The LCD is $(x - 1)(x + 1)(x + 1)$.

Two factors are needed.

Step 2

$$\frac{5}{(x - 1)(x + 1)} = \frac{5(x + 1)}{(x - 1)(x + 1)(x + 1)}$$

$$\frac{2}{(x + 1)(x + 1)} = \frac{2(x - 1)}{(x + 1)(x + 1)(x - 1)}$$

Step 3

$$\frac{5}{x^2 - 1} - \frac{2}{x^2 + 2x + 1} = \frac{5(x + 1) - 2(x - 1)}{(x - 1)(x + 1)(x + 1)}$$

Remove the parentheses and simplify in the numerator.

$$= \frac{5x + 5 - 2x + 2}{(x - 1)(x + 1)(x + 1)}$$

$$= \frac{3x + 7}{(x - 1)(x + 1)(x + 1)}$$

● ● ● **CHECK YOURSELF 5**

Add or subtract as indicated.

(**a**) $\dfrac{5}{2x + 2} + \dfrac{1}{5x + 5}$ (**b**) $\dfrac{3}{x^2 - 9} - \dfrac{1}{2x - 6}$ (**c**) $\dfrac{4}{x^2 - x - 2} - \dfrac{3}{x^2 + 4x + 3}$

Recall that

$$a - b = -(b - a)$$

Let's see how this can be used in adding or subtracting algebraic fractions.

• Example 6

Adding Fractions

Add $\dfrac{4}{x - 5} + \dfrac{2}{5 - x}$.

Rather than try a denominator of $(x - 5)(5 - x)$, let us simplify first.

Replace $5 - x$ with $-(x - 5)$.
We now use the fact that

$$\frac{a}{-b} = -\frac{a}{b}$$

$$\frac{4}{x - 5} + \frac{2}{5 - x} = \frac{4}{x - 5} + \frac{2}{-(x - 5)}$$

$$= \frac{4}{x - 5} - \frac{2}{x - 5}$$

The LCD is now $x - 5$, and we can combine the fractions as

$$\frac{4 - 2}{x - 5} = \frac{2}{x - 5}$$

● ● ●   **CHECK YOURSELF 6**

Subtract.

$$\frac{3}{x - 3} - \frac{1}{3 - x}$$

● ● ●   **CHECK YOURSELF ANSWERS**

1. (a) $\dfrac{13}{24}$, (b) $\dfrac{17}{30}$.   2. (a) $\dfrac{5x + 3}{x^3}$, (b) $\dfrac{12x - 5}{20x^2}$.   3. $\dfrac{x + 4y}{6x^2y^2}$.

4. (a) $\dfrac{8x + 10}{x(x + 2)}$, (b) $\dfrac{2x - 18}{(x + 3)(x - 3)}$.   5. (a) $\dfrac{27}{10(x + 1)}$, (b) $\dfrac{-1}{2(x + 3)}$,

(c) $\dfrac{x + 18}{(x + 1)(x - 2)(x + 3)}$.   6. $\dfrac{4}{x - 3}$.

Exercises

Add or subtract as indicated. Express your result in simplified form.

1. $\dfrac{5}{8} + \dfrac{3}{5}$

2. $\dfrac{7}{9} - \dfrac{1}{6}$

3. $\dfrac{11}{15} - \dfrac{3}{10}$

4. $\dfrac{7}{8} + \dfrac{5}{6}$

5. $\dfrac{y}{4} + \dfrac{3y}{5}$

6. $\dfrac{5x}{6} - \dfrac{2x}{3}$

7. $\dfrac{7a}{3} - \dfrac{a}{7}$

8. $\dfrac{3m}{4} + \dfrac{m}{9}$

9. $\dfrac{3}{r} - \dfrac{4}{5}$

10. $\dfrac{5}{x} + \dfrac{2}{3}$

11. $\dfrac{5}{a} + \dfrac{a}{5}$

12. $\dfrac{y}{3} - \dfrac{3}{y}$

13. $\dfrac{5}{m} + \dfrac{3}{m^2}$

14. $\dfrac{4}{x^2} - \dfrac{3}{x}$

15. $\dfrac{1}{x^2} - \dfrac{2}{3x}$

16. $\dfrac{5}{2w} + \dfrac{7}{w^3}$

17. $\dfrac{3}{5s} + \dfrac{2}{s^2}$

18. $\dfrac{7}{a^2} - \dfrac{3}{7a}$

19. $\dfrac{3}{4b^2} + \dfrac{5}{3b^3}$

20. $\dfrac{4}{5x^3} - \dfrac{3}{2x^2}$

21. $\dfrac{x}{x + 2} + \dfrac{2}{5}$

22. $\dfrac{3}{4} - \dfrac{a}{a - 1}$

23. $\dfrac{y}{y - 4} - \dfrac{3}{4}$

24. $\dfrac{m}{m + 3} + \dfrac{2}{3}$

25. $\dfrac{4}{x} + \dfrac{3}{x + 1}$

26. $\dfrac{2}{x} - \dfrac{1}{x - 2}$

27. $\dfrac{5}{a - 1} - \dfrac{2}{a}$

28. $\dfrac{4}{x + 2} + \dfrac{3}{x}$

29. $\dfrac{4}{2x - 3} + \dfrac{2}{3x}$

30. $\dfrac{7}{2y - 1} - \dfrac{3}{2y}$

31. $\dfrac{2}{x + 1} + \dfrac{3}{x + 3}$

32. $\dfrac{5}{x - 1} + \dfrac{2}{x + 2}$

33. $\dfrac{4}{y - 2} - \dfrac{1}{y + 1}$

34. $\dfrac{5}{x + 4} - \dfrac{3}{x - 1}$

35. $\dfrac{2}{b - 3} + \dfrac{3}{2b - 6}$

36. $\dfrac{4}{a + 5} - \dfrac{3}{4a + 20}$

37. $\dfrac{x}{x + 4} - \dfrac{2}{3x + 12}$

38. $\dfrac{x}{x - 3} + \dfrac{5}{2x - 6}$

39. $\dfrac{4}{3m + 3} + \dfrac{1}{2m + 2}$

40. $\dfrac{3}{5y - 5} - \dfrac{2}{3y - 3}$

41. $\dfrac{4}{5x - 10} - \dfrac{1}{3x - 6}$

42. $\dfrac{2}{3w + 3} + \dfrac{5}{2w + 2}$

43. $\dfrac{5}{4b + 8} - \dfrac{b}{5b + 10}$

44. $\dfrac{4}{3b - 9} + \dfrac{3b}{4b - 12}$

45. $\dfrac{y - 1}{y + 1} - \dfrac{y}{3y + 3}$

46. $\dfrac{x + 2}{x - 2} - \dfrac{x}{3x - 6}$

47. $\dfrac{3}{x^2 - 4} + \dfrac{2}{x + 2}$

48. $\dfrac{4}{x - 2} + \dfrac{3}{x^2 - x - 2}$

49. $\dfrac{3x}{x^2 - 3x + 2} - \dfrac{1}{x - 2}$

50. $\dfrac{a}{a^2 - 1} - \dfrac{4}{a + 1}$

51. $\dfrac{2x}{x^2 - 5x + 6} + \dfrac{4}{x - 2}$

52. $\dfrac{7a}{a^2 + a - 12} - \dfrac{4}{a + 4}$

53. $\dfrac{2}{3x - 3} + \dfrac{1}{4x + 4}$

54. $\dfrac{2}{5w + 10} - \dfrac{3}{2w - 4}$

55. $\dfrac{4}{3a - 9} - \dfrac{3}{2a + 4}$

56. $\dfrac{2}{3b - 6} + \dfrac{3}{4b + 8}$

57. $\dfrac{5}{x^2 - 16} - \dfrac{3}{x^2 - x - 12}$

58. $\dfrac{3}{x^2 + 4x + 3} - \dfrac{1}{x^2 - 9}$

59. $\dfrac{2}{y^2 + y - 6} + \dfrac{3y}{y^2 - 2y - 15}$

60. $\dfrac{2a}{a^2 - a - 12} - \dfrac{3}{a^2 - 2a - 8}$

61. $\dfrac{6x}{x^2 - 9} - \dfrac{5x}{x^2 + x - 6}$

62. $\dfrac{4y}{y^2 + 6y + 5} + \dfrac{2y}{y^2 - 1}$

63. $\dfrac{3}{a - 7} + \dfrac{2}{7 - a}$

64. $\dfrac{5}{x - 5} - \dfrac{3}{5 - x}$

65. What are the similarities between adding unlike numeric fractions and adding unlike algebraic fractions?

66. What are the differences between adding unlike numeric fractions and adding unlike algebraic fractions?

Answers

1. $\dfrac{49}{40}$ **3.** $\dfrac{13}{30}$ **5.** $\dfrac{17y}{20}$ **7.** $\dfrac{46a}{21}$ **9.** $\dfrac{15 - 4x}{5x}$ **11.** $\dfrac{25 + a^2}{5a}$ **13.** $\dfrac{5m + 3}{m^2}$ **15.** $\dfrac{3 - 2x}{3x^2}$

17. $\dfrac{3s + 10}{5s^2}$ **19.** $\dfrac{9b + 20}{12b^3}$ **21.** $\dfrac{7x + 4}{5(x + 2)}$ **23.** $\dfrac{y + 12}{4(y - 4)}$ **25.** $\dfrac{7x + 4}{x(x + 1)}$ **27.** $\dfrac{3a + 2}{a(a - 1)}$

29. $\dfrac{2(8x - 3)}{3x(2x - 3)}$ **31.** $\dfrac{5x + 9}{(x + 1)(x + 3)}$ **33.** $\dfrac{3(y + 2)}{(y - 2)(y + 1)}$ **35.** $\dfrac{7}{2(b - 3)}$ **37.** $\dfrac{3x - 2}{3(x + 4)}$

39. $\dfrac{11}{6(m + 1)}$ **41.** $\dfrac{7}{15(x - 2)}$ **43.** $\dfrac{25 - 4b}{20(b + 2)}$ **45.** $\dfrac{2y - 3}{3(y + 1)}$ **47.** $\dfrac{2x - 1}{(x - 2)(x + 2)}$

49. $\dfrac{2x + 1}{(x - 1)(x - 2)}$ **51.** $\dfrac{6}{x - 3}$ **53.** $\dfrac{5x + 11}{12(x - 1)(x + 1)}$ **55.** $\dfrac{-a + 43}{6(a - 3)(a + 2)}$

57. $\dfrac{2x + 3}{(x + 4)(x - 4)(x + 3)}$ **59.** $\dfrac{3y^2 - 4y - 10}{(y + 3)(y - 2)(y - 5)}$ **61.** $\dfrac{x}{(x - 3)(x - 2)}$ **63.** $\dfrac{1}{a - 7}$ **65.**

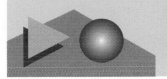

Appendix B Solving Word Problems

To make mathematics useful, one first learns to solve linear equations. The next step is to use this work in the solution of word problems. If you feel a bit uneasy about this subject, do not be too nervous. You have lots of company! To help you feel more comfortable when solving word problems, we are going to present a step-by-step approach that will, *with practice,* allow you to organize your work. Organization is the key to the solution of these problems.

To Solve Word Problems

STEP 1 Read the problem carefully. Then reread it to decide what you are asked to find.

STEP 2 Choose a letter to represent one of the unknowns in the problem. Then represent all other unknowns of the problem with expressions that use the same letter.

STEP 3 Translate the problem to the language of algebra to form an equation.

STEP 4 Solve the equation and answer the question of the original problem.

STEP 5 Verify your solution by returning to the original problem.

Step 3 is usually the hardest. We must translate words to the language of algebra. This table will help you review that translation step.

Translating Words to Algebra

Words	Algebra
The sum of x and y	$x + y$
3 plus a	$3 + a$ or $a + 3$
5 more than m	$m + 5$
b increased by 7	$b + 7$
The difference of x and y	$x - y$
4 less than a	$a - 4$
s decreased by 8	$s - 8$
The product of x and y	$x \cdot y$ or xy
5 times a	$5 \cdot a$ or $5a$
Twice m	$2m$
The quotient of x and y	$\dfrac{x}{y}$
a divided by 6	$\dfrac{a}{6}$
One-half of b	$\dfrac{b}{2}$ or $\dfrac{1}{2}b$

Now let's look at some typical examples of translating phrases to algebra.

● Example 1

Writing Algebraic Expressions

Translate each statement to an algebraic expression.

(*a*) The sum of *a* and 2 times *b*

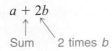

$a + 2b$

Sum   2 times *b*

(*b*) 5 times *m*, increased by 1

$5m + 1$

5 times *m*   Increased by 1

(*c*) 5 less than 3 times *x*

$3x - 5$

3 times *x*   5 less than

(*d*) The product of *x* and *y*, divided by 3

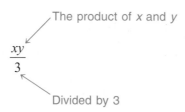

The product of *x* and *y*

$\dfrac{xy}{3}$

Divided by 3

● ● ●   **CHECK YOURSELF 1**

Translate to algebra.

(a) 2 more than twice *x*   **(b)** 4 less than 5 times *n*   **(c)** The product of twice *a* and *b*   **(d)** The sum of *s* and *t*, divided by 5

Now let's work through a complete example, using our five-step approach.

● Example 2

A Number Problem

The sum of twice a number and 5 is 17. What is the number?

Step 1 *Read carefully.* You must find the unknown number.

Step 2 *Choose letters or variables.* Let x represent the unknown number. There are no other unknowns.

Step 3 *Translate.*

The sum of
↓
$2x + 5 = 17$
↖   ↖
Twice x is

Step 4 *Solve.*

$$2x + 5 = 17$$

$$2x + 5 - 5 = 17 - 5 \qquad \text{Subtract 5.}$$

$$2x = 12$$

$$\frac{2x}{2} = \frac{12}{2} \qquad \text{Divide by 2.}$$

$$x = 6$$

So the number is 6.

Always return to the original problem to check your result.

Step 5 *Check.* Is the sum of twice 6 and 5 equal to 17? Yes ($12 + 5 = 17$). We have checked the solution.

● ● ●   **CHECK YOURSELF 2**

The sum of 3 times a number and 8 is 35. What is the number?

Consecutive integers are integers that follow one another, like 10, 11, and 12. To represent them in algebra:

If x is an integer, then $x + 1$ is the next consecutive integer, $x + 2$ is the next, and so on.

We will need this idea in Example 3.

●Example 3

A Number Problem

REMEMBER THE STEPS!

Read the problem carefully. What do you need to find?

Assign variables to the unknowns.

The sum of two consecutive integers is 41. What are the two integers?

Step 1 We want to find the two consecutive integers.

Step 2 Let x be the first integer. Then $x + 1$ must be the next integer.

Write an equation.

Step 3

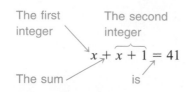

The first integer
The second integer

$$x + x + 1 = 41$$

The sum   is

Solve the equation.

Step 4

$$x + x + 1 = 41$$
$$2x + 1 = 41$$
$$2x = 40$$
$$x = 20$$

The first integer (x) is 20, and the next integer ($x + 1$) is 21.

Check.

Step 5 The sum of the two integers 20 and 21 is 41.

● ● ●   **CHECK YOURSELF 3**

The sum of three consecutive integers is 51. What are the three integers?

●Example 4

A Number Problem

There were 55 more yes votes than no votes on an election measure. If 735 votes were cast in all, how many yes votes were there? How many no votes?

What do you need to find?

Step 1 We want to find the number of yes votes and the number of no votes.

Assign variables to the unknowns.

Step 2 Let x be the number of no votes. Then

$$\underbrace{x + 55}$$

55 more than x

is the number of yes votes.

Write an equation.

Step 3

$$x + \underbrace{x + 55} = 735$$

No votes   Yes votes

Solve the equation.

Step 4

$$x + x + 55 = 735$$

$$2x + 55 = 735$$

$$2x = 680$$

$$x = 340$$

No votes (x) = 340

Yes votes ($x + 55$) = 395

Check.

Step 5 Thus 340 no votes plus 395 yes votes equals 735 total votes. The solution checks.

● ● ● **CHECK YOURSELF 4**

Francine earns $120 per month more than Rob. If they earn a total of $2680 per month, what are their monthly salaries?

Similar methods will allow you to solve a variety of word problems. Look at Example 5.

● **Example 5**

A Work Problem

Juan worked twice as many hours as Jerry. Marcia worked 3 more hours (h) than Jerry. If they worked a total of 31 h, find out how many hours each worked.

Step 1 We want to find the hours each worked, so there are three unknowns.

There are other choices for *x*, but choosing the smallest quantity will usually give the easiest equation to write and solve.

Step 2 Let x be the number of hours that Jerry worked.

Twice Jerry's hours

Then $2x$ is Juan's hours worked

3 h more than Jerry worked

and $x + 3$ is Marcia's hours.

Step 3

Jerry   Juan   Marcia

$$x \quad + 2x \quad + x + 3 = 31$$

Sum of their hours

Step 4

$$x + 2x + x + 3 = 31$$

$$4x + 3 = 31$$

$$4x = 28$$

$$x = 7$$

Jerry's hours $(x) = 7$

Juan's hours $(2x) = 14$

Marcia's hours $(x + 3) = 10$

Step 5 The sum of their hours $(7 + 14 + 10)$ is 31 h, and the solution is verified.

● ● ● **CHECK YOURSELF 5**

Lucy jogged twice as many miles (mi) as Paul but 3 mi less than Isaac. If the three ran a total of 23 mi, how far did each person run?

Many word problems involve geometric figures and measurements, as Example 6 illustrates.

●Example 6

A Geometry Problem

The perimeter of a rectangle is 46 cm. If the length is 3 cm more than the width, what are the dimensions of the rectangle?

Step 1 We want to find the dimensions (length and width) of the rectangle.

Step 2 Let x be the width. Then $\underset{\underset{\text{3 more than } x}{\uparrow}}{x + 3}$ is the length.

Always draw a sketch at this point when you can do so. It will help you form an equation in step 3.

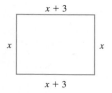

Step 3 The perimeter (the distance around) is 46 cm, so

$$\underset{\substack{\text{Width} \qquad \text{Length} \qquad \text{Width} \qquad \text{Length}}}{x + x + 3 + x + x + 3} = 46$$

Step 4

$$x + x + 3 + x + x + 3 = 46$$

$$4x + 6 = 46$$

$$4x = 40$$

$$x = 10$$

$$\text{Width } (x) = 10 \text{ cm}$$

$$\text{Length } (x + 3) = 13 \text{ cm}$$

Step 5 The perimeter is $10 + 13 + 10 + 13$, or 46 cm. The solution is verified.

● ● ● **CHECK YOURSELF 6**

The perimeter of a triangle is 55 in. If the length of the base is 5 in less than the length of the two equal legs, find the lengths of the legs and the base.

That is all we propose to do with word problems for the moment. Remember to use the five-step approach as you work on the problems for this section. We hope you find it helpful.

● ● ● **CHECK YOURSELF ANSWERS**

1. (a) $2x + 2$, (b) $5n - 4$, (c) $2ab$, (d) $\dfrac{s + t}{5}$. **2.** The equation is $3x + 8 = 35$.

The number is 9. **3.** The equation is $x + x + 1 + x + 2 = 51$. The integers are 16, 17, and 18. **4.** The equation is $x + x + 120 = 2680$. Rob's salary is $1280, and Francine's is $1400. **5.** Paul: 4 mi; Lucy: 8 mi, Isaac: 11 mi **6.** Legs: 20 in; base: 15 in.

Exercises

Solve the following word problems. Be sure to label the unknowns and to show the equation you use for the solution.

1. The sum of twice a number and 7 is 33. What is the number?

2. 3 times a number, increased by 8, is 50. Find the number.

3. 5 times a number, minus 12, is 78. Find the number.

4. 4 times a number, decreased by 20, is 44. What is the number?

5. The sum of two consecutive integers is 71. Find the two integers.

6. The sum of two consecutive integers is 145. Find the two integers.

7. The sum of three consecutive integers is 63. What are the three integers?

8. If the sum of three consecutive integers is 93, find the three integers.

9. The sum of two consecutive even integers is 66. What are the two integers? (*Hint:* Consecutive even integers such as 10, 12, and 14 can be represented by x, $x + 2$, $x + 4$, and so on.)

10. If the sum of two consecutive even integers is 86, find the two integers.

11. If the sum of two consecutive odd integers is 52, what are the two integers? (*Hint:* Consecutive odd integers such as 21, 23, and 25 can be represented by x, $x + 2$, $x + 4$, and so on.)

12. The sum of two consecutive odd integers is 88. Find the two integers.

13. The sum of three consecutive odd integers is 105. What are the three integers?

14. The sum of three consecutive even integers is 126. What are the three integers?

15. If the sum of four consecutive integers is 86, what are the four integers?

16. The sum of four consecutive integers is 62. What are the four integers?

17. 4 times an integer is 9 more than 3 times the next consecutive integer. What are the two integers?

18. 4 times an integer is 30 less than 5 times the next consecutive even integer. Find the two integers.

19. In an election, the winning candidate had 160 more votes than the losing candidate. If the total number of votes cast was 3260, how many votes did each candidate receive?

20. Jody earns $140 more per month than Frank. If their monthly salaries total $2760, what amount does each earn?

21. A washer-dryer combination costs $650. If the washer costs $70 more than the dryer, what does each appliance cost?

22. Morgan has a board that is 98 in long. He wishes to cut the board into two pieces so that one piece will be 10 in. longer than the other. What should be the length of each piece?

23. Ken is 1 year less than twice as old as his sister. If the sum of their ages is 14 years, how old is Ken?

24. Diane is twice as old as her brother Dan. If the sum of their ages is 27 years, how old are Diane and her brother?

25. José is 3 years less than 4 times as old as his daughter. If the sum of their ages is 37 years, how old is José?

26. Mrs. Jackson is 2 years more than 3 times as old as her son. If the difference between their ages is 22 years, how old is Mrs. Jackson?

27. On her vacation in Europe, Jovita's expenses for food and lodging were $60 less than twice as much as her airfare. If she spent $2400 in all, what was her airfare?

28. Rachel earns $6000 less than twice as much as Tom. If their two incomes total $48,000, how much does each earn?

29. There are 99 students registered in three sections of algebra. There are twice as many students in the 10 o'clock section as in the 8 o'clock section and 7 more students at 12 o'clock than at 8 o'clock. How many students are in each section?

30. The Randolphs used 12 more gallons (gal) of fuel oil in October than in September and twice as much oil in November as in September. If they used 132 gal for the 3 months, how much was used during each month?

31. The length of a rectangle is 5 cm more than its width. The perimeter is 98 cm. What are the dimensions? (*Hint:* Remember to draw a sketch whenever geometric figures are involved in a word problem.)

32. The length of a rectangle is 3 times its width. If the perimeter of the rectangle is 48 in, find the length and width of the rectangle.

33. The length of a rectangle is 2 ft more than 3 times its width. If the perimeter is 68 ft, what are the length and width of the rectangle?

34. The length of a rectangle is 3 cm more than twice its width. What are the dimensions of the rectangle if its perimeter is 48 cm?

35. One side of a triangle is 4 meters (m) longer than the shortest side. The third side is twice the length of the shortest side. If the perimeter of the triangle is 44 m, find the lengths of the three sides of the triangle.

36. The equal legs of an isosceles triangle are each 4 ft more than twice the length of the base. If the perimeter of the triangle is 68 ft, find the lengths of the three sides of the triangle.

37. What is the five-step approach for solving a word problem?

38. Does the solution to the equation always provide the answer to the question of the original problem? Explain your answer.

Answers

1. 13   **3.** 18   **5.** 35, 36   **7.** 20, 21, 22   **9.** 32, 34   **11.** 25, 27   **13.** 33, 35, 37

15. 20, 21, 22, 23   **17.** 12, 13   **19.** 1710 votes, 1550 votes   **21.** Washer $360, dryer $290

23. 9 years old   **25.** 29 years old   **27.** $820   **29.** 8 o'clock, 23; 10 o'clock, 46; 12 o'clock, 30

31. 22 cm, 27 cm   **33.** 26 ft, 8 ft   **35.** 10 m, 14 m, 20 m   **37.**

Answers to Self-Tests and Cumulative Review Tests

Self-Test for Chapter 1

1. b **2. (a)** D:$\{-3, 1, 2, 3, 4\}$; R:$\{-2, 0, 1, 5, 6\}$ **(b)** D: All reals; R: all reals **3. (a)** $-7x - 2$, **(b)** $-15x + 16$, **(c)** all reals

4. (a) $4x^2 + 29x - 63$, **(b)** $\dfrac{4x - 7}{x + 9}$, **(c)** $\{x \mid x \neq 9\}$ **5.** $y = 2x - 1$

6. $y = \dfrac{2}{3}x - 4$ **7.** $y = -3x + 5$ **8.** $y = 4x - 2$ **9. (a)** 1, **(b)** 7,

(c) 7 **10.** Function **11.** Not a function **12.** Function
13. Function **14.** Not a function **15.** x int.: $(5, 0)$; y int.: $(0, 7)$
16. x int.: $(12, 0)$; y int.: $(0, 4)$ **17.** D: All reals; R: all reals
18. A: $(1, 0)$; B: $(-3, -4)$ **19.** A: $(-4, -2)$; B: $(1, 2)$ **20. (a)** 3,
(b) 0, **(c)** -3 **21. (a)** -3, **(b)** -2, **(c)** 0 **22. (a)** 0, **(b)** 1, **(c)** 2
23. (a) 0, **(b)** $-2, 2$, **(c)** $-4, 4$

24.

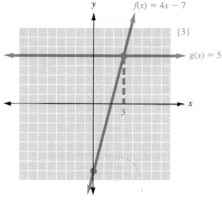

25. $y = 3.22x + 343.7$

Self-Test for Chapter 2

1. $\left\{\dfrac{4}{5}\right\}$ **2.** $\{2\}$ **3.** $\left\{\dfrac{5}{2}\right\}$ **4.** $\left\{\dfrac{14}{5}\right\}$ **5.** 3:30 PM **6.** 20,000 boxes

7.

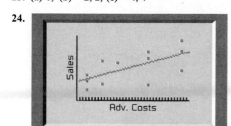

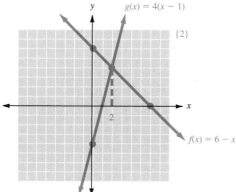

8.

9.

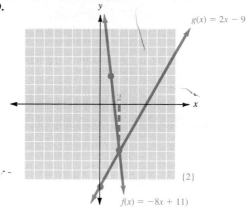

10.

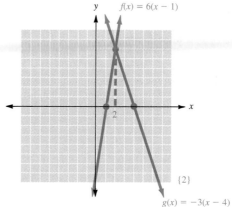

11.

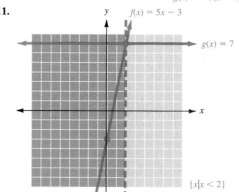

12. $2x - 1 \leq 3(x - 1)$

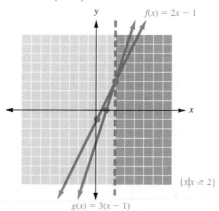

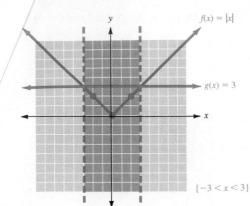

$\{-3 < x < 3\}$

14.

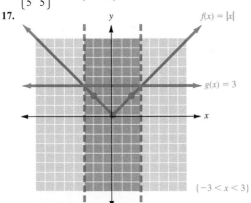

$\{-2, 7\}$

15. $\left\{\dfrac{4}{5}, \dfrac{8}{5}\right\}$ **16.** $\{-2, 6\}$

17.

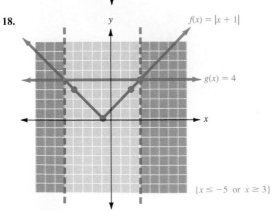

$\{-3 < x < 3\}$

18.

$\{x \le -5 \text{ or } x \ge 3\}$

19. $-3 < x < 5$ **20.** $x \le 4 \text{ or } x \ge 3$

Cumulative Test for Chapters 1–2

1. D: $\{-2, -1, 2, 3, 4\}$; R: $\{0, 1, 5, 7\}$ **2.** D: Reals; R: reals

3. $\{1\}$ **4.** $\left\{\dfrac{7}{6}\right\}$ **5.** $\{-1\}$ **6.** Function   **7.** Function

8. (a) Not a function, (b) Function

9. (a) x int.: $(6, 0)$; y int.: $(0, -4)$,

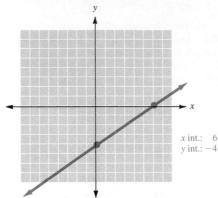

x int.:   6
y int.: -4

(b) x int.: $(7, 0)$; y int.: $(0, 3)$

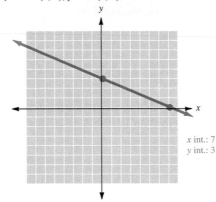

x int.: 7
y int.: 3

10. Domain: reals; range: reals   **11.** (a) -3, (b) 0, (c) 3

12. -2 **13.** $y = -2x - 7$ **14.** (a) $-4x + 3$, (b) $10x + 7$,

(c) $-21x^2 - 41x - 10$, (d) $\dfrac{3x + 5}{-7x - 2}$ **15.** $\{11\}$ **16.** $\{-1, 9\}$

17. $\{8\}$ **18.** $\{x | x \ge 4\}$ **19.** $\{x | 1 < x < 5\}$ **20.** $\{x | x < 2 \text{ or } x > 3\}$

Self-Test for Chapter 3

1. $-6x^3y^4$ **2.** $\dfrac{16m^4n^{10}}{p^6}$ **3.** x^8y^{10} **4.** $\dfrac{1}{2c^2d}$ **5.** $108x^8y^7$

6. $-128x^7y^5$ **7.** 25   **8.** (a) $6x^2 - 3x + 7$, (b) $2x^2 - 3x + 7$,

(c) 10, (d) 6, (e) 10, (f) 6   **9.** (a) $-3x^3 + 3x^2 + 5x - 9$,

(b) $-3x^3 + 7x^2 - 9x - 5$, (c) -4, (d) -10, (e) -4, (f) -10

10. $5x + 12$ **11.** $6a^2 - ab - 35b^2$ **12.** $25m^2 - 9n^2$

13. $4a^2 + 12ab + 9b^2$ **14.** $2x^3 - 13x^2 + 26x - 15$

15. $7ab(2ab - 3a + 5b)$ **16.** $(x - 3y)(x + 5)$

17. $(5c - 8d)(5c + 8d)$ **18.** $(3x - 1)(9x^2 + 3x + 1)$

19. $2a(2a + b)(4a^2 - 2ab + b^2)$ **20.** $(x - 8)(x + 6)$

21. $(5x - 2)(2x - 7)$ **22.** $3x(x + 3)(2x - 5)$ **23.** 18

24. $3x + 2; -3$ **25.** $(x - 2)(x + 4)(x - 3)$

Cumulative Test for Chapters 1–3
1. $\{x|x = 2\}$ **2.** -77 **3.** x int.: $(6, 0)$; y int.: $(0, 4)$

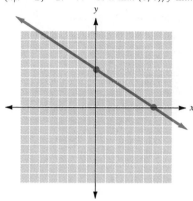

4. $y = \dfrac{5}{4}x - \dfrac{3}{4}$ **5.** $256x^{13}y^{12}$ **6.** 6 **7.** Domain: all reals; range:
all reals **8. (a)** 4, **(b)** 0, **(c)** 6 **9.** $-x^2 - 5x + 9$
10. $10x^2 + 7x - 12$ **11.** $x(3x + 2)(x - 1)$ **12.** $(4x + 5y)(4x - 5y)$
13. $(x - y)(3x + 1)$ **14. (a)** $(3x + 7)$, **(b)** $-13x - 5$,
(c) $-40x^2 - 22x + 6$, **(d)** $\dfrac{5x + 1}{8x + 6}$ **15.** $\{x|x = 5\}$ **16.** $\{x|x = -18$ or
$x = -28\}$ **17.** $\{x|x = 2\}$ **18.** $\{x|x \geq -1\}$ **19.** $\{x|-13 < x < -5\}$
20. $\{x|-6 < x < -4\}$

Self-Test for Chapter 4
1. $\dfrac{-3x^4}{4y^2}$ **2.** $\dfrac{w + 1}{w - 2}$ **3.** $\dfrac{x + 3}{x + 2}$
4.

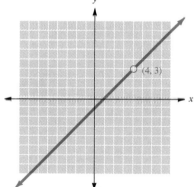

5. $\dfrac{4a^2}{7b}$ **6.** $\dfrac{m - 4}{4}$ **7.** $\dfrac{2}{x - 1}$ **8.** $\dfrac{2x + y}{x(x + 3y)}$ **9.** $\dfrac{-5(3x + 1)}{3x - 1}$
10. $\dfrac{2(2x + 1)}{x(x - 2)}$ **11.** $\dfrac{2}{x - 3}$ **12.** $\dfrac{4}{x - 2}$ **13.** $\dfrac{8x + 17}{(x - 4)(x + 1)(x + 4)}$
14. $\dfrac{y}{3y + x}$ **15.** $\dfrac{z - 1}{2(z + 3)}$ **16.** $\dfrac{1}{xy(x - y)}$ **17.** $\{x|x = 9\}$
18. $\left\{x|x = -\dfrac{1}{2} \text{ or } x = 4\right\}$
19. $-4 \leq x < 3$
20. $-3 < x \leq -1$

Cumulative Test for Chapters 1–4
1. 12 **2.** -14 **3.** $6x + 7y = -21$ **4.** x int.: $(-6, 0)$; y int.: $(0, 7)$
5. $6x^2 - 5x - 2$ **6.** $2x^3 + 5x^2 - 3x$ **7.** Domain: $x = 2$; range:

all reals **8.** 41 **9.** $x(2x + 3)(3x - 1)$ **10.** $(4x^8 + 3y^4)(4x^8 - 3y^4)$
11. $\dfrac{3}{x - 1}$ **12.** $\dfrac{1}{(x + 1)(x - 1)}$ **13.** $x^2 - 3x$
14. (a) $\dfrac{8x^2 - 34x - 29}{x - 5}$, **(b)** $\dfrac{1}{(x - 5)(8x + 6)}$,
(c) $\left\{x|x \neq 5 \text{ or } x \neq -\dfrac{3}{4}\right\}$ **15.** $\left\{x|=\dfrac{7}{2}\right\}$
16. $\left\{x|x = -\dfrac{8}{9} \text{ or } x = -\dfrac{4}{9}\right\}$ **17.** $\left\{x|x = \dfrac{30}{17}\right\}$ **18.** $\{x|x > 2\}$
19. $\left\{x|\dfrac{1}{5} < x < \dfrac{7}{5}\right\}$ **20.** $\{x|x \leq -4 \text{ or } x \geq -2\}$

Self-Test for Chapter 5
1. x^8y^{-10} **2.** $\dfrac{c^2d^7}{2}$ **3.** 4.23×10^9 **4.** 2.5×10^{-5} **5.** $7a^2$
6. $-3w^2z^3$ **7.** $\dfrac{7x}{8y}$ **8.** $\dfrac{\sqrt{10xy}}{4y}$ **9.** $x^3\sqrt{14x}$ **10.** $2x^3\sqrt{4x^2}$ **11.** 6.6
12. 1.4 **13.** $\dfrac{1}{64x^6}$ **14.** $\dfrac{9m}{n^4}$ **15.** $\dfrac{8s}{r}$ **16.** $a^2b\sqrt{a^4b}$ **17.** $5p^3q^2$
18. $3 + 10i$ **19.** $-14 + 22i$ **20.** $5 - 5i$

Cumulative Test for Chapters 1–5
1. $\{x|x = -15\}$ **2.** 5 **3.** $3x - 2y = 12$
4. $\left\{x|x = 3 \text{ or } x = \dfrac{1}{3}\right\}$ **5.** $10x^2 - 2x$ **6.** $10x^2 - 39x - 27$
7. $x(x - 1)(2x + 3)$ **8.** $9(x^2 + 2y^2)(x^2 - 2y^2)$ **9.** $(x + 2y)(4x - 5)$
10. $\dfrac{x + 5}{3x - 1}$ **11.** $\dfrac{x + 7}{(x - 5)(x - 1)}$ **12.** $\dfrac{a + 2}{2a}$ **13.** 1 **14.** $2x^4y^3\sqrt{3y}$
15. $\sqrt{6} - 5\sqrt{2} + 3\sqrt{3} - 15$
16.

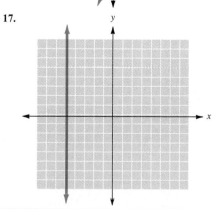

17.

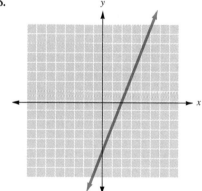

18.

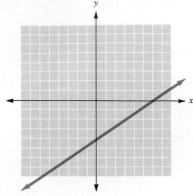

19. $15 + 18i$ **20.** $23 - 14i$

Self-Test for Chapter 6

1. $\left\{-3, -\dfrac{1}{2}\right\}$ **2.** $\left\{-\dfrac{5}{2}, \dfrac{2}{3}\right\}$ **3.** $\left\{0, \dfrac{3}{2}, -\dfrac{3}{2}\right\}$ **4.** $\{\pm\sqrt{5}\}$

5. $\{1 \pm \sqrt{10}\}$ **6.** $\left\{\dfrac{2 \pm \sqrt{23}}{2}\right\}$ **7.** $\left\{\dfrac{-3 \pm \sqrt{13}}{2}\right\}$ **8.** $\left\{\dfrac{5 \pm \sqrt{19}}{2}\right\}$

9. $\left\{\dfrac{5 \pm \sqrt{37}}{2}\right\}$ **10.** $\{-2 \pm \sqrt{11}\}$ **11.** $\left\{-\dfrac{2}{3}, 4\right\}$ **12.** 7, 9

13. 2.7 s   **14.** $x = -2, (-2, 1)$ **15.** $x = 2, (2, 9)$

16. $x = \dfrac{3}{2}, \left(\dfrac{3}{2}, \dfrac{3}{2}\right)$ **17.** $y = 3, (-2, 3)$ **18.** $y = 3, (-7, 3)$

19.

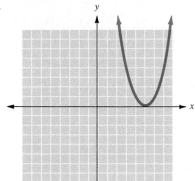

20.

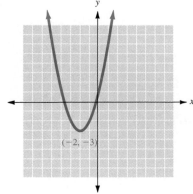

21.

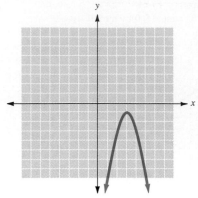

22.

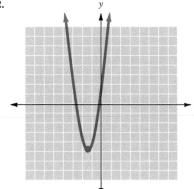

23.

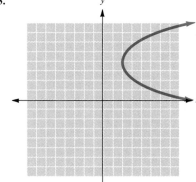

24.

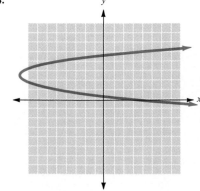

25. $x < -3$ or $x > 1$ ⬥⊢⊢⬥
                                        -3 0 1

26. $-7 < x < 2$ ⬥⊢⊢⊢⊢⊢⊢⊢⊢⬥
                       -7 0 2

27. $x \leq -3$ or $x \geq 6$ ⬥⊢⊢⊢⊢⊢⊢⊢⊢⊢⬥
                         -3 0   6

28. $-2 \leq x < \dfrac{5}{3}$

29. Center: (3, −2); radius: 6

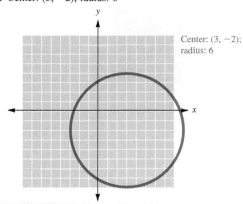

Center: (3, −2); radius: 6

30. Center: (1, −2), radius: $\sqrt{26}$

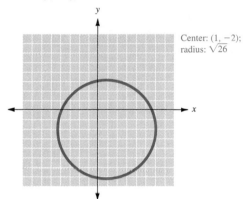

Center: (1, −2); radius: $\sqrt{26}$

31. Center: (2, −3) radius: 3

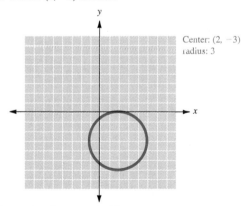

Center: (2, −3) radius: 3

32. $y = 3.2x^2 - 1.2x + 0.88$

Cumulative Test for Chapters 1–6

1.

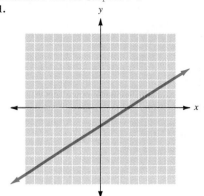

2.

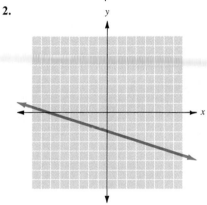

3.

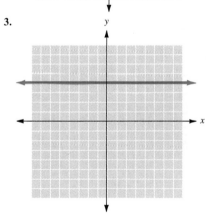

4. −3   **5.** $\dfrac{4}{3}$ **6.** 35   **7.** $f(x) = x^3 + 3x^2 - x - 3$ **8.** $x(x + 3)(x - 2)$

9. $\dfrac{-x - 4}{(x - 3)(x + 2)}$ **10.** $\sqrt{21} - \sqrt{6} + \sqrt{42} - \sqrt{3}$ **11.** $\left\{\dfrac{7}{2}\right\}$

12. {−4}   **13.** {−5, 3}   **14.** {1, 2}   **15.** {−10, 3}

16. $\left\{\dfrac{3 \pm \sqrt{21}}{2}\right\}$ **17.** $\{3 \pm \sqrt{5}\}$ **18.** −13   **19.** $-2 + 2\sqrt{97}$,
$2 + \sqrt{97}$ or 17.7 ft, and 21.7 ft   **20.** Between 21 and 59 units

Self-Test for Chapter 7

1. (−3, 4)   **2.** Dependent   **3.** Inconsistent   **4.** (−2, −5)

5. (5, 0)   **6.** $\left\{3, -\dfrac{5}{3}\right\}$ **7.** (−1, 2, 4)   **8.** $\left(2, 3, -\dfrac{1}{2}\right)$

9. Disks $2.50, ribbons $6   **10.** 60-lb jawbreakers, 40-lb
licorice   **11.** Four 5-in. sets, six 12-in. sets   **12.** $8000 savings,
$4000 bond, $2000 mutual fund   **13.** 50 by 80 ft

14.

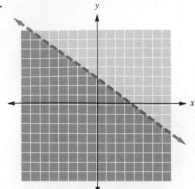

15.

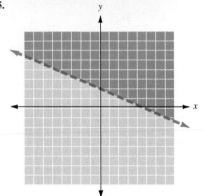

16.

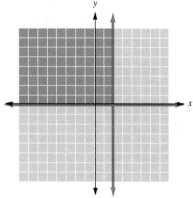

17.

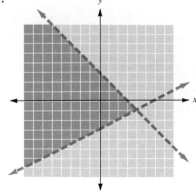

18.

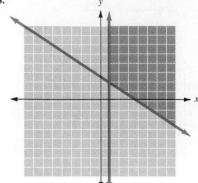

19.

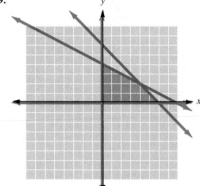

20.

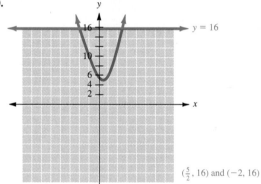

$\left(\frac{5}{2}, 16\right)$ and $(-2, 16)$

21.

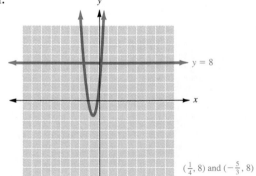

$\left(\frac{1}{4}, 8\right)$ and $\left(-\frac{5}{3}, 8\right)$

22. $\left(\frac{5}{2}, 16\right)$ and $(-2, 16)$ **23.** $\left(\frac{1}{4}, 8\right)$ and $\left(-\frac{5}{3}\right)$

24.

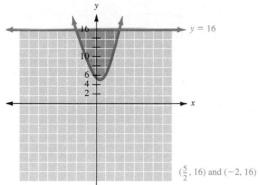

$(\frac{5}{2}, 16)$ and $(-2, 16)$

25.

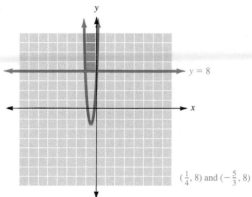

$(\frac{1}{4}, 8)$ and $(-\frac{5}{3}, 8)$

Cumulative Test for Chapters 1–7

1. $\left\{\dfrac{22}{4}\right\}$ **2.** $x > -2$ **3.** $\{-1, 4\}$ **4.** $-4 \le x \le \dfrac{2}{3}$

5. $x < -\dfrac{17}{5}$ or $x > 5$ **4.** $-4 \le x \le \dfrac{2}{3}$ **5.** $x < -\dfrac{17}{5}$ or $x > 5$

6. $x = 8$ or $x = -3$

7.

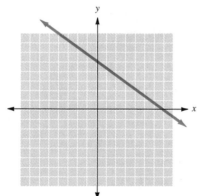

8.

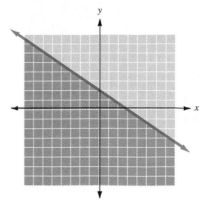

9. 24.5   **10.** 7   **11.** $f(x) = -x + 3$ **12.** $f(x) = 2x^2 - 5x - 3$

13. $f(x) = 9x^2 - 12x + 4$ **14.** $(x - 3)(x^2 - 5)$

15.

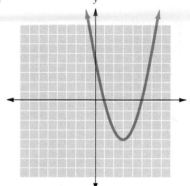

16.

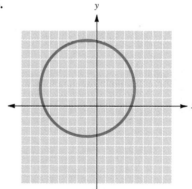

17. $\left(-10, \dfrac{26}{3}\right)$ **18.** $(2, 2, -1)$ **19.** 8 cm by 19 cm   **20.** 37

Self-Test for Chapter 8

1. (a) $f^{-1} = \left\{(x, y) \mid y = \dfrac{1}{4}x + \dfrac{1}{2}\right\}$; a function,

(b) $g^{-1} = \{(x, y) \mid y = \pm\sqrt{x - 1}\}$; not a function,

(c)

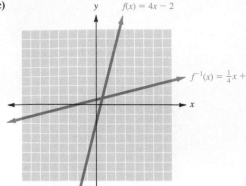

2.

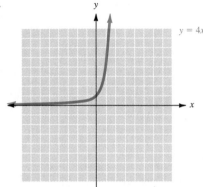

3.

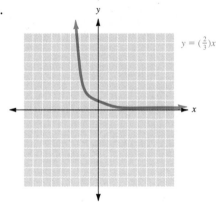

4. (a) -2, (b) $\dfrac{5}{2}$

5.

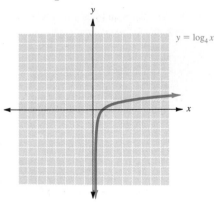

6. $\log 10{,}000 = 4$ 7. $\log_{27} 9 = \dfrac{2}{3}$ 8. $5^3 = 125$

9. $10^2 = 0.01$ 10. 6   11. 4   12. 5

13. $2 \log_b x + \log_b y + 3 \log_b z$

14. $\dfrac{1}{2}(\log_b x + 2 \log_5 y - \log_5 z)$ 15. $\log xy^2$ 16. $\log_b \sqrt[3]{\dfrac{x}{z^2}}$

17. 8   18. $\dfrac{11}{8}$ 19. 0.262   20. 2.151

Cumulative Test for Chapters 1–8

1. 11   2. $\left\{ x \,\middle|\, x < \dfrac{2}{3} \right\}$ or $\{x > 4\}$ 3. $\dfrac{10}{9}$

4.

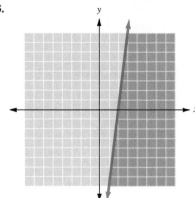

5.

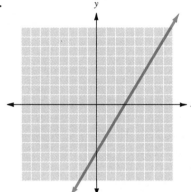

6. $6x + 5y = 7$ 7. $x \geq 10$ 8. $2x^2 - 6x + 1$ 9. $6x^2 - 13x - 5$

10. $(2x - 5)(x + 2)$ 11. $x(5x + 4y)(5x - 4y)$ 12. $\dfrac{-x + 2}{(x - 4)(x - 5)}$

13. $(x + 2)(x - 2)$ 14. $-4\sqrt{2}$ 15. $22 + 12\sqrt{2}$

16. $\dfrac{5}{3}(\sqrt{5} + \sqrt{2})$ 17. 77, 79, 81   18. $\{-2, 1\}$ 19. $\left\{ \dfrac{3 \pm \sqrt{19}}{2} \right\}$

20. $\left\{ -\dfrac{3}{2} \leq x \leq 1 \right\}$

Index